W9-CCC-917

Chemistry for Pharmacy Students

Chemistry for Pharmacy Students

General, Organic and Natural Product Chemistry

Satyajit D. Sarker
University of Ulster, Coleraine, Northern Ireland, UK

Lutfun Nahar
University of Ulster, Coleraine, Northern Ireland, UK

John Wiley & Sons, Ltd

Telephone (+44) 1243 779777

Email (for orders and customer service enquiries): cs-books@wiley.co.uk
Visit our Home Page on www.wiley.com

Other Wiley Editorial Offices

John Wiley & Sons Inc., 111 River Street, Hoboken, NJ 07030, USA

Jossey-Bass, 989 Market Street, San Francisco, CA 94103-1741, USA

Wiley-VCH Verlag GmbH, Boschstrasse 12, D-69469 Weinheim, Germany

John Wiley & Sons Australia Ltd, 42 McDougall Street, Milton, Queensland 4064, Australia

John Wiley & Sons (Asia) Pte Ltd, 2 Clementi Loop #02-01, Jin Xing Distripark, Singapore 129809

John Wiley & Sons Canada Ltd, 6045 Freemont Blvd. Mississauga, ONT, L5R 4J3, Canada

Wiley also publishes its books in a variety of electronic formats. Some content that appears in print may not be available in electronic books.

Anniversary Logo Design: Richard J. Pacifico

Library of Congress Cataloging-in-Publication Data

Sarker, Satyajit D.
Chemistry for pharmacy students: general, organic, and natural product chemistry / Satyajit D. Sarker, Lutfun Nahar.
p.; cm.
Includes bibliographical references.
ISBN 978-0-470-01780-7 (cloth : alk. paper)
1. Chemistry–Textbooks. I. Nahar, Lutfun, Ph. D. II. Title.
[DNLM: 1. Chemistry, Pharmaceutical. 2. Chemistry. QV 744 S517c 2007]
QD31.3.S377 2007
540–dc22 2007017895

British Library Cataloguing in Publication Data

A catalogue record for this book is available from the British Library

ISBN 978-0-470-01780-7 (HB) 978-0-470-01781-4 (PB)

Typeset in 11/14pt Times by Thomson Digital
Printed and bound in Great Britain by Antony Rowe Ltd., Chippenham, Wilts
This book is printed on acid-free paper responsibly manufactured from sustainable forestry in which at least two trees are planted for each one used for paper production.

This book is dedicated to
pharmacy students
from all over the world

Contents

Preface

The pharmacy profession and the role of pharmacists in the modern healthcare systems have evolved quite rapidly over the last couple of decades. The services that pharmacists provide are expanding with the introduction of supplementary prescribing, provision of health checks, patient counselling and many others. The main ethos of pharmacy profession is now as much about keeping people healthy as treating them when they are not well. The modern pharmacy profession is shifting away from a product focus and towards a patient focus. To cope with these changes, and to meet the demand of the modern pharmacy profession, the pharmacy curriculum, especially in the developed world, has evolved significantly. In the western countries, almost all registered pharmacists are employed by the community and hospital pharmacies. As a consequence, the practice, law, management, care, prescribing science and clinical aspects of pharmacy have become the main components of pharmacy curriculum. In order to incorporate all these changes, naturally, the fundamental science components, e.g. chemistry, statistics, pharmaceutical biology, microbiology, pharmacognosy and a few other topics, have been reduced remarkably. The impact of these recent changes is more innocuous in the area of pharmaceutical chemistry.

As all drugs are chemicals, and pharmacy is mainly about the study of various aspects of drugs, including manufacture, storage, actions and toxicities, metabolisms and managements, chemistry still plays a vital role in pharmacy education. However, the extent at which chemistry used to be taught a couple of decades ago has certainly changed significantly. It has been recognized that while pharmacy students need a solid foundation in chemistry knowledge the extent cannot be the same as chemistry students may need.

There are several books on general, organic and natural product chemistry available today, but all of them are written in such a manner that the level is only suitable for undergraduate Chemistry students, not for Pharmacy undergraduates. Moreover, in most modern pharmacy curricula, general, organic and natural product chemistry is taught at the first and second year undergraduate levels only. There are also a limited number of Pharmaceutical Chemistry books available to the students, but none of them can meet the demand of the recent changes in pharmacy courses in the developed countries. Therefore, there has been a pressing need for a chemistry text covering the fundamentals of general, organic and natural product chemistry written at a correct level for the Pharmacy undergraduates. Physical (Preformulation) and Analytical Chemistry (Pharmaceutical Analysis) are generally taught separately at year 2 and year 3 levels of any modern MPharm course, and there are a number of excellent and up-to-date texts available in these areas.

During our teaching careers, we have always struggled to find an appropriate book that can offer general, organic and natural product chemistry at the right level for Pharmacy undergraduate students, and address the current changes in pharmacy curricula all over the world, at least in the UK. We have always ended up recommending several books, and also writing notes for the students. Therefore, we have decided to address this issue by compiling a chemistry book for Pharmacy students, which will cover general, organic and natural product chemistry in relation to drug molecules. Thus, the aims of our book are to provide the fundamental knowledge and overview of all core topics related to general, organic and natural product chemistry currently taught in Pharmacy undergraduate courses in the UK, USA and various other developed countries, relate these topics to the better understanding of drug molecules and their development and meet the demand of the recent changes in pharmacy curricula. This book attempts to condense the essentials of general, organic and natural product chemistry into a manageable, affordable and student-friendly text, by concentrating purely on the basics of various topics without going into exhaustive detail or repetitive examples.

In Pharmacy undergraduate courses, especially in the UK, we get students of heterogeneous educational backgrounds; while some of them have very good chemistry background, the others have bare minimum or not at all. From our experience in teaching Pharmacy undergraduate students, we have been able to identify the appropriate level that is required for all these students to learn properly. While we recognise that learning styles and levels vary from student to student, we can still try to strike the balance in terms of the level and standard at a point, which is not too difficult or not too easy for any students, but will certainly be student friendly. Bearing this in mind, the contents of this book are organized and dealt with in a way that they are suitable for year 1 and year 2 levels of the pharmacy curriculum. While the theoretical aspects of various topics are covered adequately, much focus has been given to the applications of these theories in relation to drug molecules and their discovery and developments. Chapter 1 provides an overview of some general aspects of chemistry and their importance in modern life, with particular emphasis on medicinal applications, and brief discussions of various physical characteristics of drug molecules, e.g. pH, polarity and solubility. While Chapter 2 deals with the fundamentals of atomic structure and bonding, chapter 3 covers various aspects of stereochemistry. Chapter 4 incorporates organic functional groups, and various aspects of aliphatic, aromatic and heterocyclic chemistry, amino acids and nucleic acids and their pharmaceutical importance. Major organic reactions are covered adequately in Chapter 5, and various types of pharmaceutically important natural products are discussed in Chapter 6.

While the primary readership of this book is the Pharmacy undergraduate students (BPharm/MPharm), especially in their first and second years of study, the readership could also extend to the students of various other subject areas within Food Sciences, Life Sciences and Health Sciences who are not becoming chemists yet need to know the fundamentals of chemistry for their courses.

Dr Satyajit D Sarker
Dr Lutfun Nahar

1

Introduction

Learning objectives

After completing this chapter the student should be able to

- describe the role of chemistry in modern life;
- define some of the physical properties of drugs, e.g. polarity, solubility, melting point, boiling point and acid–base properties;
- explain the terms pH, pK_a, buffer and neutralization.

1.1 Role of chemistry in modern life

Chemistry is the science of the composition, structure, properties and reactions of matter, especially of atomic and molecular systems.

Life itself is full of chemistry; i.e., life is the reflection of a series of continuous biochemical processes. Right from the composition of the cell to the whole organism, the presence of chemistry is conspicuous. Human beings are constructed physically of chemicals, live in a plethora of chemicals and are dependent on chemicals for their quality of modern life. All living organisms are composed of numerous organic substances. Evolution of life begins from one single organic compound called a nucleotide. Nucleotides join together to form the building blocks of life. Our identities, heredities and continuation of generations are all governed by chemistry.

In our everyday life, whatever we see, use or consume is the gift of research in chemistry for thousands of years. In fact, chemistry is applied

Chemistry for Pharmacy Students Satyajit D Sarker and Lutfun Nahar

everywhere in modern life. From the colouring of our clothes to the shapes of our PCs, all are possible due to chemistry. It has played a major role in pharmaceutical advances, forensic science and modern agriculture. Diseases and their remedies have also been a part of human lives. Chemistry plays an important role in understanding diseases and their remedies, i.e. drugs. The focus of this section is given to the role of chemistry in modern medicine.

Medicines or drugs that we take for the treatment of various ailments are chemicals, either organic or inorganic. However, most drugs are organic molecules. Let us take aspirin as an example. It is probably the most popular and widely used analgesic drug because of its structural simplicity and low cost. Aspirin is chemically known as acetyl salicylic acid, an organic molecule. The precursor of aspirin is salicin, which is found in willow tree bark. However, aspirin can easily be synthesized from phenol using the *Kolbe reaction* (see Section 4.6.10). As we progress through various chapters of this book, we will come across a series of examples of drugs and their properties.

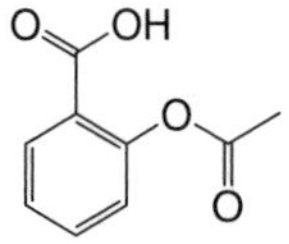

Aspirin
Acetyl salicylic acid

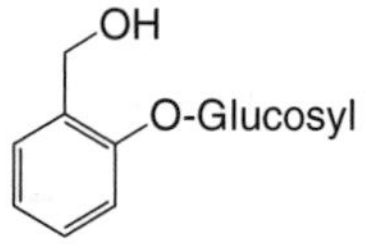

Salicin
The precursor of aspirin

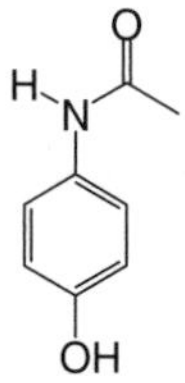

Paracetamol

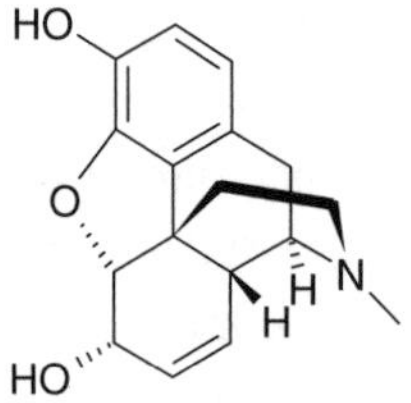

Morphine

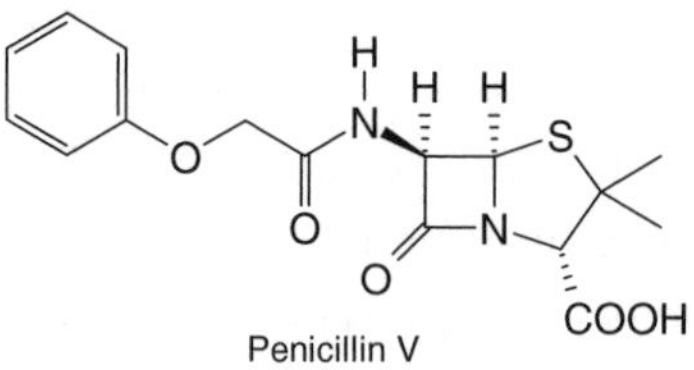

Penicillin V

In order to have a proper understanding and knowledge of these drugs and their behaviour, there is no other alternative but to learn chemistry. Everywhere, from discovery to development, from production and storage to administration, and from desired actions to adverse effects of drugs, chemistry is involved directly.

In the drug discovery stage, suitable sources are explored. Sources of drug molecules can be natural, e.g. narcotic analgesic, morphine, from *Papaver somniferum* (Poppy plant), synthetic, e.g. a popular analgesic and antipyretic, paracetamol, or semi-synthetic, e.g. semi-synthetic penicillins.

Whatever the source is, chemistry is involved in all processes in the discovery phase. For example, if a drug molecule has to be purified from a natural source, e.g. a plant, processes such as extraction, isolation and identification are used, and all these processes involve chemistry.

Similarly, in the drug development steps, especially in the pre-formulation and formulation studies, the structures and the physical properties, e.g. solubility and pH, of the drug molecules are exploited. Chemistry, particularly physical properties of drugs, is also important to determine storage conditions. Drugs having an ester functionality, e.g. aspirin, could be quite unstable in the presence of moisture, and should be kept in a dry and cool place. The chemistry of drug molecules dictates the choice of the appropriate route of administration. When administered, the action of a drug inside our body depends on its binding to the appropriate receptor, and its subsequent metabolic processes, all of which involve complex enzyme-driven biochemical reactions.

All drugs are chemicals, and pharmacy is a subject that deals with the study of various aspects of drugs. Therefore, it is needless to say that to become a good pharmacist the knowledge of the chemistry of drugs is essential. Before moving on to the other chapters, let us try to understand some of the fundamental chemical concepts in relation to the physical properties of drug molecules.

1.2 Physical properties of drug molecules

1.2.1 Physical state

Drug molecules exist in various physical states, e.g. amorphous solid, crystalline solid, hygroscopic solid, liquid or gas. The physical state of drug molecules is an important factor in the formulation and delivery of drugs.

1.2.2 Melting point and boiling point

The *melting point (m.p.)* is the temperature at which a solid becomes a liquid, and the *boiling point (b.p.)* is the temperature at which the vapour pressure of the liquid is equal to the atmospheric pressure. The boiling point of a substance can also be defined as the temperature at which it can change its state from a liquid to a gas throughout the bulk of the liquid at a given pressure. For example, the melting point of water at 1 atmosphere of pressure is 0 °C (32 °F, 273.15 K; this is also known as the *ice point*) and the boiling point of water is 100 °C.

Melting point is used to characterize organic compounds and to confirm the purity. The melting point of a pure compound is always higher than the melting point of that compound mixed with a small amount of an impurity. The more impurity is present, the lower the melting point. Finally, a minimum melting point is reached. The mixing ratio that results in the lowest possible melting point is known as the *eutectic point.*

The melting point increases as the molecular weight increases, and the boiling point increases as the molecular size increases. The increase in melting point is less regular than the increase in boiling point, because packing influences the melting point of a compound.

Packing of the solid is a property that determines how well the individual molecules in a solid fit together in a crystal lattice. The tighter the crystal lattice, the more energy is required to break it, and eventually melt the compound. Alkanes with an odd number of carbon atoms pack less tightly, which decreases their melting points. Thus, alkanes with an even number of carbon atoms have higher melting points than the alkanes with an odd number of carbon atoms. In contrast, between two alkanes having same molecular weights, the more highly branched alkane has a lower boiling point.

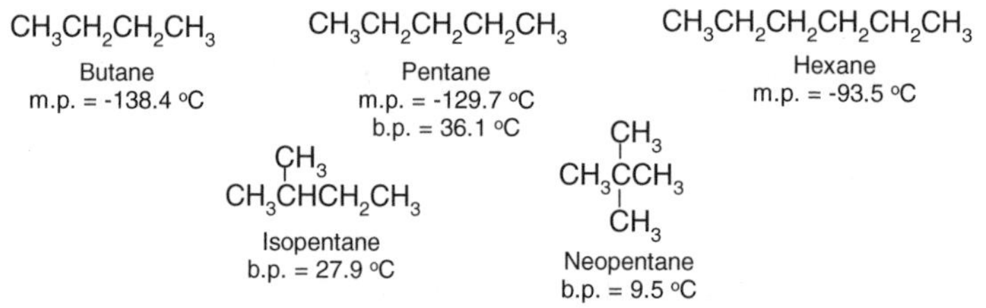

$CH_3CH_2CH_2CH_3$
Butane
m.p. = -138.4 °C

$CH_3CH_2CH_2CH_2CH_3$
Pentane
m.p. = -129.7 °C
b.p. = 36.1 °C

$CH_3CH_2CH_2CH_2CH_2CH_3$
Hexane
m.p. = -93.5 °C

$CH_3CH(CH_3)CH_2CH_3$
Isopentane
b.p. = 27.9 °C

$CH_3C(CH_3)_2CH_3$
Neopentane
b.p. = 9.5 °C

1.2.3 Polarity and solubility

Polarity is a physical property of a compound, which relates other physical properties, e.g. melting and boiling points, solubility and intermolecular interactions between molecules. Generally, there is a direct correlation between the polarity of a molecule and the number and types of polar or nonpolar covalent bond that are present. In a few cases, a molecule having polar bonds, but in a symmetrical arrangement, may give rise to a nonpolar molecule, e.g. carbon dioxide (CO_2).

The term *bond polarity* is used to describe the sharing of electrons between atoms. In a nonpolar covalent bond, the electrons are shared equally between two atoms. A polar covalent bond is one in which one atom has a greater attraction for the electrons than the other atom. When this relative attraction is strong, the bond is an ionic bond.

The polarity in a bond arises from the different electronegativities of the two atoms that take part in the bond formation. The greater the difference in electronegativity between the bonded atoms, the greater is the polarity of the bond. For example, water is a polar molecule, whereas cyclohexane is nonpolar. The bond polarity and electronegativity are discussed in Chapter 2.

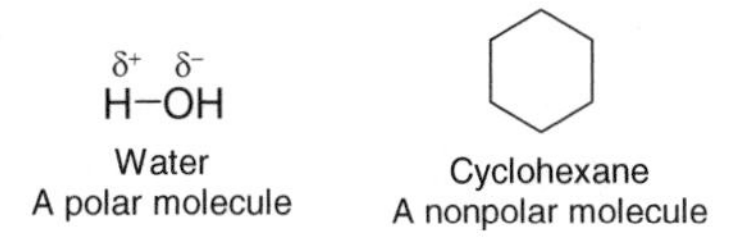

Solubility is the amount of a solute that can be dissolved in a specific solvent under given conditions. The dissolved substance is called the *solute* and the dissolving fluid is called the *solvent*, which together form a *solution*. The process of dissolving is called *solvation*, or *hydration* when the solvent is water. In fact, the interaction between a dissolved species and the molecules of a solvent is *solvation*.

The solubility of molecules can be explained on the basis of the polarity of molecules. Polar, e.g. water, and nonpolar, e.g. benzene, solvents do not mix. In general, like dissolves like; i.e., materials with similar polarity are soluble in each other. A polar solvent, e.g. water, has partial charges that can interact with the partial charges on a polar compound, e.g. sodium chloride (NaCl). As nonpolar compounds have no net charge, polar solvents are not attracted to them. Alkanes are nonpolar molecules, and are insoluble in polar solvent, e.g. water, and soluble in nonpolar solvent, e.g. petroleum ether. The hydrogen bonding and other nonbonding interactions between molecules are described in Chapter 2.

A solution at *equilibrium* that cannot hold any more solute is called a *saturated solution*. The equilibrium of a solution depends mainly on temperature. The maximum equilibrium amount of solute that can usually dissolve per amount of solvent is the *solubility* of that solute in that solvent. It is generally expressed as the maximum concentration of a saturated solution. The solubility of one substance dissolving in another is determined by the *intermolecular forces* between the solvent and solute, temperature, the entropy change that accompanies the solvation, the presence and amount of other substances and sometimes pressure or partial pressure of a solute gas. The *rate of solution* is a measure of how fast a solute dissolves in a solvent, and it depends on size of the particle, stirring, temperature and the amount of solid already dissolved.

1.2.4 Acid–base properties and pH

One of the adverse effects of aspirin is stomach bleeding, which is partly due to its acidic nature. In the stomach, aspirin is hydrolysed to salicylic acid. The carboxylic acid group (−COOH) and a phenolic hydroxyl group (−OH) present in salicylic acid make this molecule acidic. Thus, intake of aspirin increases the acidity of the stomach significantly, and if this increased acidic condition remains in the stomach for a long period, it may cause stomach bleeding. Like aspirin, there are a number of drug molecules that are acidic in nature. Similarly, there are basic and neutral drugs as well. Now, let us see what these terms *acid*, *base* and *neutral* compounds really mean, and how these parameters are measured.

Aspirin —Hydrolysis in the stomach→ Salicylic acid

Simply, an electron-deficient species that accepts an electron pair is called an *acid*, e.g. hydrochloric acid (HCl), and a species with electrons to donate is a *base*, e.g. sodium hydroxide (NaOH). A neutral species does not do either of these. Most organic reactions are either *acid–base* reactions or involve catalysis by an acid or base at some point.

Arrhenius acids and bases

According to Arrhenius's definition, an *acid* is a substance that produces hydronium ion (H_3O^+), and a base produces hydroxide ion (OH^-) in aqueous solution. An acid reacts with a base to produce salt and water.

$$HCl\ (Acid) + NaOH\ (Base) \rightleftharpoons NaCl\ (Salt) + H_2O\ (Water)$$

Brønsted–Lowry acids and bases

The Danish chemist Johannes Brønsted and the English chemist Thomas Lowry defined an *acid* as a proton (H^+) donor, and a *base* as a proton (H^+) acceptor.

$$HNO_2\ (Acid) + H_2O\ (Base) \rightleftharpoons NO_2^-\ (Conjugate\ base) + H_3O^+\ (Conjugate\ acid)$$

Each acid has a *conjugate base*, and each base has a *conjugate acid*. These conjugate pairs only differ by a proton. In the above example, HNO_2 is the acid, H_2O is the base, NO_2^- is the conjugate base, and H_3O^+ is the conjugate acid. Thus, a conjugate acid can lose an H^+ ion to form a base, and a conjugate base can gain an H^+ ion to form an acid. Water can be an acid or a base. It can gain a proton to become a hydronium ion (H_3O^+), its conjugate acid, or lose a proton to become the hydroxide ion (HO^-), its conjugate base.

When an acid transfers a proton to a base, it is converted to its conjugate base. By accepting a proton, the base is converted to its conjugate acid. In the following acid–base reaction, H_2O is converted to its conjugate base, hydroxide ion (HO^-), and NH_3 is converted to its conjugate acid, ammonium ion ($^+NH_4$). Therefore, the conjugate acid of any base always has an additional hydrogen atom, and an increase in positive charge or a decrease in negative charge. On the other hand, the conjugate base of an acid has one hydrogen atom less and an increase in negative charge or lone pair of electrons, and also a decrease in positive charge.

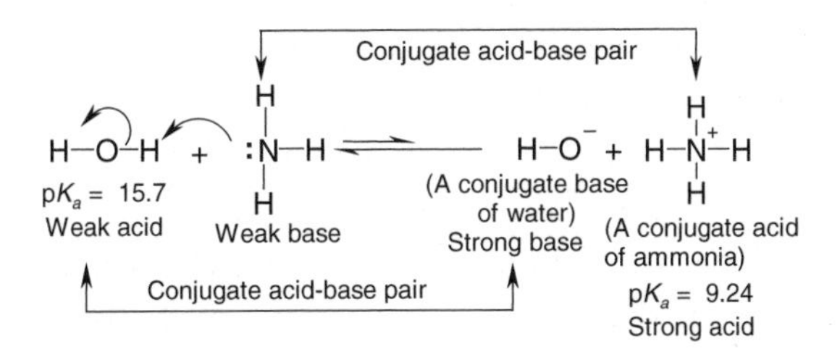

According to the Brønsted–Lowry definitions, any species that contains hydrogen can potentially act as an acid, and any compound that contains a lone pair of electrons can act as a base. Therefore, neutral molecules can also act as bases if they contain an oxygen, nitrogen or sulphur atom. Both an acid and a base must be present in a proton transfer reaction, because an acid cannot donate a proton unless a base is present to accept it. Thus, proton-transfer reactions are often called *acid–base reactions.*

For example, in the following reaction between acetic acid (CH_3CO_2H) and NH_3, a proton is transferred from CH_3CO_2H, an acid, to NH_3, a base.

Conjugate acid-base pair
$H_3C{-}C(=O){-}O{-}H + :NH_3 \rightleftharpoons H_3C{-}C(=O){-}O^- + H{-}N^+H_3$
$pK_a = 4.76$
Strong acid
Strong base
(Conjugate base of acetic acid)
Weak base
(Conjugate acid of ammonia)
$pK_a = 9.24$
Weak acid
Conjugate acid-base pair

In the above acid–base reaction, NH_3 is a base because it accepts a proton, and CH_3CO_2H is an acid because it donates a proton. In the reverse reaction, ammonium ion ($^+NH_4$) is an acid because it donates a proton, and acetate ion ($CH_3CO_2^-$) is a base because it accepts a proton. The curved arrows show the flow of electrons in an acid–base reaction.

Two half-headed arrows are used for the equilibrium reactions. A longer arrow indicates that the equilibrium favours the formation of acetate ion ($CH_3CO_2^-$) and ammonium ion ($^+NH_4$). Because acetic acid (CH_3CO_2H) is a stronger acid than ammonium ion ($^+NH_4$), the equilibrium lies towards the formation of weak acid and weak base.

Lewis theory of acids and bases

The Lewis theory of acids and bases defines an acid as an electron-pair acceptor, and a base as an electron-pair donor. Thus, a proton is only one of a large number of species that may function as a *Lewis acid.* Any molecule or ion may be an acid if it has an empty orbital to accept a pair of electrons (see Chapter 2 for orbital and Lewis theory). Any molecule or ion with a pair of electrons to donate can be a base.

Using this theory, a number of organic reactions can be considered as acid–base reactions, because they do not have to occur in solution. Lewis acids are known as *aprotic acids*, compounds that react with bases by accepting pairs of electrons, not by donating protons.

Borane (BH_3), boron trichloride (BCl_3) and boron trifluoride (BF_3) are known as *Lewis acids*, because boron has a vacant *d* orbital that accepts a pair of electrons from a donor species. For example, diethyl ether acts as a *Lewis base* towards BCl_3 and forms a complex of boron trichloride.

$$C_2H_5-\ddot{O}-C_2H_5 + BCl_3 \longrightarrow C_2H_5-\overset{+}{\ddot{O}}(C_2H_5)-\overset{-}{B}Cl_3$$

Diethyl ether (Lewis base) | Boron trichloride (Lewis acid) | A complex of diethyl ether and boron trichloride

Acid–base properties of organic functional groups

Let us see the acid–base properties of some molecules having different functional groups. The most common examples are carboxylic acids, amines, alcohols, amides, ethers and ketones. Drug molecules also contain various types of functional group, and these functional groups contribute to the overall acidity or basicity of drug molecules. Organic compounds with nonbonding electrons on nitrogen, oxygen, sulphur, or phosphorus can act as *Lewis bases* or *Brønsted bases*. They react with Lewis acids or Brønsted acids. Lewis acids may be either protic or aprotic acids. Brønsted acids are also called *protic acids*.

The most common organic acids are carboxylic acids. They are moderately strong acids having pK_a values ranging from about 3 to 5. Acetic acid (p$K_a = 4.76$) can behave as an acid and donate a proton, or as a base and accept a proton. A protonated acetic acid (p$K_a = -6.1$) is a strong acid. Equilibrium favours reaction of the stronger acid and stronger base to give the weaker acid and weaker base.

$$H_3C{-}C({=}O){-}O{-}H + HO^- \rightleftharpoons H_3C{-}C({=}O){-}O^- + HO{-}H$$

$H_3C{-}CO{-}OH$: pK_a = 4.76, Strong acid; HO^-: Strong base; $H_3C{-}CO{-}O^-$: (A conjugate base) Weak base; HO–H: (A conjugate acid) pK_a = 15.7, Weak acid

$$H_3C{-}C({=}O){-}OH + H{-}SO_3OH \rightleftharpoons H_3C{-}C({=}O^+{-}H){-}OH + HSO_4^-$$

$H_3C{-}CO{-}OH$: Weak base; $H{-}SO_3OH$: pK_a = -5.2, Weak acid; $H_3C{-}C(=O^+H){-}OH$: (A conjugate acid) pK_a = -6.1, Strong acid; HSO_4^-: (A conjugate base) Strong base

Amines are the most important organic bases as well as weak acids. Thus, an amine can behave as an acid and donate a proton, or as a base and accept a proton.

$$H_3C{-}NH_2 + HO^- \rightleftharpoons H_3C{-}NH^- + HO{-}H$$

$H_3C{-}NH_2$: pK_a = 40, Weak acid; HO^-: Weak base; $H_3C{-}NH^-$: (A conjugate base) Strong base; HO–H: (A conjugate acid) pK_a = 15.7, Strong acid

$$H_3C{-}NH_2 + H{-}SO_3OH \rightleftharpoons H_3C{-}NH_3^+ + HSO_4^-$$

$H_3C{-}NH_2$: Strong base; $H{-}SO_3OH$: pK_a = -5.2, Strong acid; $H_3C{-}NH_3^+$: (A conjugate acid) pK_a = 10.64, Weak acid; HSO_4^-: (A conjugate base) Weak base

An alcohol can behave like an acid and donate a proton. However, alcohols are much weaker organic acids, with pK_a values close to 16. Alcohol may also behave as a base; e.g., ethanol is protonated by sulphuric acid and gives ethyloxonium ion ($C_2H_5OH_2^+$). A protonated alcohol (p$K_a = -2.4$) is a strong acid.

$$C_2H_5{-}O{-}H + HO^- \rightleftharpoons C_2H_5{-}O^- + HO{-}H$$

$C_2H_5{-}O{-}H$: pK_a = 15.9, Weak acid; HO^-: Weak base; $C_2H_5{-}O^-$: (A conjugate base) Strong base; HO–H: pK_a = 15.7, Strong acid

$$C_2H_5{-}OH + H{-}SO_3OH \rightleftharpoons C_2H_5{-}O^+H_2 + HSO_4^-$$

$C_2H_5{-}OH$: Strong base; $H{-}SO_3OH$: pK_a = -5.2, Strong acid; $C_2H_5{-}O^+H_2$: (A conjugate acid) pK_a = -2.4, Weak acid; HSO_4^-: (A conjugate base) Weak base

Some organic compounds have more than one atom with nonbonding electrons, thus more than one site in such a molecule can react with acids. For example, acetamide has nonbonding electrons on both nitrogen and oxygen atoms, and either may be protonated. However, generally the reaction stops when one proton is added to the molecule.

Both acetamide and acetic acid are more readily protonated at the carbonyl oxygen than the basic site. The protonation of the nonbonding electrons on the oxygen atom of a carbonyl or hydroxyl group is an important first step in the reactions under acidic conditions of compounds such as acetamide, acetic acid and alcohols. The conjugate acids of these compounds are more reactive towards Lewis bases than the unprotonated forms are. Therefore, acids are used as catalysts to enhance reactions of organic compounds.

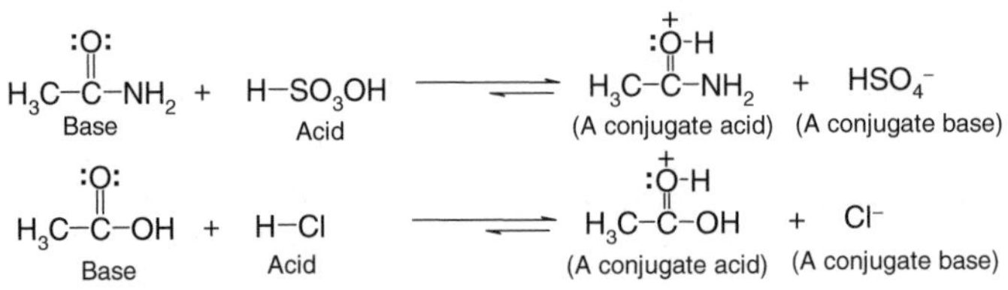

$$\underset{\text{Base}}{H_3C{-}\overset{:O:}{\overset{\|}{C}}{-}NH_2} + \underset{\text{Acid}}{H{-}SO_3OH} \rightleftharpoons \underset{\text{(A conjugate acid)}}{H_3C{-}\overset{:\overset{+}{O}{-}H}{\overset{\|}{C}}{-}NH_2} + \underset{\text{(A conjugate base)}}{HSO_4^-}$$

$$\underset{\text{Base}}{H_3C{-}\overset{:O:}{\overset{\|}{C}}{-}OH} + \underset{\text{Acid}}{H{-}Cl} \rightleftharpoons \underset{\text{(A conjugate acid)}}{H_3C{-}\overset{:\overset{+}{O}{-}H}{\overset{\|}{C}}{-}OH} + \underset{\text{(A conjugate base)}}{Cl^-}$$

The reaction of diethyl ether with concentrated hydrogen chloride (HCl) is typical of that of an oxygen base with a protic acid. Just like water, organic oxygenated compounds are protonated to give oxonium ions, e.g. protonated ether.

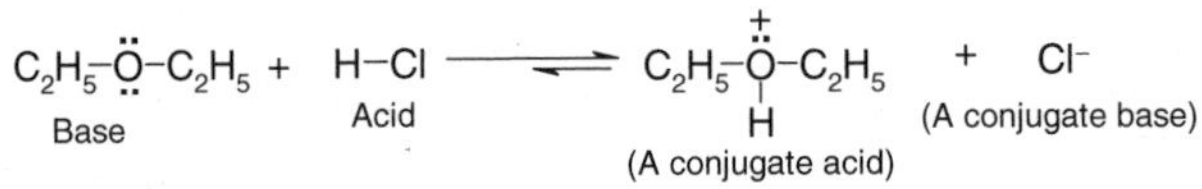

$$\underset{\text{Base}}{C_2H_5{-}\ddot{\underset{\cdot\cdot}{O}}{-}C_2H_5} + \underset{\text{Acid}}{H{-}Cl} \rightleftharpoons \underset{\text{(A conjugate acid)}}{C_2H_5{-}\overset{+}{\ddot{\underset{\overset{|}{H}}{O}}}{-}C_2H_5} + \underset{\text{(A conjugate base)}}{Cl^-}$$

Ketones can behave as bases. Acetone donates electrons to boron trichloride, a Lewis acid, and forms a complex of acetone and boron trichloride.

$$\underset{\substack{\text{Acetone}\\ \text{(Lewis base)}}}{H_3C{-}\overset{:O:}{\overset{\|}{C}}{-}CH_3} + \underset{\substack{\text{Boron trichloride}\\ \text{(Lewis acid)}}}{BCl_3} \longrightarrow \underset{\substack{\text{A complex of acetone}\\ \text{and boron trichloride}}}{H_3C{-}\overset{:\ddot{O}}{\overset{\|+}{\underset{\underset{CH_3}{|}}{C}}}{-}\overset{-}{B}Cl_3}$$

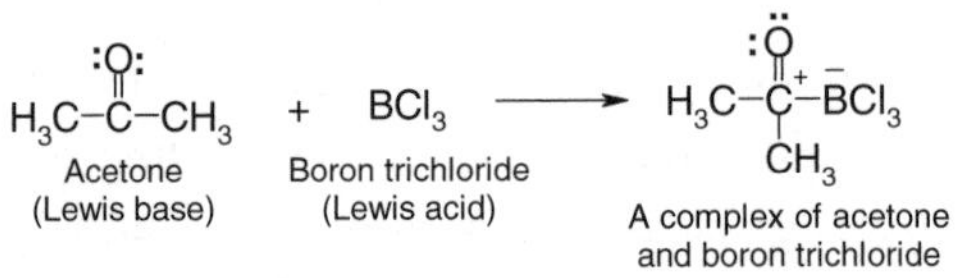

The reaction of an organic compound as an acid depends on how easily it can lose a proton to a base. The acidity of the hydrogen atom depends on the electronegativity of the bonded central atom. The more electronegative the bonded central atom, the more acidic are the protons. Carbon is less electronegative than nitrogen and oxygen. Thus, carbon attracts and holds electrons less strongly than nitrogen and oxygen do. For example, ethane, in which the hydrogen atoms are bonded to carbon atoms, is a very weak acid. Nitrogen is less electronegative than oxygen. Thus, nitrogen attracts and holds the electrons less strongly than oxygen does. For example, in methylamine, the hydrogen atoms on nitrogen are acidic, but the hydrogen atom bonded to the oxygen atom in methanol is even more acidic. Weak acids produce strong conjugate bases. Thus, ethane gives a stronger conjugate base than methylamine and methanol. The conjugate bases of ethane, methylamine and methanol are shown below.

CH_3CH_3 (Ethane) → CH_3NH_2 (Methylamine) → CH_3OH (Methanol)
(Increasing acidity of hydrogen bonded to carbon, nitrogen and oxygen)
CH_3O^- (Methoxide anion) → CH_3NH^- (Methylamide anion) →
$CH_3CH_2^-$ (Ethyl anion)
(Increasing basicity of the conjugate base)

pH and pK_a values

The pH value is defined as the negative of the logarithm to base 10 of the concentration of the hydrogen ion. The acidity or basicity of a substance is defined most typically by the pH value.

$$pH = -\log_{10}[H_3O^+]$$

The acidity of an aqueous solution is determined by the concentration of H_3O^+ ions. Thus, the pH of a solution indicates the concentration of hydrogen ions in the solution. The concentration of hydrogen ions can be indicated as $[H^+]$ or its solvated form in water as $[H_3O^+]$. Because the $[H_3O^+]$ in an aqueous solution is typically quite small, chemists have found an equivalent way to express $[H_3O^+]$ as a positive number whose value normally lies between 0 and 14. The lower the pH, the more acidic is the solution. The pH of a solution can be changed simply by adding acid or base to the solution. Do not confuse pH with pK_a. The pH scale is used to describe the acidity of a solution. The pK_a is characteristic of a particular compound, and it tells how readily the compound gives up a proton.

The pH of the salt depends on the strengths of the original acids and bases as shown below.

Acid	Base	Salt pH
Strong	Strong	7
Weak	Strong	>7
Strong	Weak	<7
Weak	Weak	Depends on which one is stronger

At equilibrium the concentration of H^+ is 10^{-7}, so we can calculate the pH of water at equilibrium as

$pH = -\log_{10}[H^+] = -\log[10^{-7}] = 7$. Solutions with a pH of 7 are said to be *neutral*, while those with pH values below 7 are defined as acidic, and those above pH of 7 as being basic. The pH of blood plasma is around 7.4, whereas that of the stomach is around 1.

Strong acids, e.g. HCl, HBr, HI, H_2SO_4, HNO_3, $HClO_3$ and $HClO_4$, completely ionize in solution, and are always represented in chemical equations in their ionized form. Similarly, *strong bases*, e.g. LiOH, NaOH, KOH, RbOH, $Ca(OH)_2$, $Sr(OH)_2$ and $Ba(OH)_2$, completely ionize in solution and are always represented in their ionized form in chemical equations. A *salt* is formed when an acid and a base are mixed and the acid releases H^+ ions while the base releases OH^- ions. This process is called *hydrolysis*. The conjugate base of a strong acid is very weak and cannot undergo hydrolysis. Similarly, the conjugate acid of a strong base is very weak and likewise does not undergo hydrolysis.

Acidity and basicity are described in terms of equilibria. Acidity is the measure of how easily a compound gives up a proton, and basicity is a measure of how well a compound shares its electrons with a proton. A strong acid is one that gives up its proton easily. This means that its conjugate base must be weak because it has little affinity for a proton. A weak acid gives up its proton with difficulty, indicating that its conjugate base is strong because it has a high affinity for a proton. Thus, the stronger the acid, the weaker is its conjugate base.

When a strong acid, e.g. hydrochloric acid (an inorganic or mineral acid), is dissolved in water, it dissociates almost completely, which means that the products are favoured at equilibrium. When a much weaker acid, e.g. acetic acid (an organic acid), is dissolved in water, it dissociates only to a small extent, so the reactants are favoured at equilibrium.

$$\underset{\substack{pK_a = -7 \\ \text{Strong acid}}}{H{-}Cl} + \underset{\text{Strong base}}{H_2\ddot{O}:} \rightleftharpoons \underset{\substack{\text{(A conjugate acid)} \\ pK_a = -1.74 \\ \text{Weak acid}}}{H_3\ddot{O}^+} + \underset{\substack{\text{(A conjugate base)} \\ \text{Weak base}}}{Cl^-}$$

$$\underset{\substack{pK_a = 4.76 \\ \text{Weak acid}}}{H_3C{-}C({=}O){-}\ddot{O}{-}H} + \underset{\text{Weak base}}{H_2\ddot{O}:} \rightleftharpoons \underset{\substack{\text{(A conjugate acid)} \\ pK_a = -1.74 \\ \text{Strong acid}}}{H_3\ddot{O}^+} + \underset{\substack{\text{(A conjugate base)} \\ \text{Strong base}}}{H_3C{-}C({=}O){-}\ddot{O}:^-}$$

Whether a reversible reaction favours reactants or products at equilibrium is indicated by the equilibrium constant of the reaction (K_{eq}). Remember that square brackets are used to indicate concentration in moles/litre = molarity (M). The degree to which an acid (HA) dissociates is described by its acid dissociation constant (K_a). The acid dissociation constant is obtained by multiplying the equilibrium constant (K_{eq}) by the concentration of the solvent in which the reaction takes place.

$$K_a = K_{eq}[H_2O] = \frac{[H_3O^+][A]}{[HA]}$$

The larger the acid dissociation constant, the stronger is the acid. Hydrochloric acid has an acid dissociation constant of 10^7, whereas acetic acid has an acid dissociation constant of only 1.74×10^{-5}. For convenience, the strength of an acid is generally indicated by its pK_a value rather than its K_a value. The pK_a of hydrochloric acid, strong acid, is -7, and the pK_a of acetic acid, much weaker acid, is 4.76.

$$pK_a = -\log K_a$$

Very strong acids	$pK_a < 1$	Moderately strong acids	$pK_a = 1-5$
Weak acids	$pK_a = 5-15$	Extremely weak acids	$pK_a > 15$

Buffer

A *buffer* is a solution containing a weak acid and its conjugate base (e.g. CH_3COOH and CH_3COO^-) or a weak base and its conjugate acid (e.g. NH_3 and NH_4^+).

The most important application of acid–base solutions containing a common ion is buffering. Thus, a buffer solution will maintain a relatively constant pH even when acidic or basic solutions are added to it. The most important practical example of a buffered solution is human blood, which can absorb the acids and bases produced by biological reactions without changing its pH. The normal pH of human blood is 7.4. A constant pH for blood is vital, because cells can only survive this narrow pH range around 7.4.

A buffered solution may contain a weak acid and its salt, e.g. acetic acid and acetate ion, or a weak base and its salt, e.g. NH_3 and NH_4Cl. By choosing the appropriate components, a solution can be buffered at virtually any pH. The pH of a buffered solution depends on the ratio of the concentrations of buffering components. When the ratio is least affected by adding acids or bases, the solution is most resistant to a change in pH. It is more effective when the acid–base ratio is equal to unity. The pK_a of the weak acid selected for the buffer should be as close as possible to the desired pH, because it follows the following equation:

$$pH = pK_a$$

The role of a buffer system in the body is important, because it tends to resist any pH changes as a result of metabolic processes. Large fluctuation in pH would denature most enzymes and hence interfere with the body metabolism. Carbon dioxide from metabolism combines with water in blood plasma to produce carbonic acid (H_2CO_3). The amount of H_2CO_3 depends on the amount of CO_2 present. The following system acts as a buffer, since carbonic acid can neutralize any base:

$$CO_2 + H_2O \rightleftharpoons H_2CO_3$$
$$H_2CO_3 + H_2O \rightleftharpoons H_3O^+ + HCO_3^-$$

Acid–base titration: neutralization

The process of obtaining quantitative information on a sample using a fast chemical reaction by reacting with a certain volume of reactant whose concentration is known is called *titration*. Titration is also called *volumetric analysis*, which is a type of quantitative chemical analysis. Generally, the *titrant* (the known solution) is added from a burette to a known quantity of the analyte (the unknown solution) until the reaction is complete. From the added volume of the titrant, it is possible to determine the concentration of the unknown. Often, an indicator is used to detect the end of the reaction, known as the *endpoint*.

An *acid–base titration* is a method that allows quantitative analysis of the concentration of an unknown acid or base solution. In an *acid–base titration*, the base will react with the weak acid and form a solution that contains the weak acid and its conjugate base until the acid is completely neutralized. The following equation is used frequently when trying to find the pH of buffer solutions.

$$\mathrm{pH} = \mathrm{p}K_a + \log[\mathrm{base}]/[\mathrm{acid}]$$

where pH is the log of the molar concentration of the hydrogen, $\mathrm{p}K_a$ is the equilibrium dissociation constant for the acid, [base] is the molar concentration of the basic solution and [acid] is the molar concentration of the acidic solution.

For the titration of a strong base with a weak acid, the equivalence point is reached when the pH is greater than 7. The half equivalence point is when half of the total amount of base needed to neutralize the acid has been added. It is at this point that the $\mathrm{pH} = \mathrm{p}K_a$ of the weak acid. In acid–base titrations, a suitable acid–base indicator is used to detect the endpoint from the change of colour of the indicator used. An acid–base indicator is a weak acid or a weak base. The following table contains the names and the pH range of some commonly used acid–base indicators.

Indicator	pH range	Quantity to be used per 10 mL	Colour in acid	Colour in base
Bromophenol blue	3.0–4.6	1 drop of 0.1% aq. solution	Yellow	Blue–violet
Methyl orange	3.1–4.4	1 drop of 0.1% aq. solution	Red	Orange
Phenolphthalein	8.0–10.0	1–5 drops of 0.1% solution in 70% alcohol	Colourless	Red
Thymol blue	1.2–2.8	1–2 drops of 0.1% aq. solution	Red	Yellow

Recommended further reading

Ebbing, D. D. and Gammon, S. D. *General Chemistry*, 7th edn, Houghton Mifflin, New York, 2002.

2

Atomic structure and bonding

Learning objectives

After completing this chapter the student should be able to

- describe the fundamental concepts of atomic structure;
- explain various aspects of chemical bonding;
- discuss the relevance of chemical bonding in drug molecules and drug–receptor interactions.

2.1 Atoms, elements and compounds

The basic building block of all matter is called an *atom*. Atoms are a collection of various subatomic particles containing negatively charged *electrons*, positively charged *protons* and neutral particles called *neutrons*. Each element has its own unique number of protons, neutrons and electrons. Both protons and neutrons have mass, whereas the mass of electrons is negligible. Protons and neutrons exist at the centre of the atom in the nucleus.

Electrons move around the nucleus, and are arranged in shells at increasing distances from the nucleus. These shells represent different energy levels, the outermost shell being the highest energy level.

Chemistry for Pharmacy Students Satyajit D Sarker and Lutfun Nahar

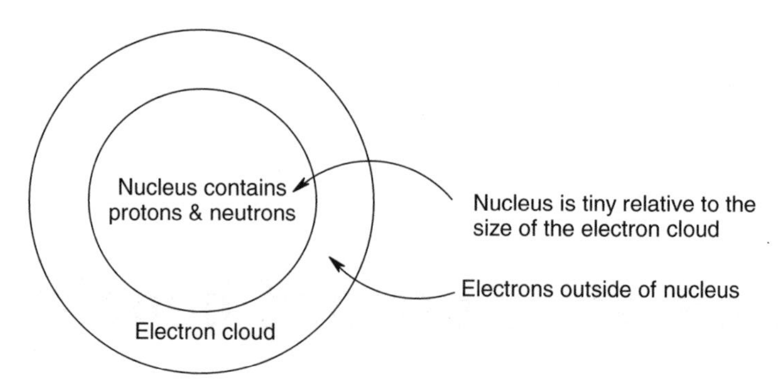

The number of protons that an atom has in its nucleus is called the *atomic number*. The total number of protons and neutrons in the nucleus of an atom is known as the *mass number*. For example, a carbon atom containing six protons and six neutrons has a mass number of 12.

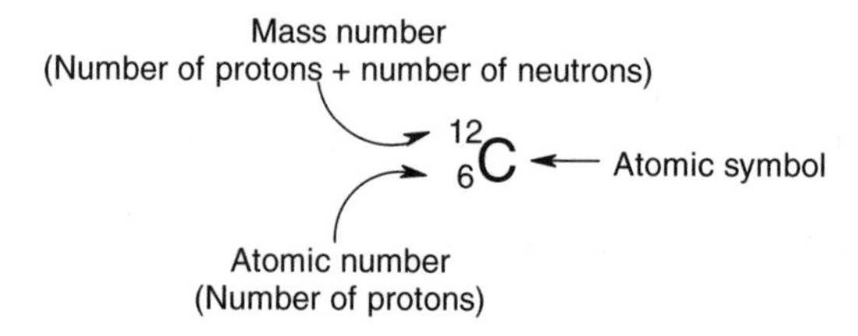

Elements are substances containing atoms of one type only, e.g. O_2, N_2 and Cl_2.

Compounds are substances formed when atoms of two or more elements join together, e.g. NaCl, H_2O and HCl. Although 109 elements exist naturally, some of them are extremely rare (check out the periodic table).

2.2 Atomic structure: orbitals and electronic configurations

It is important to understand the location of electrons, as it is the arrangement of the electrons that creates the bonds between the atoms, and chemical reactions are just that to form new bonds. Electrons are involved in the chemical bonding and reactions of an atom. Electrons are said to occupy *orbitals* in an atom.

An *orbital* is a region of space that can hold two electrons. Electrons do not move freely in the space around the nucleus but are confined to regions of space called *shells*. Each shell can contain up to $2n^2$ electrons, where n is the number of the shell. Each shell contains subshells known as *atomic orbitals*. The first shell contains a single orbital known as the 1*s* orbital. The second shell contains one 2*s* and three 2*p* orbitals. These three 2*p* orbitals are designated as $2p_x$, $2p_y$ and $2p_z$. The third shell contains one 3*s* orbital, three 3*p* orbitals and five 3*d* orbitals. Thus, the first shell can hold only two electrons, the second shell eight electrons and the third shell up to 18 electrons, and so on. As the number of electrons goes up, the shell numbers also increase. Therefore, electron shells are identified by the principal quantum number, $n = 1$, 2, 3 and so on.

Shell	Total number of shell electrons	Relative energies of shell electrons
4	32	Higher
3	16	↑
2	8	|
1	2	Lower

The electronic configuration of an atom describes the number of electrons that an atom possesses, and the orbitals in which these electrons are placed. The arrangements of electrons in orbitals, subshells and shells are called *electronic configurations*. Electronic configurations can be represented by using noble gas symbols to show some of the inner electrons, or by using Lewis structures in which the valence electrons are represented by dots.

Valence is the number of electrons an atom must lose or gain to attain the nearest noble gas or inert gas electronic configuration. Electrons in the outer shells that are not filled are called *valence electrons*.

The *ground-state electronic configuration* is the lowest energy, and the *excited-state electronic configuration* is the highest energy orbital. If energy is applied to an atom in the ground state, one or more electrons can jump into a higher energy orbital. Thus, it takes a greater energy to remove an electron from the first shell of an atom than from any other shells. For example, the sodium atom has electronic configuration of two, eight and one. Therefore, to attain the stable configuration, the Na atom must lose one electron from its outermost shell and become the nearest noble gas configuration, i.e. the configuration of neon, which has the electronic configuration of two and eight. Thus, sodium has a valence of 1. Since all other elements of Group IA in the periodic table have one electron in their outer shells, it can be said that Group IA elements have a valence of 1.

At the far end on the right hand side of the periodic table, let us take another example, chlorine, which has the electronic configuration of two, eight and seven, and the nearest noble gas is argon, which has the electronic configuration of two, eight and eight. To attain the argon electronic configuration chlorine must gain one electron. Therefore, chlorine has a valence of 1. Since all other elements of Group 7A in the periodic table have seven electrons in their outermost shells and they can gain one electron, we can say that the Group 7A elements have a valence of 1.

Shell	Number of orbitals contained each shell
4	$4s$, $4p_x$, $4p_y$, $4p_z$, five $4d$, seven $4f$
3	$3s$, $3p_x$, $3p_y$, $3p_z$, five $3d$
2	$2s$, $2p_x$, $2p_y$, $2p_z$
1	$1s$

Each atom has an infinite number of possible electronic configurations. We are here only concerned with the ground-state electronic configuration, which has the lowest energy. The ground-state electronic configuration of an atom can be determined by the following *three* principles.

- The *Aufbau principle* states that the orbitals fill in order of increasing energy, from lowest to highest. Because a $1s$ orbital is closer to the nucleus it is lower in energy than a $2s$ orbital, which is lower in energy than a $3s$ orbital.

- The *Pauli exclusion principle* states that no more than two electrons can occupy each orbital, and if two electrons are present, their spins must be paired. For example, the two electrons of a helium atom must occupy the $1s$ orbital in opposite spins.

- *Hund's rule* explains that when *degenerate orbitals* (orbitals that have same energy) are present but not enough electrons are available to fill all the shell completely, then a single electron will occupy an empty orbital first before it will pair up with another electron. This is understandable, as it takes energy to pair up electrons. Therefore, the six electrons in the carbon atom are filled as follows: the first four electrons will go to the $1s$ and $2s$ orbitals, a fifth electron goes to the $2p_x$, the sixth electron to the $2p_y$ orbital and the $2p_z$ orbital will remain empty.

The ground-state electronic configurations for elements 1–18 are listed below (electrons are listed by symbol, atomic number and ground-state electronic configuration).

First period			Second period			Third period		
H	1	$1s^1$	Li	3	[He] $2s^1$	Na	11	[Ne] $3s^1$
He	2	$1s^2$	Be	4	[He] $2s^2$	Mg	12	[Ne] $3s^2$
			B	5	[He] $2s^2\ 2p^1$	Al	13	[Ne] $3s^2\ 3p^1$
			C	6	[He] $2s^2\ 2p^2$	Si	14	[Ne] $3s^2\ 3p^2$
				7	[He] $2s^2\ 2p^3$	P	15	[Ne] $3s^2\ 3p^3$
				8	[He] $2s^2\ 2p^4$	S	16	[Ne] $3s^2\ 3p^4$
				9	[He] $2s^2\ 2p^5$	Cl	17	[Ne] $3s^2\ 3p^5$
				10	[He] $2s^2\ 2p^6$	Ar	18	[Ne] $3s^2\ 3p^6$

Let us see how we can write the ground-state electronic configurations for oxygen, chlorine, nitrogen, sulphur and carbon showing the occupancy of each *p* orbital. Oxygen has the atomic number 8, and the ground-state electronic configuration for oxygen can be written as $1s^2\ 2s^2\ 2p_x^{\ 2}\ 2p_y^{\ 1}\ 2p_z^{\ 1}$. Similarly, we can write the others as follows:

Chlorine (atomic number 17): $1s^2\ 2s^2\ 2p_x^{\ 2}\ 2p_y^{\ 2}\ 2p_z^{\ 2}\ 3s^2\ 3p_x^{\ 2}\ 3p_y^{\ 2}\ 3p_z^{\ 1}$

Nitrogen (atomic number 7): $1s^2\ 2s^2\ 2p_x^{\ 1}\ 2p_y^{\ 1}\ 2p_z^{\ 1}$

Sulphur (atomic number 16): $1s^2\ 2s^2\ 2p_x^{\ 2}\ 2p_y^{\ 2}\ 2p_z^{\ 2}\ 3s^2\ 3p_x^{\ 2}\ 3p_y^{\ 1}\ 3p_z^{\ 1}$

Carbon (atomic number 6): $1s^2\ 2s^2\ 2p_x^{\ 1}\ 2p_y^{\ 1}\ 2p_z^{\ 0}$

2.3 Chemical bonding theories: formation of chemical bonds

Atoms form bonds in order to obtain a stable electronic configuration, i.e. the electronic configuration of the nearest noble gas. All noble gases are inert, because their atoms have a stable electronic configuration in which they have eight electrons in the outer shell except helium (two electrons). Therefore, they cannot donate or gain electrons.

One of the driving forces behind the bonding in an atom is to obtain a stable valence electron configuration. A filled shell is also known as a noble gas configuration. Electrons in filled shells are called *core electrons*. The core electrons do not participate in chemical bonding. Electrons in shells that are not completely filled are called *valence electrons*, also known as *outer-shell electrons*, and the energy level in which they are found is also known as the *valence shell*. Carbon, for example, with the ground-state electronic configuration $1s^2\ 2s^2\ 2p^2$, has four outer-shell electrons. We generally use the Lewis structure to represent the outermost electrons of an atom.

2.3.1 Lewis structures

Lewis structures provide information about what atoms are bonded to each other, and the total electron pairs involved. According to the *Lewis theory*, an atom will give up, accept or share electrons in order to achieve a filled outer shell that contains eight electrons. The Lewis structure of a covalent molecule shows all the electrons in the valence shell of each atom; the bonds between atoms are shown as shared pairs of electrons. Atoms are most stable if they have a filled valence shell of electrons. Atoms transfer or share electrons in such a way that they can attain a filled shell of electrons. This stable configuration of electrons is called an *octet*. Except for hydrogen and helium, a filled valence shell contains eight electrons.

Lewis structures help us to track the valence electrons and predict the types of bond. The number of valence electrons present in each of the elements is to be considered first. The number of valence electrons determines the number of electrons needed to complete the octet of eight electrons. Simple ions are atoms that have gained or lost electrons to satisfy the octet rule. However, not all compounds follow the octet rule.

Elements in organic compounds are joined by *covalent bonds*, a sharing of electrons, and each element contributes one electron to the bond. The number of electrons necessary to complete the octet determines the number of electrons that must be contributed and shared by a different element in a bond. This analysis finally determines the number of bonds that each element may enter into with other elements. In a *single bond* two atoms share one pair of electrons and form a σ bond. In a *double bond* they share two pairs of electrons and form a σ bond and a π bond. In a *triple bond* two atoms share three pairs of electrons and form a σ bond and two π bonds.

Sodium (Na) loses a single electron from its $3s$ orbital to attain a more stable neon gas configuration ($1s^2\ 2s^2\ 2p^6$) with no electron in the outer shell. An atom having a filled valence shell is said to have a *closed shell configuration*. The total number of electrons in the valence shell of each atom can be determined from its group number in the periodic table. The shared electrons are called the *bonding electrons* and may be represented by a line or lines between two atoms. The valence electrons that are not being shared are the *nonbonding electrons* or *lone pair electrons*, and they are shown in the Lewis structure by dots around the symbol of the atom. A species that has an unpaired electron are called *radical*s. Usually they are very reactive, and are believed to play significant roles in aging, cancer and many other ailments.

In neutral organic compounds, C forms four bonds, N forms three bonds (and a lone pair), O forms two bonds (and two lone pairs) and H forms one bond. The number of bonds an atom normally forms is called the *valence*.

Lewis structure shows the connectivity between atoms in a molecule by a number of dots equal to the number of electrons in the outer shell of an atom of that molecule. A pair of electrons is represented by two dots, or a dash. When drawing Lewis structures, it is essential to keep track of the number of electrons available to form bonds and the location of the electrons. The number of valence electrons of an atom can be obtained from the periodic table because it is equal to the group number of the atom. For example, hydrogen (H) in Group 1A has one valence electron, carbon (C) in Group 4A has four valence electrons, and fluorine (F) in Group 7A has seven valence electrons.

To write the Lewis formula of CH_3F, first of all, we have to find the total number of valence electrons of all the atoms involved in this structure, i.e. C, H and F, having four, one and seven valence electrons, respectively.

$$\underset{\text{C}}{4} + \underset{\text{3H}}{3(1)} + \underset{\text{H}}{7} = 14$$

The carbon atom bonds with three hydrogen atoms and one fluorine atom, and it requires four pairs of electrons. The remaining six valence electrons are with the fluorine atom in the three nonbonding pairs.

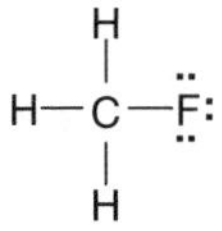

```
     H
     |   ..
  H—C—F:
     |   ..
     H
```

In the periodic table, the period 2 elements C, N, O, and F have valence electrons that belong to the second shell (2*s* and three 2*p*). The shell can be completely filled with eight electrons. In period 3, elements Si, P, S and Cl have the valence electrons that belong to the third shell (3*s*, three 3*p* and five 3*d*). The shell is only partially filled with eight electrons in 3*s* and three 3*p*, and the five 3*d* orbitals can accommodate an additional ten electrons. For these differences in valence shell orbitals available to elements of the second and third periods, we see significant differences in the covalent bonding of oxygen and sulphur, and of nitrogen and phosphorus. Although oxygen and nitrogen can accommodate no more than eight electrons in their valence shells, many phosphorus-containing compounds have 10 electrons in the valence shell of phosphorus, and many sulphur-containing compounds have 10 and even 12 electrons in the valence shell of sulphur.

So, to derive Lewis structures for most molecules the following sequence should be followed.

(a) Draw a tentative structure. The element with the least number of atoms is usually the central element.

(b) Calculate the number of valence electrons for all atoms in the compound.

(c) Put a pair of electrons between each symbol.

(d) Place pairs of electrons around atoms beginning with the outer atom until each has eight electrons, except for hydrogen. If an atom other than hydrogen has fewer than eight electrons then move unshared pairs to form multiple bonds.

If the structure is an ion, electrons are added or subtracted to give the proper charge. Lewis structures are useful as they show what atoms are bonded together, and whether any atoms possess lone pairs of electrons or have a formal charge. A *formal charge* is the difference between the number of valence electrons an atom actually has when it is not bonded to any other atoms, and the number of nonbonding electrons and half of its bonding electrons. Thus, a positive or negative charge assigned to an atom is called a formal charge. The decision as to where to put the charge is made by calculating the formal charge for each atom in an ion or a molecule. For example, the hydronium ion (H_3O^+) is positively charged and the oxygen atom has a formal charge of +1.

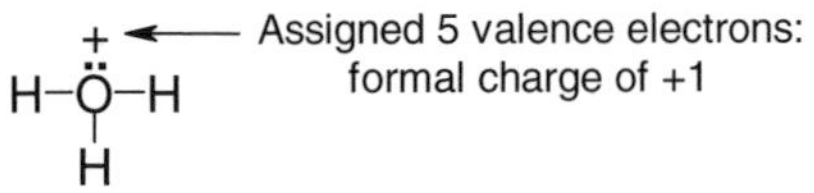

H–Ö–H (with a third H bonded below the O; + on O) ← Assigned 5 valence electrons: formal charge of +1

$$\begin{aligned}\text{So, formal charge} &= (\text{group number}) - (\text{nonbonding electrons}) - 1/2\ (\text{shared electrons})\\ &= 6 - 2 - 1/2(6)\\ &= 1.\end{aligned}$$

An uncharged oxygen atom must have six electrons in its valence shell. In the hydronium ion, oxygen bonds with three hydrogen atoms. So, only five electrons effectively belong to oxygen, which is one less than the valence electrons. Thus, oxygen bears a formal charge of +1. Elements of the second period, including carbon, nitrogen, oxygen and fluorine, cannot accommodate more than eight electrons as they have only four orbitals ($2s$, $2p_x$, $2p_y$ and $2p_z$) in their valence shells.

2.3.2 Various types of chemical bonding

A *chemical bond* is the attractive force that holds two atoms together. Valence electrons take part in bonding. An atom that gains electrons becomes an *anion*, a negatively charged ion, and an atom that loses electrons becomes a *cation*, a positively charged ion. Metals tend to lose electrons and nonmetals tend to gain electrons. While cations are smaller than atoms, anions are larger. Atoms decrease in size as they go across a period, and increase in size as they go down a group and increase the number of shells to hold electrons.

The energy required for removing an electron from an atom or ion in the gas phase is called *ionization energy.* Atoms can have a series of ionization energies, since more than one electron can always be removed, except for hydrogen. In general, the first ionization energies increase across a period and decrease down the group. Adding more electrons is easier than removing electrons. It requires a vast amount of energy to remove electrons.

Ionic bonds

Ionic bonds result from the transfer of one or more electrons between atoms. The more electronegative atom gains one or more valence electrons and hence becomes an anion. The less electronegative atom loses one or more valence electrons and becomes a cation. A single-headed arrow indicates a single electron transfer from the less electronegative element to the more electronegative atom. Ionic compounds are held together by the attraction of opposite charges. Thus, ionic bonds consist of the electrostatic attraction between positively and negatively charged ions. Ionic bonds are commonly formed between reactive metals, electropositive elements (on the left hand side of the periodic table), and nonmetals, electronegative elements (on the right hand side of the periodic table). For example, Na (electronegativity 0.9) easily gives up an electron, and Cl (electronegativity 3.0) readily accepts an electron to form an ionic bond. In the formation of ionic compound Na^+Cl^-, the single 3*s* valence electron of Na is transferred to the partially filled valence shell of chlorine.

$$\mathrm{Na}([\mathrm{Ne}]3s^1) + \mathrm{Cl}([\mathrm{Ne}]3s^2 3p^5) \rightarrow \mathrm{Na}^+([\mathrm{Ne}]3s^0) + \mathrm{Cl}^-([\mathrm{Ne}]3s^2 3p^6) \rightarrow \mathrm{Na}^+\mathrm{Cl}^-$$

Covalent bonds

Covalent bonds result from the sharing of electrons between atoms. In this case, instead of giving up or acquiring electrons, an atom can obtain a filled valence shell by sharing electrons. For example, two chlorine atoms can achieve a filled valence shell of 18 electrons by sharing their unpaired valence electrons.

$$:\ddot{\underset{..}{\mathrm{Cl}}}:\ddot{\underset{..}{\mathrm{Cl}}}: \longrightarrow \mathrm{Cl{-}Cl}$$

Similarly, hydrogen and fluorine can form a covalent bond by sharing electrons. By doing this, hydrogen fills its only shell and fluorine achieves its valence shell of eight electrons.

$$\mathrm{H}:\ddot{\underset{..}{\mathrm{F}}}: \longrightarrow \mathrm{H{-}F}$$

Nonpolar and polar covalent bonds In general, most bonds within organic molecules, including various drug molecules, are covalent. The exceptions are compounds that possess metal atoms, where the metal atoms should be treated as ions. If a bond is covalent, it is possible to identify whether it is a polar or nonpolar bond. In a *nonpolar* covalent bond, the electrons are shared equally between two atoms, e.g. H—H and F—F. Bonds between different atoms usually result in the electrons being attracted to one atom more strongly than the other. Such an unequal sharing of the pair of bonding electrons results in a polar covalent bond.

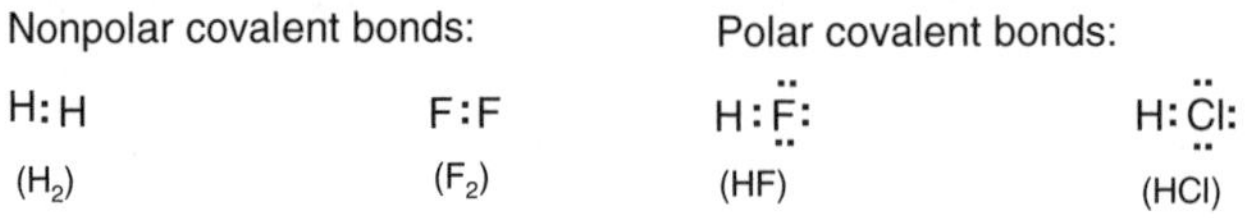

Nonpolar covalent bonds:

H:H (H_2) F:F (F_2)

Polar covalent bonds:

H:F: (HF) H:Cl: (HCl)

In a *polar covalent bond*, one atom has a greater attraction for the electrons than the other atom, e.g. chloromethane (CH_3Cl). When chlorine is bonded to carbon, the bonding electrons are attracted more strongly to chlorine. In other words, in a polar covalent bond, the electron pair is not shared equally. This results in a small partial positive charge on the carbon, and an equal but opposite partial negative charge on the chlorine. Bond polarity is measured by dipole moment (μ, which for chloromethane is 1.87). The dipole moment is measured in a unit called the debye (D). Generally, the C—H bond is considered nonpolar.

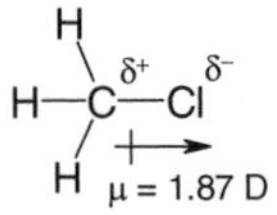

H
H—C—Cl (δ^+ on C, δ^- on Cl)
H
μ = 1.87 D

Chemists use two parameters, *bond lengths* and *bond angles*, to describe the 3D structures of covalent compounds. A bond length is the average distance between the nuclei of the atoms that are covalently bonded together. A bond angle is the angle formed by the interaction of two covalent bonds at the atom common to both.

Covalent bonds are formed when atomic orbitals overlap. The overlap of atomic orbitals is called *hybridization*, and the resulting atomic orbitals are called *hybrid orbitals*. There are two types of orbital overlap, which form sigma (σ) and pi (π) bonds. Pi bonds never occur alone without the bonded atoms also being joined by a σ bond. Therefore, a double bond consists of a σ bond and a π bond, whereas a triple bond consists of a σ bond and two π bonds. A sigma overlap occurs when there is one bonding interaction that results from the overlap of two *s* orbitals or an *s* orbital overlaps a *p* orbital or two *p* orbitals overlap head to head. A π overlap occurs only when two bonding interactions result from the sideways overlap of two parallel *p* orbitals. The *s* orbital is spherical in shape and *p* orbitals are in dumbbell shapes.

Sigma overlap of a *s* orbital with a *p* orbital

Pi overlap of two parallel *p* orbitals

Let us consider the formation of σ overlap in the hydrogen molecule (H_2), from two hydrogen atoms. Each hydrogen atom has one electron, which occupies the 1*s* orbital. The overlap of two *s* orbitals, one from each of two hydrogen atoms, forms a σ bond. The electron density of a σ bond is greatest along the axis of the bond. Since *s* orbitals are spherical in shape, two hydrogen atoms can approach one another from any direction resulting in a strong σ bond.

H· ·H ⟶ H:H = H:H ⟶ H_2

Hydrogen atoms, each contains 1*s* atomic orbital

Hydrogen atoms, formation of bonding molecular orbital

Hydrogen molecule

2.4 Electronegativity and chemical bonding

Electronegativity is the ability of an atom that is bonded to another atom or atoms to attract electrons strongly towards it. This competition for electron density is scaled by electronegativity values. Elements with higher electronegativity values have greater attraction for bonding electrons. Thus, the electronegativity of an atom is related to bond polarity. The difference in electronegativity between two atoms can be used to measure the polarity of the bonding between them. The greater the difference in electronegativity between the bonded atoms, the greater is the polarity of the bond. If the difference is great enough, electrons are transferred from the less electronegative atom to the more electronegative one, hence an ionic bond is formed. Only if the two atoms have exactly the same electronegativity is a nonpolar bond formed. Electronegativity increases from left to right and bottom to top in the periodic table as shown below (electronegativity is shown in parentheses).

1A	2A	3A	4A	5A	6A	7A
H (2.2)						
Li (1.0)	Be (1.6)	B (1.8)	C (2.5)	N (3.0)	O (3.4)	F (4.0)
Na (0.9)	Mg (1.3)	Al (1.6)	Si (1.9)	P (2.2)	S (2.6)	Cl (3.2)
						Br (3.0)
						I (2.7)

In general, if the electronegativity difference is equal to or less than 0.5 the bond is nonpolar covalent, and if the electronegativity difference between bonded atoms is 0.5–1.9 the bond is polar covalent. If the difference in electronegativities between the two atoms is 2.0 or greater, the bond is ionic. Some examples are shown below.

Bond	Difference in electronegativity	Types of bond
C—Cl	$3.0 - 2.5 = 0.5$	Polar covalent
P—H	$2.1 - 2.1 = 0$	Nonpolar covalent
C—F	$4.0 - 2.5 = 1.5$	Polar covalent
S—H	$2.5 - 2.1 = 0.4$	Nonpolar covalent
O—H	$3.5 - 2.1 = 1.4$	Polar covalent

Electrons in a polar covalent bond are unequally shared between the two bonded atoms, which results in partial positive and negative charges. The separation of the partial charges creates a *dipole*. The word dipole means two poles, the separated partial positive and negative charges. A polar molecule results when a molecule contains polar bonds in an unsymmetrical arrangement. Nonpolar molecules whose atoms have equal or nearly equal electronegativities have zero or very small dipole moments, as do molecules that have polar bonds but the molecular geometry is symmetrical, allowing the bond dipoles to cancel each other.

2.5 Bond polarity and intermolecular forces

Bond polarity is a useful concept for describing the sharing of electrons between atoms. The shared electron pairs between two atoms are not necessarily shared equally and this leads to a *bond polarity*. Atoms, such as nitrogen, oxygen and halogens, that are more electronegative than carbon have a tendency to have partial negative charges. Atoms such as carbon and hydrogen have a tendency to be more neutral or have partial positive charges. Thus, bond polarity arises from the difference in electronegativities of two atoms participating in the bond formation. This also depends on the attraction forces between molecules, and these interactions are called *intermolecular interactions or forces*. The physical properties, e.g. boiling points, melting points and solubilities of the molecules are determined, to a large extent, by intermolecular nonbonding interactions.

There are three types of nonbonding intermolecular interaction: *dipole–dipole interactions*, *van der Waals forces* and *hydrogen bonding*. These interactions increase significantly as the molecular weights increase, and also increase with increasing polarity of the molecules.

2.5.1 Dipole–dipole interactions

The interactions between the positive end of one dipole and the negative end of another dipole are called *dipole–dipole interactions*. As a result of dipole–dipole interactions, polar molecules are held together more strongly than nonpolar molecules. Dipole–dipole interactions arise when electrons are not equally shared in the covalent bonds because of the difference in electronegativity. For example, hydrogen fluoride has a dipole moment of 1.98 D, which lies along the H—F bond. As the fluorine atom has greater electronegativity than the hydrogen atom, the electrons are pulled towards fluorine, as shown below.

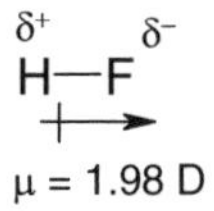

$$\overset{\delta^+}{H}-\overset{\delta^-}{F}$$
$$\mu = 1.98\ D$$

The arrow indicates the electrons are towards the more electronegative atom fluorine. The δ^+ and δ^- symbols indicate partial positive and negative charges.

2.5.2 van der Waals forces

Relatively weak forces of attraction that exist between nonpolar molecules are called *van der Waals forces* or *London dispersion forces*. Dispersion forces between molecules are much weaker than the covalent bonds within molecules. Electrons move continuously within bonds and molecules, so at any time one side of the molecule can have more electron density than the other side, which gives rise to a temporary dipole. Because the dipoles in the molecules are induced, the interactions between the molecules are also called *induced dipole–induced dipole interactions*.

van der Waals forces are the weakest of all the intermolecular interactions. Alkenes are nonpolar molecules, because the electronegativities of carbon and hydrogen are similar. Consequently, there are no significant partial charges on any of the atoms in an alkane. Therefore, the size of the van der Waals forces that hold alkane molecules together depends on the area of contact between the molecules. The greater the area of contact, the stronger are the van der Waals forces, and the greater is the amount of energy required to overcome these forces. For example, isobutane (b.p. $-10.2\,°C$) and butane (b.p. $-0.6\,°C$), both with the molecular formula C_4H_{10}, have different boiling points. Isobutane is a more compact molecule than butane. Thus, butane molecules have a greater surface area for interaction with each other than isobutane. The stronger interactions that are possible for *n*-butane are reflected in its boiling point, which is higher than the boiling point of isobutane.

Isobutane
b.p. -10.2

n-Butane
b.p. - 0.6

2.5.3 Hydrogen bonding

Hydrogen bonding is the attractive force between the hydrogen attached to an electronegative atom of one molecule and an electronegative atom of the same (*intramolecular*) or a different molecule (*intermolecular*). It is an unusually strong force of attraction between highly polar molecules in which hydrogen is covalently bonded to nitrogen, oxygen or fluorine. Therefore, a hydrogen bond is a special type of interaction between atoms. A hydrogen bond is formed whenever a polar covalent bond involving a hydrogen atom is in close proximity to an electronegative atom such as O or N. The attractive forces of hydrogen bonding are usually indicated by a dashed line rather than the solid line used for a covalent bond. For example, water molecules form intermolecular hydrogen bonding.

Hydrogen bond
Donor
Acceptor

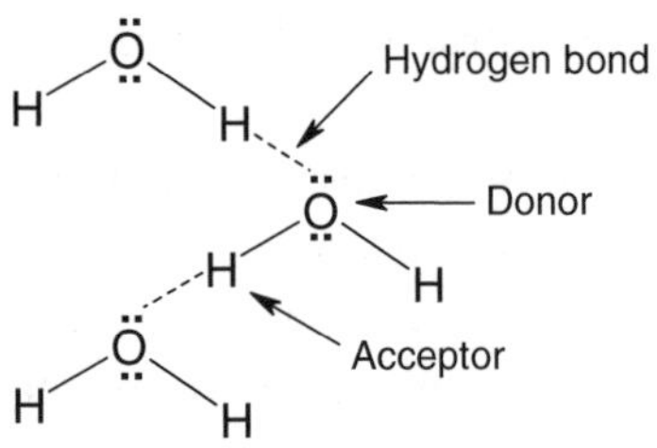

The above diagram shows a cluster of water molecules in the liquid state. Water is a polar molecule due to the electronegativity difference between hydrogen and oxygen atoms. The polarity of the water molecule with the attraction of the positive and negative partial charges is the basis for the hydrogen bonding. Hydrogen bonding is responsible for certain characteristics of water, e.g. surface tension, viscosity and vapour pressure.

Hydrogen bonding occurs with hydrogen atoms covalently bonded to oxygen, fluorine or nitrogen, but not with chlorine, which has larger atom size. The strength of a hydrogen bond involving an oxygen, a fluorine or a nitrogen atom ranges from 3 to 10 kcal/mol, making hydrogen bonds the strongest known type of intermolecular interaction. The intermolecular hydrogen bonding in water is responsible for the unexpectedly high boiling point of water (b.p. 100 °C). Hydrogen bonds are interactions between molecules and should not be confused with covalent bonds to hydrogen within a molecule. Hydrogen bonding is usually stronger than normal dipole forces between molecules, but not as strong as normal ionic or covalent bonds.

The hydrogen bond is of fundamental importance in biology. The hydrogen bond is said to be the 'bond of life'. The double helix structure of DNA is formed and held together with hydrogen bonds (see Section 4.8.2). The nature of the hydrogen bonds in proteins dictates their properties and behaviour. *Intramolecular hydrogen bonds* (within the molecule) in proteins result in the formation of globular proteins, e.g. enzymes or hormones. On the other hand, *intermolecular hydrogen bonds* (between different molecules) tend to give insoluble proteins such as fibrous protein. Cellulose, a polysaccharide, molecules are held together through hydrogen bonding, which provides plants with rigidity and protection (see Section 6.3.10). In drug–receptor binding, hydrogen bonding often plays an important role.

2.6 Significance of chemical bonding in drug–receptor interactions

Most drugs interact with receptor sites localized in macromolecules that have protein-like properties and specific three-dimensional shapes. A *receptor* is the specific chemical constituents of the cell with which a drug interacts to produce its pharmacological effects. One may consider that every protein that acts as the molecular target for a certain drug should be called a receptor. However, this term mainly incorporates those proteins that play an important role in the intercellular communication via chemical messengers. As such, enzymes, ion channels and carriers are usually not classified as receptors. The term receptor is mostly reserved for those protein structures that serve as intracellular antennas for chemical messengers. Upon recognition of the appropriate chemical signal (known as the *ligand*), the receptor proteins transmit the signal into a biochemical change in the target cell via a wide variety of possible pathways.

A minimum three-point attachment of a drug to a receptor site is essential for the desired effect. In most cases, a specific chemical structure is required for the receptor site and a complementary drug structure. Slight changes in the molecular structure of the drug may drastically change specificity, and thus the efficacy. However, there are some drugs that act exclusively by physical means outside cells, and do not involve any binding to the receptors. These sites include external surfaces of skin and gastrointestinal tract. Drugs also act outside cell membranes by chemical interactions, e.g. neutralization of stomach acid by antacids.

$$\text{Drug} + \text{Receptor} \rightarrow \text{Drug–Receptor Complex} \rightarrow \text{Altered Function}$$

The drug–receptor interaction, i.e. the binding of a drug molecule to its receptor, is governed by various types of chemical bonding that have been discussed earlier. A variety of chemical forces may result in a temporary binding of the drug to its receptor. Interaction takes place by utilizing the same bonding forces as involved when simple molecules interact, e.g. covalent (40–140 kcal/mol), ionic (10 kcal/mol), ion–dipole (1–7 kcal/mol), dipole–dipole (1–7 kcal/mol), van der Waals (0.5–1 kcal/mol), hydrogen bonding (1–7 kcal/mol) and hydrophobic interactions (1 kcal/mol). However, most useful drugs bind through the use of multiple weak bonds (ionic and weaker).

Covalent bonds are strong, and practically irreversible. Since the drug–receptor interaction is a reversible process, covalent bond formation is rather rare except in a few situations. Some drugs that interfere with DNA function by chemically modifying specific nucleotides are mitomycin C, cisplatin and anthramycin. Mitomycin C is a well characterized antitumour agent, which forms a covalent interaction with DNA after reductive activation, forming a cross-linking structure between guanine bases on adjacent strands of DNA, thereby inhibiting single strand formation. Similarly, anthramycin is another antitumour drug, which binds covalently to N-2 of guanine located in the minor groove of DNA. Anthramycin has a preference for purine–G–purine sequences (purines are adenine and guanine) with bonding to the middle G. Cisplatin, an anticancer drug, is a transition metal complex, *cis-diamine-dichloro-platinum*. The effect of the drug is due to the ability to platinate the N-7 of guanine on the major groove site of the DNA double helix. This chemical modification of the platinum atom cross-links two adjacent guanines on the same DNA strand, interfering with the mobility of DNA polymerases (see Section 4.8.2 for nucleic acid structures).

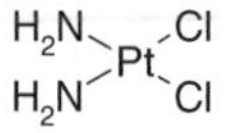

Cisplatin
An anticancer drug

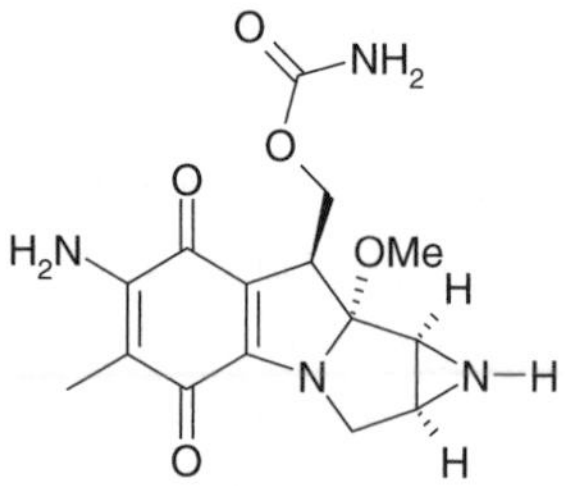

Mitomycin C
An antitumour agent

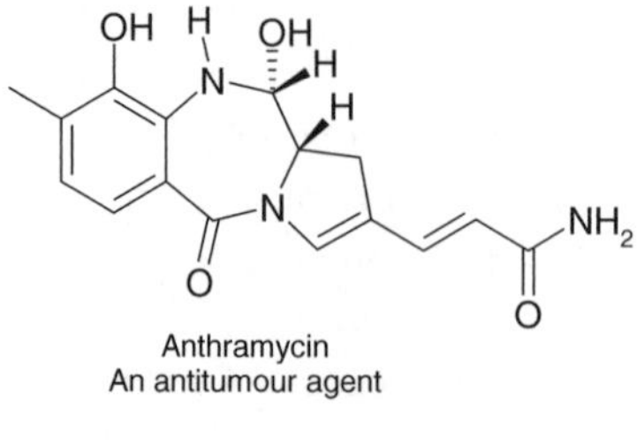

Anthramycin
An antitumour agent

Many drugs are acids or amines, easily ionized at physiological pH, and able to form ionic bonds by the attraction of opposite charges in the receptor site, for example the ionic interaction between the protonated amino group on salbutamol or the quaternary ammonium on acetylcholine and the dissociated carboxylic acid group of its receptor site. Similarly, the dissociated carboxylic group on the drug can bind with amino groups on the receptor. Ion–dipole and dipole–dipole bonds have similar interactions, but are more complicated and are weaker than ionic bonds.

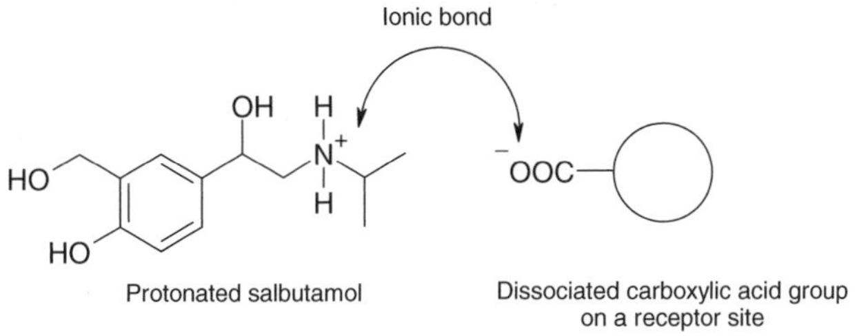

Ionic bond

Protonated salbutamol

Dissociated carboxylic acid group on a receptor site

Polar–polar interaction, e.g. hydrogen bonding, is also an important binding force in drug–receptor interaction, because the drug–receptor interaction is basically an exchange of the hydrogen bond between a drug molecule, surrounding water and the receptor site.

Formation of hydrophobic bonds between nonpolar hydrocarbon groups on the drug and those in the receptor site is also common. Although these bonds are not very specific, the interactions take place to exclude water molecules. Repulsive forces that decrease the stability of the drug–receptor interaction include repulsion of like charges and *steric hindrance*.

Recommended further reading

Clayden, J., Greeves, N., Warren, S. and Wothers, P. *Organic Chemistry*, Oxford University Press, Oxford, 2001.

Ebbing, D. D. and Gammon, S. D. *General Chemistry*, Houghton Mifflin, Boston, MA, 2002.

3 Stereochemistry

Learning objectives

After completing this chapter the student should be able to

- define stereochemistry;
- outline different types of isomerism;
- distinguish between conformational isomers and configurational isomers;
- discuss conformational isomerism in alkanes;
- explain the terms torsional energy, torsional strain, angle strain, enantiomers, chirality, specific rotation, optical activity, diastereomers, meso compounds and racemic mixture;
- designate the configuration of enantiomers using the D and L system and the (*R*) and (*S*) system;
- explain geometrical isomerism in alkenes and cyclic compounds;
- outline the synthesis of chiral molecules;
- explain resolution of racemic mixtures;
- discuss the significance of stereoisomerism in determining drug action and toxicity.

Chemistry for Pharmacy Students Satyajit D Sarker and Lutfun Nahar

3.1 Stereochemistry: definition

Stereochemistry is the chemistry of molecules in *three dimensions*. A clear understanding of stereochemistry is crucial for the study of complex molecules that are biologically important, e.g. proteins, carbohydrates and nucleic acids, and also drug molecules, especially in relation to their behaviour and pharmacological actions. Before we go into further detail, let us have a look at different types of isomerism that may exist in organic molecules.

3.2 Isomerism

Compounds with the same molecular formula but different structures are called *isomers*. For example, 1-butene and 2-butene have the same molecular formula, C_4H_8, but structurally they are different because of the different positions of the double bond. There are two types of isomer: *constitutional isomers* and *stereoisomers*.

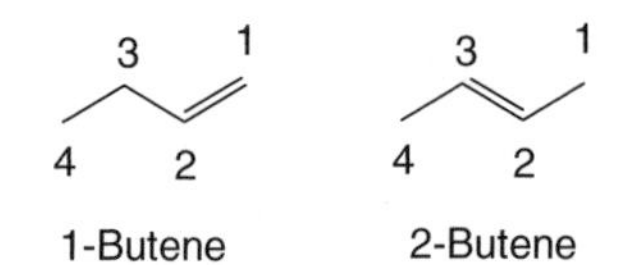

3.2.1 Constitutional isomers

When two different compounds have the same molecular formula but differ in the nature or sequence of bonding, they are called *constitutional isomers*. For example, ethanol and dimethylether have same molecular formula, C_2H_6O, but they differ in the sequence of bonding. Similarly, butane and isobutane are two constitutional isomers. Constitutional isomers generally have different physical and chemical properties.

3.2.2 Stereoisomers

Stereoisomers are compounds where the atoms are connected in the same order but with different geometries, i.e. they differ in the three-dimensional arrangements of groups or atoms in space. For example, in α-glucose and β-glucose, the atoms are connected in the same order, but the three dimensional orientation of the hydroxyl group at C–1 is different in each case. Similarly, *cis*- and *trans*-cinnamic acid only differ in the three dimensional orientation of the atoms or groups.

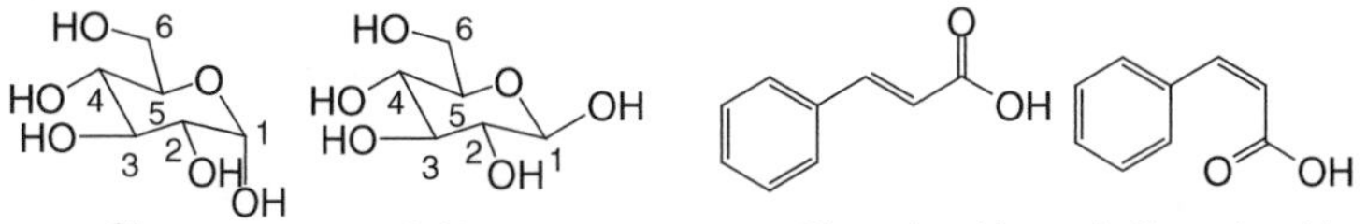

α-Glucose β-Glucose *trans*-Cinnamic acid *cis*-Cinnamic acid

There are two major types of stereoisomer: *conformational isomers* and *configurational isomers*. Configurational isomers include optical isomers, geometrical isomers, enantiomers and diastereomers.

Conformational isomers

Atoms within a molecule move relative to one another by rotation around single bonds. Such rotation of covalent bonds gives rise to different conformations of a compound. Each structure is called a *conformer* or *conformational isomer.* Generally, conformers rapidly interconvert at room temperature.

Conformational isomerism can be presented with the simplest example, ethane (C_2H_6), which can exist as an infinite number of conformers by the rotation of the C–C σ bond. Ethane has two sp^3-hybridized carbon atoms, and the tetrahedral angle about each is 109.5°. The most significant conformers of ethane are the *staggered* and *eclipsed* conformers. The staggered conformation is the most stable as it has the lowest energy.

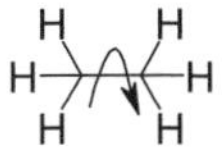

Rotation about the C-C bond in ethane

Visualization of conformers There are four conventional methods for visualization of three-dimensional structures on paper. These are the ball and stick method, the sawhorse method, the wedge and broken line method and the Newman projection method. Using these methods, the staggered and eclipsed conformers of ethane can be drawn as follows.

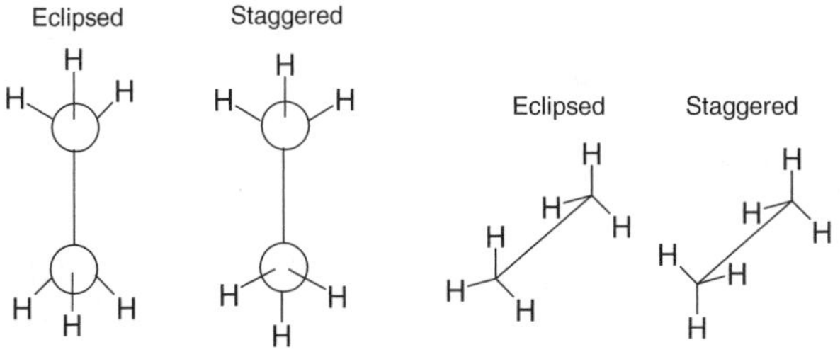

Eclipsed Staggered

Ball and stick method

Eclipsed Staggered

Sawhorse method

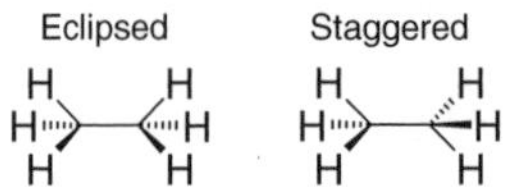

Eclipsed Staggered

Wedge and broken line method

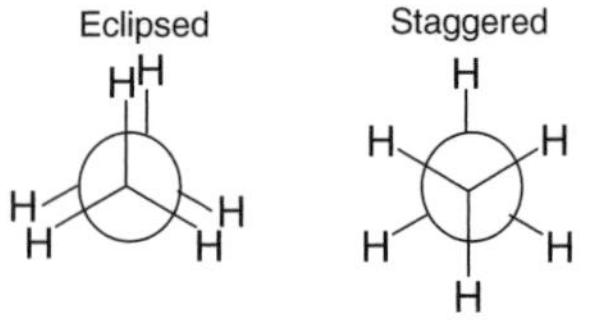

Eclipsed Staggered

Newman projection method

Staggered and eclipsed conformers In the staggered conformation, the H atoms are as far apart as possible. This reduces repulsive forces between them. This is why staggered conformers are stable. In the eclipsed conformation, H atoms are closest together. This gives higher repulsive forces between them. As a result, eclipsed conformers are unstable. At any moment, more molecules will be in staggered form than any other conformation.

Torsional energy and torsional strain Torsional energy is the energy required for rotating about the C–C σ bond. In ethane, this is very low (only 3 kcal). Torsional strain is the strain observed when a conformer rotates away from the most stable conformation (i.e. the staggered form). Torsional strain is due to the slight repulsion between electron clouds in the C–H bonds as they pass close by each other in the eclipsed conformer. In ethane, this is also low.

Conformational isomerism in propane Propane is a three-carbon- (sp^3-hybridized) atom-containing linear alkane. All are tetrahedrally arranged. When a hydrogen atom of ethane is replaced by a methyl (CH_3) group, we have propane. There is rotation about two C–C σ bonds.

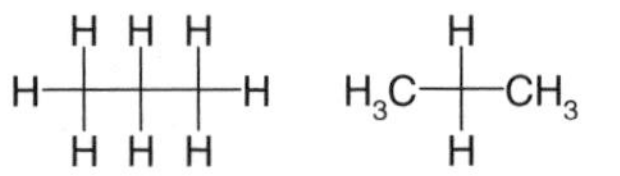

Propane

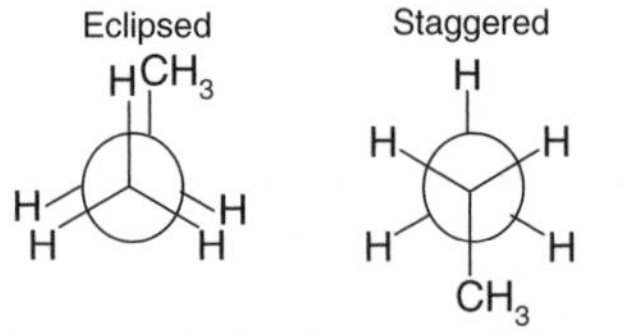

Eclipsed Staggered

Newman projection of propane conformers

In the eclipsed conformation of propane, we now have a larger CH_3 close to H atom. This results in increased repulsive force or increased steric strain. The energy difference between the eclipsed and staggered forms of propane is greater than that of ethane.

Conformational isomerism in butane Butane is a four-carbon- (sp^3-hybridized) atom-containing linear alkane. All are tetrahedrally arranged. When a hydrogen atom of propane is replaced by a methyl (CH_3) group, we have butane. There is rotation about two C–C σ bonds, but the rotation about C_2–C_3 is the most important.

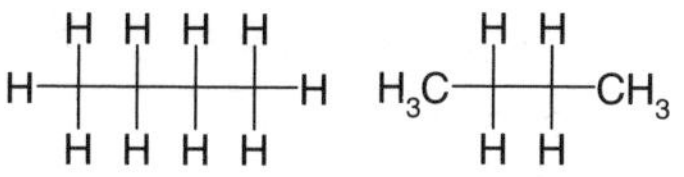

Butane

Among the conformers, the least stable is the first eclipsed structure, where two CH_3 groups are totally eclipsed, and the most stable is the first staggered conformer, where two CH_3 groups are staggered, and far apart from each other. When two bulky groups are staggered we get the *anti* conformation, and when they are at 60° to each other, we have the *gauche* conformer. In butane, the torsional energy is even higher than in propane. Thus, there is slightly restricted rotation about the C_2–C_3 bond in butane. The order of stability (from the highest to the lowest) among the following conformers is anti → gauche → another eclipsed → eclipsed. The most stable conformer has the lowest steric strain and torsional strain.

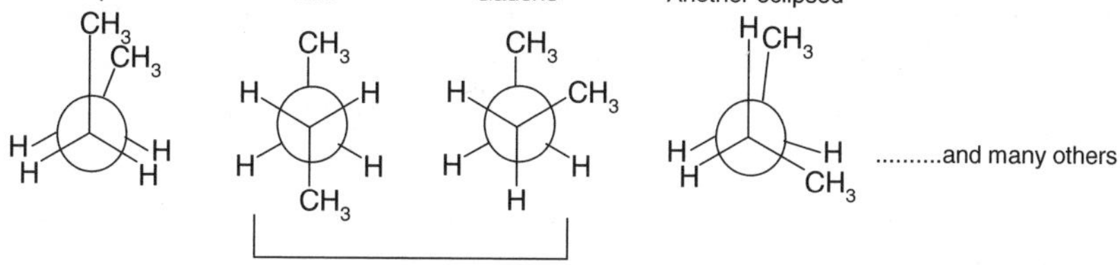

Eclipsed Anti Gauche Another eclipsed ....and many others

Staggered

Newman projection of butane conformers

Conformational isomerism in cyclopropane Cyclopropane is the first member of the cycloalkane series, and composed of three carbons and six hydrogen atoms (C_3H_6). The rotation about C–C bonds is quite restricted in cycloalkanes, especially in smaller rings, e.g. cyclopropane.

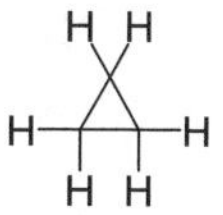

Cyclopropane

In cyclopropane, each C atom is still sp^3-hybridized, so we should have a bond angle of 109.5°, but each C atom is at the corner of an equilateral triangle, which has angles of 60°! As a result, there is considerable *angle strain*. The sp^3 hybrids still overlap but only just! This gives a very unstable and weak structure. The *angle strain* can be defined as the strain induced in a molecule when bond angle deviates from the ideal tetrahedral value. For example, this deviation in cyclopropane is from 109.5 to 60°.

Conformational isomerism in cyclobutane Cyclobutane comprises four carbons and eight hydrogen atoms (C_4H_8). If we consider cyclobutane to have a flat or planar structure, the bond angles will be 90°, so the angle strain (cf. 109.5°) will be much less than that of cyclopropane. However, cyclobutane in its planar form will give rise to torsional strain, since all H atoms are eclipsed.

Cyclobutane as a planar molecule
All H atoms are eclipsed
Bond angle = 90°

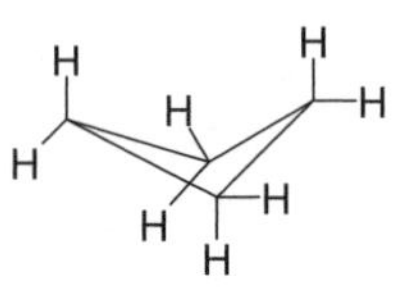

Most stable folded conformation of cyclobutane
H atoms are not eclipsed

Cyclobutane, in fact, is not a planar molecule. To reduce torsional strain, this compound attains the above nonplanar folded conformation. Hydrogen atoms are not eclipsed in this conformation and torsional strain is much less than in the planar structure. However, in this form angles are less than 90°, which means a slight increase in angle strain.

Conformational isomerism in cyclopentane Cyclopentane is a five-carbon cyclic alkane. If we consider cyclopropane as a planar and regular pentagon, the angles are 108°. Therefore, there is very little or almost no angle strain (cf. 109.5° for sp^3 hybrids). However, in this form the torsional strain is very large, because most of its hydrogen atoms are eclipsed. Thus, to reduce torsional strain, cyclopentane twists to adopt a puckered or envelope shaped, nonplanar conformation that strikes a balance between increased angle strain and decreased torsional strain. In this conformation, most of the hydrogen atoms are almost staggered.

Cyclopentane as a planar molecule
Bond angle = 108°

Most stable puckered conformation of cyclopentane
Most H atoms are nearly staggered

Conformational isomerism in cyclohexane Cyclohexane (C_6H_{12}) is a six-carbon cyclic alkane that occurs extensively in nature. Many pharmaceutically important compounds possess cyclohexane rings, e.g. steroidal molecules. If we consider cyclohexane as a planar and regular hexagon, the angles are 120° (cf. 109.5° for sp^3 hybrids).

Cyclohexane as a planar molecule
Bond angle 120°

Most stable chair conformation of cyclohexane
All neighbouring C-H bonds are staggered

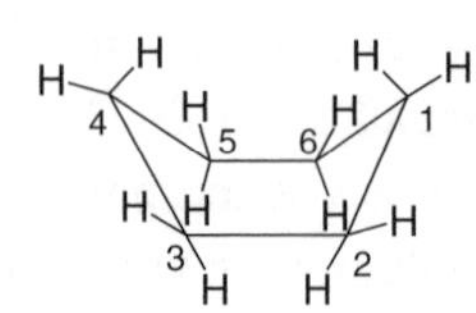

Boat conformation of cyclohexane

Again, in reality, cyclohexane is not a planar molecule. To strike a balance between torsional strain and angle strain, and to achieve more stability, cyclohexane attains various conformations, among which the *chair* and *boat conformations* are most significant. At any one moment 99.9 per cent of cyclohexane molecules will have the chair conformation.

The chair conformation of cyclohexane is the most stable conformer. In the chair conformation, the C–C–C angles can reach the strain free tetrahedral value (109.5°), and all neighbouring C–H bonds are staggered. Therefore, this conformation does not have any angle strain or torsional strain.

Another conformation of cyclohexane is the *boat conformation*. Here the H atoms on C_2–C_3 and C_5–C_6 are eclipsed, which results in an increased torsional strain. Also, the H atoms on C_1 and C_4 are close enough to produce steric strain.

In the chair conformation of cyclohexane, there are two types of position for the substituents on the ring, *axial* (perpendicular to the ring, i.e. parallel to the ring axis) and *equatorial* (in the plane of the ring, i.e. around the ring equator) positions. Six hydrogen atoms are in the axial positions and six others in the equatorial positions. Each carbon atom in the cyclohexane chair conformation has an axial hydrogen and an equatorial hydrogen atom, and each side of the ring has three axial and three equatorial hydrogen atoms.

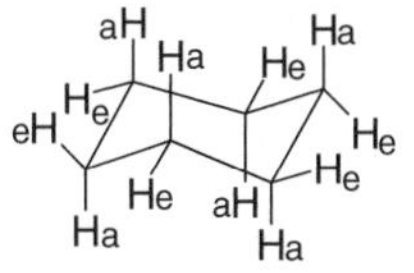

Chair conformation of cyclohexane
Six axial (a) and six equatorial (e) hydrogen atoms

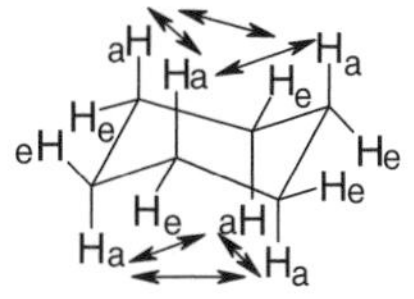

Chair conformation of cyclohexane
Diaxial interaction

When all 12 substituents are hydrogen atoms, there is no steric strain. The presence of any groups larger than H changes the stability by increasing the steric strain, especially if these groups are present in axial positions. When axial, *diaxial interaction* can cause steric strain. In the equatorial case, there is more room and less steric strain. Bulky groups always preferably occupy equatorial positions.

Because of axial and equatorial positions in the chair conformation of cyclohexane, one might expect to see two isomeric forms of a monosubstituted cyclohexane. However, in reality, only one monosubstituted form exists, because cyclohexane rings are conformationally mobile at room temperature. Different chair conformations interconvert, resulting in the exchange of axial and equatorial positions. This interconversion of chair conformations is known as a *ring-flip*. During ring-flip, the middle four carbon atoms remain in place, while the two ends are folded in opposite directions. As a result, an axial substituent in one chair form of cyclohexane becomes an equatorial substituent in the ring-flipped chair form, and vice versa.

R = Any substituent group or atom other than H

R in axial position ⇌ R in equatorial position

Ring-flip in chair conformation of monosubstituted cyclohexane

Configurational isomers

Configurational isomers differ from each other only in the arrangement of their atoms in space, and cannot be converted from one into another by rotations about single bonds within the molecules. Before we look into the details of various configurational isomers, we need to understand the concept of *chirality.*

Chirality Many objects around us are handed. For example, our left and right hands are mirror images of each other, and cannot be superimposed on each other. Other chiral objects include shoes, gloves and printed pages. Many molecules are also handed, i.e. they cannot be superimposed on their mirror images. Such molecules are called *chiral* molecules. Many compounds that occur in living organisms, e.g. carbohydrates and proteins, are chiral

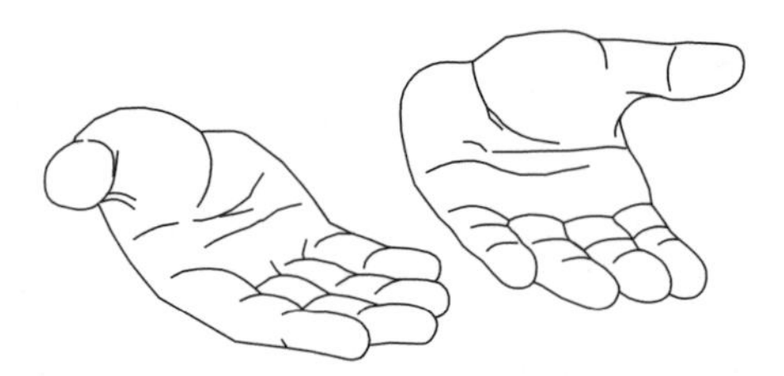

The most common feature in *chiral molecules* is a tetrahedral (i.e. sp^3-hybridized) carbon atom with four different atoms or groups attached. Such a carbon atom is called a *chiral carbon* or an *asymmetric carbon*. Chiral molecules do not have a plane of symmetry.

X, W—C*—Y, Z
Chiral carbon
Four different groups/atoms present

X, W—C—Z, Z
Achiral carbon
At least two same groups/atoms (**Z**) present

When there are two or more atoms/groups that are the same, the carbon is called *achiral*. Achiral molecules often have a *plane of symmetry*. If a molecule can be divided by a plane into two equal halves that are mirror images of each other, the plane is a *plane of symmetry*, and the molecule is not chiral. With achiral molecules, the compound and its mirror image are the same, i.e. they can be superimposed.

If you rotate the mirror image through 180°, it is identical to the original structure.

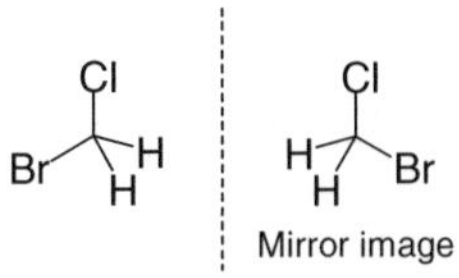

Mirror image

Achiral molecule
Superimposable mirror image

Enantiomers The Greek word *enantio* means 'opposite'. A chiral molecule and its mirror image are called *enantiomers* or an *enantiomeric pair.* They are nonsuperimposable. The actual arrangement or orientation (in space) of atoms/groups attached to the chiral carbon (stereogenic centre or stereocentre) is called the *configuration* of a compound.

The arrangement of W, X, Y and Z is configuration

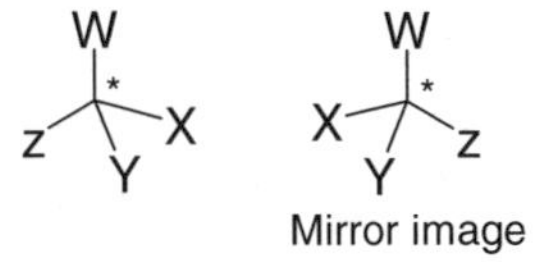

Mirror image

Enantiomers
Not superimposable

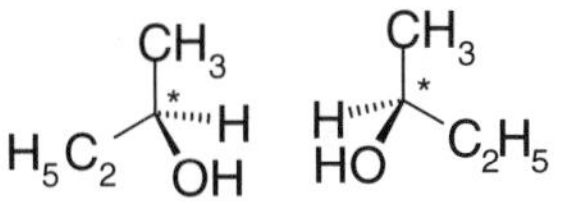

(-)-2-Butanol (+)-2-Butanol

Enantiomers of 2-butanol

Properties of enantiomers Enantiomers share same physical properties, e.g. melting points, boiling points and solubilities. They also have same chemical properties. However, they differ in their activities with *plane polarized light*, which gives rise to *optical isomerism*, and also in their pharmacological actions.

Drawing a chiral molecule (enantiomer) On a plane paper, chiral molecules can be drawn using *wedge bonds*. There are also a few other methods that use horizontal bonds representing bonds pointing out of the paper and vertical bonds pointing into the paper. Some examples are given below.

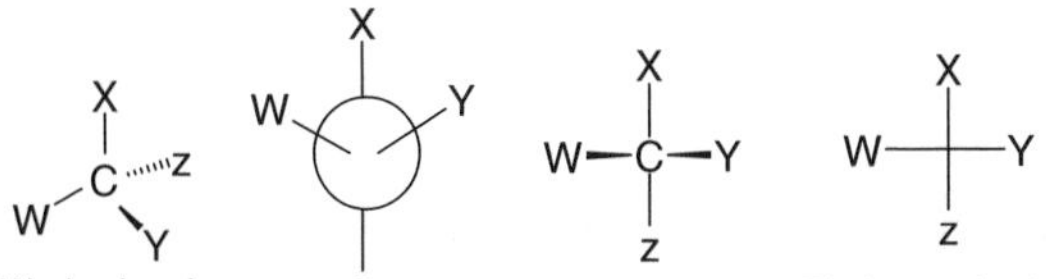

Wedge bonds

Fischer projection

Optical isomers Light consists of waves that are composed of electrical and magnetic vectors (at right angles). If we looked 'end-on' at light as it travels, we would see it oscillates in all directions. When a beam of ordinary light is passed through a *polarizer*, the polarizer interacts with the electrical vector in such a way that the light emerging from it oscillates in one direction or plane. This is called *plane polarized light.*

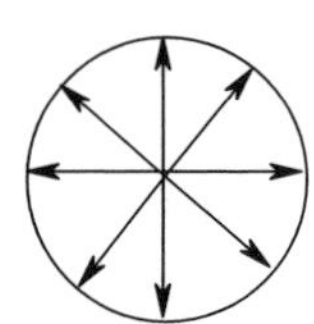
Light oscillates in all directions

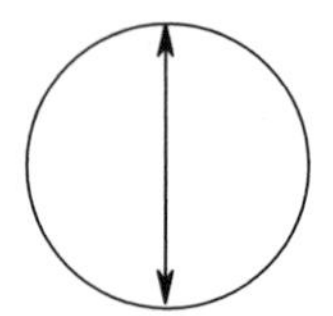
The plane of oscillation in plane-polarized light

When plane-polarized light passes through a solution of an enantiomer, the plane of light rotates. Any compounds that rotate plane-polarized light are called *optically active*. If the rotation is in a clockwise direction, the enantiomer is said to be *dextrorotatory* and is given the (+) sign in front of its name. Anticlockwise rotation gives an enantiomer which is known as *levorotatory* and is given the sign (−) in front of its name.

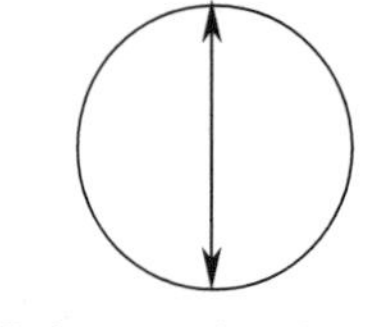
Before passing through a solution of an enantiomer

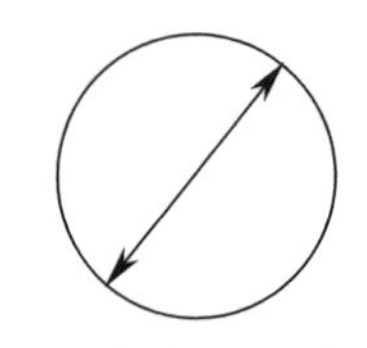
Rotates clock-wise (right) Dextrorotatory (+)

or

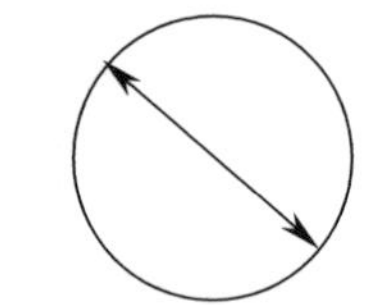
Rotates anti-clock-wise (left) Levorotatory (–)

After passing through a solution of an enantiomer

The amount of rotation can be measured with an instrument called a *polarimeter.* A solution of optically active molecule (enantiomer) is placed in a sample tube, plane-polarized light is passed through the tube and a rotation of the polarization plane takes place. The light then goes through a second polarizer called an analyser. By rotating the analyser until the light passes through it, the new plane of polarization can be found, and the extent of rotation that has taken place can be measured.

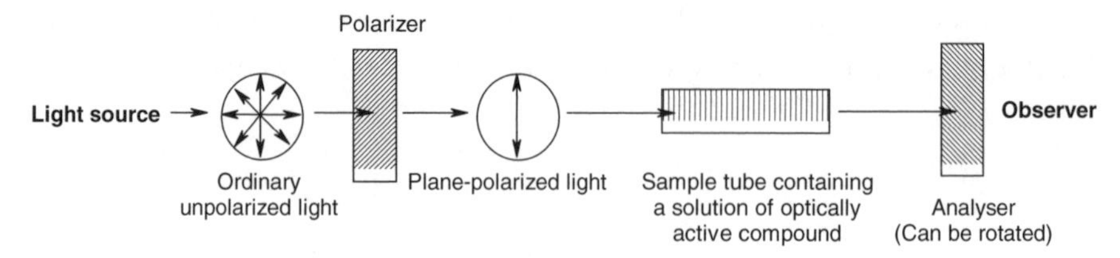

Enantiomers are optically active, and are called *optical isomers*, with one being (+) and the other (−).

When we have a pair of enantiomers, each rotates the plane-polarized light by the same amount, but in the opposite direction. A mixture of enantiomers with the same amount of each is called a *racemic mixture*. Racemic mixtures are optically inactive (i.e. they cancel each other out) and are denoted by (±).

Specific rotations The more molecules (optically active) the light beam encounters, the greater the observed rotation. Thus, the amount of rotation depends on both sample concentration and sample path length.

If we keep the concentration constant but double the length of the sample tube, the observed rotation doubles. The amount of rotation also depends on the temperature and the wavelength of the light used.

Therefore, to obtain a meaningful optical rotation data, we have to choose standard conditions, and here the concept of *specific rotation* comes in.

The *specific rotation* of a compound, designated as $[\alpha]_D$, is defined as the observed rotation, α, when the sample path length l is 1 dm, the sample concentration C is 1g/mL and light of 599.6 nm wavelength (the D line of a sodium lamp, which is the yellow light emitted from common sodium lamps) is used.

$$[\alpha]_D = \frac{\text{Observed rotation } \alpha \text{ in degrees}}{\text{Pathlength, } l(\text{dm}) \times \text{Concentration, } C(\text{g/mL})} = \frac{\alpha}{l \times C}$$

As the specific rotation also depends on temperature, the temperature at which the rotation is measured is often shown in the equation. A specific rotation measured at 25 °C is denoted more precisely as

$$[\alpha]_D^{25}$$

When optical rotation data are expressed in this standard way, the specific rotation, $[\alpha]_D$, is a physical constant, characteristic of a given optically active compound. For example, the specific rotation of morphine is −132°, i.e.

$$[\alpha]_D^{25} = -132^\circ$$

This means that the D line of a sodium lamp ($l = 599.6$ nm) was used for light, that a temperature of 25 °C was maintained and that a sample containing 1.00 g/mL of the optically active morphine, in a 1 dm tube, produced a rotation of 132° in an anti-clockwise direction.

How to designate the configuration of enantiomers We have already seen that a (+) or (−) sign indicates the optical activity of an enantiomer.

However, the optical activity does not tell us the actual configuration of an enantiomer. It only gives us the information whether an enantiomer rotates the plane-polarized light clockwise or anti-clockwise.

Let us look at the example of glyceraldehyde, an optically active molecule. Glyceraldehyde can exist as enantiomers, i.e. (+) and (−) forms, but the sign does not describe the exact configuration.

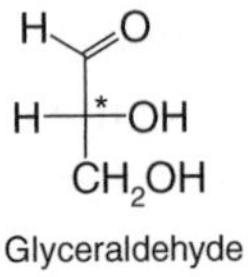

Glyceraldehyde

There are two systems to designate configuration of enantiomers: the D and L system, and the (*R*) and (*S*) system (also known as the Cahn–Ingold–Prelog system).

D and L system Emil Fischer used glyceraldehyde as a standard for the D and L system of designating configuration. He arbitrarily took the (+)-glyceraldehyde enantiomer and assigned this as D-glyceraldehyde. The other enantiomer is the (−)-glyceraldehyde and this was assigned as L-glyceraldehyde. We can easily identify the only difference in the following structures, which is the orientation of the hydroxyl group at the chiral centre. In the case of D-glyceraldehyde the –OH group on the chiral carbon is in on the right hand side, whereas in L-glyceraldehyde it is on the left. In the D and L system, structures that are similar to glyceraldehyde (at chiral carbon) are compared, for example 2,3-dihydroxypropanoic acid.

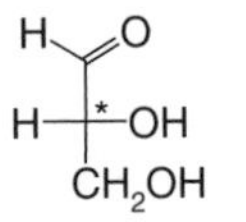

(+)-D-Glyceraldehyde 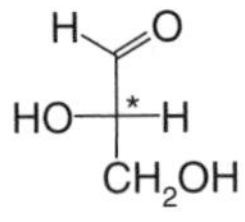 (−)-L-Glyceraldehyde

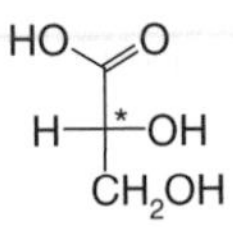

D-2,3-dihydroxypropanoic acid
The -OH on the chiral carbon (*) is on the right

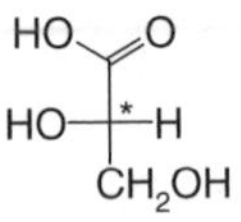

D-2,3-dihydroxypropanoic acid
The -OH on the chiral carbon (*) is on the left

One must remember that there is no correlation between D and L configurations, and (+) and (−) rotations. The D-isomer does not have to have a (+) rotation, and similarly the L-isomer does not have to have a (−) rotation. For some compounds the D-isomer is (+), and for others the L-isomer may be (+). Similarly, for some we may have L (−) and others may have D

(−). This D and L system is common in biology/biochemistry, especially with sugars and amino acids (with amino acids, the $-NH_2$ configuration is compared with the –OH of glyceraldehyde). This system is particularly used to designate various carbohydrate or sugar molecules, e.g. D-glucose, L-rhamnose and L-alanine.

(*R*) and (*S*) system (Cahn–Ingold–Prelog system) Three chemists, R. S. Cahn (UK), C. K. Ingold (UK) and V. Prelog (Switzerland), devised a system of nomenclature that can describe the configuration of enantiomers more precisely. This system is called the (*R*) and (*S*) system, or the *Cahn–Ingold–Prelog system.*

According to this system, one enantiomer of 2-hexanol should be designated (*R*)-2-hexanol, and the other (*S*)-2-hexanol. *R* and *S* came from the Latin words *rectus* and *sinister*, meaning right and left, respectively.

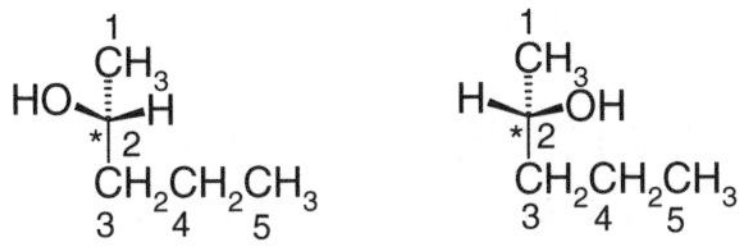

Enantiomers of 2-hexanol

The following rules or steps are applied for designating any enantiomer as *R* or *S*.

(a) Each of the four groups attached to the chiral carbon is assigned **1–4** (or **a–d**) in terms of order of priority or preference, **1** being the highest and **4** being the lowest priority. Priority is first assigned on the basis of the atomic number of the atom that is directly attached to the chiral carbon.

Higher atomic number gets higher priority. This can be shown by the structure of 2-hexanol.

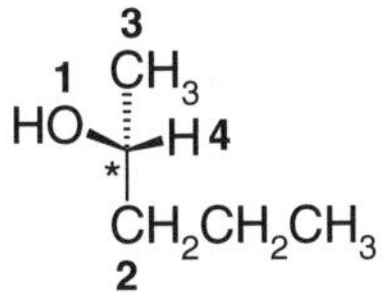

Priority determination in 2-hexanol

(b) When a priority cannot be assigned on the basis of the atomic numbers of the atoms that are directly attached to the chiral carbon, the next set of atoms in the unassigned groups is examined. This process is continued to the first point of difference.

In 2-hexanol, there are two carbon atoms directly attached to the chiral carbon; one is of the methyl group, and the other is of the propyl group. In this case, on the basis of the atomic number of the carbon atom alone, one cannot assign the priority of these two carbon atoms, and must consider the next set of atoms attached to these carbon atoms. When we examine the methyl group of the enantiomer, we find that the next set of atoms consists of three H atoms. On the other hand, in the propyl group, the next set of atoms consists of one C and two H atoms. Carbon has a higher atomic number than H, so the $CH_3CH_2CH_2$– group receives priority over CH_3.

(c) Groups containing double or triple bonds (π bonds) are assigned priorities as if both atoms were duplicated or triplicated. For example

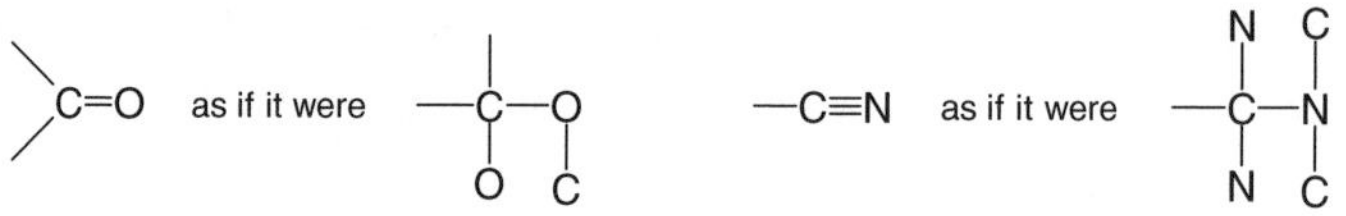

>C=O as if it were —C(O)—O(C)

—C≡N as if it were —C(N)(N)—N(C)(C)

(d) Having decided on the priority of the four groups, one has to arrange (rotate) the molecule in such a way that group 4, i.e. the lowest priority, is pointing away from the viewer.

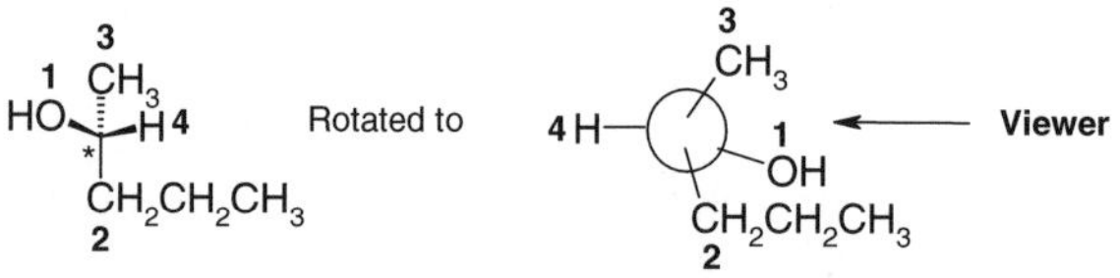

Rotated to

Viewer

The least priority H atom (**4**) is away from the viewer

Then an arrow from group **1** → **2** → **3** is to be drawn. If the direction is clockwise, it is called an (*R*)-isomer. If it is anti-clockwise, it is called an (*S*)-isomer. With enantiomers, one will be the (*R*)-isomer and the other the (*S*)-isomer. Again, there is no correlation between (*R*) and (*S*), and (+) and (−).

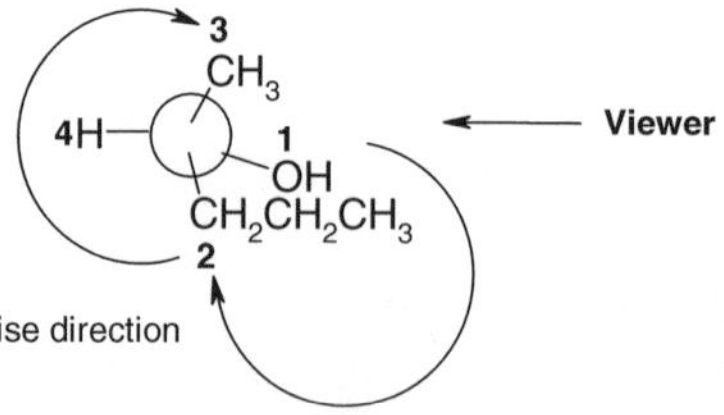

Viewer

(*R*)-2-Hexanol
Priority order **1** to **4** is in clock-wise direction

Following the *Cahn–Ingold–Prelog system,* it is now possible to draw the structures of (*R*)- and (*S*)-enantiomers of various chiral molecules, for example 2,3-dihydroxypropanoic acid, where the priorities are 1 = OH, 2 = COOH, 3 = CH_2OH and 4 = H.

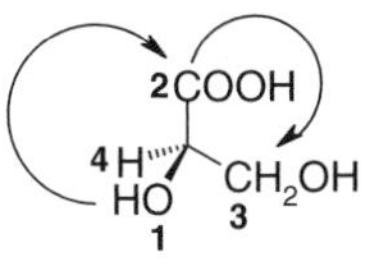

(R)-2,3-Dihydroxypropanoic acid
Priority order **1** to **4** is in clockwise direction

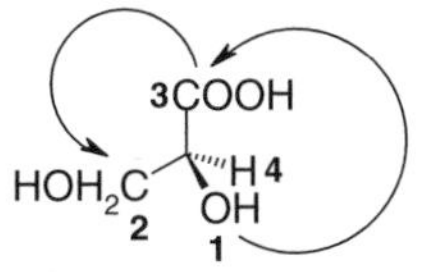

(S)-2,3-Dihydroxypropanoic acid
Priority order **1** to **4** is in clockwise direction

When there is more than one stereocentre (chiral carbon) present in a molecule, it is possible to have more than two stereoisomers. It is then necessary to designate all these stereoisomers using the (R) and (S) system. In 2,3,4-trihydroxybutanal, there are two chiral carbons. The chiral centres are at C-2 and C-3. Using the (R) and (S) system, one can designate these isomers as follows.

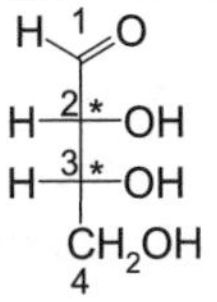

2,3,4-Trihydroxybutanal
A molecule with two chiral centres (*)

C-2	C-3	Designation of stereoisomer
R	R	(2R, 3R)
R	S	(2R, 3S)
S	R	(2S, 3R)
S	S	(2S, 3S)

Stereoisomerism in compounds with two stereo centres: diastereomers and meso structure In compounds whose stereoisomerism is due to tetrahedral stereocentres, the total number of stereoisomers will not exceed 2^n, where n is the number of tetrahedral stereocentres. For example, in 2,3,4-trihydroxybutanal, there are two chiral carbons. The chiral centres are at C-2 and C-3. Therefore, the maximum number of possible isomers will be $2^2 = 4$. All four stereoisomers of 2,3,4-trihydroxybutanal (**A**–**D**) are optically active, and among them there are two enantiomeric pairs, **A** and **B**, and **C** and **D**, as shown in the structures below.

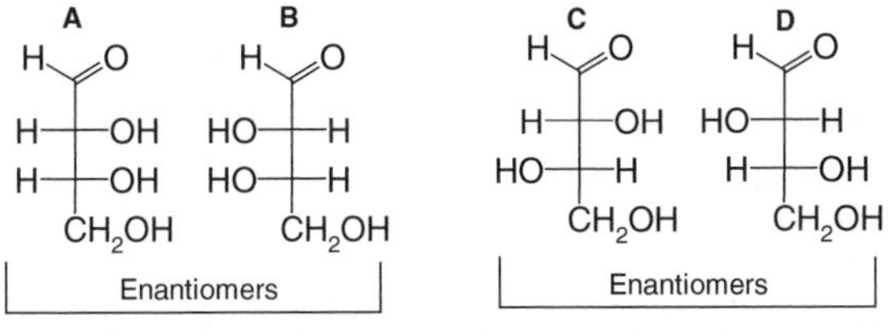

Four possible stereoisomers of 2,3,4-trihydroxybutanal

If we look at structures **A** and **C** or **B** and **D**, we have stereoisomers, but not enantiomers. These are called *diastereomers*. Diastereomers have different physical properties (e.g. melting point). Other pairs of diastereomers among the stereoisomers of 2,3,4-trihydroxybutanal are **A** and **D**, and **B** and **C**.

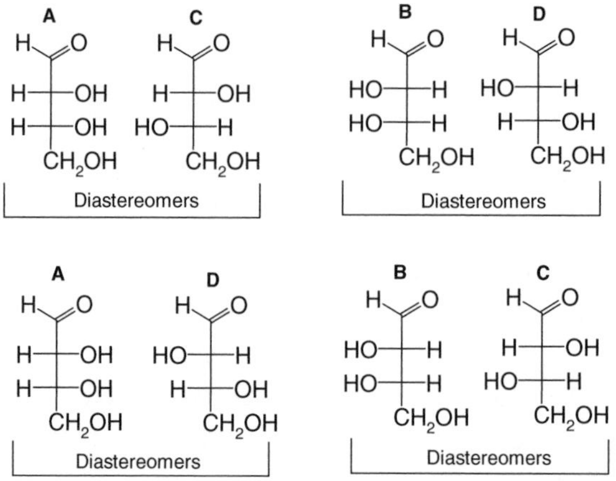

A C Diastereomers

B D Diastereomers

A D Diastereomers

B C Diastereomers

Now, let us consider another similar molecule, tartaric acid, where there are two chiral carbons. In tartaric acid, four isomeric forms are theoretically expected ($2^2 = 4$). However, because one half of the tartaric acid molecule is a mirror image of the other half, we get a *meso* structure. This means this compound and its mirror image are superimposable, i.e. they are the same compound. Thus, instead of four, we obtain only three stereoisomers for tartaric acid.

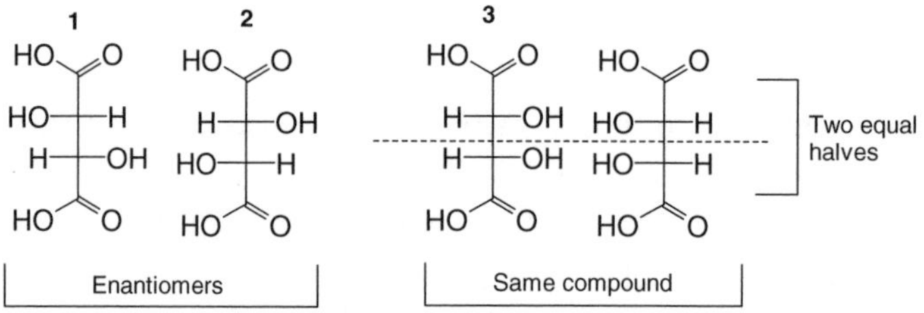

1 2 Enantiomers

3 Same compound

Two equal halves

Stereoisomers of tartaric acid

Structures **1** and **2** are enantiomers, and both are optically active. In structures **3** and **4**, there is a plane of symmetry, i.e., there is a mirror image within a single molecule. Such a structure is called a *meso* structure. Structures **3** and **4** are superimposable, and essentially are the same compound. Hence, we have a *meso*-tartaric acid and it is achiral (since it has a plane of symmetry, and it is superimposable on its mirror image). *Meso*-tartaric acid is optically inactive. Therefore, for tartaric acid, we have (+), (−) and *meso*-tartaric acid.

Cyclic compounds Depending on the type of substitution on a ring, the molecule can be chiral (optically active) or achiral (optically inactive). For example, 1,2-dichlorocyclohexane can exists as meso compounds (optically inactive) and enantiomers (optically active). If the two groups attached to the ring are different, i.e. no plane of symmetry, there will be four isomers.

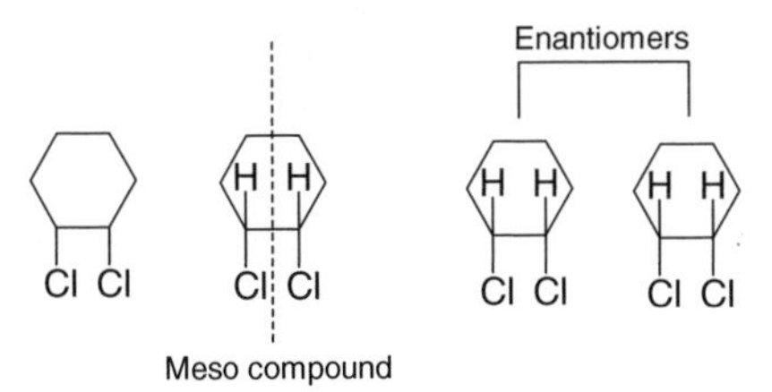

Stereoisomerism in1,2-dichlorocyclohexane

Geometrical isomers

Geometrical isomerism is found in alkenes and cyclic compounds. In alkenes, there is restricted rotation about the double bond. When there are substituent groups attached to the double bond, they can bond in different ways, resulting in *trans* (opposite side) and *cis* (same side) isomers. These are called *geometrical isomers*. They have different chemical and physical properties. Each isomer can be converted to another when enough energy is supplied, e.g. by absorption of UV radiation or being heated to temperatures around 300 °C. The conversion occurs because the π bond breaks when energy is absorbed, and the two halves of the molecule can then rotate with respect to each other before the π bond forms again.

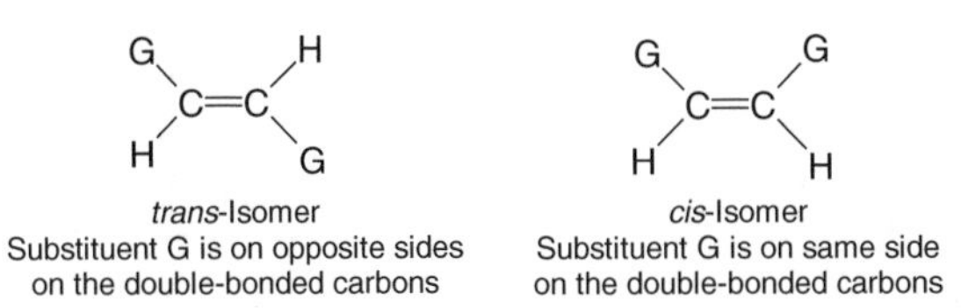

trans-Isomer
Substituent G is on opposite sides on the double-bonded carbons

cis-Isomer
Substituent G is on same side on the double-bonded carbons

When there is the same substituent attached to the double bonded carbons, as in the above example, it is quite straightforward to designate *trans* or *cis*. However, if there are more than one different groups or atoms present, as in the following examples, the situation becomes a bit more complicated for assigning *cis* and *trans*.

Alkenes with different substituents on the double-bonded carbons

To simplify this situation, the *E/Z* system is used for naming geometrical isomers. *Z* stands for German *zusammen*, which means the same side, and *E* for German *entgegen,* meaning on the opposite side.

In the *E* and *Z* system, the following rules or steps are followed.

(a) On each C atom of the double bond, we have to assign the priority of the atoms bonded. The priority should be on the same basis as the (*R*)/(*S*) system (i.e. on the basis of atomic number).

(b) If the two higher priority groups of the two C atoms are on the *same* side of the double bond, it is called the (*Z*)-isomer.

(c) If the two higher priority groups of the two C atoms are on *opposite* sides of the double bond, it is called the (*E*)-isomer.

Let us take a look at 1-bromo-1,2-dichloroethene as an example. In this molecule, atoms attached are Cl and Br on C-1, and Cl and H on C-2. Atomic numbers of these substituents are in the order of Br > Cl > H. So, once the priorities are assigned, we can easily draw the (*E*)- and (*Z*)-isomers of 1-bromo-1,2-dichloroethene in the following way.

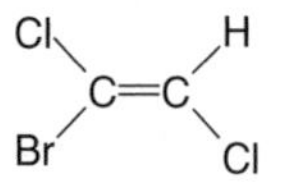

(*Z*)-1-Bromo-1,2-dichloroethene
The two higher priority groups are on the same side

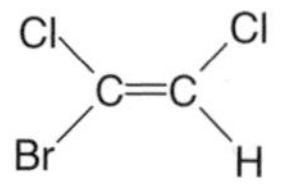

(*E*)-1-Bromo-1,2-dichloroethene
The two higher priority groups are on the opposite side

Now, let us have a look at the cyclic compounds. We can use this (*E*) and (*Z*) system for a cyclic compound when two or more groups are attached to a ring. For example, if in the following substituted cyclopentane A and B are different groups, each C atom attached to A and B is a chiral carbon or stereocentre.

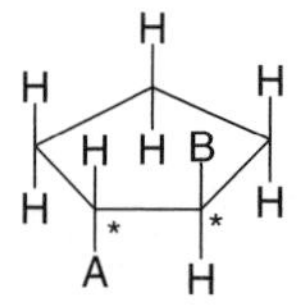

(*E*)-form
The two higher priority groups (A or B > H) are on the opposite side

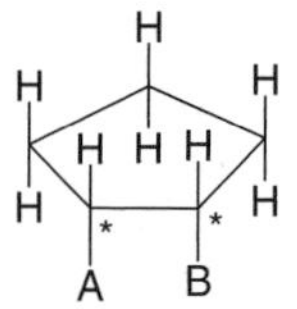

(*Z*)-form
The two higher priority groups (A or B > H) are on the same side

In 1-bromo-2-chlorocyclopentane, there are two chiral centres. Therefore, four possible stereoisomers can be expected ($2^2 = 4$).

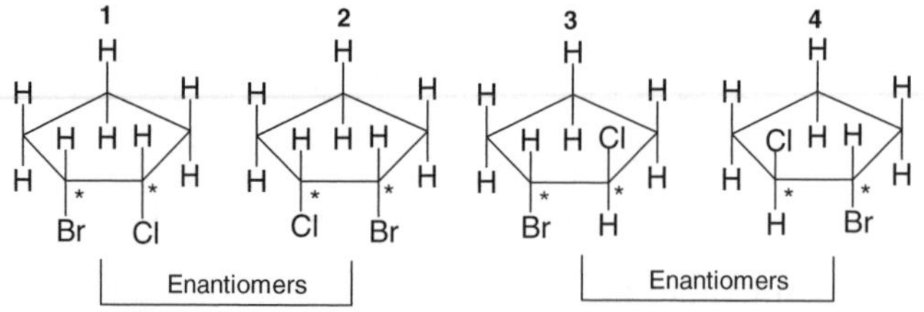

Enantiomers Enantiomers

Four possible isomers of 1-bromo-2-chlorocyclopentane

The isomers are (+)-*cis*-2-bromo-1-chlorocyclopentane (**1**), (−)-*cis*-1-bromo-2-chlorocyclopentane (**2**), (+)-*trans*-2-bromo-1-chlorocyclopentane (**3**) and (−)-*trans*-1-bromo-2-chlorocyclopentane (**4**).

However, when A = B, i.e. two substituents are the same, as in 1,2-dihydroxycyclopentane, only three isomers are possible, because of the presence of a plane of symmetry with this molecule. In this case, we have a *meso* structure.

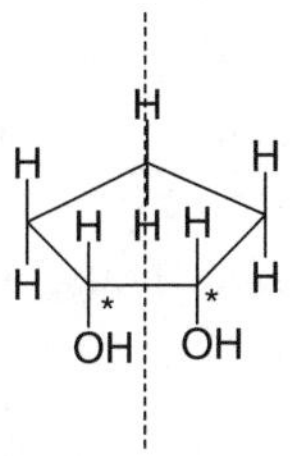

1,2-Dihydroxycyclopentane
There is a plane of symmetry within the molecule

In 1,2-dihydroxycyclohexane, there exists a plane of symmetry within the molecule, and instead of four, it produces three isomers as follows.

One is an optically inactive *meso* isomer (*cis* or (*Z*)-isomer) and two are optically active *trans* or (*E*)-isomers. With cyclohexane, we can have equatorial and axial bonds. Thus, with *trans* structure, we obtain di-axial and di-equatorial bonds, and with *cis* structure we obtain axial–equatorial bonds.

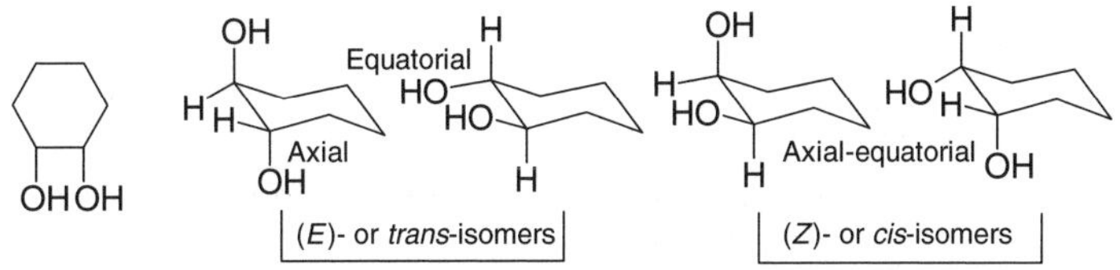

(*E*)- or *trans*-isomers | (*Z*)- or *cis*-isomers

Four possible isomers of 1,2-dichlorocyclohexane

3.3 Significance of stereoisomerism in determining drug action and toxicity

Pharmacy is a discipline of science that deals with various aspects of drugs. All drugs are chemical entities, and a great majority (30–50 per cent) of them contain stereocentres, show stereoisomerism and exist as enantiomers. Moreover, the current trend in drug markets is a rapid increase of the sales of chiral drugs at the expense of the achiral ones. In the next few years, chiral drugs, whether enantiomerically pure or sold as a racemic mixture, will dominate drug markets. It is therefore important to understand how drug chirality affects its interaction with drug targets and to be able to use proper nomenclature in describing the drugs themselves and the nature of forces responsible for those interactions.

Most often only one form shows correct physiological and pharmacological action. For example, only one enantiomer of morphine is active as an analgesic, only one enantiomer of glucose is metabolized in our body to give energy and only one enantiomeric form of adrenaline is a neurotransmitter.

One enantiomeric form of a drug may be active, and the other may be inactive, less active or even toxic. Not only drug molecules, but also various other molecules that are essential for living organisms exist in stereoisomeric forms, and their biological properties are often specific to one stereoisomer. Most of the molecules that make up living organisms are chiral, i.e. show stereoisomerism. For example, all but one of the 20 essential amino acids are chiral. Thus, it is important to understand stereochemistry for a better understanding of drug molecules, their action and toxicity.

Ibuprofen is a popular analgesic and anti-inflammatory drug. There are two stereoisomeric forms of ibuprofen. This drug can exist as (*S*)- and (*R*)-stereoisomers (enantiomers). Only the (*S*)-form is active. The (*R*)-form is completely inactive, although it is slowly converted in the body to the active (*S*)-form. The drug marketed under the trade names, commercially known as Advil®, Anadin®, Arthrofen®, Brufen®, Nurofen®, Nuprin®, Motrin® etc., is a racemic mixture of (*R*)- and (*S*)-ibuprofen.

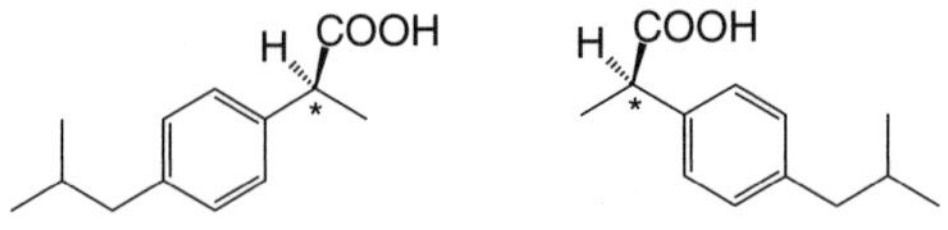

(*S*)-Ibuprofen
Active stereoisomer

(*R*)-Ibuprofen
Inactive stereoisomer

In the early 1950s, Chemie Grunenthal, a German pharmaceutical company, developed a drug called thalidomide. It was prescribed to prevent nausea or morning sickness in pregnant women. The drug, however, caused severe adverse effects on thousands of babies who were exposed to this drug while their mothers were pregnant. The drug caused 12 000 babies to be born with severe birth defects, including limb deformities such as missing or stunted limbs. Later, it was found that thalidomide molecule can exist in two stereoisomeric forms; one form is active as a sedative, but the other is responsible for its *teratogenic* activity (the harmful effect on the foetus).

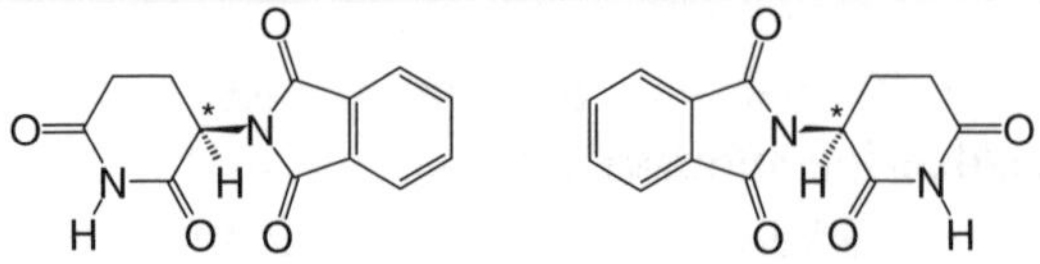

Sedative

Teratogenic

Thalidomide stereoisomers

Limonene is a monoterpene that occurs in citrus fruits. Two enantiomers of limonene produce two distinct flavours: (−)-limonene is responsible for the flavour of lemons and (+)-limonene for orange. Similarly, one enantiomeric form of carvone is the cause of caraway flavour, while the other enantiomer has the essence of spearmint.

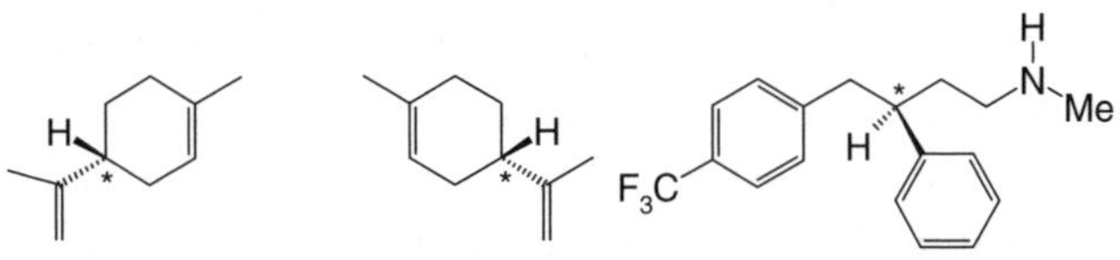

(+)-Limonene (in orange)   (−)-Limonene (in lemon)   (*S*)-Fluoxetine (prevents migraine)

Fluoxetine, commonly known as Prozac®, as a racemic mixture is an antidepressant drug, but has no effect on migraine. The pure *S*-enantiomer works remarkably well in the prevention of migraine and is now under clinical evaluation.

3.4 Synthesis of chiral molecules

3.4.1 Racemic forms

On many occasions, a reaction carried out with *achiral* reactants results in the formation of a chiral product. In the absence of any chiral influence, the outcome of such reactions is the formation of a racemic form. For example, hydrogenation of ethylmethylketone yields a racemic mixture of 2-hydroxybutane.

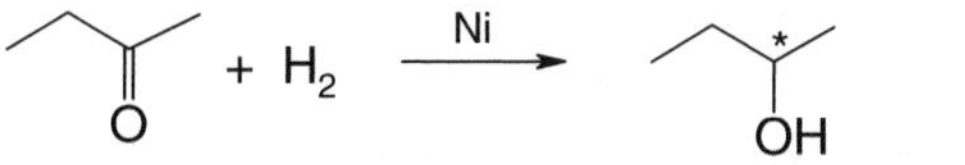

Ethylmethylketone + H_2 —Ni→ (±)-2-Hydroxybutane

Similarly, the addition of HBr to 1-butene produces a racemic mixture of 2-bromobutane.

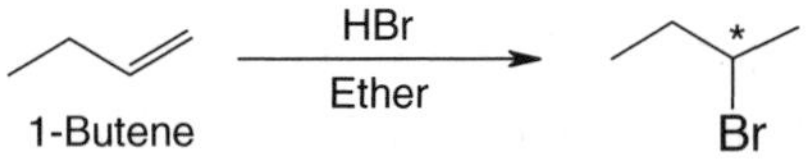

1-Butene —HBr, Ether→ (±)-2-Bromobutane

3.4.2 Enantioselective synthesis

A reaction that produces a predominance of one enantiomer over other is known as *enantioselective synthesis*. To carry out an enantioselective reaction, a chiral reagent, solvent, or catalyst must assert an influence on the course of the reaction. In nature, most of the organic or bioorganic reactions are enantioselective, and the chiral influence generally comes from various *enzymes*. Enzymes are chiral molecules, and they possess an active site where the reactant molecules are bound momentarily during the reaction. The active site in any enzyme is chiral, and allows only one enantiomeric form of a chiral reactant to fit in properly. Enzymes are also used to carry out enantioselective reactions in the laboratories. *Lipase* is one such enzyme used frequently in the laboratories.

Lipase catalyses a reaction called *hydrolysis*, where esters react with a molecule of water and are converted to a carboxylic acid and an alcohol.

The use of lipase allows the hydrolysis to be used to prepare almost pure enantiomers.

Ethyl (±)-2-fluorohexanoate —Lipase, H-OH→ Ethyl (*R*)-(+)-2-fluorohexanoate (>99%) + (*S*)-(−)-2-Fluorohexanoic acid (>69%) + EtOH

3.5 Separation of stereoisomers: resolution of racemic mixtures

A number of compounds exist as racemic mixtures (±), i.e. a mixture of equal amounts of two enantiomers, (−) and (+). Often, one enantiomer shows medicinal properties. Therefore, it is important to purify the racemic mixture so that active enantiomer can be obtained. The separation of a mixture of enantiomers is called the *resolution of a racemic mixture*.

Through luck, in 1848, Louis Pasteur was able to separate or resolve racemic tartaric acid into its (+) and (−) forms by *crystallization*. Two enantiomers of the sodium ammonium salt of tartaric acid give rise to two distinctly different types of chiral crystal that can then be separated easily. However, only a very few organic compounds crystallize into separate crystals (of two enantiomeric forms) that are visibly chiral as are the crystals of the sodium ammonium salt of tartaric acid. Therefore, Pasteur's method of separation of enantiomers is not generally applicable to the separation of enantiomers.

One of the current methods for resolution of enantiomers is the reaction of a racemic mixture with a single enantiomer of some other compound.

This reaction changes a racemic form into a mixture of diastereomers. Diastereomers have different b.p., m.p. and solubilities, and can be separated by conventional means, e.g. recrystallization and chromatography.

Resolution of a racemic mixture can also be achieved by using an enzyme. An enzyme selectively converts one enantiomer in a racemic mixture to another compound, after which the unreacted enantiomer and the new compound are separated. For example, lipase is used in the hydrolysis of chiral esters as shown above.

Among the recent instrumental methods, *chiral chromatography* can be used to separate enantiomers. The most commonly used chromatographic technique is chiral high performance liquid chromatography (HPLC).

Diastereomeric interaction between molecules of the racemic mixture and the chiral chromatography medium causes enantiomers of the racemate to move through the stationary phase at different rates.

3.6 Compounds with stereocentres other than carbon

Silicon (Si) and germanium (Ge) are in the same group of the periodic table as carbon, and they form tetrahedral compounds as carbon does. When four different groups are situated around the central atom in silicon, germanium and nitrogen compounds, the molecules are chiral. Sulphoxides, where one of the four groups is a nonbonding electron pair, are also chiral.

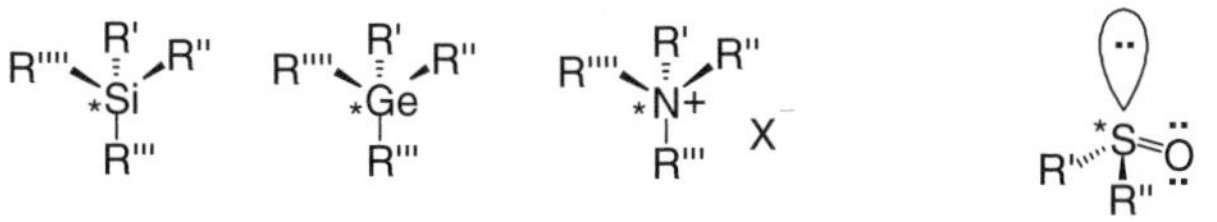

Chiral compounds with silicon, germanium, and nitrogen stereocentres Chiral sulphoxide

3.7 Chiral compounds that do not have a tetrahedral atom with four different groups

A molecule is chiral if it is not superimposable on its mirror image. A tetrahedral atom with four different groups is just one of the factors that confer chirality on a molecule. There are a number of molecules where a tetrahedral atom with four different groups is not present, yet they are not superimposable, i.e. chiral. For example, 1,3-dichloroallene is a chiral molecule, but it does not have a tetrahedral atom with four different groups.

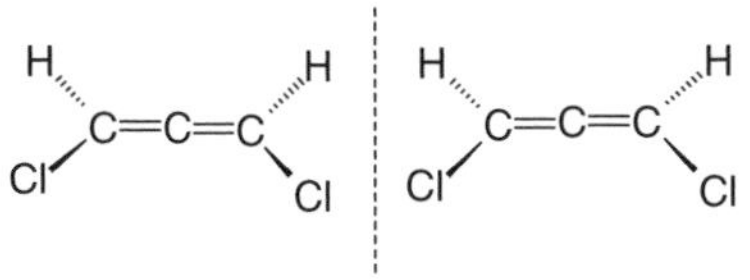

1,3-Dichloroallene

An allene is a hydrocarbon in which one atom of carbon is connected by double bonds with two other atoms of carbon. Allene is also the common name for the parent compound of this series, 1,2-propadiene. The planes of the π bonds of allenes are perpendicular to each other. This geometry of the π bonds causes the groups attached to the end carbon atoms to lie in perpendicular planes. Because of this geometry, allenes with different substitutents on the end carbon atoms are chiral. However, allenes do not show *cis–trans* isomerism.

Recommended further reading

Robinson, M. J. T. *Organic Stereochemistry*, Oxford University Press, Oxford, 2002.

4

Organic functional groups

Learning objectives:

After completing this chapter the student should be able to

- recognize various organic functional groups;
- discuss the importance of organic functional groups in determining drug action and toxicity;
- describe the significance of organic functional groups in determining stability of drug molecules;
- outline the preparation and reactions of alkanes, alkenes, alkynes and their derivatives;
- define aromaticity, recognize aromatic compounds and describe the preparation and reactions of various aromatic compounds;
- provide an overview of heterocyclic aromatic chemistry;
- classify amino acids, describe the properties of amino acids and discuss the formation of peptides;
- explain the fundamentals of the chemistry of nucleic acids.

Chemistry for Pharmacy Students Satyajit D Sarker and Lutfun Nahar

4.1 Organic functional groups: definition and structural features

All organic compounds are grouped into classes based on characteristic features called *functional groups*. A *functional group* is an atom or a group of atoms within a molecule that serves as a site of chemical reactivity. Carbon combines with other atoms such as H, N, O, S and halogens to form functional groups. A *reaction* is the process by which one compound is transformed into a new compound. Thus, functional groups are important in chemical reactions. It is important that you are able to recognize these functional groups because they dictate the physical, chemical and other properties of organic molecules, including various drug molecules. The most important functional groups are shown in the following table, with the key structural elements and a simple example.

Name	General structure	Example
Alkane	R–H	$CH_3CH_2CH_3$ Propane
Alkene	>C=C<	$CH_3CH_2CH{=}CH_2$ 1-Butene
Alkyne	—C≡C—	HC≡CH Ethyne
Aromatic	(benzene ring) C_6H_5– = Ph = Ar	(benzene ring)–CH_3 Toluene
Haloalkane	R-Cl, R-Br, R-I, R-F	CH–Cl_3 Chloroform
Alcohol	R–OH (R is never H)	CH_3CH_2–OH Ethanol (ethyl alcohol)
Thiol (Mercaptan)	R–SH (R is never H)	$(CH_3)_3C$–SH *tert*-Butyl marcaptan
Sulfide	R–S–R (R is never H)	CH_3CH_2–S–CH_3 Ethyl methyl sulphide
Ether	R–OR (R is never H)	CH_3CH_2–O–CH_2CH_3 Diethyl ether
Amine	RNH_2, R_2NH, R_3N	$(CH_3)_2$–NH Dimethyl amine
Aldehyde	R–C(=O)–H RCHO	H_3C–C(=O)–H Acetaldehyde
Ketone	R–C(=O)–R RCOR	H_3C–C(=O)–CH_3 Acetone
Carboxylic acid	R–C(=O)–OH RCO_2H	H_3C–C(=O)–OH Acetic acid

(*Continued*)

Name	General structure	Example
Ester	R–C(=O)–OR RCO_2R	H_3C–C(=O)–C_2H_5 Ethyl acetate
Anhydride	R–C(=O)–O–C(=O)–R $(ROC)_2O$	H_3C–C(=O)–O–C(=O)–CH_3 Acetic anhydride
Amide	R–C(=O)–NH_2 $RCONH_2$	C_2H_5–C(=O)–NH_2 Propanamide
Nitrile	R–C≡N RCN	H_3C–C≡N Acetonitrile

R = hydrocarbon group such as methyl or ethyl and can sometimes be H or phenyl. Where two R groups are shown in a single structure, they do not have to be the same, but they can be.

4.2 Hydrocarbons

Hydrocarbons are compounds that only contain carbon and hydrogen atoms, and they can be classified as follows depending on the bond types that are present within the molecules.

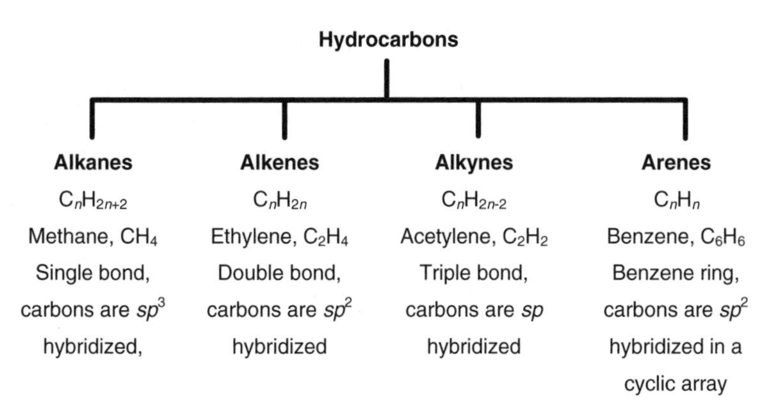

4.3 Alkanes, cycloalkanes and their derivatives

4.3.1 Alkanes

Hydrocarbons having no double or triple bond are classified as *alkanes*. The carbon atoms of the molecule are arranged in chains (*alkanes*) or in rings (*cycloalkanes*). All alkanes have the general molecular formula C_nH_{2n+2} and are called saturated hydrocarbons. *Saturated hydrocarbons* contain only single bonds, and are also commonly referred to as *aliphatic* or *acyclic* alkanes (alkanes without rings). Thus, the alkane family is characterized by the presence of tetrahedral carbon (sp^3) atoms. Methane (CH_4) and ethane (C_2H_6) are the first two members of the alkane family. A group derived from an alkane by removal of one of its hydrogen atoms is known as an *alkyl group*, for example the methyl group (CH_3—) from methane (CH_4) and the ethyl group (CH_3CH_2—) from ethane (CH_3CH_3).

The IUPAC nomenclature of alkanes

In general, organic compounds are given systematic names by using the order prefix–parent–suffix, where prefix indicates how many branching groups are present, parent indicates how many carbons are in the longest chain and suffix indicates the name of the family. Common names as well as systematic names are used for alkanes and their derivatives. However, it is advisable to use systematic names or the IUPAC (*International Union of Pure and Applied Chemistry*) nomenclature, which can be derived from a simple set of rules.

The IUPAC naming of the alkanes is based on a prefix indicating the number of carbon atoms in the chain (as shown below) followed by the suffix *-ane*. For example, if a chain contains three carbons the parent name is *propane*, if four carbons the parent name is *butane* and so on. The remaining parts of the structure are treated as substituents on the chain. Numbers are used to indicate the positions of the substituents on the parent carbon chain.

Prefix	Number of carbon atoms	Prefix	Number of carbon atoms
Meth-	1	Hept-	7
Eth-	2	Oct-	8
Prop-	3	Non-	9
But-	4	Dec-	10
Pent-	5	Undec-	11
Hex-	6	Dodec-	12

First, one has to identify and name the groups attached to the chain, and number the chain so that the substituent gets the lowest possible number. For example, one of the isomers of pentane is 2-methylbutane, where the parent chain is a four-carbon butane chain, and is numbered starting from the end nearer the substituent group (methyl group). Therefore, the methyl group is indicated as being attached to carbon atom number 2.

CH_3
$H_3C-C-CH_2-CH_3$
1 2 3 4
H

Systematic name: 2-Methylbutane

CH_3
$H_3C-C-CH_3$
1 2 3
H

Systematic name: 2-Methylpropane
Common name: Isobutane

Similarly, *isobutane* is the common name of one of the structural isomers of C_4H_{10} (butane). The longest continuous chain of carbons consists of three atoms in length, so the systematic name is based on propane. Finally, since a methyl group appears on the second carbon, the correct name is 2-methylpropane.

When more than one substituent is present, the location of each substituent should be designated by an appropriate name and number. The presence of two or more identical substituents is indicated by the prefixes di-, tri-, tetra- and so on, and the position of each substituent is indicated by a number in the prefix. A number and a word are separated by a hyphen, and numbers are separated by comma. For example, in 2,2-dimethylbutane, both methyl groups are attached to carbon atom 2 of a butane chain. The names of the substituents are arranged in alphabetical order, not numerical order, e.g. 3-ethyl-2-methylhexane, *not* 2-methyl-3-ethylhexane.

CH_3
$H_3C-C-CH_2-CH_3$
1 2 3 4
CH_3

Systematic name: 2,2-Dimethylbutane

CH_3
$CH_3CH-CH-CH_2CH_2CH_3$
1 2 3 4 5 6
C_2H_5

Systematic name: 3-Ethyl-2-methylhexane

Isomerism and physical properties

Compounds that differ from each other in their molecular formulas by the unit $-CH_2-$ are called members of *homologous series*. Thus, methane and ethane belong to a homologous series of saturated hydrocarbons. Compounds that have same molecular formula but different order of attachment of their atoms are called *constitutional isomers* (see Section 3.2.1). For the molecular formulas CH_4, C_2H_6 and C_3H_8, only one order of attachment of atoms is possible. The molecular formula C_4H_{10} gives rise to two different structural formulas in which four carbon atoms and 10 hydrogen atoms can be connected to each other in the following ways. These structures also can be drawn using line drawings, where *zigzag* lines represent carbon chains.

n-Butane and isobutane (2-methylpropane) are constitutional isomers. Their structures differ in connectivity, and they are different compounds. They have different physical properties, e.g. different boiling points.

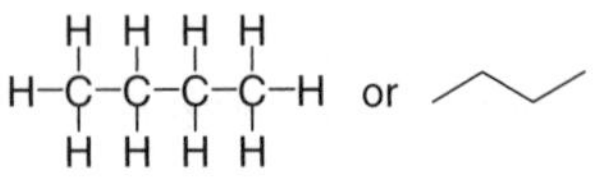

n-Butane (bp −0.6 °C)

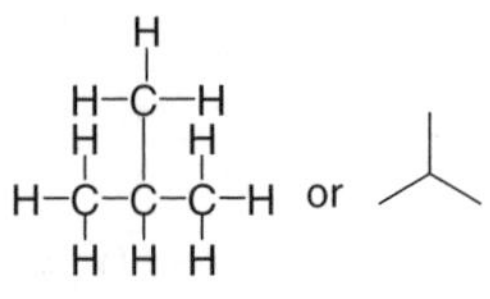

Isobutane (bp −10.2 °C)

Alkanes have similar chemical properties, but their physical properties vary with molecular weight and the shape of the molecule. The low polarity of all the bonds in alkanes means that the only intermolecular forces between molecules of alkanes are the weak *dipole–dipole forces* (see 2.5.1), which are easily overcome. As a result, compared with other functional groups, alkanes have low melting and boiling points, and low solubility in polar solvents, e.g. water, but high solubility in nonpolar solvents, e.g. hexane and dichloromethane. Most cycloalkanes also have low polarity.

Name	Number of carbons	Molecular formula	Condensed structure	bp (°C)	mp (°C)
Methane	1	CH_4	CH_4	−164	−182.5
Ethane	2	C_2H_6	CH_3CH_3	−88.6	−183.3
Propane	3	C_3H_8	$CH_3CH_2CH_3$	−42.1	−189.7
Butane	4	C_4H_{10}	$CH_3(CH_2)_2CH_3$	−0.60	−138.4
Pentane	5	C_5H_{12}	$CH_3(CH_2)_3CH_3$	36.1	−129.7
Hexane	6	C_6H_{14}	$CH_3(CH_2)_4CH_3$	68.9	−93.5
Heptane	7	C_7H_{16}	$CH_3(CH_2)_5CH_3$	98.4	−90.6
Octane	8	C_8H_{18}	$CH_3(CH_2)_6CH_3$	125.7	−56.8
Nonane	9	C_9H_{20}	$CH_3(CH_2)_7CH_3$	150.8	−51.0
Decane	10	$C_{10}H_{22}$	$CH_3(CH_2)_8CH_3$	174.1	−29.7
Undecane	11	$C_{11}H_{24}$	$CH_3(CH_2)_9CH_3$	196	−26
Dodecane	12	$C_{12}H_{26}$	$CH_3(CH_2)_{10}CH_3$	216	−10

The boiling points of alkanes increase steadily with increasing molecular weights, as shown in the above table. Alkanes from methane to butane are gases at room temperature.

Structure and conformation of alkanes

Alkanes have only sp^3-hybridized carbons. The conformation of alkanes is discussed in Chapter 3 (see Section 3.2.2). Methane (CH_4) is a nonpolar molecule, and has four covalent carbon–hydrogen bonds. In methane, all four C—H bonds have the same length (1.10 Å), and all the bond angles (109.5°) are the same. Therefore, all four covalent bonds in methane are identical. Three different ways to represent a methane molecule are shown here. In a perspective formula, bonds in the plane of the paper are drawn as solid lines, bonds sticking out of the plane of the paper towards you are drawn as solid wedges, and those pointing back from the plane of the paper away from you are drawn as broken wedges.

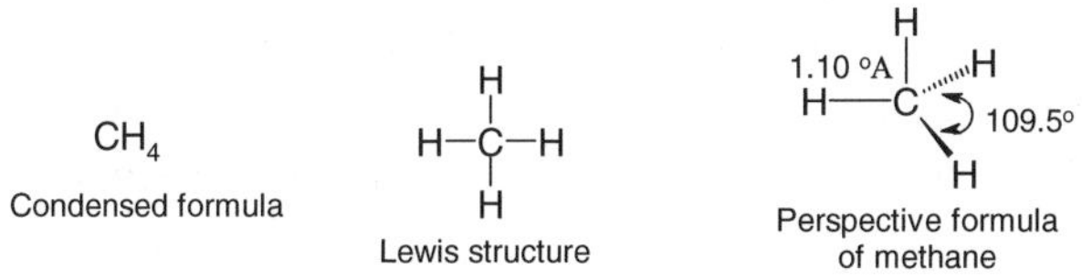

CH_4
Condensed formula

Lewis structure

1.10 °A 109.5°
Perspective formula of methane

One of the hydrogen atoms in CH_4 is replaced by another atom or group to give a new derivative, such as alkyl halide or alcohol. Chloromethane (CH_3Cl) is a compound in which one of the hydrogen atoms in CH_4 is substituted by a Cl atom. Chloromethane (methyl chloride) is an alkyl halide, where the hydrocarbon part of the molecule is a methyl group (CH_3—). Similarly, in methanol (CH_3OH), one of the hydrogen atoms of CH_4 is replaced by an OH group.

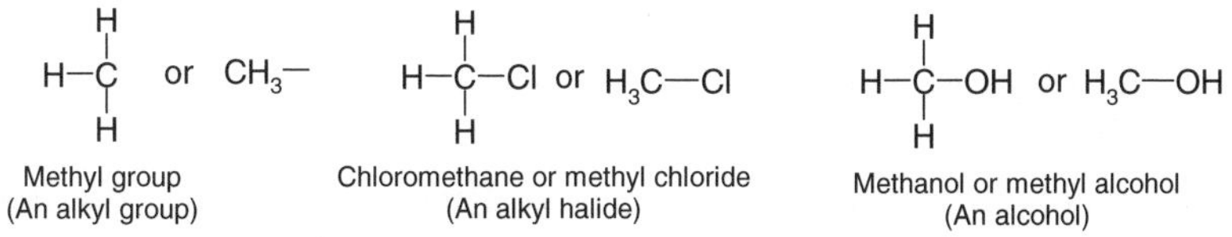

or CH_3—
Methyl group (An alkyl group)

or H_3C—Cl
Chloromethane or methyl chloride (An alkyl halide)

or H_3C—OH
Methanol or methyl alcohol (An alcohol)

In the above examples, the first name given for each compound is its systematic name, and the second is the common name. The name of an alkyl group is obtained by changing the suffix *–ane* to *–yl*. Thus, the methyl group (CH_3—) is derived from methane (CH_4), the ethyl group (C_2H_5—) from ethane (C_2H_6) and so on. Sometimes, an alkane is represented by the symbol RH; the corresponding alkyl group is symbolized by R—.

Classification of carbon substitution

A carbon atom is classified as primary (1°), secondary (2°), tertiary (3°) and quaternary (4°) depending on the number of carbon atoms bonded to it. A carbon atom bonded to only one carbon atom is known as 1°; when bonded to two carbon atoms, it is 2°; when bonded to three carbon atoms, it is 3°, and when bonded to four carbon atoms, it is known as 4°. Different types of carbon atom are shown in the following compound.

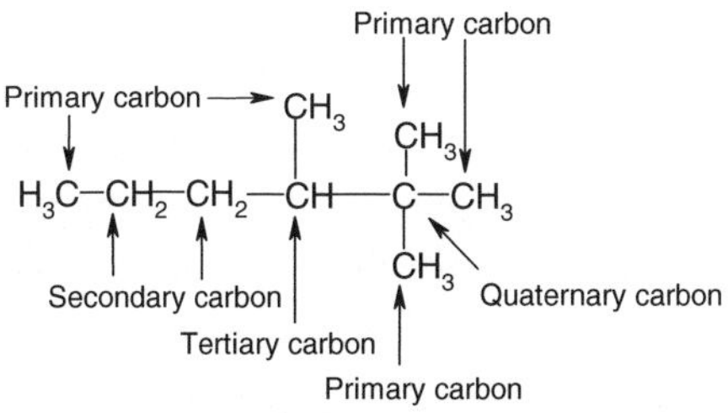

Primary carbon
Primary carbon
Secondary carbon
Tertiary carbon
Quaternary carbon
Primary carbon

4.3.2 Cycloalkanes

Cycloalkanes are alkanes that are cyclic with the general formula C_nH_{2n}. The simplest members of this class consist of a single, unsubstituted carbon ring, and these form a homologous series similar to the unbranched alkanes. The C_3 to C_6 cycloalkanes with their structural representations are shown below.

Name	Molecular formula	Structural formula	Name	Molecular formula	Structural formula
Cyclopropane	C_3H_6	△	Cyclopentane	C_5H_{10}	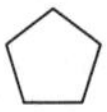 ⬠
Cyclobutane	C_4H_8	□	Cyclohexane	C_6H_{12}	 ⬡

Nomenclature of cycloalkanes

The nomenclature of cycloalkanes is almost the same as that for alkanes, with the exception that the prefix *cyclo-* is to be added to the name of the alkane. When a substituent is present on the ring, the name of the substituent is added as a prefix to the name of the cycloalkane. No number is required for rings with only one substituent.

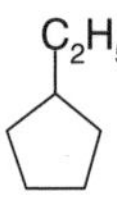

C_2H_5 Ethylcyclopentane

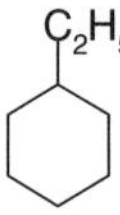

C_2H_5 Ethyl cyclohexane

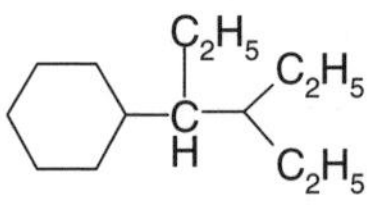

C_2H_5 C H C_2H_5 C_2H_5 (1,1-Diethylbutyl)-cyclohexane

However, if two or more substituents are present on the ring, numbering starts from the carbon that has the group of alphabetical priority, and proceeds around the ring so as to give the second substituent the lowest number.

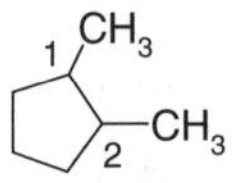

1 CH_3 2 CH_3
1,2-Dimethylcyclopentane
not 1,5-dimethylcyclopentane

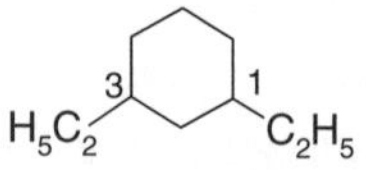

H_5C_2 3 1 C_2H_5
1,3-Diethylcyclohexane
not 1,5-diethylcyclohexane

When the number of carbons in the ring is greater than or equal to the number of carbons in the longest chain, the compound is named as a cycloalkane. However, if an alkyl chain of the cycloalkane has a greater number of carbons, then the alkyl chain is used as the parent, and the cycloalkane as a cycloalkyl substituent.

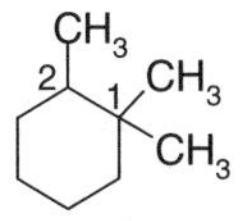

1,1,2-Trimethylcyclohexane
not 1,2,2-Trimethylcyclohexane

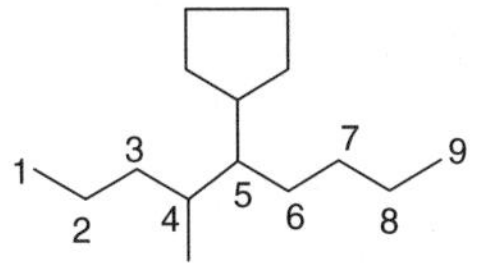

5-Cyclopentyl-4-methylnonane
not 5-Cyclopentyl-6-methylnonane

Geometric isomerism in cycloalkanes

Butane (C_4H_{10}) can exist in two different isomeric forms, e.g. *n*-butane and isobutane (2-methylpropane). Open chain alkanes have free rotation about their C—C bonds, but cycloalkanes cannot undergo free rotation, so substituted cycloalkanes can give rise to *cis* and *trans* isomers (see Section 3.2.2).

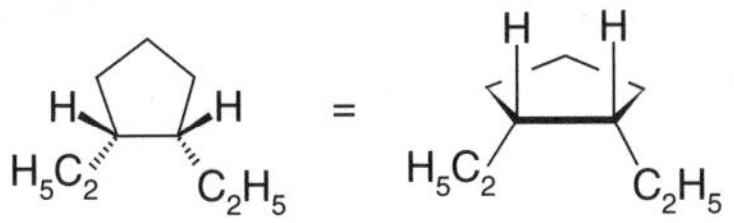

cis-1,2-Diethylcyclopentane

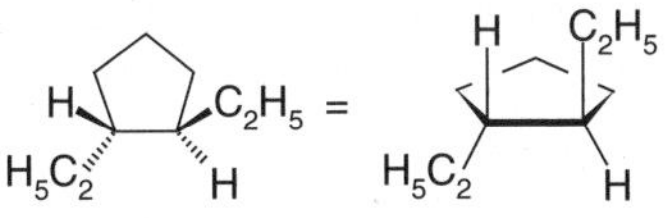

trans-1,2-Diethylcyclopentane

Physical properties of cycloalkenes

Cycloalkenes are nonpolar molecules like alkanes. As a result, they tend to have low melting and boiling points compared with other functional groups.

4.3.3 Sources of alkanes and cycloalkanes

The principal source of alkanes is petroleum and natural gas, which contain only the more volatile alkanes. Therefore, low molecular weight alkanes, e.g. methane and small amounts of ethane, propane and other higher alkanes can be obtained directly from natural gas. Another fossil fuel, coal, is a potential second source of alkanes. Usually alkanes are obtained through refinement or hydrogenation of petroleum and coal.

Cycloalkanes of ring sizes ranging from three to 30 are found in nature. Compounds containing five-membered rings (cyclopentane) and six-membered rings (cyclohexane) are especially common.

4.3.4 Preparation of alkanes and cycloalkanes

Alkanes are prepared simply by *catalytic hydrogenation* of alkenes or alkynes (see Section 5.3.1).

$$RHC{=}CHR \text{ (Alkene)} \xrightarrow[\text{Pt-C or Pd-C}]{H_2} R{-}CH_2{-}CH_2{-}R \text{ (Alkane)} \xleftarrow[\text{Pt-C or Pd-C}]{2\,H_2} R{-}C{\equiv}C{-}R \text{ (Alkyne)}$$

$$\text{Cyclohexene} \xrightarrow[\text{Pt-C or Pd-C}]{H_2} \text{Cyclohexane}$$

Alkanes can also be prepared from alkyl halides by *reduction*, directly with Zn and acetic acid (AcOH) (see Section 5.7.14) or via the Grignard reagent formation followed by hydrolytic work-up (see Section 5.7.15). The coupling reaction of alkyl halides with Gilman reagent (R'_2CuLi, lithium organocuprates) also produces alkanes (see Section 5.5.2).

$$\underset{\text{Alkane}}{R{-}H} \xleftarrow[\text{i. Mg, Dry ether; ii. } H_2O]{\text{Zn, AcOH or}} \underset{\text{Alkyl halide}}{R{-}X} \xrightarrow[\text{Ether}]{R'_2CuLi} \underset{\text{Alkane}}{R{-}R'}$$

Selective reduction of aldehydes or ketones, either by *Clemmensen reduction* (see Section 5.7.17) or *Wolff–Kishner reduction* (see Section 5.7.18) yields alkanes.

$$\underset{\substack{Y = H \text{ or } R \\ \text{Aldehyde or ketone}}}{R{-}C({=}O){-}Y} \xrightarrow[NH_2NH_2,\ NaOH]{\text{Zn(Hg) in HCl or}} \underset{\text{Alkane}}{R{-}CH_2{-}Y}$$

4.3.5 Reactions of alkanes and cycloalkanes

Alkanes contain only strong σ bonds, and all the bonds (C—C and C—H) are nonpolar. As a result, alkanes and cycloalkanes are quite unreactive towards most reagents. In fact, it is often convenient to regard the hydrocarbon framework of a molecule as an unreactive support for the more reactive functional groups. More branched alkanes are more stable and less reactive than linear alkanes. For example, isobutane is more stable than *n*-butane. Alkanes and cycloalkanes react with O_2 under certain conditions. They also react with halogens under UV light or at high temperatures, and the reaction is called a *free radical chain reaction* (see Section 5.2). Catalytic hydrogenation of smaller cycloalkanes produces open chain alkanes.

Combustion or oxidation of alkanes

Alkanes undergo combustion reaction with oxygen at high temperatures to produce carbon dioxide and water. This is why alkanes are good fuels. Oxidation of saturated hydrocarbons is the basis for their use as energy sources for heat, e.g. natural gas, liquefied petroleum gas (LPG) and fuel oil, and for power, e.g. gasoline, diesel fuel and aviation fuel.

$$CH_4 + 2\ O_2 \longrightarrow CO_2 + 2\ H_2O \qquad CH_3CH_2CH_3 + 2\ O_2 \longrightarrow 3\ CO_2 + 4\ H_2O$$

Methane Propane

Reduction of smaller cycloalkanes

Cyclopropane and cyclobutane are unstable due to their ring strain compared with the larger cycloalkanes, e.g. cyclopentane and cyclohexane. The two smaller cycloalkanes react with hydrogen, even though they are not alkenes. In the presence of a nickel catalyst, the rings open up, and form corresponding acyclic (open chain) alkanes. Cyclobutanes require higher temperature than cyclopentane for ring opening.

$$\text{Cyclopropane} \xrightarrow[120\ ^{\circ}C]{H_2/Ni} CH_3CH_2CH_3 \ (\text{Propane}) \qquad \text{Cyclobutane} \xrightarrow[200\ ^{\circ}C]{H_2/Ni} CH_3CH_2CH_2CH_3 \ (n\text{-Butane})$$

4.3.6 Alkyl halides

Alkyl halides (haloalkanes) are a class of compounds where a halogen atom or atoms are attached to a tetrahedral carbon (sp^3) atom. The functional group is —X, where —X may be —F, —Cl, —Br or —I. Two simple members of this class are methyl chloride (CH_3Cl) and ethyl chloride (CH_3CH_2Cl).

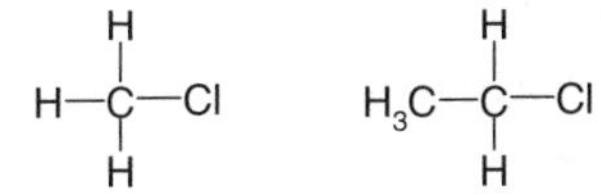

Chloromethane
Methyl chloride

Chloroethane
Ethyl chloride

Based on the number of alkyl groups attached to the C—X unit, alkyl halides are classed as primary (1°), secondary (2°) or tertiary (3°).

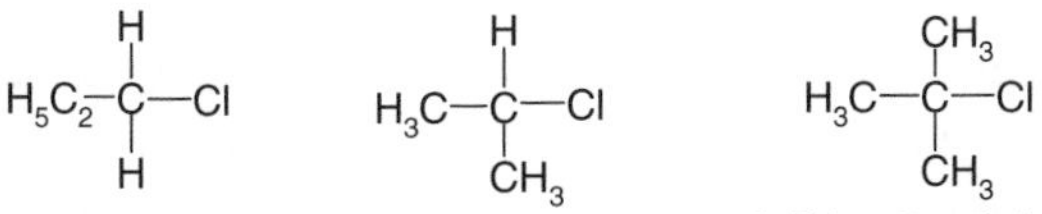

Chloropropane
(Propyl chloride)
(1° halide)

2-Chloropropane
(Isopropyl chloride)
(2° halide)

2-Chloro-2-methyl propane
(*tert*-Butyl chloride)
(3° halide)

A *geminal* (*gem*)-dihalide has two halogen atoms on the same carbon, and a *vicinal* (*vic*)-dihalide has halogen atoms on adjacent carbon atoms.

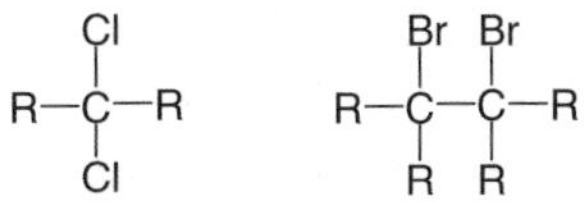

gem-Dichloride

vic-Dibromide

Nomenclature of alkyl halides

According to the IUPAC system, alkyl halides are treated as alkanes with a halogen substituent. The halogen prefixes are fluoro-, chloro-, bromo- and iodo-. An alkyl halide is named as a haloalkane with an alkane as the parent structure.

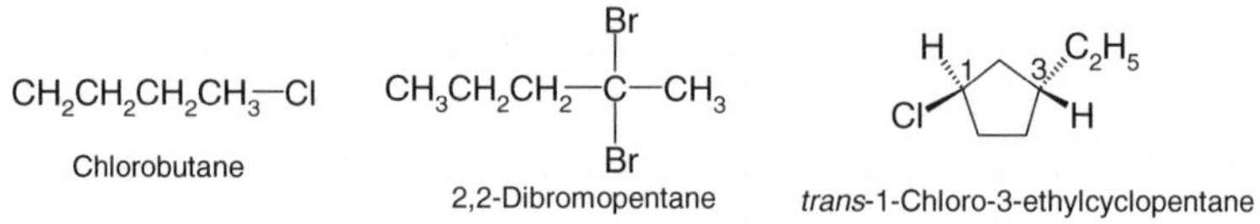

$CH_2CH_2CH_2CH_3$–Cl
Chlorobutane

$CH_3CH_2CH_2$–C(Br)(Br)–CH_3
2,2-Dibromopentane

trans-1-Chloro-3-ethylcyclopentane

Often compounds of CH_2X_2 type are called methylene halides, e.g. methylene chloride (CH_2Cl_2), CHX_3 type compounds are called *haloforms*, e.g. chloroform ($CHCl_3$), and CX_4 type compounds are called carbon tetrahalides, e.g. carbon tetrachloride (CCl_4). Methylene chloride (dichloromethane, DCM), chloroform and carbon tetrachloride are extensively used in organic synthesis as nonpolar solvents.

Physical properties of alkyl halides

Alkyl halides have considerably higher melting and boiling points compared with analogous alkanes. The boiling points also increase with increasing atomic weight of the halogen atom. Thus, alkyl fluoride has the lowest boiling point and alkyl iodide has the highest boiling point. Alkyl halides are insoluble in water as they are unable to form hydrogen bonds, but are soluble in nonpolar solvents, e.g. ether and chloroform.

Preparation of alkyl halides

Alkyl halides are almost always prepared from corresponding alcohols by the use of hydrogen halides (HX) or phosphorus halides (PX_3) in ether (see Section 5.5.3). Alkyl chlorides are also obtained by the reaction of alcohols with thionyl chloride ($SOCl_2$) in triethylamine (Et_3N) or pyridine (see Section 5.5.3).

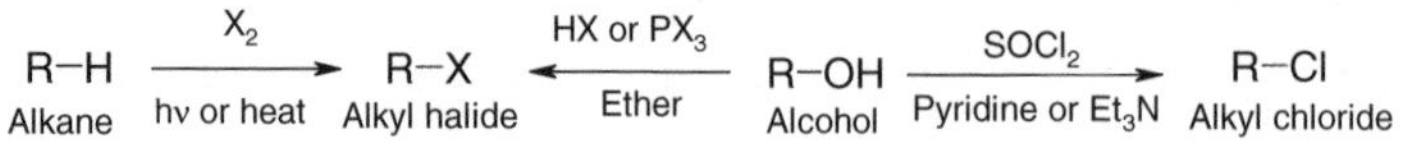

$$\underset{\text{Alkane}}{\text{R–H}} \xrightarrow[\text{hν or heat}]{X_2} \underset{\text{Alkyl halide}}{\text{R–X}} \xleftarrow[\text{Ether}]{\text{HX or } PX_3} \underset{\text{Alcohol}}{\text{R–OH}} \xrightarrow[\text{Pyridine or } Et_3N]{SOCl_2} \underset{\text{Alkyl chloride}}{\text{R–Cl}}$$

Other methods for the preparation of alkyl halides are electrophilic addition of hydrogen halides (HX) to alkenes (see Section 5.3.1) and free radical halogenation of alkanes (see Section 5.2).

$$\underset{\text{Alkene}}{RHC{=}CH_2} \xrightarrow{HX} \underset{\substack{\text{Alkyl halide} \\ \text{Markovnikov addition}}}{R{-}CHX{-}CH_3}$$

Reactivity of alkyl halides

The alkyl halide functional group consists of an sp^3-hybridized carbon atom bonded to a halogen atom via a strong σ bond. The C—X bonds in alkyl halides are highly polar due to the higher electronegativity and polarizability of the halogen atoms. Halogens (Cl, Br and I) are good leaving groups in the nucleophilic substitution reactions. The electronegativity of halides decreases and the polarizability increases in the order of: F > Cl > Br > I.

$$H_3\overset{\delta+}{C}{-}\overset{\delta-}{Cl} \quad (\mu \rightarrow)$$

Chloromethane

Reactions of alkyl halides

Alkyl halides undergo not only *nucleophilic substitution* but also *elimination*, and both reactions are carried out in basic reagents. Often substitution and elimination reactions occur in competition with each other. In general, most nucleophiles can also act as bases, therefore the preference for elimination or substitution is determined by the reaction conditions and the alkyl halide used.

$$RCH_2CH_2{-}X + Y{:}^- \xrightarrow{\text{Substitution}} RCH_2CH_2{-}Y + X{:}^-$$

$$RCH_2CH_2{-}X + Y{:}^- \xrightarrow{\text{Elimination}} RCH{=}CH_2 + H{-}Y + X{:}^-$$

Alkyl halides are most commonly converted to *organometallic compounds* that contain carbon–metal bonds, usually Mg or Li.

$$H_3\overset{\delta-}{C}{-}\overset{\delta+}{Mg} \qquad H_3\overset{\delta-}{C}{-}\overset{\delta+}{Li}$$

Organometallic compounds When a compound has a covalent bond between a carbon and a metal, it is called an *organometallic compound*. Carbon–metal bonds vary widely in character from covalent to ionic depending on the metal. The greater the ionic character of the bond, the more reactive is the compound. The most common types of organometallic compound are *Grignard reagents*, *organolithium reagents* and *Gilman reagents* (*lithium organocuprates*, R_2CuLi). A carbon–metal bond is polarized, with significant negative charge on the carbon, because metals are electropositive. These compounds have nucleophilic carbon atoms, and therefore are strong bases.

Organometallic reagents react readily with hydrogen atoms attached to oxygen, nitrogen or sulphur, in addition to other acidic hydrogen atoms. They react with terminal alkynes to form alkynides (alkynyl Grignard reagent and alkynyllithium) by acid–base reactions (see Section 4.5.3). Alkynides are useful nucleophiles for the synthesis of a variety of other compounds (see Section 4.5.3). Organolithium reagents react similarly to Grignard reagents, but they are more reactive than Grignard reagents. Gilman reagents are weaker organometallic reagents. They react readily with acid chlorides, but do not react with aldehydes, ketones, esters, amides, acid anhydrides or nitriles. Gilman reagents also undergo coupling reactions with alkyl halides.

Grignard reagents These reagents are prepared by the reaction of organic halides with magnesium turnings, usually in dry ether. An ether solvent is used, because it forms a complex with the Grignard reagent, which stabilizes it. This reaction is versatile: primary, secondary and tertiary alkyl halides can be used, and also vinyl, allyl and aryl halides.

$$R{-}X + Mg \xrightarrow{\text{Dry ether}} RMgX \qquad C_2H_5{-}Br + Mg \xrightarrow{\text{Dry ether}} C_2H_5MgBr$$

Organolithium reagents These reagents are prepared by the reaction of alkyl halides with lithium metals in an ether solvent. Unlike Grignard reagents, organolithiums can also be prepared using a variety of hydrocarbon solvents e.g. hexane, and pentane.

$$R{-}X + 2\,Li \xrightarrow[\text{Hexane}]{\text{Ether or}} R{-}Li + LiX$$

Gilman reagents or lithium organocuprates The most useful Gilman reagents are lithium organocuprates (R_2CuLi). They are easily prepared by the reaction of two equivalents of the organolithium reagent with copper (I) iodide in ether.

$$2\,R\text{-}X + Cu\text{-}I \xrightarrow{\text{Ether}} R_2CuLi + Li\text{-}I$$

Alkanes can be prepared from alkyl halides by *reduction*, directly with Zn and acetic acid (AcOH) (see Sections 4.3.4 and 5.7.14) or via the Grignard reagent formation followed by hydrolytic work-up (see Sections 4.3.4 and 5.7.15). The coupling reaction of alkyl halides with Gilman reagent (R'_2CuLi, lithium organocuprates) also produces alkanes (see Section 5.5.2). Base-catalysed dehydrohalogenation of alkyl halides is an important reaction for the preparation of alkenes (see Section 5.4).

$$\underset{\text{Alkyl halide}}{\text{H–C–C–X}} \xrightarrow[\text{Heat}]{\text{NaOH}} \underset{\text{Alkene}}{\text{C=C}}$$

Alkyl halide reacts with triphenylphospine to give a phosphonium salt, which is an important intermediate for the preparation of phosphorus ylide (see Section 5.3.2).

$$\underset{\text{Alkyl halide}}{RCH_2–X} + \underset{\text{Triphenylphosphine}}{(Ph)_3P:} \longrightarrow \underset{\text{Phosphonium salt}}{RCH_2–\overset{+}{P}(Ph)_3\overset{-}{X}}$$

Alkyl halides undergo S_N2 reactions with a variety of nucleophiles, e.g. metal hydroxides (NaOH or KOH), metal alkoxides (NaOR or KOR) or metal cyanides (NaCN or KCN), to produce alcohols, ethers or nitriles, respectively. They react with metal amides ($NaNH_2$) or NH_3, 1° amines and 2° amines to give 1°, 2° or 3° amines, respectively. Alkyl halides react with metal acetylides (R′C≡CNa), metal azides (NaN_3) and metal carboxylate ($R'CO_2Na$) to produce internal alkynes, azides and esters, respectively. Most of these transformations are limited to primary alkyl halides (see Section 5.5.2). Higher alkyl halides tend to react via elimination.

$$RCH_2–X \text{ (Alkyl halide)} \xrightarrow{\text{R'C?CNa}} RCH_2–C{\equiv}CR' \text{ (Internal alkyne)}$$

$$RCH_2–X \xrightarrow{NaN_3} RCH_2–N_3 \text{ (Primary azide)}$$

$$RCH_2–X \xrightarrow{NaNH_2 \text{ or } NH_3} RCH_2–NH_2 \text{ (Primary amine)}$$

$$RCH_2–X \xrightarrow{RNH_2} RCH_2–NHR \text{ (Secondary amine)}$$

$$RCH_2–X \xrightarrow{KOH} RCH_2–OH \text{ (Alcohol)}$$

$$RCH_2–X \xrightarrow{NaOR} RCH_2–OR \text{ (Ether)}$$

$$RCH_2–X \xrightarrow{KCN} RCH_2–C{\equiv}N \text{ (Nitrile)}$$

$$RCH_2–X \xrightarrow{R'CO_2Na} RCH_2–CO_2R' \text{ (Ester)}$$

4.3.7 Alcohols

The functional group of an *alcohol* is the hydroxyl (—OH) group. Therefore, an alcohol has the general formula ROH. The simplest and most common alcohols are methyl alcohol (CH_3OH) and ethyl alcohol (CH_3CH_2OH).

Methanol
(Methyl alcohol)

Ethanol
(Ethyl alcohol)

An alcohol may be acyclic or cyclic. It may contain a double bond, a halogen atom or additional hydroxyl groups. Alcohols are usually classified as primary (1°), secondary (2°) or tertiary (3°). When a hydroxyl group is linked directly to an aromatic ring, the compound is called a *phenol* (see Section 4.6.10), which differs distinctly from alcohols.

Propanol
Propyl alcohol (1°)

Isopropanol
2-Propanol (2°)

tert-Butanol
2-Methyl-2-propanol (3°)

Phenol

Nomenclature of alcohols

Generally, the name of an alcohol ends with *-ol*. An alcohol can be named as an alkyl alcohol, usually for small alkyl groups e.g. methyl alcohol and ethyl alcohol. The longest carbon chain bearing the —OH group is used as the parent; the last *-e* from this alkane is replaced by an *-ol* to obtain the root name. The longest chain is numbered starting from the end nearest to the —OH group, and the position of the —OH group is numbered. Cyclic alcohols have the prefix *cyclo-*, and the —OH group is deemed to be on C-1.

1-Bromo-3-methyl-2-butanol
(The -OH is at C-2 of butane)

1-Chloro-3-pentanol
(The -OH is at C-3 of pentane)

1-Propylcyclopentanol

Alcohols with double or triple bonds are named using the *-ol* suffix on the alkene or alkyne name. Numbering gives the hydroxyl group the lowest possible number. When numbers are also given for the multiple bond position, the position of the hydroxyl can be written immediately before the *-ol* prefix. If the hydroxyl group is only a minor part of the structure, it may be named as a hydroxy- substituent.

3-Hydroxypentanoic acid

Pent-4-en-2-ol

3-Bromo-5-chlorocyclohexanol

Diols are compounds with two hydroxyl groups. They are named as for alcohols except that the suffix *-diol* is used and two numbers are required to locate the hydroxyls. 1,2-diols are called glycols. The common names for glycols usually arise from the name of the alkene from which they are prepared.

1,2-Ethane diol (Ethylene glycol)

1,2-propane diol (Propylene glycol)

3,4-Hexanediol

Physical properties of alcohols

Alcohols can be considered as organic analogues of water. Both the C—O and O—H bonds are polarized due to the electronegativity of the oxygen atom. The highly polar nature of the O—H bond results in the formation of hydrogen bonds with other alcohol molecules or other hydrogen bonding systems, e.g. water and amines. Thus, alcohols have considerably higher boiling point due to the hydrogen bonding between molecules (*intermolecular hydrogen bonding*). They are more polar than hydrocarbons, and are better solvents for polar molecules.

Hydrogen bonds between alcohol molecules

Hydrogen bonds to water in aqueous solution

The hydroxyl group is *hydrophilic* (water loving), whereas the alkyl (hydrocarbon) part is *hydrophobic* (water repellent or fearing). Small alcohols are miscible with water, but solubility decreases as the size of the alkyl group increases. Similarly, the boiling point of alcohols increases with the increase in the alkyl chain length as shown in the following table. It is interesting to note that isopentanol has a lower b.p. than its isomer *n*-pentanol, and this pattern is observed in all isomeric alkanols (alkyl alcohols).

Name	Molecular formula	Molecular weight	Solubility in water (g/100 g water)	Boiling point (°C)
Methanol	CH_3OH	32	Infinite	64.5
Ethanol	C_2H_5OH	46	Infinite	78.3
n-propanol	C_3H_7OH	60	Infinite	97.0
Isopropanol	$CH_3CHOHCH_3$	60	Infinite	82.5
n-butanol	C_4H_9OH	74	8.0	118.0
Isobutanol	$(CH_3)_2CHCH_2OH$	74	10.0	108.0
n-pentanol	$C_5H_{11}OH$	88	2.3	138.0
Isopentanol	$(CH_3)_2CH(CH_3)_2OH$	88	2.0	132.0

Acidity and basicity of alcohols

Alcohols resemble water in their acidity and basicity. They are stronger acids than terminal alkynes and primary or secondary amines. However, they are weaker acids than HCl, H_2SO_4 and even acetic acid. They dissociate in water and form alkoxides (RO^-) and hydronium ion (H_3O^+).

$$R{-}OH + H_2O \longrightarrow \underset{\text{Alkoxide ion}}{RO^-} + \underset{\text{Hydronium ion}}{H{-}\overset{+}{O}(H){-}H}$$

They are considerably acidic and react with active metals to liberate hydrogen gas. Thus, an *alkoxide* (RO^-) can be prepared by the reaction of an alcohol with Na or K metal.

Like hydroxide ion (HO^-), alkoxide ions are strong bases and nucleophiles. Halogens increase the acidity, but acidity decreases as the alkyl group increases.

$$R{-}OH + Na \longrightarrow \underset{\text{Sodium alkoxide}}{R{-}O^-Na^+} + 1/2\ H_2$$

Alcohols are basic enough to accept a proton from strong acids, e.g. HCl and H_2SO_4, and able to dissociate completely in acidic solution. Sterically hindered alcohols, e.g. *tert*-butyl alcohol, are strongly basic (higher pK_a values), and react with strong acids to give oxonium ions (ROH_2^+).

$$R{-}OH + H_2SO_4 \longrightarrow \underset{\text{Oxonium ion}}{R{-}\overset{+}{O}(H){-}H} + HSO_4^-$$

Alcohol	Molecular formula	pK_a	Alcohol	Molecular formula	pK_a
Methanol	CH_3OH	15.5	Cyclohexanol	$C_6H_{11}OH$	18.0
Ethanol	C_2H_5OH	15.9	Phenol	C_6H_5OH	10.0
2-chloroethanol	ClC_2H_4OH	14.3	Water	HOH	15.7
2,2,2-trifluoroethanol	CF_3CH_2OH	12.4	Acetic acid	CH_3COOH	4.76
t-butanol	$(CH_3)_3COH$	19.0	Hydrochloric acid	HCl	−7

Preparation of alcohols

Alcohols can be prepared conveniently from the hydration of alkenes (see Section 5.3.1).

$$\underset{\substack{\text{Alcohol}\\ \textit{anti}\text{-Markovnikov addition}}}{R_2CH{-}CH(OH)R} \xleftarrow[\text{ii. } H_2O_2,\ NaOH]{\text{i. } BH_3.THF} \underset{\text{Alkene}}{R_2C{=}CHR} \xrightarrow[\text{i. } Hg(OAc)_2,\ THF,\ H_2O;\ \text{ii. } NaBH_4,\ NaOH]{H_2O,\ H_2SO_4 \text{ or}} \underset{\substack{\text{Alcohol}\\ \text{Markovnikov addition}}}{R_2C(OH){-}CH_2R}$$

However, the most important methods for preparing alcohols are catalytic hydrogenation (H_2/Pd–C) or metal hydride ($NaBH_4$ or $LiAlH_4$) reduction of aldehydes, ketones, carboxylic acids, acid chlorides and esters (see Sections 5.7.15 and 5.7.16), and nucleophilic addition of organometallic reagents (RLi and RMgX) to aldehydes, ketones, acid chlorides and esters (see Sections 5.3.2 and 5.5.5).

$$\underset{1^\circ \text{ or } 2^\circ \text{ Alcohol}}{R{-}CH(OH){-}Y} \xleftarrow[NaBH_4 \text{ or } LiAlH_4]{H_2/Pd\text{-}C \text{ or}} \underset{\substack{Y = H \text{ or } R\\ \text{Aldehyde or ketone}}}{R{-}C(=O){-}Y} \xrightarrow[\text{ii. } H_3O^+]{\text{i. } R'MgX \text{ or } R'Li} \underset{2^\circ \text{ or } 3^\circ \text{ Alcohol}}{R{-}C(OH)(R'){-}Y}$$

$$\underset{1^\circ \text{ Alcohol}}{R{-}CH_2OH} \xleftarrow[\text{ii. } H_3O^+]{\text{i. } LiAlH_4,\ \text{dry ether}} \underset{Y = Cl \text{ or } OR}{R{-}C(=O){-}Y} \xrightarrow[\text{ii. } H_3O^+]{\text{i. } 2\ R'MgX \text{ or } 2\ R'Li} \underset{3^\circ \text{ Alcohol}}{R{-}C(OH)(R'){-}R'}$$

Alcohols are obtained from epoxides by acid-catalysed cleavage of H_2O or base-catalysed cleavage by Grignard reagents (RMgX, RLi), metal acetalides or alkynides (RC≡CM), metal hydroxides (KOH or NaOH) and $LiAlH_4$ (see Section 5.5.4).

$$\underset{\text{2-Substituted alcohol}}{R{-}CH(OH){-}CH_2{-}Nu} \xleftarrow[(Nu:^- = R^-,\ RC{\equiv}C^-,\ C{\equiv}N^-,\ N_3^-,\ HO^-,\ H^-)]{\text{Base-catalysed cleavage}} \underset{\text{Epoxide}}{R\text{-epoxide}} \xrightarrow[Nu: = H_2O]{\text{Acid-catalysed cleavage}} \underset{\text{1-Substituted alcohol}}{Nu{-}CH(R){-}CH_2{-}OH}$$

Reactivity of alcohols

The hydroxyl (—OH) group in alcohol is polarized due to the electronegativity difference between atoms. The oxygen of the —OH group can react as either a base or a nucleophile in the nucleophilic substitution reactions.

Reactions of alcohols

Alcohol itself cannot undergo nucleophilic substitution reaction, because the hydroxyl group is strongly basic and a poor leaving group. Therefore, it needs to be converted to a better leaving group, e.g. water, a good leaving group. Only weakly basic nucleophiles (halides) can be used. Moderately (ammonia, amines) and strongly basic nucleophiles (alkoxides, cyanides) would be protonated in the acidic solution, resulting in the total loss or significant decrease of their nucleophilicity.

Alkyl halides are almost always prepared from corresponding alcohols by the use of hydrogen halides (HX) or phosphorus halides (PX_3) in ether (see Section 5.5.3). Alkyl chlorides are also obtained by the reaction of alcohols with $SOCl_2$ in pyridine or Et_3N (see Section 5.5.3).

$$\underset{\text{Alkyl halide}}{R{-}X} \xleftarrow[\text{Ether}]{\text{HX or } PX_3} \underset{\text{Alcohol}}{R{-}OH} \xrightarrow[\text{Pyridine or } Et_3N]{SOCl_2} \underset{\text{Alkyl chloride}}{R{-}Cl}$$

Alkenes are obtained by dehydration of alcohols via elimination reactions (see Section 5.4.3), and esters are prepared conveniently by the acid-catalysed reaction of alcohols and carboxylic acids (see Section 5.5.5).

$$\underset{\text{Ester}}{RCH_2CH_2{-}OR'} \xleftarrow[\text{HCl}]{R'CO_2H} \underset{\text{Alcohol}}{RCH_2CH_2{-}OH} \xrightarrow{H_2SO_4,\ \text{Heat}} \underset{\text{Alkene}}{RCH{=}CH_2}$$

Symmetrical ethers are obtained from the dehydration of two molecules of alcohol with H_2SO_4 (see Section 5.5.3). Alcohols react with *p*-toluenesulphonyl chloride (tosyl chloride, TsCl), also commonly known as *sulphonyl chloride*, in pyridine or Et_3N to yield alkyl tosylates (see Section 5.5.3). Carboxylic acids, aldehydes and ketones are prepared by the oxidation of 1° and 2° alcohols (see Sections 5.7.9 and 5.7.10). Tertiary alcohols cannot undergo oxidation, because they have no hydrogen atoms attached to the oxygen bearing carbon atom.

$$\underset{\text{Carboxylic acid or Aldehyde}}{RCOH \text{ or } RCO_2H} \xleftarrow[1^\circ\text{ alcohol}]{\text{Oxidation of}} \underset{\text{Alcohol}}{R{-}OH} \xrightarrow[140\ ^\circ C]{ROH,\ H_2SO_4} \underset{\text{Ether}}{R{-}O{-}R}$$

$$\underset{\text{Ketone}}{RCOR} \xleftarrow[2^\circ\text{ alcohol}]{\text{Oxidation of}} \underset{\text{Alcohol}}{R{-}OH} \xrightarrow[\text{Pyridine}]{TsCl} \underset{\text{Alkyl tosylate}}{R{-}OTs}$$

4.3.8 Thiols

Thiols, general formula RSH, are the sulphur analogues of alcohols. The functional group of a thiol is −SH. The simplest members of this class are methanethiol (CH_3SH), ethanethiol (C_2H_5SH) and propanethiol (C_3H_7SH).

$$\underset{\text{Methanethiol}}{H_3C{-}SH} \qquad \underset{\text{Ethanethiol}}{C_2H_5{-}SH} \qquad \underset{\text{Propanethiol}}{C_2H_5{-}CH_2{-}SH}$$

Nomenclature of thiols

The nomenclature is similar to alcohols, except that they are named using the suffix *-thiol*, and as a substituent as *mercapto-*.

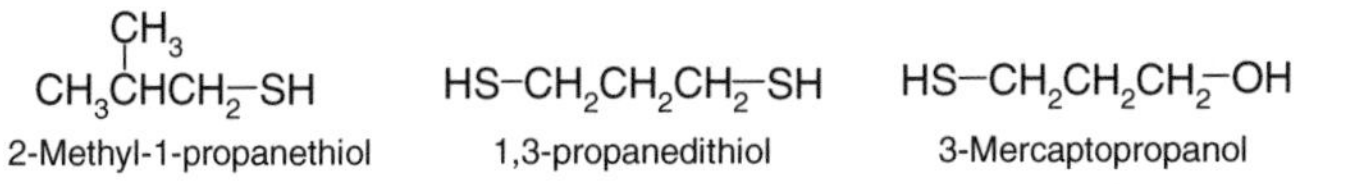

$CH_3CH(CH_3)CH_2$–SH
2-Methyl-1-propanethiol

HS–$CH_2CH_2CH_2$–SH
1,3-propanedithiol

HS–$CH_2CH_2CH_2$–OH
3-Mercaptopropanol

Physical properties of thiols

The S—H bond in thiols is less polar than O—H bond in alcohols, since sulphur is less electronegative than the oxygen atom. Thus, thiols form much weaker hydrogen bonding than alcohols, and have lower boiling points than analogous alcohols.

Acidity and basicity of thiols

Thiols are much more acidic than analogous alcohols, e.g. RSH ($pK_a = 10$) versus ROH ($pK_a = 16$–19), and more nucleophilic than analogous alcohols. In fact, RSH is about as nucleophilic as RO^-.

Preparation of thiols

Thiols are prepared from alkyl halides and sodium hydrosulphide (Na^+SH^-) by S_N2 reaction. A large excess of Na^+SH^- is used with unhindered alkyl halide to prevent dialkylation (R—S—R).

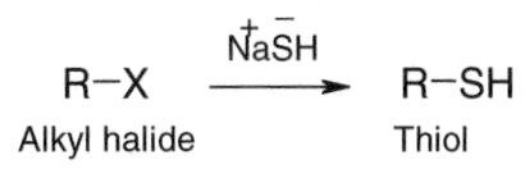

$$\underset{\text{Alkyl halide}}{R{-}X} \xrightarrow{\overset{+}{Na}\overset{-}{S}H} \underset{\text{Thiol}}{R{-}SH}$$

Reactions of thiols

Thiols are easily oxidized to disulphides, an important feature of protein structure. Vigorous oxidation with $KMnO_4$, HNO_3 or sodium hypochlorite (NaOCl) produces sulphonic acids.

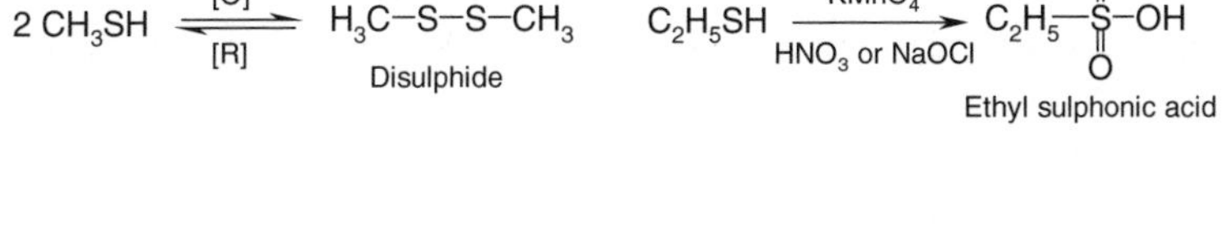

$$2\,CH_3SH \underset{[R]}{\overset{[O]}{\rightleftharpoons}} \underset{\text{Disulphide}}{H_3C{-}S{-}S{-}CH_3}$$

$$C_2H_5SH \xrightarrow[HNO_3 \text{ or } NaOCl]{KMnO_4} \underset{\text{Ethyl sulphonic acid}}{C_2H_5{-}S(=O)_2{-}OH}$$

4.3.9 Ethers

Ethers are also organic relatives of water, where alkyl groups replace both hydrogen atoms. Thus, ethers have two hydrocarbons bonded to an oxygen atom. The simplest and most common ethers are diethyl ether and tetrahydrofuran (THF), which is a cyclic ether.

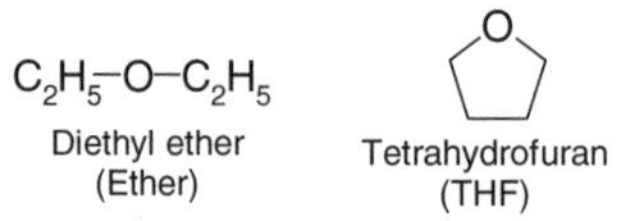

$C_2H_5-O-C_2H_5$
Diethyl ether
(Ether)

Tetrahydrofuran
(THF)

Ethers are relatively unreactive towards most reagents, so they are frequently used as solvents in organic reactions. A few other common ether solvents are shown below.

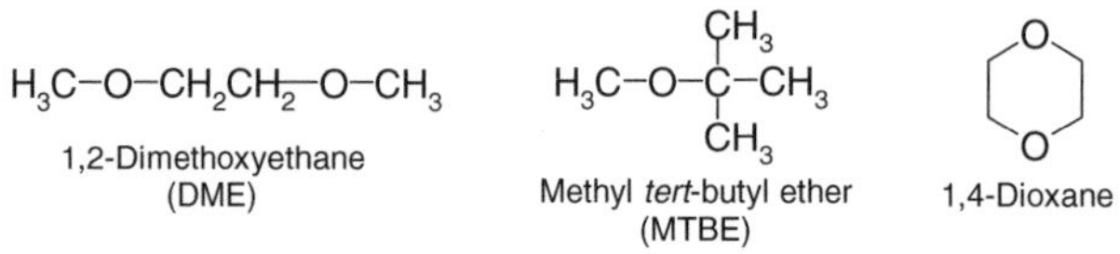

$H_3C-O-CH_2CH_2-O-CH_3$
1,2-Dimethoxyethane
(DME)

$H_3C-O-C(CH_3)_2-CH_3$
Methyl *tert*-butyl ether
(MTBE)

1,4-Dioxane

Nomenclature of ethers

Ethers can be symmetrical, where the two alkyl groups are the same, or unsymmetrical, where the two alkyl groups are different. While diethyl ether is symmetrical, ethyl methyl ether is unsymmetrical. The common name of an unsymmetrical ether is obtained by quoting the names of the two alkyl groups in alphabetical order followed by the prefix ether.

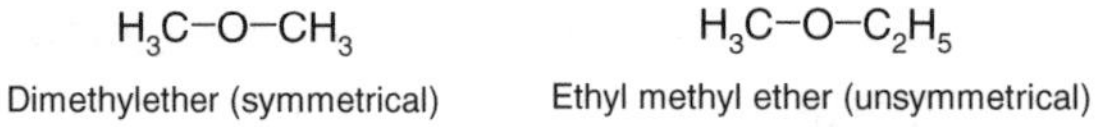

$H_3C-O-CH_3$
Dimethylether (symmetrical)

$H_3C-O-C_2H_5$
Ethyl methyl ether (unsymmetrical)

In the nomenclature of ethers, either the suffix *-ether* or the prefix *alkoxy-* is used. For example, diethyl ether can be called ethoxyethane, and methyl *t*-butyl ether can be named as 2-methyl-2-methoxypropane.

$C_2H_5-O-C_2H_5$
Diethylether
(Ethoxyethane)

$H_3C-O-C(CH_3)_2-CH_3$
Methyl *t*-butyl ether
(2-Methyl-2-methoxypropane)

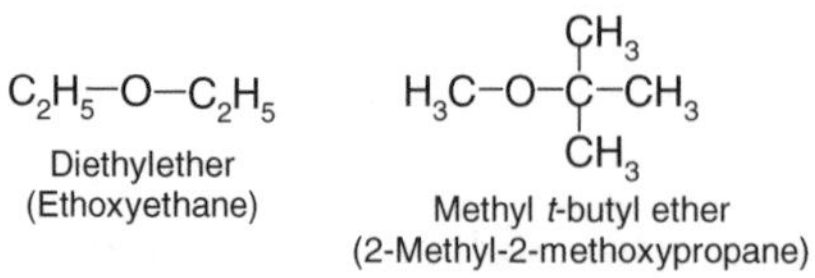

Three-membered cyclic ethers are known as *epoxides*. They are just a subclass of ethers containing a three-membered oxirane ring (C—O—C unit). Cyclic ethers have the prefix *epoxy-* and suffix *-alkene oxide*. Five-membered and six-membered cyclic ethers are known as oxolane and oxane, respectively.

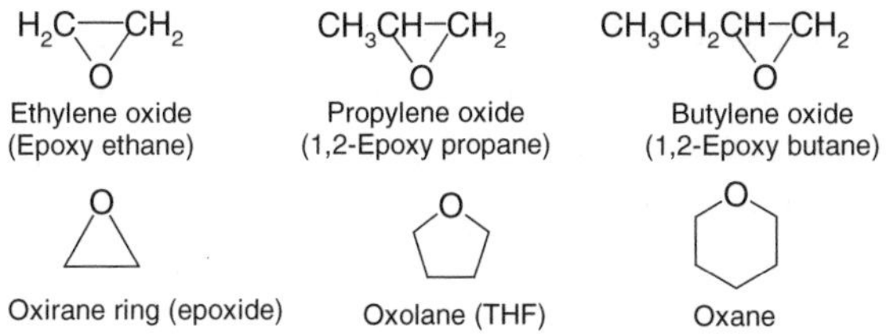

Ethylene oxide (Epoxy ethane) | Propylene oxide (1,2-Epoxy propane) | Butylene oxide (1,2-Epoxy butane)

Oxirane ring (epoxide) | Oxolane (THF) | Oxane

Physical properties of ethers

An ether cannot form hydrogen bonds with other ether molecules since there is no H to be donated as there is no —OH group, but can be involved in hydrogen bonding with hydrogen bonding systems, e.g. water, alcohols and amines. Ethers have much lower m.p. and b.p., and less water solubility than analogous alcohols. They are fairly unreactive, and this makes them useful as good polar protic solvents to carry out many organic reactions. For example, diethyl ether and THF are common solvents used in the Grignard reaction. Ethers often form complexes with molecules that have vacant orbitals, e.g. THF complexes with borane (BH_3.THF), which is used in the *hydroboration–oxidation* reaction (see Section 5.3.1).

Preparation of ethers

Ethers are prepared from alkyl halides by the treatment of metal alkoxide. This is known as Williamson ether synthesis (see Sections 4.3.6 and 5.5.2). *Williamson ether synthesis* is an important laboratory method for the preparation of both symmetrical and unsymmetrical ethers. Symmetrical ethers are prepared by dehydration of two molecules of primary alcohols and H_2SO_4 (see Sections 4.3.7 and 5.5.3). Ethers are also obtained from alkenes either by acid-catalysed addition of alcohols or alkoxymercuration–reduction (see Section 5.3.1).

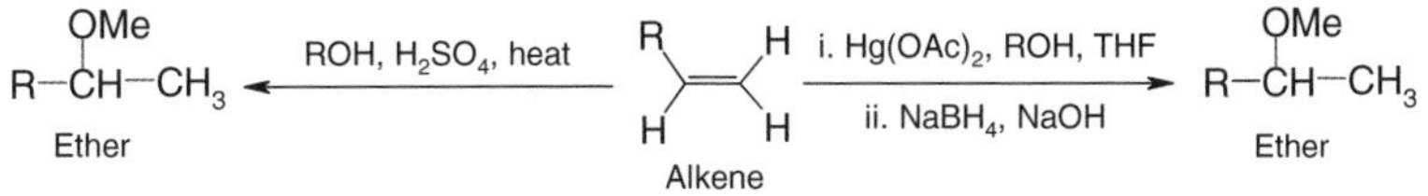

R–CH(OMe)–CH3 (Ether) ← ROH, H2SO4, heat — RHC=CH2 (Alkene) — i. Hg(OAc)2, ROH, THF; ii. NaBH4, NaOH → R–CH(OMe)–CH3 (Ether)

Conversion of alkenes to epoxides The simplest epoxide, ethylene dioxide, is prepared by catalytic oxidation of ethylene, and alkenes are also oxidized to other epoxides by peracid or peroxy acid (see Section 5.7.2).

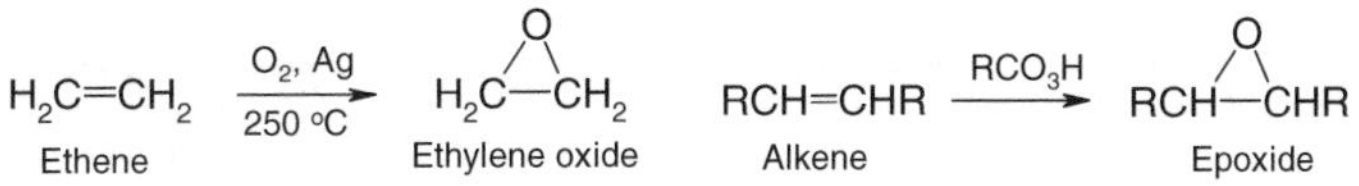

H2C=CH2 (Ethene) —O2, Ag, 250 °C→ Ethylene oxide   RCH=CHR (Alkene) —RCO3H→ Epoxide

Alkenes are converted to *halohydrins* by the treatment of halides and water. When halohydrins are treated with a strong base (NaOH), an intramolecular cyclization occurs and epoxides are formed. For example, 1-butene can be converted to butylene oxide via butylene chlorohydrin.

$$CH_3CH_2CH{=}CH_2 \xrightarrow{Cl_2,\ H_2O} C_2H_5CH(Cl){-}CH_2(OH) \xrightarrow{NaHO} C_2H_5CH{-}CH_2\ (\text{epoxide, }O) + NaCl + H_2O$$

1-Butene — Butylene chlorohydrin — Butylene oxide

Reactions of ethers

Simple ethers (acyclic) are relatively unreactive towards bases, oxidizing agents and reducing agents. Ethers cannot undergo nucleophilic substitution reactions, except with haloacids (usually HBr or HI) at high temperatures, where ethers are protonated to undergo nucleophilic substitution reactions to form corresponding alkyl halides (see Section 5.5.4).

$$R{-}O{-}R \xrightarrow[X = Br\ or\ I]{HX,\ heat} RX$$

Ethers — Alkyl halides

Epoxides are much more reactive than simple ethers due to ring strain, and are useful intermediates because of their chemical versatility. They undergo nucleophilic substitution reactions with both acids and bases to produce alcohols (see Sections 4.3.7 and 5.5.4).

4.3.10 Amines

Amines are nitrogen-containing compounds, where the functional group is an amino group ($-NH_2$). They are organic relatives of ammonia, where one or more of the hydrogen atoms of ammonia are replaced by alkyl group(s). Thus, an amine has the general formula RNH_2, R_2NH or R_3N. The simplest and most common amines are methylamine (CH_3NH_2) and ethylamine ($CH_3CH_2NH_2$).

$H_3C{-}NH_2$ Methylamine

$C_2H_5{-}NH_2$ Ethylamine

Amines are classified as primary (1°), secondary (2°), tertiary (3°) or quaternary (4°) depending on how many alkyl groups are attached to the N atom. Quaternary amines, $(CH_3)_4N^+$, are known as ammonium cations.

$CH_3CH_2CH_2{-}NH_2$ Propylamine (1° amine)

$H_3C{-}NH{-}CH_3$ Dimethylamine (2° amine)

$H_3C{-}N(CH_3){-}CH_3$ Trimethylamine (3° amine)

$H_3C{-}N^+(CH_3)_2{-}CH_3$ Tetramethylamine (4° amine)

Nomenclature of amines

Aliphatic amines are named according to the alkyl group or groups attached to nitrogen with the suffix *-amine*. If there is more than one alkyl group bonded to the nitrogen atom, the prefixes *di-* and *tri-* are used to indicate the presence of two or three alkyl groups of the same kind. Often the prefix *amino-* is used with the name of the parent chain for more complex amines. If other substituents are attached to the nitrogen atom, they are indicated by the prefix *N-* before the name of the substituents. The simplest aromatic amine is aniline ($C_6H_5NH_2$), where nitrogen is attached directly to a benzene ring.

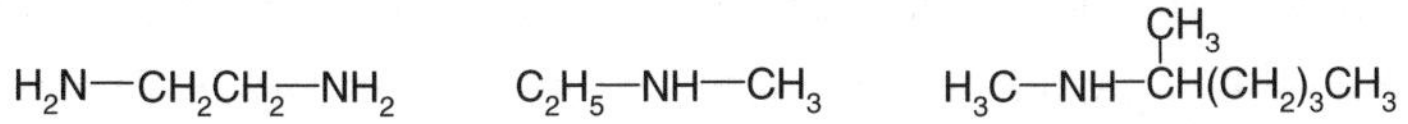

$H_2N—CH_2CH_2—NH_2$
Ethylenediamine
(1° amine)

$C_2H_5—NH—CH_3$
Methylethylamine
(2° amine)

$H_3C—NH—CH(CH_3)(CH_2)_3CH_3$
2-(*N*-Methylamino)hexane
(2° amine)

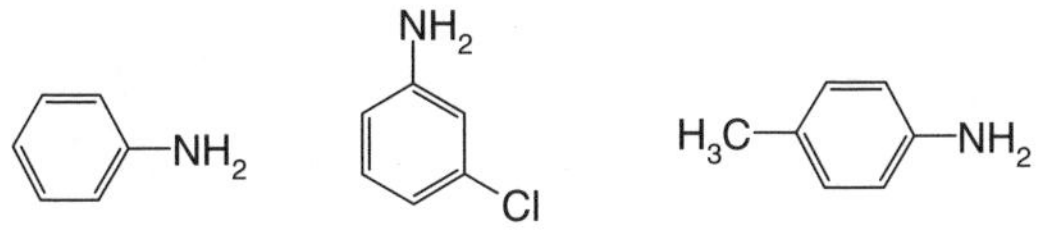

Aniline

o-Chloroaniline

P-Toluidine

Physical properties of amines

In amines, both the C—N and the N—H bonds are polar owing to the electronegativity of the nitrogen atom. The polar nature of the N—H bond results in the formation of hydrogen bonds with other amine molecules or other hydrogen-bonding systems, e.g. water and alcohols. Thus, amines have higher m.p. and b.p., and are more soluble in aqueous media than analogous alkanes.

Basicity and reactivity of amines

The nitrogen atom of amines has a lone pair of electrons, and they can react as either bases or nucleophiles. Thus, the basicity and the nucleophilicity of amines (NH_2) are quite similar to those of ammonia (NH_3). Amines are more basic than analogous alcohols and ethers. The —NH group is a poor leaving group like the —OH group, and needs to be converted to a better leaving group before substitution can occur. The anion derived from the deprotonation of an amine is the amide ion, NH_2^-, and should not be confused with the carboxylic acid derivative amide, $RCONH_2$. Amide ions are important bases in organic reactions.

Preparation of amines

Amines are prepared by *aminolysis* of alkyl halides, and also *reductive amination* (reduction in the presence of ammonia) of aldehydes and ketones (see Section 5.7.19). They are obtained conveniently from *Hofmann rearrangement* of amides.

Aminolysis of halides Amines are prepared from primary alkyl halides by treatment with an aqueous or alcoholic solution of ammonia. This reaction is known as *aminolysis*; it is not product specific, and produces more than one class of amine. Therefore, it is difficult to prepare pure primary amines using this method. The primary amines can be separated by distillation from these by-products, but the yield is poor. However, this can be avoided using a large excess of ammonia.

$$\underset{\text{Excess}}{R{-}X + NH_3} \xrightarrow[\text{Alcohol}]{H_2O\text{ or}} \underset{1^\circ\text{ Amine}}{RCH_2NH_2} \xleftarrow[\text{NaBH}_3\text{CN or H}_2\text{/Pd-C}]{NH_3} \underset{\substack{Y = H \text{ or } R \\ \text{Aldehyde or ketone}}}{R{-}\overset{O}{\overset{\|}{C}}{-}Y}$$

Hofmann rearrangement In this reaction, amines (with one less carbon) are prepared from amides by the treatment of halides (Br_2 or Cl_2) in aqueous sodium or potassium hydroxide (NaOH or KOH).

$$\underset{tert\text{-Butylamide}}{(CH_3)_3C{-}\overset{O}{\overset{\|}{C}}{-}NH_2} \xrightarrow[H_2O]{Br_2,\ NaOH} \underset{tert\text{-Butylamine}}{(CH_3)_3C{-}NH_2}$$

Catalytic hydrogenation or $LiAlH_4$ reduction of amides, azides or nitriles produces amines (see Section 5.7.23).

$$\underset{1^\circ\text{ Amide}}{R{-}\overset{O}{\overset{\|}{C}}{-}NH_2} \xrightarrow[\text{i. LiAlH}_4\text{ ii. H}_2\text{O}]{H_2/Pd\text{-}C\text{ or}} \underset{1^\circ\text{ Amine}}{RCH_2{-}NH_2}$$

$$\underset{\text{Alkyl azide}}{RCH_2{-}N{=}\overset{+}{N}{=}\overset{-}{N}} \xrightarrow[\text{i. LiAlH}_4\text{ ii. H}_2\text{O}]{H_2/Pd\text{-}C\text{ or}} \underset{1^\circ\text{ Amine}}{RCH_2{-}NH_2}$$

$$\underset{\text{Nitrile}}{R{-}C{\equiv}N} \xrightarrow[\text{i. LiAlH}_4\text{ ii. H}_2\text{O}]{2\ H_2/Pd\text{-}C\text{ or}} \underset{1^\circ\text{ Amine}}{RCH_2{-}NH_2}$$

Reactions of amines

Primary (RNH_2) and secondary (R_2NH) amines undergo nucleophilic acyl substitution with acid chlorides and anhydrides in pyridine or Et_3N to give 2° and 3° amides (see Section 5.5.5). Primary amines (RNH_2) react with aldehydes and ketones, followed by loss of water to give imines, also known as *Schiff bases* (see Section 5.3.2). Secondary amines (R_2NH) react similarly to give enamines, after loss of water and tautomerization (see Section 5.3.2).

$$\underset{\text{Imine}}{R'CH_2-C(=NCH_2R)-Y} \xleftarrow[Y = H \text{ or } R]{R'CH_2CO-Y} \underset{1^\circ \text{ Amine}}{RCH_2-NH_2} \xrightarrow[(R'CO)_2O]{R'COCl \text{ or}} \underset{\substack{2^\circ \text{ amide} \\ N\text{-substituted amide}}}{R'CONHCH_2R}$$

$$\underset{\text{Enamine}}{R'CH=C(NR_2)-Y} \xleftarrow[Y = H \text{ or } R]{R'CH_2CO-Y} \underset{2^\circ \text{ Amine}}{R_2-NH} \xrightarrow[(R'CO)_2O]{R'COCl \text{ or}} \underset{\substack{3^\circ \text{ amide} \\ N,N\text{-disubstituted amide}}}{R'CONR_2}$$

The amides derived from sulphonic acids are called *sulphonamides*. They are obtained from amines by the reaction with sulphonyl chloride ($R'SOCl_2$) in pyridine.

$$\underset{1^\circ \text{ Amine}}{RCH_2-NH_2} \xrightarrow[\text{Pyridine}]{R'SO_2Cl} \underset{\substack{N\text{-substituted} \\ \text{sulphonamide } (2^\circ)}}{R'SO_2NHCH_2R} \qquad \underset{2^\circ \text{ Amine}}{R_2-NH} \xrightarrow[\text{Pyridine}]{R'SO_2Cl} \underset{\substack{N,N\text{-disubstituted} \\ \text{sulphonamide } (3^\circ)}}{R'SO_2NR_2}$$

Base-catalysed quaternary ammonium salts give alkenes and 3° amines. This reaction is known as *Hofmann elimination* or *Hofmann degradation.* Amines can readily be converted to quaternary ammonium salt by the treatment of excess primary alkyl halides, and then Ag_2O and H_2O. Quaternary ammonium salts undergo E2 elimination, when heated with NaOH to give alkenes and tertiary amines. Thermal decomposition of a quaternary ammonium salt by NaOH to an alkene is known as *Hofmann elimination.*

$$\underset{1^\circ \text{ Amine}}{H-C-C-NH_2} \xrightarrow[\text{ii. } Ag_2O/H_2O]{\text{i. Excess RX}} \underset{\text{Quarternary ammonium salt}}{H-C-C-\overset{+}{N}R_3} \xrightarrow[\text{E2}]{\text{NaOH, heat}} \underset{\text{Alkene}}{>C=C<} + \underset{3^\circ \text{ Amine}}{NR_3}$$

4.3.11 Aldehydes and ketones

A carbonyl functional group (C=O) is a carbon double bonded to an oxygen atom. An acyl functional group (R−C=O) consists of a carbonyl group attached to an alkyl or an aryl group. Carbonyl-group-containing compounds can be classified into two broad classes: one group includes compounds that have hydrogen and carbon atoms bonded to the carbonyl carbon, and the other group contains an electronegative atom bonded to the carbonyl carbon (see Section 4.3.13).

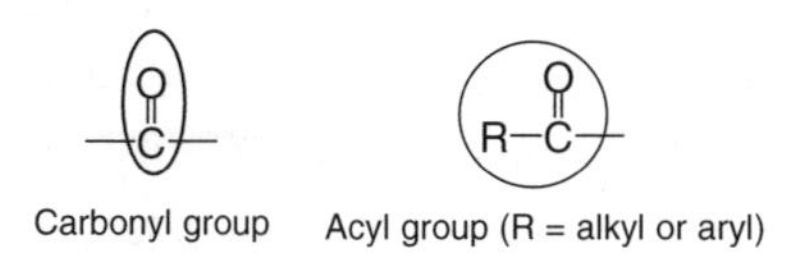

Carbonyl group Acyl group (R = alkyl or aryl)

Aldehydes have an acyl group with a hydrogen atom bonded to the carbonyl carbon. The most abundant natural aldehyde is glucose. The simplest aldehyde is formaldehyde (CH_2O), where the carbonyl carbon is bonded to two hydrogen atoms. In all other aldehydes, the carbonyl carbon is bonded to one hydrogen atom and one alkyl or aryl group, e.g. acetaldehyde (CH_3CHO).

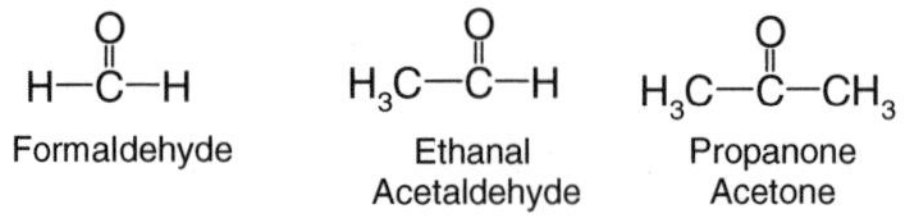

Formaldehyde | Ethanal Acetaldehyde | Propanone Acetone

Ketones have an acyl group with another alkyl or aryl group connected to the carbonyl carbon. Many steroid hormones contain ketone functionality, e.g. testosterone and progesterone. The simplest ketone is acetone (CH_3COCH_3), where the carbonyl carbon is bonded to two methyl groups.

Nomenclature of aldehydes and ketones

The common names of aldehydes are derived from the corresponding carboxylic acids by replacing *-ic acid* by *-aldehyde*; e.g., formic acid gives formaldehyde, and acetic acid gives acetaldehyde. The simplest ketone has the common name of acetone. In the IUPAC nomenclature of aldehydes, the *-e* of the alkane is replaced with *-al*, e.g. ethan*al* (the parent alkane is ethan*e*). Similarly, ketones are named by replacing the *-e* ending of the alkyl name with *-one*, e.g. propan*one* (the parent alkane is propan*e*). The longest chain carrying the carbonyl group is considered the parent structure. The carbonyl carbon is the first carbon atom of the chain. Other substituents are named using prefixes and their positions are indicated by numbers relative to the carbonyl group. If the aldehyde group is a substituent on a ring, the suffix *-carbaldehyde* is used in the name.

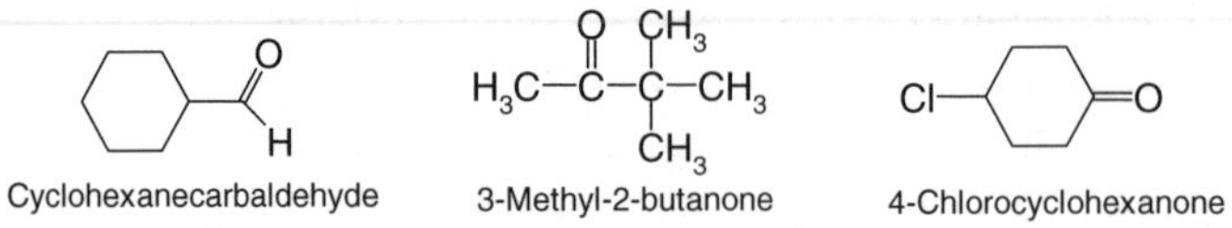

Cyclohexanecarbaldehyde | 3-Methyl-2-butanone | 4-Chlorocyclohexanone

In certain polyfunctional compounds, an aldehyde or ketone group can also be named as a substituent on a molecule with another functional group as its root. The aldehyde carbonyl is given the prefix *formyl-*, and the ketone group is named *oxo-* with a number to show its position in the molecule. Compounds with both an aldehyde and ketone are named as aldehydes, because aldehydes have functional group priority over ketones. A ketone containing a benzene ring is named as a *-phenone*.

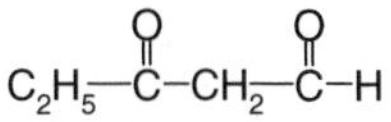

$C_2H_5-C(=O)-CH_2-C(=O)-H$
3-Oxopentanal

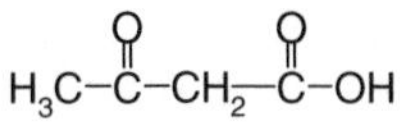

$H_3C-C(=O)-CH_2-C(=O)-OH$
3-Oxobutanoic acid

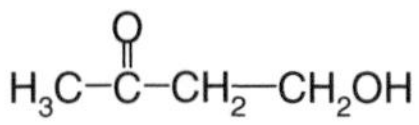

$H_3C-C(=O)-CH_2-CH_2OH$
4-Hydroxy-2-butanone

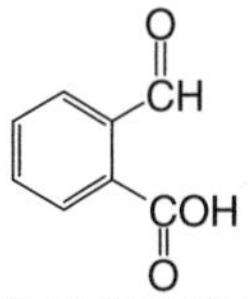

2-Formylbenzoic acid

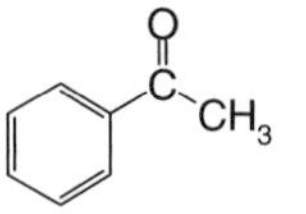

Acetophenone

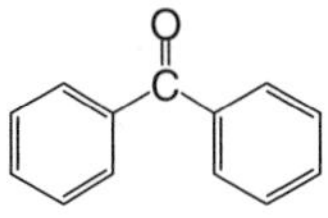

Benzophenone

Physical properties of aldehydes and ketones

The carbonyl oxygen atom is a Lewis base (see Sections 1.2.3 and 1.2.4), and can be readily protonated in the presence of an acid. The polar nature of the C=O group is due to the electronegativity difference of the carbon and oxygen atoms. The C=O group cannot form intermolecular hydrogen bonding, but it can accept hydrogen from hydrogen bond donors, e.g. water, alcohols and amines. Therefore, aldehydes and ketones have higher melting and boiling points compared with analogous alkanes, and much lower boiling points than analogous alcohols. They are much more soluble than alkanes but less soluble than analogous alcohols in aqueous media; e.g., acetone and acetaldehyde are miscible with water.

Preparation of aldehydes and ketones

Aldehydes are prepared by the hydroboration-oxidation of alkynes (see Section 5.3.1) or selective oxidation of primary alcohols (see Section 5.7.9), and partial reduction of acid chlorides (see Section 5.7.21) and esters (see Section 5.7.22) or nitriles (see Section 5.7.23) with lithium tri-*tert*-butoxyaluminium hydride [$LiAlH(O\text{-}^tBu)_3$] and diisobutylaluminium hydride (DIBAH), respectively.

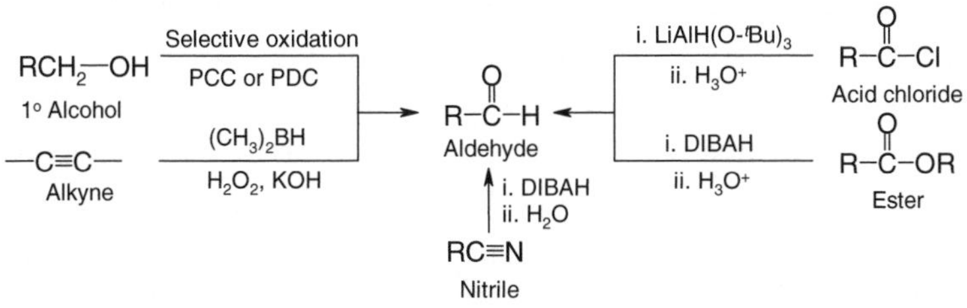

RCH_2-OH (1° Alcohol) —Selective oxidation, PCC or PDC→ $R-C(=O)-H$ (Aldehyde)

$-C{\equiv}C-$ (Alkyne) —$(CH_3)_2BH$, H_2O_2, KOH→ Aldehyde

$R-C(=O)-Cl$ (Acid chloride) —i. $LiAlH(O\text{-}^tBu)_3$, ii. H_3O^+→ Aldehyde

$R-C(=O)-OR$ (Ester) —i. DIBAH, ii. H_3O^+→ Aldehyde

$RC{\equiv}N$ (Nitrile) —i. DIBAH, ii. H_2O→ Aldehyde

Ketones are prepared by the oxidation of secondary alcohols (see Section 5.7.10) and partial reduction of acid chlorides by the treatment of Gilman reagents (organocopper reagents, R'_2CuLi) followed by hydrolytic work-up (see Section 5.5.5) or by the reaction with nitriles and organometallic reagents (R' MgBr or $R'Li$) followed by the acidic work-up (see Section 5.3.2)

$$\underset{2^\circ \text{ Alcohol}}{R\text{-}CH(R')OH} \xrightarrow[\text{KMnO}_4 \text{ or } K_2Cr_2O_7 \text{ or PCC or PDC}]{\text{Any oxidizing agent}} \underset{\text{Ketone}}{R\text{-}CO\text{-}R'} \xleftarrow[\text{ii. } H_2O]{\text{i. } R'_2CuLi\text{, ether}} \underset{\text{Acid chloride}}{R\text{-}CO\text{-}Cl}$$

$$\underset{\text{Nitrile}}{RC{\equiv}N} \xrightarrow[\text{ii. } H_3O^+]{\text{i. R'MgBr, ether}} \text{Ketone}$$

Aldehydes and ketones are obtained by ozonolysis of alkenes (see Section 5.7.6) and hydration of alkynes (see Section 5.3.1).

$$\underset{\text{Alkene}}{>C{=}C<} \xrightarrow[\text{ii. Zn, AcOH}]{\text{i. } O_3\text{, } CH_2Cl_2} \underset{\text{Aldehyde or ketone}}{>C{=}O \ + \ O{=}C<}$$

$$\underset{\text{Alkyne}}{\text{—}C{\equiv}C\text{—}} \xrightarrow[\text{HgSO}_4\text{, } H_2SO_4]{H_2O} \underset{\substack{Y = H \text{ or } R \\ \text{Aldehyde or ketone}}}{\text{—}CH_2\text{—}CO\text{—}Y}$$

Structure and reactivity

The carbonyl group of aldehydes and ketones is highly polarized, because carbon is less electronegative than oxygen. The carbonyl carbon bears a partial positive charge (δ^+), while the oxygen bears a partial negative charge (δ^-). Therefore, the carbonyl group can function as both a nucleophile and an electrophile. Aldehydes and ketones cannot undergo substitution reactions, because they do not have a leaving group. Thus, the common carbonyl group reactions are nucleophilic additions.

$$\text{—}\overset{\delta^+}{C}\text{—}{=}\overset{\delta^-}{\ddot{O}}\text{:}$$

Nucleophilic oxygen

Electrophilic carbon

Aldehydes are more reactive than ketones. Two factors that make aldehydes more reactive than ketones are electronic and steric effects. Ketones have two alkyl groups, whereas aldehydes have only one. Because alkyl groups are electron donating, ketones have their effective partial positive charge reduced more than aldehydes. The electrophilic carbon is the site where the nucleophile approaches for reaction to occur. In ketones, two alkyl groups create more steric hindrance than one in aldehydes. As a result, ketones offer more steric resistance toward the nucleophilic attack than aldehydes.

Reactions of aldehydes and ketones: nucleophilic addition

Carbonyl compounds are of central importance in organic chemistry because of their unique ability to form a range of other derivatives. As shown earlier, alcohols can be prepared from aldehydes and ketones by nucleophilic addition of organometallic reagents (see Section 5.3.2), catalytic hydrogenation (H_2/Pd–C) and metal hydride reduction, e.g. sodium borohydride ($NaBH_4$) or lithium aluminium hydride ($LiAlH_4$) (see Section 5.7.16). Aldehydes and ketones are selectively reduced to alkanes by *Clemmensen reduction* (see Section 5.7.17) and *Wolff-Kishner reduction* (see Section 5.7.18), and to amines by reductive amination (see Section 5.7.19).

RCH_2NH_2 (1° Amine) ← NH_3, $NaBH_3CN$ — R–C(=O)–Y (Y = H or R; Aldehyde or ketone) — Zn(Hg) in HCl or NH_2NH_2, NaOH → $R–CH_2–Y$ (Alkane)

R–C(OH)(R')–Y (2° or 3° Alcohol) ← i. R'MgX or R'Li; ii. H_3O^+ — R–C(=O)–Y — H_2/Pd-C or i. $NaBH_4$ or $LiAlH_4$; ii. H_2O or H_3O^+ → R–CH(OH)–Y (1° or 2° Alcohol)

One of the most important reactions of aldehydes and ketones is the *Aldol condensation.* In this reaction, an enolate anion is formed from the reaction between an aldehyde or a ketone and an aqueous base, e.g. NaOH. The enolate anion reacts with another molecule of aldehyde or ketone to give β-hydroxyaldehyde or β-hydroxyketone, respectively (see Section 5.3.2).

RCH_2–C(=O)–Y + RCH_2–C(=O)–Y $\xrightarrow{NaOH,\ H_2O}$ RCH_2–C(OH)(Y)–CH(R)–C(=O)–Y

Y = H or R
Aldehyde or ketone

β-Hydroxyaldehyde or ketone

4.3.12 Carboxylic acids

Carboxylic acid is an organic acid that has an acyl group (R–C=O) linked to a hydroxyl group (–OH). In a condensed structural formula, a carboxyl group may be written as $-CO_2H$, and a carboxylic acid as RCO_2H.

–C(=O)–OH
$-CO_2H$ or –COOH
Carboxyl group

R–C(=O)–OH
RCO_2H or RCOOH
Carboxylic acid

Carboxylic acids are classified as aliphatic acids, where an alkyl group is bonded to the carboxyl group, and aromatic acids, where an aryl group is bonded to the carboxyl group. The simplest carboxylic acids are formic acid (HCO_2H) and acetic acid (CH_3CO_2H).

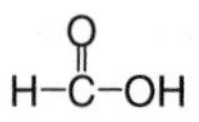

H–C(=O)–OH
Methanoic acid
Formic acid

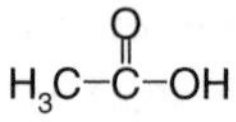

H_3C–C(=O)–OH
Ethanoic acid
Acetic acid

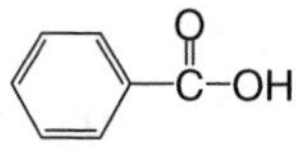

C_6H_5–C(=O)–OH
Benzenecarboxylic acid
Benzoic acid

Nomenclature of carboxylic acids

The root name is based on the longest continuous chain of carbon atoms bearing the carboxyl group. The *-e* is replaced by *-oic acid*. The chain is numbered starting with the carboxyl carbon atom. The carboxyl group takes priority over any other functional groups as follows: carboxylic acid > ester > amide > nitrile > aldehyde > ketone > alcohol > amine > alkene > alkyne.

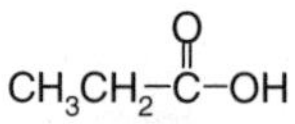

CH_3CH_2–C(=O)–OH
Propanoic acid
Propionic acid

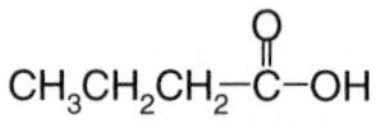

$CH_3CH_2CH_2$–C(=O)–OH
Butanoic acid
Butyric acid

$CH_3CH_2CH_2CH_2$–C(=O)–OH
Pentanoic acid
Valeric acid

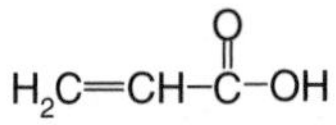

$H_2C{=}CH$–C(=O)–OH
Propenoic acid
Acrylic acid

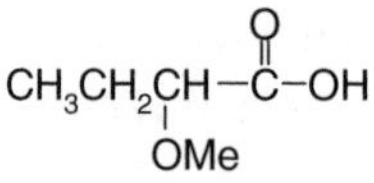

$CH_3CH_2CH(OMe)$–C(=O)–OH
2-Methoxybutanoic acid

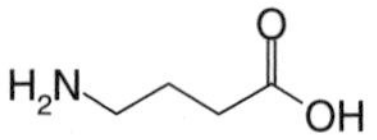

H_2N–$CH_2CH_2CH_2$–C(=O)OH
4-Aminobutanoic acid

Cycloalkanes with carboxyl substituents are named as cycloalkanecarboxylic acids. Unsaturated acids are named using the name of the alkene with *-e* replaced with *-oic acid*. The chain is numbered starting with the carboxyl group, a number designates the location of the double bond and *Z* or *E* is used.

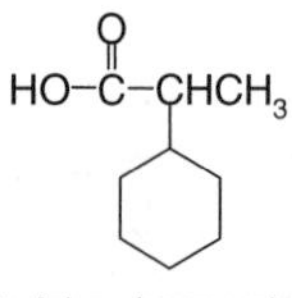

HO–C(=O)–CHCH$_3$ (cyclohexyl)
2-Cyclohexylpropanoic acid

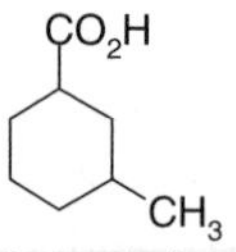

CO_2H, CH_3
3-Methylcyclohexanecarboxylic acid

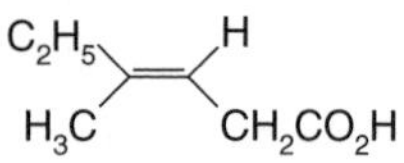

C_2H_5, H, H_3C, CH_2CO_2H
(*E*)-4-Methyl-3-hexenoic acid

Aromatic acids are named as derivatives of benzoic acids, with *ortho*, *meta* and *para* indicating the location relative to the carboxyl group.

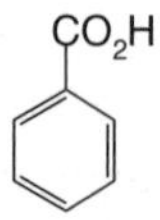

CO_2H
Benzoic acid
Benzene carrboxylic acid

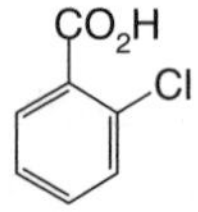

CO_2H, Cl
o-Chlorobenzoic acid

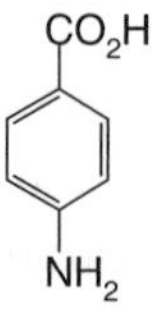

CO_2H, NH_2
p-Aminobenzoic acid

Aliphatic dicarboxylic acids are named by simply adding the suffix *-dioic acid* to the root name. The root name comes from the longest carbon chain containing both carboxyl groups. Numbering starts at the end closest to a substituent.

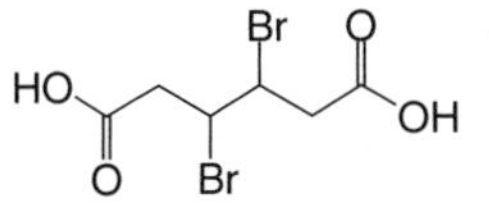

3,4-Dibromohexanedioic acid

Structure of the carboxyl group

The most stable conformation of a carboxyl group is a planar arrangement of the molecule. The carbon is sp^2-hybridized, and the O—H bond lies in the plane, eclipsing the C=O double bond. This unexpected geometric arrangement can be explained by resonance. The following resonance forms can be written for a carboxyl group.

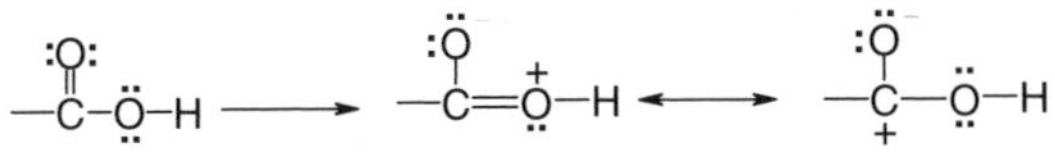

Resonance forms of carboxyl group

Acidity of carboxylic acids

Although carboxylic acids are much weaker acids than the strong mineral acids, e.g. HCl, H_2SO_4 and HNO_3, they can still dissociate in aqueous solution and form carboxylate ions (RCO_2^-). The equilibrium constant for this process is $K_a = \sim 10^{-5}$ (p$K_a = \sim 5$). Carboxylic acids are more acidic than analogous alcohols. For example, the pK_a values of ethanoic acid and ethanol are, respectively, 4.74 and ~15.9.

$$R{-}C(=O){-}OH + H_2O \rightleftharpoons R{-}C(=O){-}O^- + H_3O^+$$

Substituent effects on the acidity of carboxylic acids

Any substituent that stabilizes a negative charge enhances the dissociation process, i.e. increases the acidity. Electronegative elements can enhance the acid strength through inductive effects. The closer the substituent to the anion, the more profound the effects are.

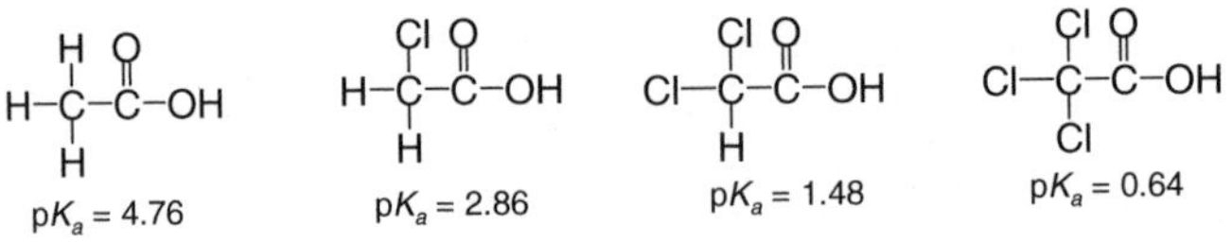

CH_3COOH, $pK_a = 4.76$; $ClCH_2COOH$, $pK_a = 2.86$; $Cl_2CHCOOH$, $pK_a = 1.48$; Cl_3CCOOH, $pK_a = 0.64$

Salts of carboxylic acids

Carboxylic acids are more acidic than alcohols and acetylene. Strong aqueous bases can completely deprotonate carboxylic acids, and salts of carboxylic acids are formed. Strong aqueous mineral acids readily convert the salt back to the carboxylic acids. Salts are soluble in water but insoluble in nonpolar solvents, e.g. hexane or dichloromethane.

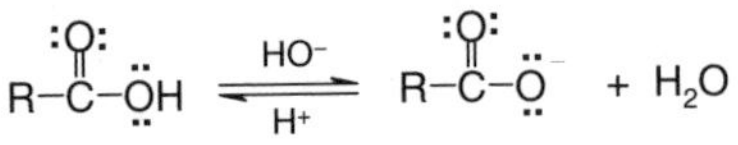

$$R{-}COOH \underset{H^+}{\overset{HO^-}{\rightleftharpoons}} R{-}COO^- + H_2O$$

Physical properties of carboxylic acids

Carboxylic acids are polar molecules due to the polar nature of both the O—H and C=O functionalities. They form strong hydrogen bonds with other carboxylic acid molecules or water. Therefore, carboxylic acids have higher m.p. and b.p. than analogous alcohols. They are highly soluble in aqueous media. The hydrogen atom of RCO_2H has a p$K_a \sim 5$.

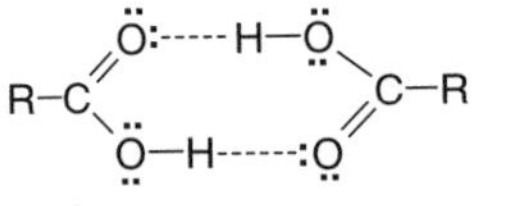

Hydrogen bonds between two molecules of carboxylic acid

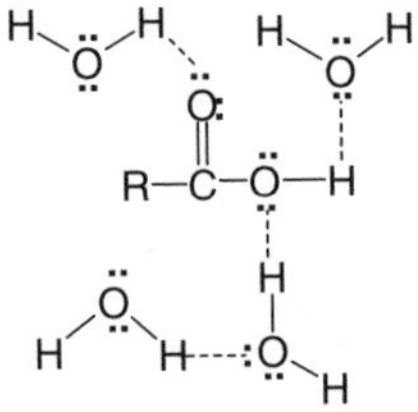

Hydrogen bonds to water in aqueous solution

Preparation of carboxylic acids

Acetic acid, the most important carboxylic acid, can be prepared by catalytic air oxidation of acetaldehyde.

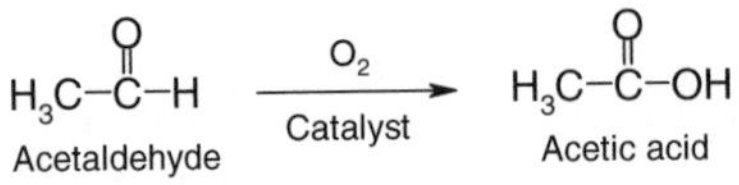

$$\underset{\text{Acetaldehyde}}{H_3C{-}CHO} \xrightarrow[\text{Catalyst}]{O_2} \underset{\text{Acetic acid}}{H_3C{-}COOH}$$

Carboxylic acids can be obtained from oxidation of alkenes (see Section 5.7.2), alkynes (see Section 5.7.7), 1° alcohols (see Section 5.7.9) and aldehydes (see Section 5.7.11), ozonolysis of alkenes and alkynes with oxidative work-up (see Sections 5.7.6 and 5.7.8) and carbonation of Grignard reagents (see Section 5.3.2).

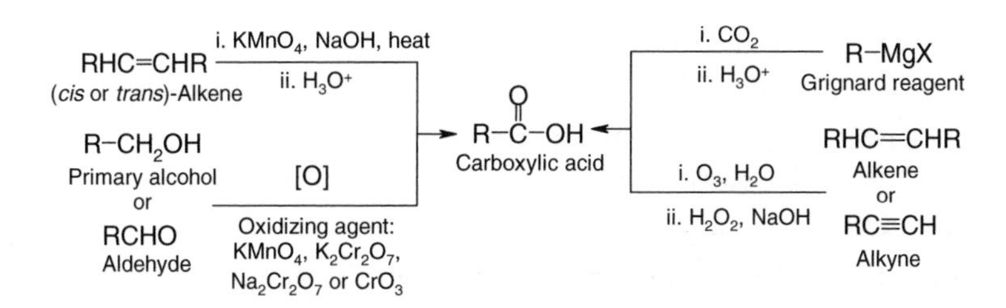

Carboxylic acids are prepared by the hydrolysis of acid chlorides and acid anhydrides, and acid- or base-catalysed hydrolysis (see Section 5.6.1) of esters, primary amides and nitriles (see Section 5.6.1).

R-C(=O)-Cl (Acid) —H_2O→ R-C(=O)-OH (Carboxylic acid)
R-C(=O)-O-C(=O)-R (Acid anhydride) —H_2O→ R-C(=O)-OH
R-C(=O)-OR (Ester) —H_3O^+, heat or i. HO^-, heat; ii. H_3O^+→ R-C(=O)-OH
R-C(=O)-NH_2 (1° Amide) —6M HCl or 40% NaOH→ R-C(=O)-OH

RC≡N (Nitrile) —H_3O^+, Heat→ R-C(=O)-NH_2 (1° Amide) —H_3O^+, heat, Long period→ R-C(=O)-OH (Carboxylic acid)
RC≡N —i. HO^-, 50 °C→ R-C(=O)-NH_2 (1° Amide) —i. HO^-, 200 °C; ii. H_3O^+→ R-C(=O)-OH

Reactions of carboxylic acids

The most important reactions of carboxylic acids are the conversions to various carboxylic acid derivatives, e.g. acid chlorides, acid anhydrides and esters. Esters are prepared by the reaction of carboxylic acids and alcohols. The reaction is acid catalysed and is known as *Fischer esterification* (see Section 5.5.5). Acid chlorides are obtained from carboxylic acids by the treatment of thionyl chloride ($SOCl_2$) or oxalyl chloride [$(COCl)_2$], and acid anhydrides are produced from two carboxylic acids. A summary of the conversion of carboxylic acid is presented here. All these conversions involve nucleophilic acyl substitutions (see Section 5.5.5).

R-C(=O)-OH (Carboxylic acid) —R'OH, H^+/heat→ R-C(=O)-OR' (Ester)
R-C(=O)-OH —HO^-, Heat→ R-C(=O)-O^- (Carboxylate)
R-C(=O)-OH —$SOCl_2$, Pyridine→ R-C(=O)-Cl (Acyl chloride)
R-C(=O)-OH —$R'CO_2Na$, Heat→ R-C(=O)-O-C(=O)-R' (Acid anhydride)

4.3.13 Carboxylic acid derivatives

Carboxylic acid derivatives are compounds that possess an acyl group (R—C=O) linked to an electronegative atom, e.g. —Cl, $-CO_2$ R, —OR or $-NH_2$. They can be converted to carboxylic acids via simple acidic or basic hydrolysis. The important acid derivatives are acid chlorides, acid anhydrides, esters and amides. Usually nitriles are also considered as carboxylic acid derivatives. Although nitriles are not directly carboxylic acid derivatives, they are conveniently hydrolysed to carboxylic acids by acid or base catalysts. Moreover, nitriles can be easily prepared through dehydration of amides, which are carboxylic acid derivatives.

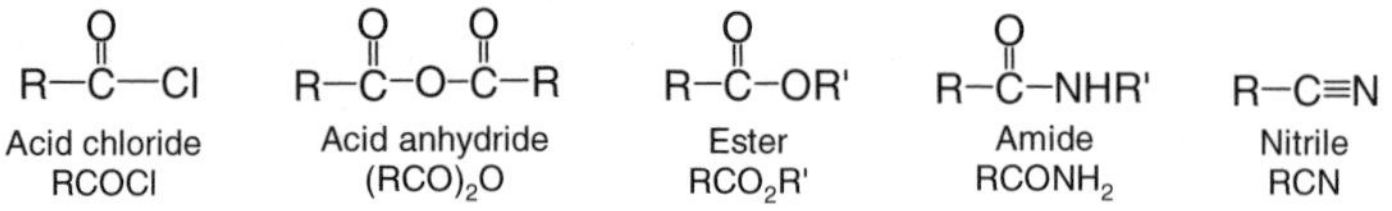

R—C(=O)—Cl	R—C(=O)—O—C(=O)—R	R—C(=O)—OR'	R—C(=O)—NHR'	R—C≡N
Acid chloride RCOCl	Acid anhydride $(RCO)_2O$	Ester RCO_2R'	Amide $RCONH_2$	Nitrile RCN

Reactivity of carboxylic acid derivatives

The carboxylic acid derivatives vary greatly in their reactivities in the acyl substitution reactions. In general, it is easy to convert more reactive derivatives into less reactive derivatives. Therefore, an acid chloride is easily converted to an anhydride, ester or amide, but an amide can only be hydrolysed to a carboxylic acid. Acid chlorides and acid anhydrides are hydrolysed easily, whereas amides are hydrolysed slowly in boiling alkaline water.

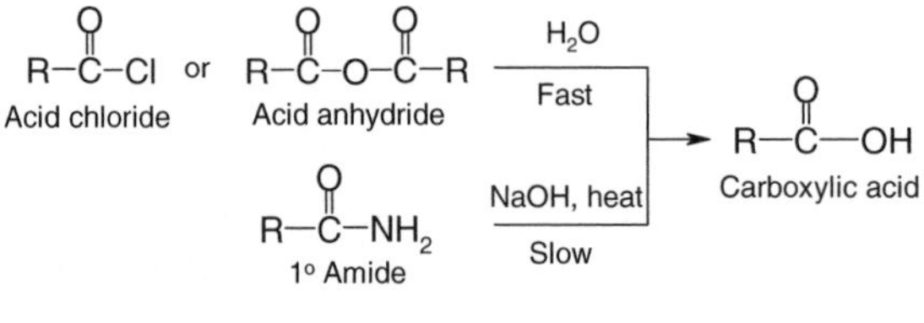

R—C(=O)—Cl (Acid chloride) or R—C(=O)—O—C(=O)—R (Acid anhydride) —[H_2O, Fast]→ R—C(=O)—OH (Carboxylic acid)

R—C(=O)—NH_2 (1° Amide) —[NaOH, heat, Slow]→ R—C(=O)—OH (Carboxylic acid)

The reactivity of carboxylic acid derivatives depends on the basicity of the substituent attached to the acyl group. Therefore, the less basic the substituent, the more reactive is the derivative. In other words, strong bases make poor leaving groups. Carboxylic acid derivatives undergo a variety of reactions under both acidic and basic conditions, and almost all involve the nucleophilic acyl substitution mechanism (see Section 5.5.5).

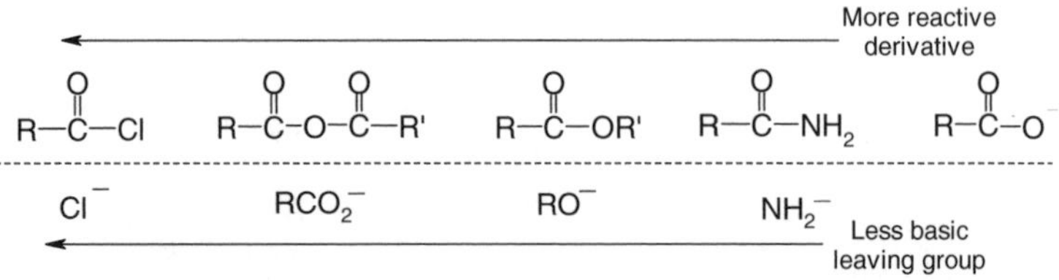

← More reactive derivative

R—C(=O)—Cl	R—C(=O)—O—C(=O)—R'	R—C(=O)—OR'	R—C(=O)—NH_2	R—C(=O)—O^-
Cl^-	RCO_2^-	RO^-	NH_2^-	

← Less basic leaving group

4.3.14 Acid chlorides

The functional group of an acid chloride, also known as acyl chloride, is an acyl group bonded to a chlorine atom. The simplest member of this family is acetyl chloride (CH_3COCl), where the acyl group is bonded to a chlorine atom.

$H_3C\text{-}C(=O)\text{-}Cl$ Ethanoyl chloride Acetyl chloride

$CH_3CH_2\text{-}C(=O)\text{-}Cl$ Propanoyl chloride

$CH_3CH_2CH_2\text{-}C(=O)\text{-}Cl$ Butanoyl chloride

Nomenclature of acid chlorides

Acid chlorides are named by replacing the *-ic* acid ending with *-yl choride* or replacing the carboxylic acid ending with *-carbonyl chloride.*

$CH_3CH_2CH_2CH_2\text{-}C(=O)\text{-}Cl$ Pentanoyl chloride

$CH_3CH_2CH(Br)\text{-}C(=O)\text{-}Cl$ 3-Bromobutanoyl chloride

Cyclohexanecarbonyl chloride

Benzoyl chloride

Preparation of acid chlorides

Acid chlorides are prepared from the corresponding carboxylic acids, most commonly from the reaction with thionyl chloride or oxalyl chloride (see Section 5.5.5).

$$\underset{\text{Carboxylic acid}}{R\text{-}C(=O)\text{-}OH} \xrightarrow[(COCl)_2]{SOCl_2 \text{ or}} \underset{\text{Acid chloride}}{R\text{-}C(=O)\text{-}Cl}$$

Reactions of acid chlorides

Acid chlorides are the most reactive carboxylic acid derivatives, and easily converted to acid anhydrides, esters and amides via nucleophilic acyl substitutions (see Section 5.5.5). Acid chlorides are sufficiently reactive with H_2O, and quite readily hydrolysed to carboxylic acid (see Section 5.6.1).

$R\text{-}C(=O)\text{-}Cl$ (Acid chloride) $\xrightarrow[Et_3N]{R'NH_2}$ $R\text{-}C(=O)\text{-}NHR'$ (2° Amide)

$R\text{-}C(=O)\text{-}Cl$ (Acid chloride) $\xrightarrow[H_2O]{\text{Hydrolysis}}$ $R\text{-}C(=O)\text{-}OH$ (Carboxylic acid)

$R\text{-}C(=O)\text{-}Cl$ (Acid chloride) $\xrightarrow[\text{Ether}]{R'CO_2Na}$ $R\text{-}C(=O)\text{-}O\text{-}C(=O)\text{-}R'$ (Acid anhydride)

$R\text{-}C(=O)\text{-}Cl$ (Acid chloride) $\xrightarrow[\text{Pyridine}]{R'OH}$ $R\text{-}C(=O)\text{-}OR'$ (Ester)

Acid chlorides are easily converted to 1° alcohols and aldehydes (see Section 5.7.21) and 3° alcohols and ketones through the choice of appropriate metal hydride and organometallic reagents (see Section 5.5.5). Acid chloride reacts with benzene in the presence of Lewis acid ($AlCl_3$) in Friedel–Crafts acylation (see Section 5.5.6).

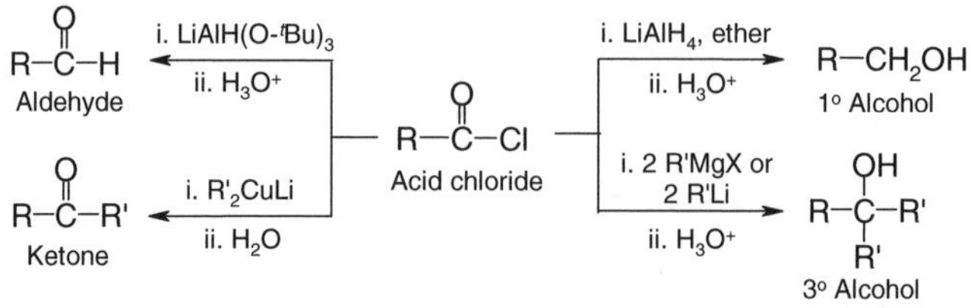

R–C(=O)–H Aldehyde ← i. $LiAlH(O\text{-}^tBu)_3$ ii. H_3O^+ — R–C(=O)–Cl Acid chloride — i. $LiAlH_4$, ether ii. H_3O^+ → $R{-}CH_2OH$ 1° Alcohol

R–C(=O)–R' Ketone ← i. R'_2CuLi ii. H_2O — R–C(=O)–Cl Acid chloride — i. 2 R'MgX or 2 R'Li ii. H_3O^+ → R–C(OH)(R')–R' 3° Alcohol

4.3.15 Acid anhydrides

The functional group of an acid anhydride is two acyl groups bonded to an oxygen atom. These compounds are called *acid anhydrides* or *acyl anhydrides*, because they are condensed from two molecules of carboxylic acid by the loss of a water molecule. An acid anhydride may be symmetrical, where two acyl groups are identical, or it may be mixed, where two different acyl groups are bonded to an oxygen atom. The simplest member of this family is acetic anhydride, $(CH_3CO)_2O$, where the acyl group (CH_3CO) is bonded to an acetate group (CH_3CO_2).

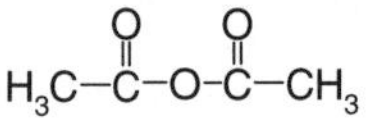

$H_3C{-}C(=O){-}O{-}C(=O){-}CH_3$
Ethanoic anhydride
Acetic anhydride

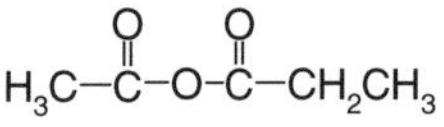

$H_3C{-}C(=O){-}O{-}C(=O){-}CH_2CH_3$
Acetic propanoic anhydride

Nomenclature of acid anhydrides

Symmetrical acid anhydrides are named by replacing the *-acid* suffix of the parent carboxylic acids with the word *anhydride*. Mixed anhydrides that consist of two different acid-derived parts are named using the names of the two individual acids with an alphabetical order.

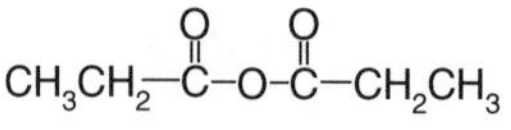

$CH_3CH_2{-}C(=O){-}O{-}C(=O){-}CH_2CH_3$
Propanoic anhydride

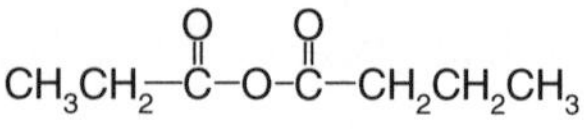

$CH_3CH_2{-}C(=O){-}O{-}C(=O){-}CH_2CH_2CH_3$
Butanoic propanoic anhydride

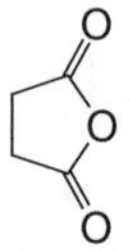

Butanedioicanhydride
Succinic anhydride

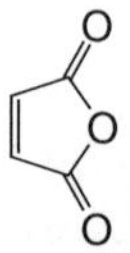

2-Butenedioic anhydride
Maleic anhydride

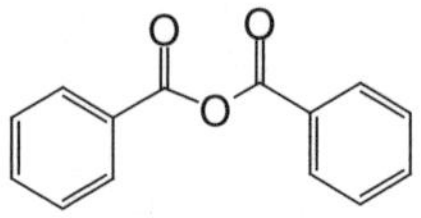

Benzoic anhydride

Preparation of acid anhydrides

Anhydrides are produced most commonly by the reaction of an acid chloride and a carboxylic acid or carboxylate salt (see Sections 4.3.14 and 5.5.5). Five- or six-membered cyclic anhydrides are prepared by heating dicarboxylic acids at high temperatures.

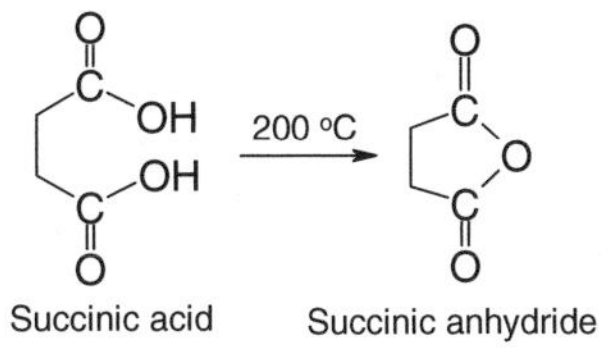

200 °C

Succinic acid Succinic anhydride

Reactions of acid anhydrides

Acid anhydrides are the second most reactive of the carboxylic acid derivatives. They are fairly readily converted to the other less reactive carboxylic acid derivatives, e.g. esters, carboxylic acids and amides. Acid anhydrides undergo many reactions similar to those of acid chlorides, and they can often be used interchangeably.

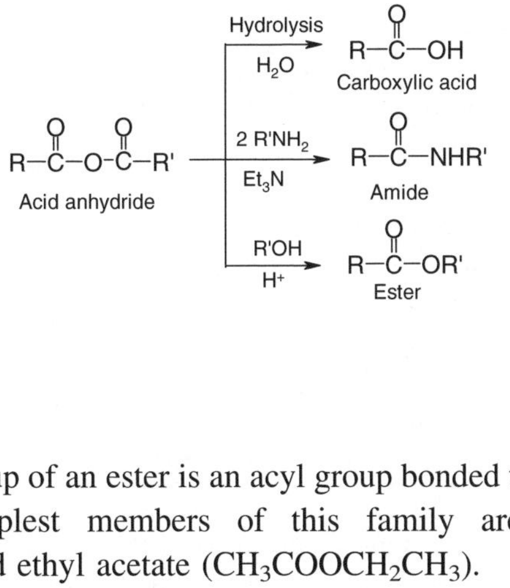

R–C(=O)–O–C(=O)–R' (Acid anhydride)

Hydrolysis, H_2O → R–C(=O)–OH (Carboxylic acid)

2 $R'NH_2$, Et_3N → R–C(=O)–NHR' (Amide)

R'OH, H^+ → R–C(=O)–OR' (Ester)

4.3.16 Esters

The functional group of an ester is an acyl group bonded to an alkoxy group (–OR). The simplest members of this family are methyl acetate (CH_3COOCH_3) and ethyl acetate ($CH_3COOCH_2CH_3$).

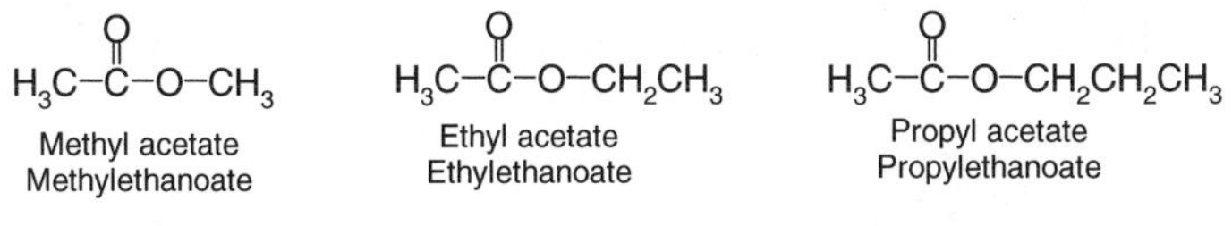

H_3C–C(=O)–O–CH_3 Methyl acetate Methylethanoate

H_3C–C(=O)–O–CH_2CH_3 Ethyl acetate Ethylethanoate

H_3C–C(=O)–O–$CH_2CH_2CH_3$ Propyl acetate Propylethanoate

Nomenclature of esters

The names of esters originate from the names of the compounds that are used to prepare them. The first word of the name comes from the alkyl group of the alcohol, and the second part comes from the carboxylate group of the carboxylic acid used. A cyclic ester is called a *lactone*, and the IUPAC names of lactones are derived by adding the term *lactone* at the end of the name of the parent carboxylic acid.

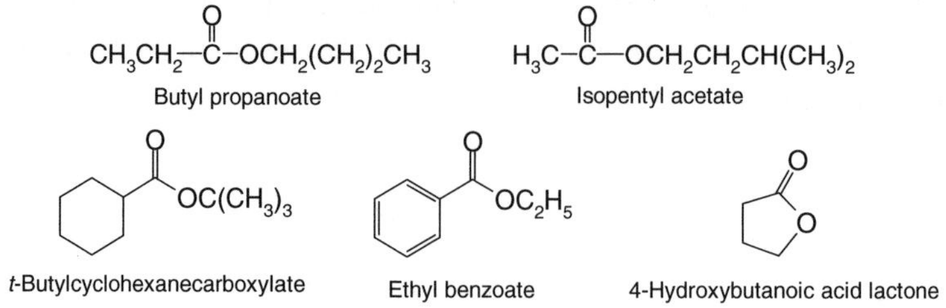

$CH_3CH_2-C(=O)-OCH_2(CH_2)_2CH_3$
Butyl propanoate

$H_3C-C(=O)-OCH_2CH_2CH(CH_3)_2$
Isopentyl acetate

$OC(CH_3)_3$
t-Butylcyclohexanecarboxylate

OC_2H_5
Ethyl benzoate

4-Hydroxybutanoic acid lactone

Preparation of esters

Esters are produced by acid-catalysed reaction of carboxylic acids with alcohols, known as *Fischer esterification*. They are also obtained from acid chlorides, acid anhydrides and other esters. The preparation of esters from other esters in the presence of an acid or a base catalyst is called *transesterification*. All these conversions involve nucleophilic acyl substitutions (see Section 5.5.5).

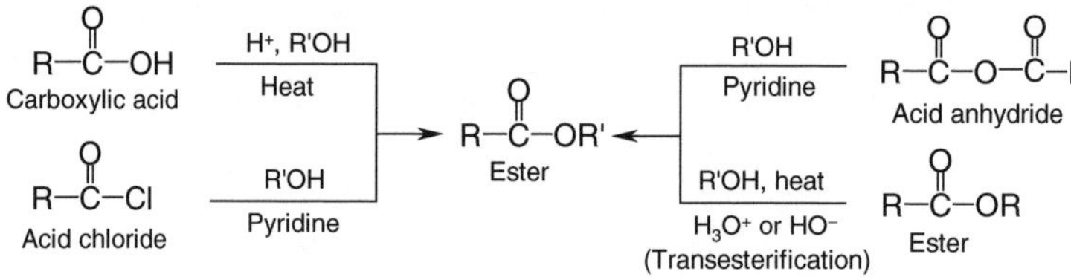

R–C(=O)–OH Carboxylic acid — H^+, R'OH, Heat
R–C(=O)–Cl Acid chloride — R'OH, Pyridine
R–C(=O)–OR' Ester
R–C(=O)–O–C(=O)–R Acid anhydride — R'OH, Pyridine
R–C(=O)–OR Ester — R'OH, heat, H_3O^+ or HO^- (Transesterification)

Lactones are made from the *Fischer esterification*, where the hydroxyl and carboxylic acid groups are present in the same molecule.

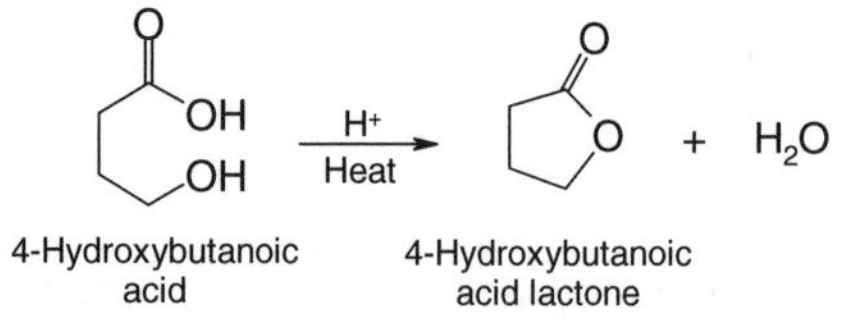

4-Hydroxybutanoic acid —(H^+, Heat)→ 4-Hydroxybutanoic acid lactone + H_2O

Alcohols react with inorganic acids to form esters, e.g. tosylate esters (see Section 5.5.3) and phosphate esters. Phosphate esters are important in nature since they link the nucleotide bases together in DNA (see Section 4.8).

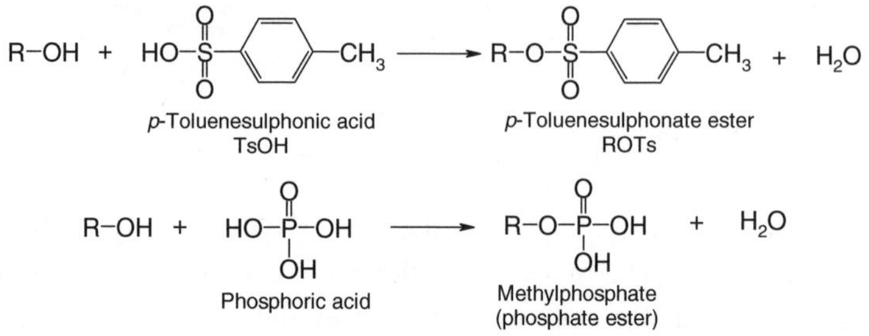

R–OH + HO–SO_2–C_6H_4–CH_3 → R–O–SO_2–C_6H_4–CH_3 + H_2O
p-Toluenesulphonic acid TsOH
p-Toluenesulphonate ester ROTs

R–OH + HO–P(=O)(OH)–OH → R–O–P(=O)(OH)–OH + H_2O
Phosphoric acid
Methylphosphate (phosphate ester)

Reactions of esters

Esters are less reactive than acid chlorides and acid anhydrides. They are converted to carboxylic acid by acid or base hydrolysis, and to another ester by acid or base alcoholysis (transesterification). The 1°, 2° or 3° amides are obtained from esters by treatment with ammonia or 1° or 2° amines, respectively.

R–C(=O)–NHR' (2° Amide) ← $R'NH_2$, Et_3N — R–C(=O)–OR (Ester) — H+, heat or i. HO−, heat ii. H_3O^+ → R–C(=O)–OH (Carboxylic acid)

R–C(=O)–O–C(=O)–R' (Acid anhydride) ← $R'CO_2H$, Pyridine — R–C(=O)–OR (Ester) — R'OH, heat, H_3O^+ or HO− → R–C(=O)–OR' (Ester) (Transesterification)

Primary and tertiary alcohols are obtained conveniently from esters by the reduction of $LiAlH_4$ and two molar equivalents of organometallic reagents (R'MgX or R'Li), respectively (see Sections 5.7.22 and 5.5.5). A less powerful reducing agent, diisobutylaluminium hydride (DIBAH), can reduce an ester to an aldehyde (see Section 5.7.22).

R–C(OH)(R')–R' (3° Alcohol) ← i. 2 R'MgX or 2 R'Li, ii. H_3O^+ — R–C(=O)–OR (Ester) — i. $LiAlH_4$, ether, ii. H_3O^+ → $R–CH_2OH$ (1° Alcohol)

R–C(=O)–OR (Ester) — i. DIBAH, ii. H_2O → R–C(=O)–H (Aldehyde)

Another important reaction of esters is the *Claisen condensation*. In this reaction, an enolate anion is formed from the reaction between an ester and a strong base, e.g. sodium ethoxide (NaOEt in EtOH). The enolate anion reacts with another molecule of ester to produce β-ketoester (see Section 5.5.5).

RCH_2–C(=O)–OR' (Ester) + RCH_2–C(=O)–OR' (Ester) — i. NaOEt, EtOH, ii. H_3O^+ → R–C(=O)–CH(R)–C(=O)–OR' (β-Ketoester)

4.3.17 Amides

The functional group of an amide is an acyl group bonded to a nitrogen atom. The simplest members of this family are formamide ($HCONH_2$) and acetamide (CH_3CONH_2).

H–C(=O)–NH_2 (Methanamide, Formamide) H_3C–C(=O)–NH_2 (Ethanamide, Acetamide) C_2H_5–C(=O)–NH_2 (Propanamide)

Amides are usually classified as primary (1°) amide, secondary (2°) or *N*-substituted amide, and tertiary (3°) or *N,N*-disubstituted amide.

$R-C(=O)-NH_2$ Primary amide (1°)

$R-C(=O)-NHR'$ Secondary (2°) or *N*-substituted amide

$R-C(=O)-NR'_2$ Tertiary (3°) or *N,N*-disubstituted amide

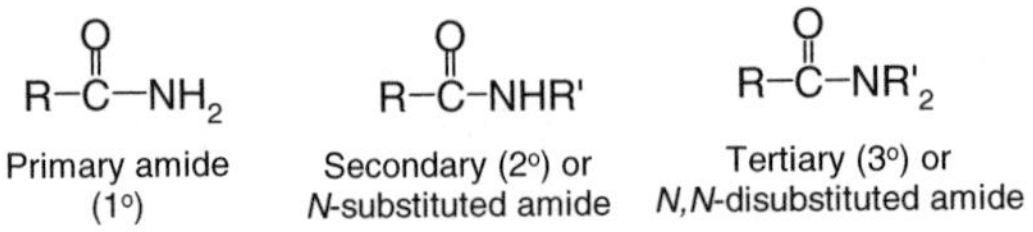

Nomenclature of amides

Amides are named by replacing the *-oic* acid or *-ic* acid suffix of the parent carboxylic acids with the suffix *-amide*, or by replacing the *-carboxylic acid* ending with *-carboxamide*. Alkyl groups on nitrogen atoms are named as substituents, and are prefaced by *N*-or *N,N*-, followed by the name(s) of the alkyl group(s).

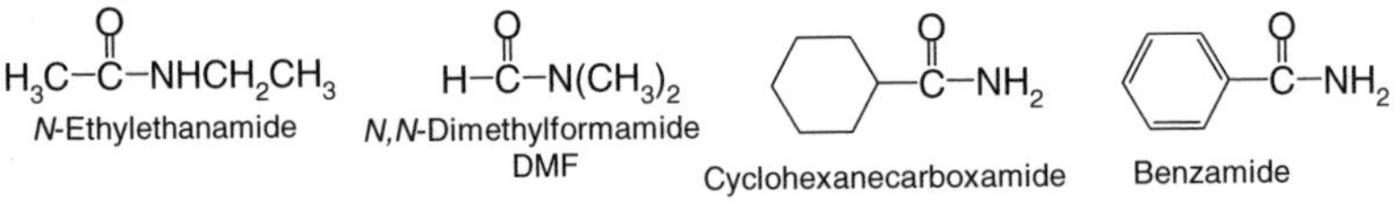

$H_3C-C(=O)-NHCH_2CH_3$ *N*-Ethylethanamide

$H-C(=O)-N(CH_3)_2$ *N,N*-Dimethylformamide DMF

Cyclohexanecarboxamide

Benzamide

If the substituent on the nitrogen atom of an amide is a phenyl group, the ending *-amide* is changed to *-anilide*. Cyclic amides are known as *lactams*, and the IUPAC names are derived by adding the term *lactam* at the end of the name of the parent carboxylic acid.

Acetanilide

Benzanilide

4-Aminobutanoic acid lactam

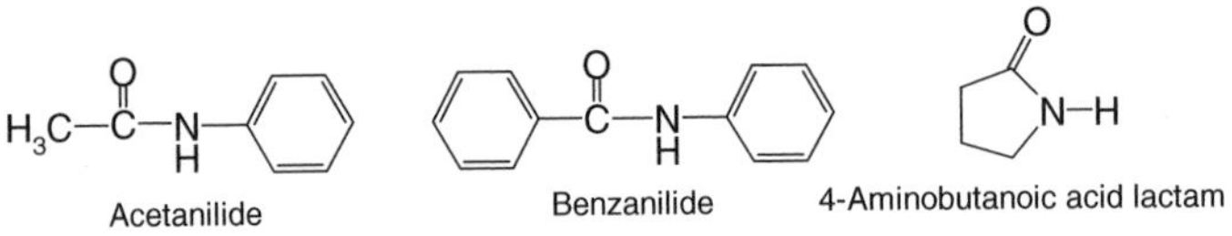

Physical properties of amides

Amides are much less basic than their parent amines. The lone pair of electrons on the nitrogen atom is delocalized on the carbonyl oxygen, and in the presence of a strong acid the oxygen is protonated first. Amides have high b.p. because of their ability to form intermolecular hydrogen bonding. The borderline for solubility in water ranges from five to six carbons for the amides.

Preparation of amides

Amides are the least reactive carboxylic acid derivatives, and are easily obtained from any of the other carboxylic acid derivatives. Carboxylic acids react with ammonia and 1° and 2° amines to give 1°, 2° and 3° amides, respectively (see Section 5.5.5). The amides derived from sulphonic acids are called *sulphonamides*. In general, sulphonamides are obtained by the reaction of amines and sulphonyl chlorides (see Section 4.3.10).

R–C(=O)–Cl (Acid chloride) —R'NH$_2$, Pyridine→
R–C(=O)–O–C(=O)–OR (Acid anhydride) —2 R'NH$_2$, Et$_3$N→ R–C(=O)–NHR' (2° Amide)
R–C(=O)–OR (Ester) —R'NH$_2$, Et$_3$N→

Lactams are produced from amino acids, where the amino and the carboxylic acid groups of the same molecule react to form an amide linkage. β-lactams are the active functionality in modern antibiotics, e.g. penicillin V.

4-Aminobutanoic acid —Heat→ 4-Aminobutanoic acid lactam + H_3O^+

Reactions of amides

Amides are the least reactive of the carboxylic acid derivatives, and undergo acid or base hydrolysis to produce the parent carboxylic acids, and reduction to appropriate amines (see Section 4.3.10). They can also be dehydrated to nitriles, most commonly with boiling acetic anhydride, $(AcO)_2O$, sulphonyl chloride ($SOCl_2$) or phosphorus oxychloride ($POCl_3$) (see Section 4.3.18). Amines (with one less carbon) are prepared from amides by the treatment of halides (Br_2 or Cl_2) in aqueous NaOH or KOH. This reaction is known as *Hofmann rearrangement* (see Section 4.3.10).

R–C(=O)–NH$_2$ (Amide):
—H^+, heat or i. HO^-, heat ii. H_3O^+→ R–C(=O)–OH (Carboxylic acid)
—i. LiAlH$_4$, ether ii. H$_2$O→ RCH$_2$–NH$_2$ (1° Amine)
—(AcO)$_2$O or SOCl$_2$ or POCl$_3$→ R–C≡N (Nitrile)
—X$_2$, NaOH, H$_2$O→ R–NH$_2$ (1° Amine (one carbon less than the amide))

4.3.18 Nitriles

Nitriles are organic compounds that contain a triple bond between a carbon and a nitrogen atom. The functional group in nitriles is the cyano (–C≡N) group, and they are often named as cyano compounds. Nitriles are not carbonyl compounds, but are often included with them because of the similarities in nitrile and carbonyl chemistry. Nitriles are considered to be acid derivatives, because they can be hydrolysed to form amides and carboxylic acids. The nitriles related to acetic acid and benzoic acid are called acetonitrile and benzonitrile.

$H_3C-C\equiv N$ Ethanenitrile Acetonitrile

$CH_3CH_2-C\equiv N$ Propanenitrile

Nomenclature of nitriles

The IUPAC requires nitriles to be named on the basis of the name of the alkanes, with the suffix *-nitrile*.

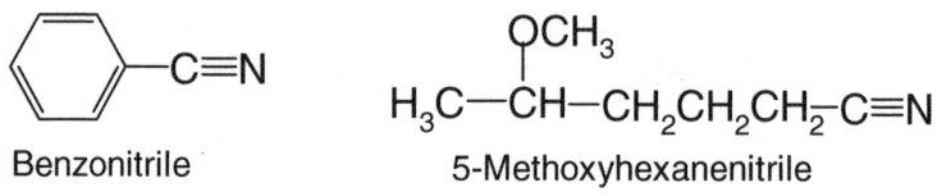

$C_6H_5-C\equiv N$ Benzonitrile

$H_3C-CH(OCH_3)-CH_2CH_2CH_2-C\equiv N$ 5-Methoxyhexanenitrile

Preparation of nitriles

Nitriles are most commonly prepared via the conversion of carboxylic acids to primary amides, followed by dehydration with boiling acetic anhydride, or other commonly employed dehydration reagents, e.g. $SOCl_2$ or $POCl_3$. This is a useful synthesis for amide, because it is not limited by steric hindrance. Alkyl nitriles can be prepared by the action of metal cyanides on alkyl halides (see Section 5.5.2).

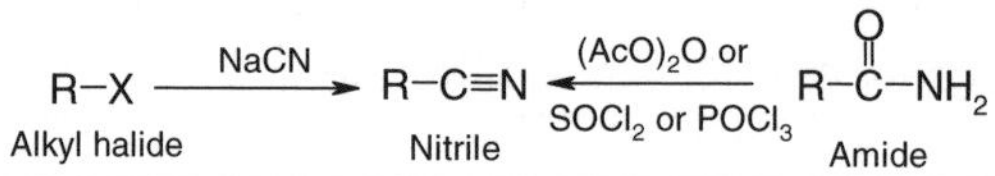

$$\underset{\text{Alkyl halide}}{R-X} \xrightarrow{\text{NaCN}} \underset{\text{Nitrile}}{R-C\equiv N} \xleftarrow[\text{SOCl}_2 \text{ or POCl}_3]{(\text{AcO})_2\text{O or}} \underset{\text{Amide}}{R-C(=O)-NH_2}$$

Reactions of nitriles

In nitriles, as nitrogen is more electronegative than carbon, the triple bond is polarized towards the nitrogen, similar to the C=O bond. Therefore, nucleophiles can attack the electrophilic carbon of the nitrile group. Nitriles undergo hydrolysis to primary amides, and then to carboxylic acids (see Section 5.6.1). Reduction of nitriles by $LiAlH_4$ and catalytic hydrogenation gives primary amines (see Section 5.7.23), and reaction with Grignard reagent or organolithium produces ketones, after the acidic hydrolysis (see Section 5.3.2).

R–CH$_2$NH$_2$ (1° Amine) ← 2 H$_2$/Pd-C or Raney Ni — R–C≡N (Nitrile) — i. R'MgX or R'Li, ii. H$_3$O$^+$ → R–C(=O)–R' (Ketone)

R–CH$_2$NH$_2$ (1° Amine) ← i. LiAlH$_4$, ether, ii. H$_2$O — R–C≡N (Nitrile) — H$^+$, heat or HO$^-$, heat → R–C(=O)–NH$_2$ (Amide)

R–C(=O)–NH$_2$ (Amide) — H$^+$, heat or i. HO$^-$, heat, ii. H$_3$O$^+$ → R–C(=O)–OH (Carboxylic acid)

4.4 Alkenes and their derivatives

Alkenes (olefins) are unsaturated hydrocarbons that contain carbon–carbon double bonds. A double bond consists of a σ bond and a π bond. A π bond is weaker than a σ bond, and this makes π bonds more reactive than σ bonds. Thus, π bond is considered to be a functional group. Alkenes form a homologous series with general molecular formula C_nH_{2n}. The simplest members of the series are ethene (C_2H_4), propene (C_3H_6), butene (C_4H_8) and pentene (C_5H_{10}).

Ethene (Ethylene) — Propene (Propylene) — 1-Butene (Butylene) — 1-Pentene (Pentylene)

Among the cycloalkenes, cyclobutene, cyclopropene and cylcohexene are most common. Cyclobutene is about 4 kcal/mol more strained than cyclopentene. The smaller bond angles mean more deviation from 120°, and this makes cyclobutene more reactive than cyclopentene.

Cyclobutene — Cyclopentene — Cyclohexene

4.4.1 Nomenclature of alkenes

The systematic name of an alkene originates from the name of the alkane corresponding to the longest continuous chain of carbon atoms that contains the double bond. When the chain is longer than three carbons, the atoms are numbered starting from the end nearest to the double bond. The functional group suffix is *-ene*.

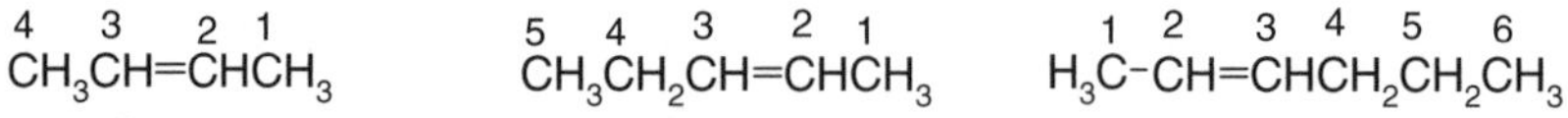

$\overset{4}{C}H_3\overset{3}{C}H{=}\overset{2}{C}H\overset{1}{C}H_3$
2-Butene
(*cis* or *trans*)

$\overset{5}{C}H_3\overset{4}{C}H_2\overset{3}{C}H{=}\overset{2}{C}H\overset{1}{C}H_3$
2-Pentene
(*cis* or *trans*)

$H_3\overset{1}{C}{-}\overset{2}{C}H{=}\overset{3}{C}H\overset{4}{C}H_2\overset{5}{C}H_2\overset{6}{C}H_3$
2-Hexene
(*cis* or *trans*)

For branches, each alkyl group is given a number, but the double bond still gets preference when numbering the chain.

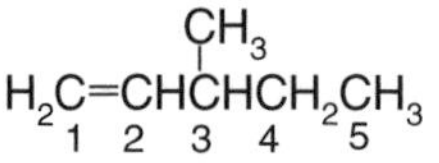

$H_2\underset{1}{C}{=}\underset{2}{C}H\underset{3}{C}H(CH_3)\underset{4}{C}H_2\underset{5}{C}H_3$
3-Methyl-1-pentene

A cyclic alkene is named by a prefix *cyclo-* to the name of the acyclic alkene. Double bonded carbons are considered to occupy positions 1 and 2.

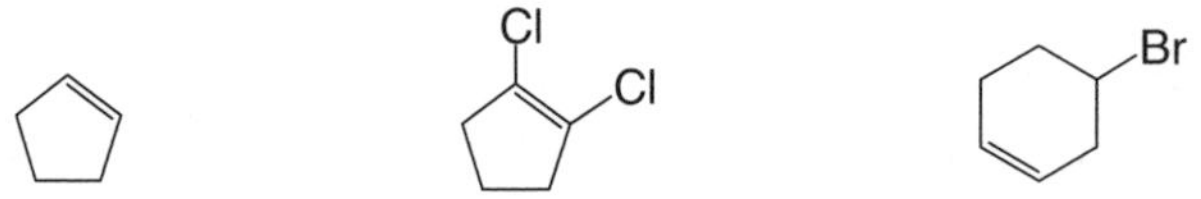

Cl Cl Br
Cyclopentene 1,2-Dicholorocyclopentene 4-Bromo-1,2-cyclohexene

When a geometric isomer is present, a prefix *cis* (*Z*) or *trans* (*E*) is added. Because of the double bonds, alkenes cannot undergo free rotation. Thus, the rigidity of a π bond gives rise to *geometric isomers*. Simple 1,2-alkenes can be described as *cis*- or *trans*-alkenes. When similar groups are present on the same side of the double bond, the alkene is said to be *cis*. When similar groups are present on opposite sides of the double bond, it is said to be *trans*. More complex alkenes are best described as *E*- or *Z*- based on the Cahn–Ingold–Prelog priority rules (see Section 3.2.2).

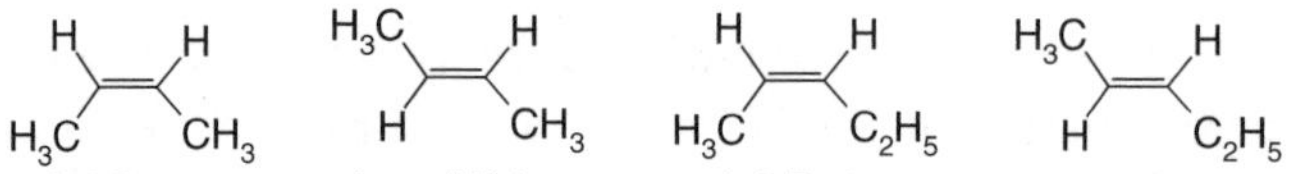

H H H_3C CH_3 — *cis*-2-Butene
H_3C H H CH_3 — *trans*-2-Butene
H H H_3C C_2H_5 — *cis*-2-Pentene
H_3C H H C_2H_5 — *trans*-2-Pentene

Compounds with two double bonds are called *dienes*, three double bonds are *trienes* and so on. Where geometric isomerism exists, each double bond is specified with numbers indicating the positions of all the double bonds.

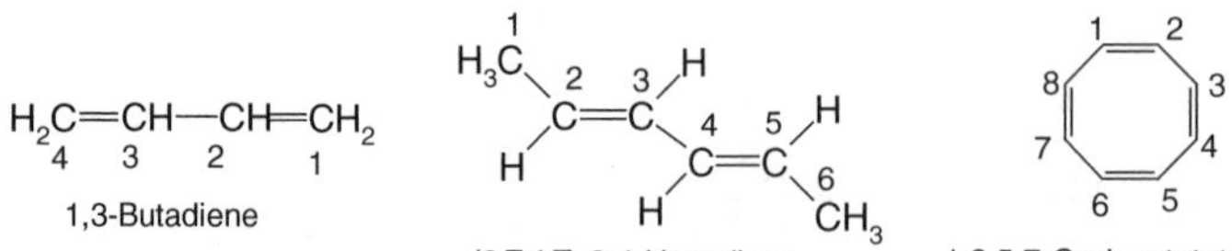

$H_2\underset{4}{C}{=}\underset{3}{C}H{-}\underset{2}{C}H{=}\underset{1}{C}H_2$
1,3-Butadiene

$H_3\overset{1}{C}$ $\overset{2}{C}{=}\overset{3}{C}$ H H $\overset{4}{C}{=}\overset{5}{C}$ H H $\overset{6}{C}H_3$
(2*E*,4*E*)-2,4-Hexadiene

1 2 3 4 5 6 7 8
1,3,5,7-Cyclooctatetraene

The sp^2 carbon of an alkene is called *vinylic carbon*, and an sp^3 carbon that is adjacent to a vinylic carbon is called an *allylic carbon*. The two unsaturated groups are called the *vinyl group* ($CH_2{=}CH{-}$) and the *allyl group* ($CH_2CHCH_2{-}$).

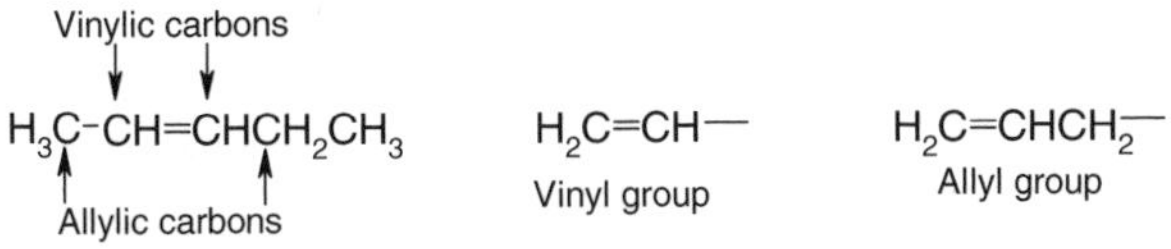

Vinylic carbons

$H_3C{-}CH{=}CHCH_2CH_3$

Allylic carbons

$H_2C{=}CH{-}$
Vinyl group

$H_2C{=}CHCH_2{-}$
Allyl group

Cycloalkenes must have eight or more carbons before they are large enough to incorporate a *trans* double bond. Thus, cycloalkenes are considered to be *cis* unless otherwise specified. A bridged bicyclic (two ring) compound cannot have a double bond at a bridgehead position unless one of the rings contains at least eight carbon atoms. A bridgehead carbon is part of both rings. A bridged bicyclic compound has at least one carbon in each of the three links between the bridgehead atoms.

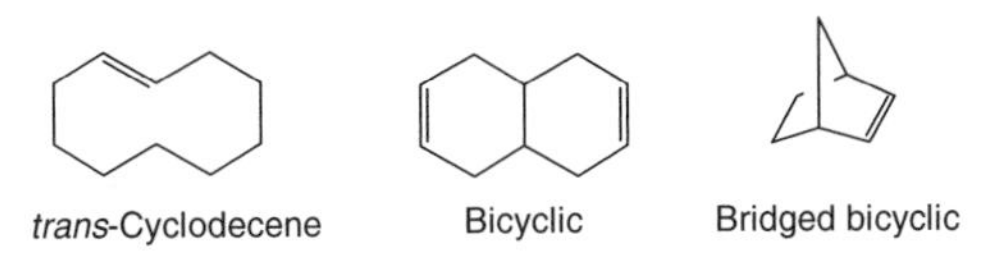

trans-Cyclodecene   Bicyclic   Bridged bicyclic

4.4.2 Physical properties of alkenes

As with alkanes, the boiling points and melting points of alkenes decrease with increasing molecular weight, but show some variations that depend on the shape of the molecule. Alkenes with the same molecular formula are isomers of one another if the position and the stereochemistry of the double bond differ. For example, there are four different acyclic structures that can be drawn for butene (C_4H_8). They have different b.p. and m.p. as follows.

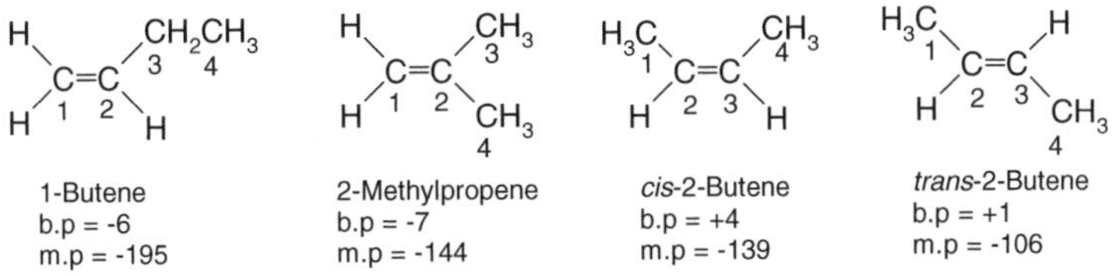

1-Butene ($H_2C{=}CHCH_2CH_3$)
b.p = -6
m.p = -195

2-Methylpropene ($H_2C{=}C(CH_3)_2$)
b.p = -7
m.p = -144

cis-2-Butene ($H_3CCH{=}CHCH_3$)
b.p = +4
m.p = -139

trans-2-Butene ($H_3CCH{=}CHCH_3$)
b.p = +1
m.p = -106

4.4.3 Structure of alkenes

In ethene (C_2H_4), each carbon atom has three σ-bonding electron pairs in its *p* orbitals to form three σ bonds, one with the carbon and the other two with two hydrogen atoms. The π bond in C_2H_4 is formed from the sideways overlap of a parallel *p* orbital on each carbon atom. The C$-$C bond in ethene is shorter and stronger than in ethane, partly because of the sp^2–sp^2 overlap being stronger than sp^3–sp^3, but especially because of the extra π bond in ethene.

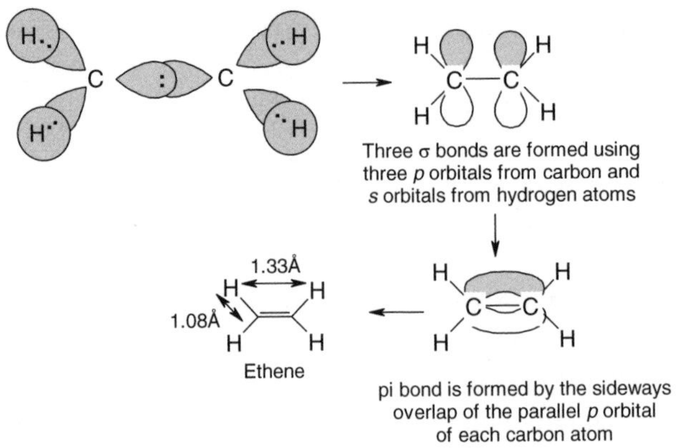

Three σ bonds are formed using three p orbitals from carbon and s orbitals from hydrogen atoms

1.33Å

1.08Å

Ethene

pi bond is formed by the sideways overlap of the parallel p orbital of each carbon atom

4.4.4 Industrial uses of alkenes

Alkenes are useful intermediates in organic synthesis, but their main commercial use is as precursors for polymers. For example, styrene polymerizes to polystyrene.

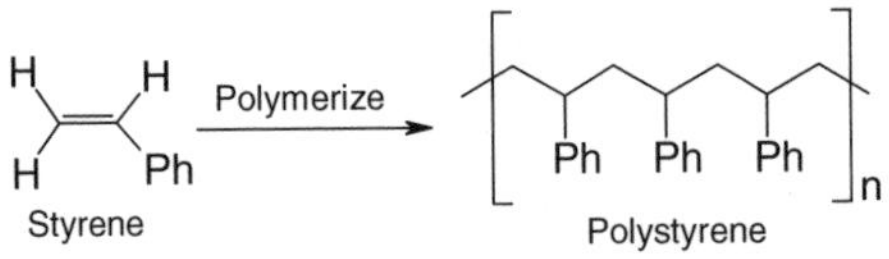

Styrene —Polymerize→ Polystyrene

4.4.5 Preparation of alkenes and cycloalkenes

Alkenes are obtained by the transformation of various functional groups, e.g. dehydration of alcohols (see Section 5.4.3), dehalogenation of alkyl halides (see Section 5.4.5) and dehalogenation or reduction of alkyl dihalides (see Section 5.4.5). These reactions are known as *elimination reactions*. An elimination reaction results when a proton and a leaving group are removed from adjacent carbon atoms, giving rise to a π bond between the two carbon atoms.

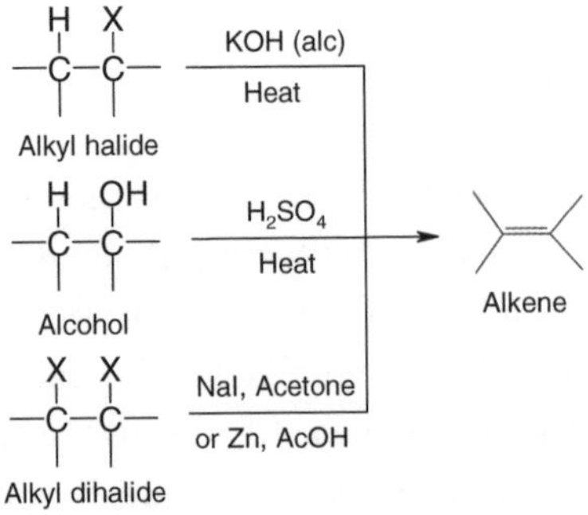

Alkyl halide — KOH (alc), Heat

Alcohol — H_2SO_4, Heat

Alkyl dihalide — NaI, Acetone or Zn, AcOH

→ Alkene

Alkenes are obtained from selective hydrogenation of alkynes (see Section 5.3.1), and the reaction of a phosphorus ylide (Wittig reagent) with an aldehyde or a ketone (see Section 5.3.2).

RC≡CR (Alkyne) $\xrightarrow[\text{Quinoline, } CH_3OH]{H_2\text{, Pd/BaSO}_4}$ *cis*-Alkene, *syn* addition

RC≡CR (Alkyne) $\xrightarrow[\text{Liq. } NH_3\text{, } -78\ ^{\circ}C]{Na}$ *trans*-Alkene, *anti* addition

RYC=O (Y = H or R) $\xrightarrow[\text{Phosphorus yilde}]{Ph_3\overset{+}{P}-\overset{-}{C}HR'}$ RYC=CHR' (Alkene) + $Ph_3P=O$ (Triphenylphosphine oxide)

4.4.6 Reactivity and stability of alkenes

Alkenes are typically nucleophiles, because the double bonds are electron rich and electrons in the π bond are loosely held. Electrophiles are attracted to the π electrons. Thus, alkenes generally undergo addition reactions, and addition reactions are typically exothermic. The following three factors influence the stability of alkenes.

(a) The degree of substitution: more highly alkylated alkenes are more stable. Thus, the stability follows the order tetra > tri > di > mono-substituted. This is because the alkyl groups stabilize the double bond. The stability arises because the alkyl groups are electron donating (hyperconjugation) and so donate electron density into the π bond. Moreover, a double bond (sp^2) carbon separates bulky groups better than an sp^3 carbon, thus reducing steric hindrance.

(b) The stereochemistry: *trans* > *cis* due to reduced steric interactions when R groups are on opposite sides of the double bond.

(c) The conjugated alkenes are more stable than isolated alkenes.

4.4.7 Reactions of alkenes and cycloalkenes

Alkenes are electron-rich species. The double bond acts as a nucleophile, and attacks the electrophile. Therefore, the most important reaction of alkenes is *electrophilic addition* to the double bond (see Section 5.3.1). An outline of the electrophilic addition reactions of alkenes is presented here.

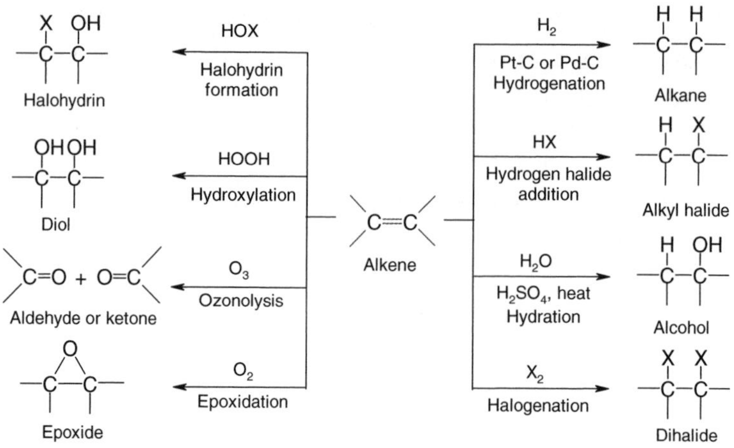

Halohydrin ← HOX, Halohydrin formation
Diol ← HOOH, Hydroxylation
Aldehyde or ketone ← O_3, Ozonolysis
Epoxide ← O_2, Epoxidation

Alkene

H_2, Pt-C or Pd-C, Hydrogenation → Alkane
HX, Hydrogen halide addition → Alkyl halide
H_2O, H_2SO_4, heat, Hydration → Alcohol
X_2, Halogenation → Dihalide

4.5 Alkynes and their derivatives

Alkynes are hydrocarbons that contain a carbon–carbon triple bond. A triple bond consists of a σ bond and two π bonds. The general formula for the alkynes is C_nH_{2n-2}. The triple bond possesses two elements of unsaturation. Alkynes are commonly named as substituted acetylenes. Compounds with triple bonds at the end of a molecule are called *terminal alkynes*. Terminal —CH groups are called acetylenic hydrogens. If the triple bond has two alkyl groups on both sides, it is called an *internal alkyne*.

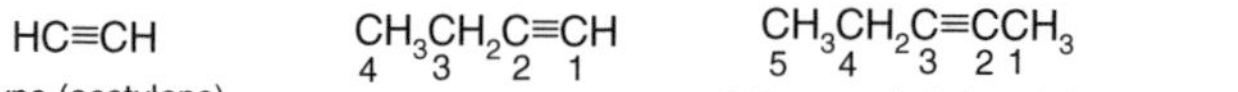

HC≡CH
Ethyne (acetylene)

$CH_3CH_2C{\equiv}CH$ (4 3 2 1)
1-Butyne (ethylacetylene)
Terminal alkyne

$CH_3CH_2C{\equiv}CCH_3$ (5 4 3 2 1)
2-Pentyne (ethylmethylacetylene)
Internal alkyne

4.5.1 Nomenclature of alkynes

The IUPAC nomenclature of alkynes is similar to that for alkenes, except the *–ane* ending is replaced with *–yne*. The chain is numbered from the end closest to the triple bond. When additional functional groups are present, the suffixes are combined.

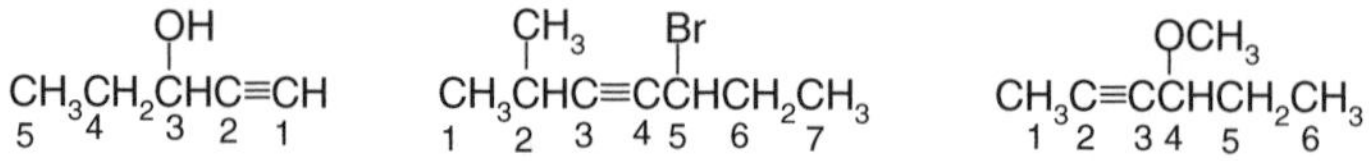

$CH_3CH_2CH(OH)C{\equiv}CH$ (5 4 3 2 1)
1-Pentyn-3-ol

$CH_3CH(CH_3)C{\equiv}CCH(Br)CH_2CH_3$ (1 2 3 4 5 6 7)
5-Bromo-2-methyl-3-heptyne

$CH_3C{\equiv}CCH(OCH_3)CH_2CH_3$ (1 2 3 4 5 6)
4-Methoxy-2-hexyne

4.5.2 Structure of alkynes

The triple bond consists of one σ bond and two π bonds. Each carbon is bonded to two other atoms, and there are no nonbonding electrons. Carbon requires two hybrid orbitals to bond to the atoms, thus *sp* hybrids are used. The *sp* orbitals are linear and oriented at 180°. The C—C bond is formed from *sp–sp* overlap. The C—H bond is formed from *sp–s* overlap. The formation of *sp* hybrids leaves two free *p* orbitals; these contribute to the formation of the two other π bonds. The C—C bond length for ethyne is 1.20 Å, which is shorter than ethane (1.54 Å) and ethene (1.33 Å). The C—H bond length in ethyne is 1.06 Å, which is also shorter than in ethane (1.09 Å) or ethene (1.08 Å). This is because the C—H bond contains more *s* character ($sp^3 \rightarrow sp^2 \rightarrow sp$), which gives stronger bonds.

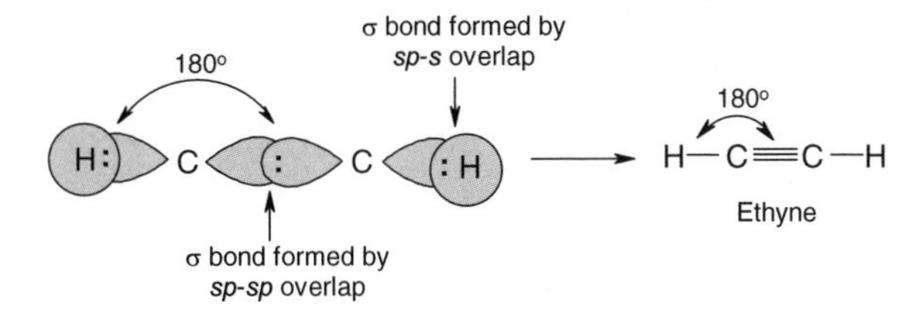

4.5.3 Acidity of terminal alkynes

Terminal alkynes are acidic, and the end hydrogen can be removed as a proton by strong bases (e.g. organolithiums, Grignard reagents and $NaNH_2$) to form metal acetylides and alkynides. They are strong nucleophiles and bases, and are protonated in the presence of water and acids. Therefore, metal acetylides and alkynides must be protected from water and acids.

$$RC{\equiv}CH \text{ (Terminal alkyne)} \xrightarrow{CH_3CH_2Li} RC{\equiv}\bar{C}\text{–}\overset{+}{Li} + CH_3CH_3 \text{ (Lithium acetylide)}$$

$$RC{\equiv}CH \xrightarrow{CH_3CH_2MgBr} RC{\equiv}\bar{C}\text{–}\overset{+}{Mg}Br + CH_3CH_3 \text{ (Alkynyl Grignard reagent)}$$

$$RC{\equiv}CH \xrightarrow{NaNH_2} RC{\equiv}\bar{C}\text{–}\overset{+}{Na} + NH_3 \text{ (Sodium acetylide)}$$

4.5.4 Heavy metal acetylides: test for terminal alkynes

The position of the triple bond can alter the reactivity of the alkynes. Acidic alkynes react with certain heavy metal ions, e.g. Ag^+ and Cu^+, to form precipitation. Addition of an alkyne to a solution of $AgNO_3$ in alcohol forms a precipitate, which is an indication of hydrogen attached to the triple bonded carbon. Thus, this reaction can be used to differentiate terminal alkynes from internal alkynes.

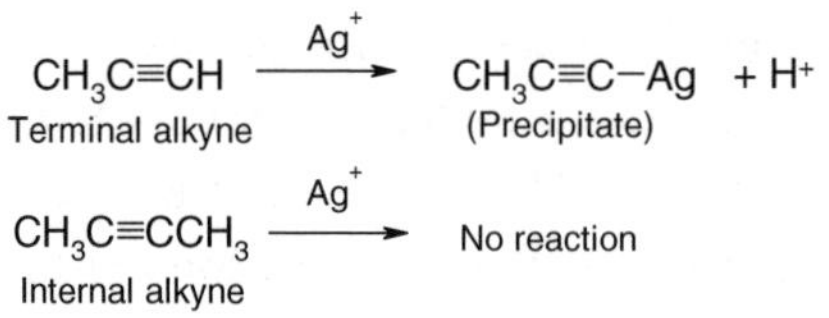

$$CH_3C{\equiv}CH \xrightarrow{Ag^+} CH_3C{\equiv}C{-}Ag + H^+$$

Terminal alkyne (Precipitate)

$$CH_3C{\equiv}CCH_3 \xrightarrow{Ag^+} \text{No reaction}$$

Internal alkyne

4.5.5 Industrial uses of alkynes

Ethynes are industrially used as a starting material for polymers, e.g. vinyl flooring, plastic piping, Teflon and acrylics. Polymers are large molecules, which are prepared by linking many small monomers. Polyvinyl chloride, also commonly known as PVC, is a polymer produced from the polymerization of vinyl chloride.

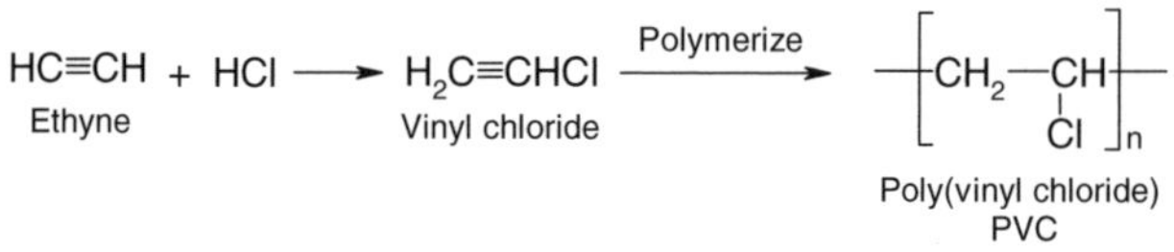

$$HC{\equiv}CH + HCl \longrightarrow H_2C{=}CHCl \xrightarrow{\text{Polymerize}} {-}[CH_2{-}CH(Cl)]_n{-}$$

Ethyne; Vinyl chloride; Poly(vinyl chloride) PVC

4.5.6 Preparation of alkynes

Alkynes are prepared from alkyl dihalides via elimination of atoms or groups from adjacent carbons. Dehydrohalogenation of *vicinal*- or *geminal*-dihalides is a particularly useful method for the preparation of alkynes (see Section 5.4.5).

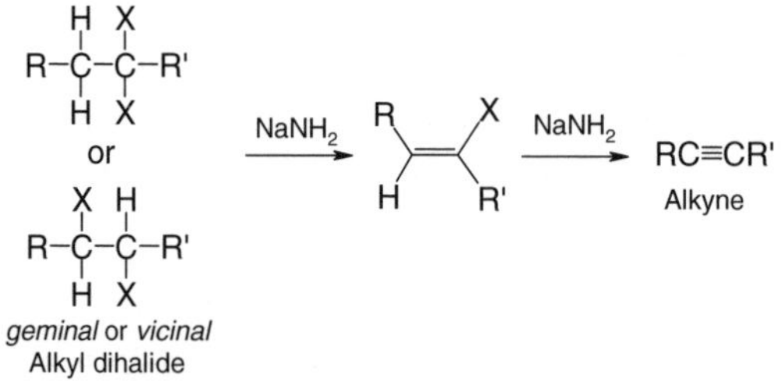

$$\text{R-CH}_2\text{-CX}_2\text{-R' or R-CHX-CHX-R'} \xrightarrow{NaNH_2} \text{RCH=CXR'} \xrightarrow{NaNH_2} RC{\equiv}CR'$$

geminal or *vicinal* Alkyl dihalide; Alkyne

Metal acetylides or alkynides react with primary alkyl halides or tosylates to prepare alkynes (see Sections 5.5.2 and 5.5.3).

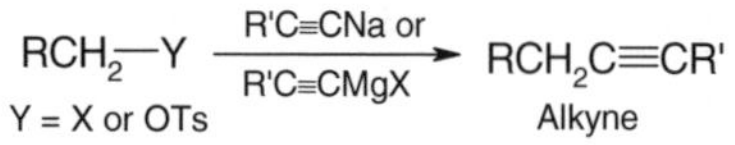

$$RCH_2{-}Y \xrightarrow[R'C{\equiv}CMgX]{R'C{\equiv}CNa \text{ or}} RCH_2C{\equiv}CR'$$

Y = X or OTs; Alkyne

4.5.7 Reactions of alkynes

Alkynes are electron-rich reagents. The triple bond acts as a nucleophile, and attacks the electrophile. Therefore, alkynes undergo *electrophilic addition reactions*, e.g. hydrogenation, halogenation and hydrohalogenation, in the same way as alkenes, except that two molecules of reagent are needed for each triple bond for the total addition. It is possible to stop the reaction at the first stage of addition for the formation of alkenes. Therefore, two different halide groups can be introduced in each stage. A summary of electrophilic addition reactions of alkynes (see Section 5.3.1) is presented here.

Alkyne ($-C{\equiv}C-$):
- H_2O, H_2SO_4, $HgSO_4$ (Hydration) → Ketone
- $2\,X_2$ (Halogenation) → Alkyl tetrahalide
- $2\,HX$ (Hydrogen halide addition) → Alkyl dihalide
- $2\,H_2$, Pt-C or Pd-C (Hydrogenation) → Alkane
- H_2 (*Syn* addition) → *cis*-Alkene
- Na or Li, NH_3 (*Anti* addition) → *trans*-Alkene
- O_2 (Epoxidation) → Epoxide

4.5.8 Reactions of acetylides and alkynides

Besides electrophilic addition, terminal alkynes also perform acid–base type reaction due to acidic nature of the terminal hydrogen. The formation of acetylides and alkynides (alkynyl Grignard reagent and alkylnyllithium) are important reactions of terminal alkynes (see Section 4.5.3). Acetylides and alkynides undergo *nucleophilic addition* with aldehydes and ketones to produce alcohols (see Section 5.3.2).

$$R{-}CO{-}Y \xrightarrow[\text{ii. } H_3O^+]{\text{i. } R'C{\equiv}\bar{C}{-}\overset{+}{M}} R{-}C(OH)(Y){-}C{\equiv}CR' \xleftarrow[\text{ii. } H_3O^+]{\text{i. } R'C{\equiv}\bar{C}{-}\overset{+}{M}gBr} R{-}CO{-}Y$$

Aldehyde or ketone (Y = H or R); M+ = Na, Li; Alcohol

They react with alkyl halides to give internal alkynes (see Section 5.5.2) via *nucleophilic substitution* reactions. This type of reaction also is known as *alkylation*. Any terminal alkyne can be converted to acetylide and alkynide, and then alkylated by the reaction with alkyl halide to produce an internal alkyne. In these reactions, the triple bonds are available for electrophilic additions to a number of other functional groups.

$$R{-}X \xrightarrow[\text{ii. } H_3O^+]{\text{i. } R'C{\equiv}\bar{C}{-}\overset{+}{M}} RC{\equiv}CR' \xleftarrow[\text{ii. } H_3O^+]{\text{i. } R'C{\equiv}\bar{C}{-}\overset{+}{M}gBr} R{-}X$$

Alkyl halide; M+ = Na, Li; Internal alkyne; Alkyl halide

4.6 Aromatic compounds and their derivatives

All drugs are chemicals and many of them are aromatic compounds. Therefore, in order to understand the chemical nature, physical properties, stability, pharmacological actions and toxicities of a majority of drug molecules, the knowledge of aromatic chemistry is extremely important. Before we look into the specific examples of various drugs that belong to this aromatic class, let us try to understand what *aromaticity* really is.

Generally, the term 'aromatic compounds' means fragrant substances. Later, benzene and its structural relatives were termed as aromatic. However, there are a number of other nonbenezenoid compounds that can be classified as aromatic compounds.

4.6.1 History

In 1825, Michael Faraday discovered benzene, and he named it 'bicarburet of hydrogen' because of the equal number of carbon and hydrogen atoms. He isolated benzene from a compressed illuminating gas that had been made by pyrolysing whale oil. In 1834, Eilhardt Mitscherlich synthesized benzene by heating benzoic acid with calcium oxide. In the late 19th century, August Kekulé first noticed that all early aromatic compounds contain a six-carbon unit that is retained through most chemical transformation and degradation.

4.6.2 Definition: Hückel's rule

An aromatic compound has a molecular structure containing cyclic clouds of delocalized π electrons above and below the plane of the molecule, and the π clouds contain a total of $(4n + 2)$ π electrons (where n is a whole number). This is known as *Hückel's rule* (introduced first by Erich Hückel in 1931). For example, benzene is an aromatic compound.

Benzene

If $n = 1$, we have $4 \times 1 + 2 = 6$, which means that any compound containing a total number of six π electrons is an aromatic compound. In the above structure of benzene, there are three double bonds and six π electrons, and it is a planar molecule. Thus, benzene follows Hückel's rule, and is an aromatic compound.

4.6.3 General properties of aromatic compounds

Aromatic compounds have the following general properties.

(a) They have a high degree of unsaturation, but are resistant to addition reactions.

(b) They favour electrophilic substitution reactions.

(c) These compounds are unusually stable.

(d) They have a low heat of hydrogenation and low heat of combustion.

(e) They are cyclic compounds.

(f) These compounds are flat and planar.

4.6.4 Classification of aromatic compounds

Benzene and its monocylic derivatives

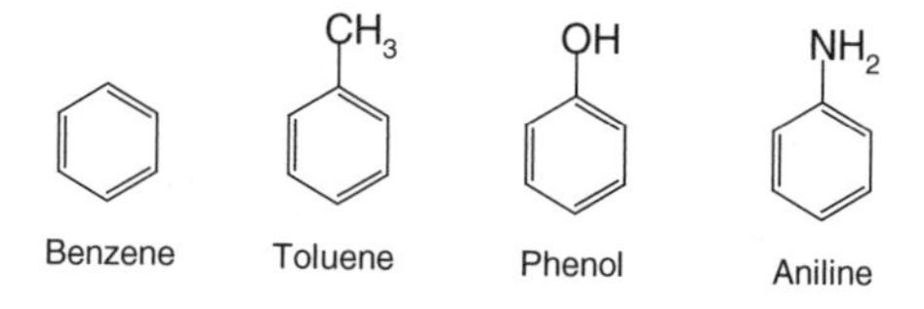

Polycyclic benzenoids These aromatic compounds have two or more benzene rings fused together, e.g. naphthalene and anthracene.

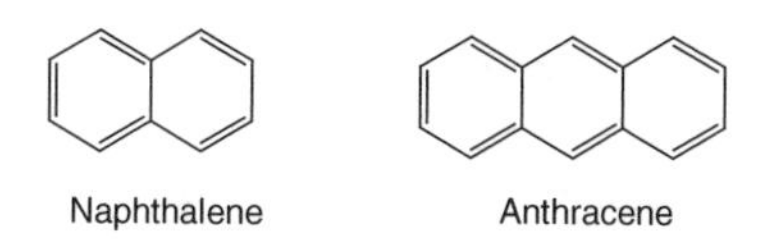

Nonbenzenoids These compounds generally have two or more rings fused together, but none of the rings is a benzene structure, and they conform to Hückel's rule, i.e. they have $(4n+2)$ π electrons, and are aromatic compounds, e.g. azulene.

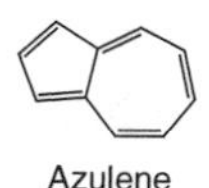

In the above structure of azulene, there are five conjugated double bonds and ten π electrons, which means that it follows *Hückel's rule* $(4 \times 2 + 2 = 10)$.

Macrocyclic These are monocyclic nonbenzene structures, and the ring sizes are quite big. There is an adequate number of double bonds and π electrons to conform to Hückel's rule; e.g., [14] annulene obeys Hückel's rule and is aromatic.

[14] Annulene

Heterocyclic These are compounds having at least one hetero atom (any other atom but carbon, e.g. O, N and S) within the ring, and conforming to Hückel's rule. The aromaticity of heterocyclic compounds, e.g. pyridine and pyrrole, can be explained as follows.

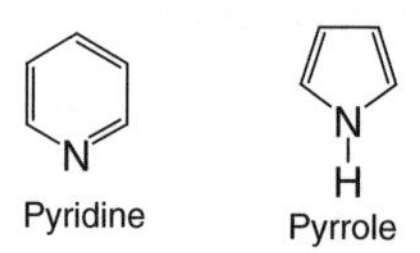

Pyridine has π electron structure similar to that of benzene. Each of the five sp^2-hybridized carbons has a *p* orbital perpendicular to the plane of the ring. Each *p* orbital has one π electron. The nitrogen atom is also sp^2-hybridized and has one electron in the *p* orbital. So, there are six π electrons in the ring. The nitrogen lone pair electrons are in an sp^2 orbital in the plane of the ring and are not a part of the aromatic π system.

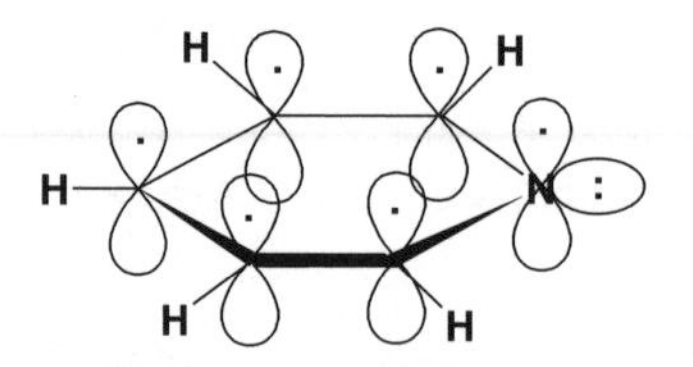

Structure of pydridine with *p* orbitals

The situation in pyrrole is slightly different. Pyrrole has a π electron system similar to that of the cyclopentadienyl anion. It has four sp^2-hybridized carbons, each of which has a *p* orbital perpendicular to the ring and contributes one π electron. The nitrogen atom is also sp^2-hybridized and its lone pair electrons occupies a *p* orbital. Therefore, there is a total of six π electrons, which makes pyrrole an aromatic compound.

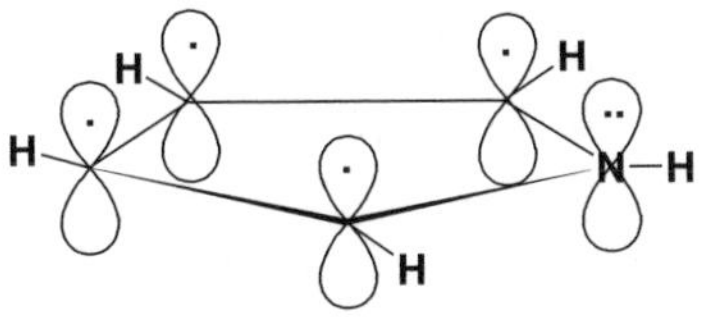

Structure of pyrrole with *p* orbitals

4.6.5 Pharmaceutical importance of aromatic compounds: some examples

There are numerous examples of aromatic compounds that are pharmaceutically important as drugs or pharmaceutical additives. Just a few examples of pharmaceutically important aromatic compounds are cited here. Aspirin, a well known non-narcotic analgesic and antipyretic drug, is a classic example of a pharmaceutically important benzene derivative. Morphine, an aromatic alkaloid, is a narcotic (habit-forming) analgesic that is used extensively for the management of post-operative pain. Aromatic compound valium is prescribed as a tranquillizer and ibuprofen as an anti-inflammatory, and sulpha drugs, e.g. sulphamethoxazole, are used as antimicrobial agents. Taxol, one of the best selling anticancer drugs of modern time, also belongs to the class of aromatic compounds. Saquinavir and crixivan, two anti-HIV drugs (protease inhibitors), also possess aromatic characters.

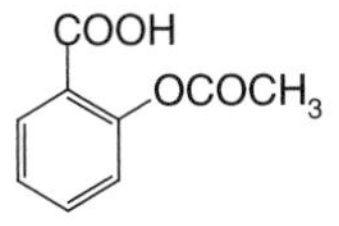

Aspirin
Acetyl salicylic acid

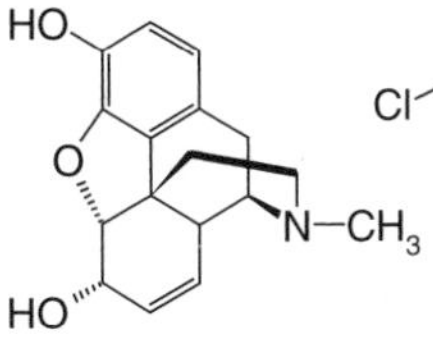

Morphine

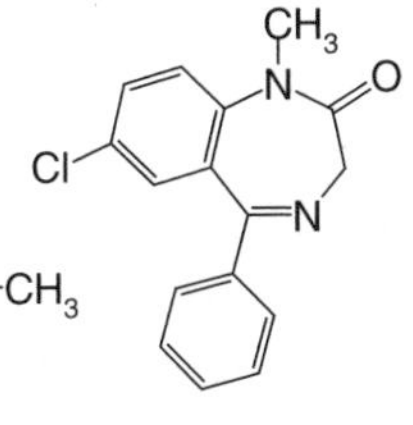

Valium

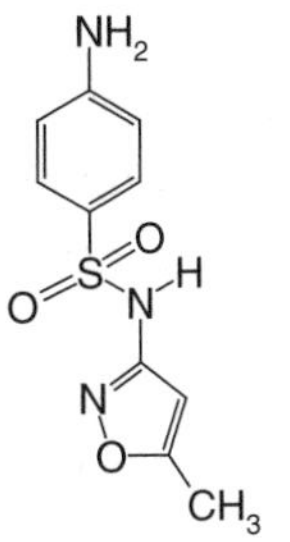

Sulphamethoxazole

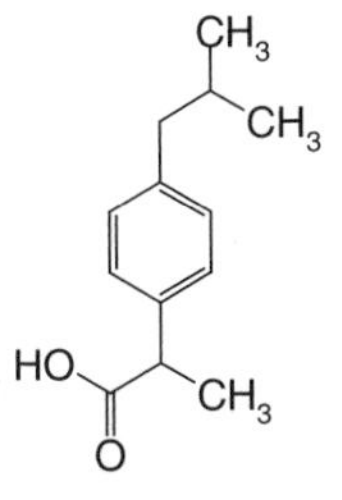

Ibuprofen

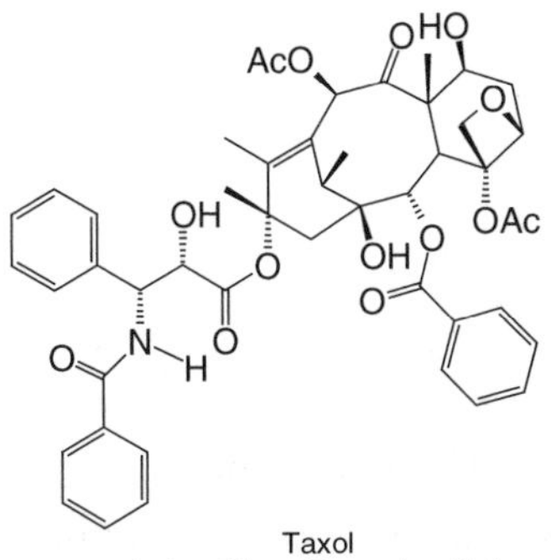

Taxol
Isolated from *Taxus brevifolia*

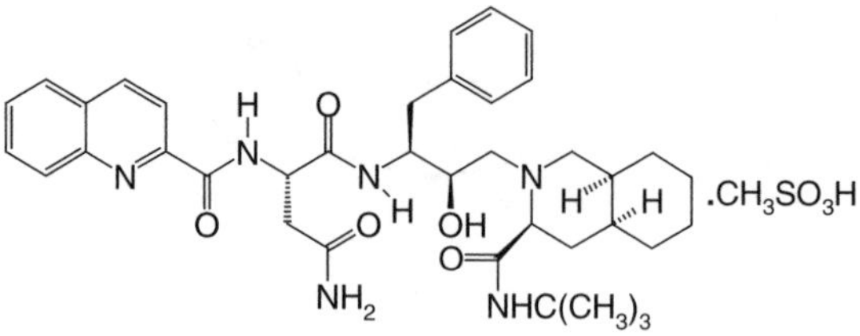

Saquinavir

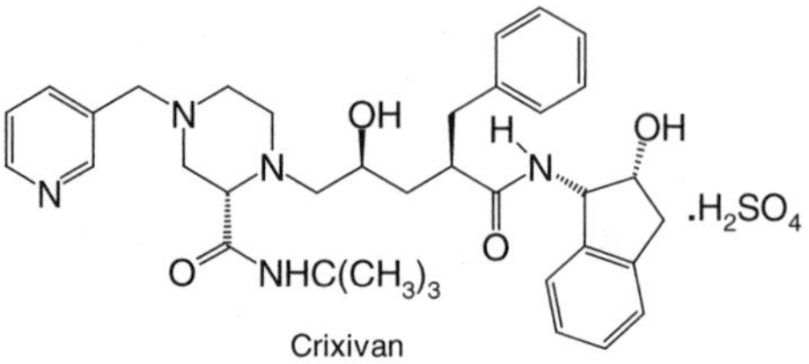

Crixivan

4.6.6 Structure of benzene

Kekulé structure of benzene

In 1865, August Kekulé proposed the structure of benzene (C_6H_6). According to his proposals, in benzene

(a) all six carbon atoms are in a ring;

(b) all carbon atoms are bonded to each other by alternating single and double bonds;

(c) one hydrogen atom is attached to each carbon atom;

(d) all hydrogen atoms are equivalent.

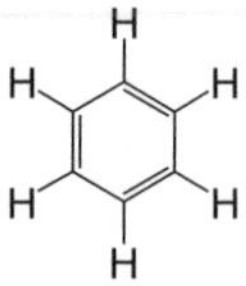

Kekulé structure of benzene

Limitations of Kekulé structure The Kekulé structure predicts that there should be two different 1,2-dibromobenzenes. In practice, only one 1,2-dibromobenzene has ever been found. Kekulé proposed that these two forms are in equilibrium, which is established so rapidly that it prevents isolation of the separate compounds. Later, this proposal was proved to be incorrect, because no such equilibrium exists!

Two different 1,2-dibromobenzenes as suggested by Kekulé

Benzene cannot be represented accurately by either individual Kekulé structure, and does not oscillate back and forth between two. The Kekulé structure also cannot explain the stability of benzene.

The resonance explanation of the structure of benzene

The resonance theory can be applied successfully to explain the structure of benzene. First of all, let us have a look at the resonance theory. According to this theory

(a) resonance forms are imaginary, not real;

(b) resonance structures differ only in the positions of their electrons;

(c) different resonance forms do not have to be equivalent;

(d) the more resonance structures there are, the more stable the molecule is;

(e) whenever it is possible to draw two or more resonance structures of a molecule, none of the structures will be in complete agreement with the compound's chemical and physical properties;

(f) the actual molecule or ion is better represented by a hybrid of these structures;

(g) whenever an equivalent resonance structure can be drawn for a molecule, the molecule (or hybrid) is much more stable than any of the resonance structures could be individually if they could exist.

If we consider the Kekulé structure of benzene, it is evident that the two proposed structures differ only in the positions of the electrons. Therefore, instead of being two separate molecules in equilibrium, they are indeed two resonance contributors to a picture of the real molecule of benzene.

Two structures of benzene as suggested by Kekulé

If we think of a hybrid of these two structures, then the C—C bonds in benzene are neither single bonds nor double bonds. They should have a bond order between a single (1.47 Å) and a double bond (1.33 Å). It has actually been proven that benzene is a planar molecule, and all of its C—C bonds are of equal length (1.39 Å). The bond order (1.39 Å) is indeed in between a single and a double bond! Thus, instead of drawing the benzene structure using alternative single and double bonds, a hybrid structure can be drawn as follows.

Hybrid structure of benzene

The hybrid structure of benzene is represented by inscribing a circle in the hexagon as depicted above. With benzene, the circle represents the six electrons that are delocalized about the six carbon atoms of the benzene ring.

The resonance theory accounts for the much greater stability of benzene (*resonance energy*) when compared with the hypothetical 1,3,5-cyclohexatriene. It also explains why there is only one 1,2-dibromobenzene rather than two. Therefore, the structure of benzene is not really a 1,3,5-cyclohexatriene, but a hybrid structure as shown above.

The molecular orbital explanation of the structure of benzene

The bond angles of the carbon atoms in benzene are 120°. All carbon atoms are sp^2-hybridized, and each carbon atom has a single unhybridized *p* orbital perpendicular to the plane of the ring. The carbon sp^2-hybridized orbitals overlap to form the ring of the benzene molecule. Because the C—C bond lengths are 1.39 Å, the *p* orbitals are close enough to overlap efficiently and equally all round the ring.

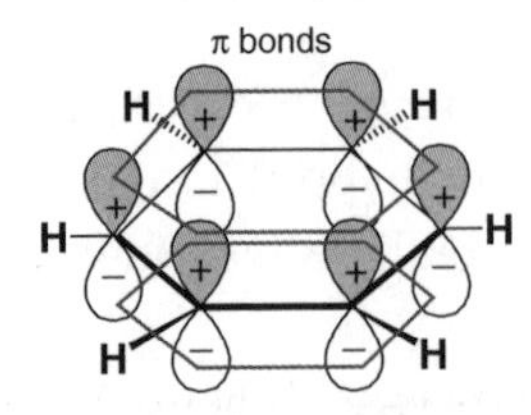

Benzene structure in the light of molecular orbital theory

The six overlapping *p* orbitals overlap to form a set of six *p* molecular orbitals. Six π electrons are completely delocalized around the ring, and form two doughnut-shaped clouds of π electrons, one above and one below the ring.

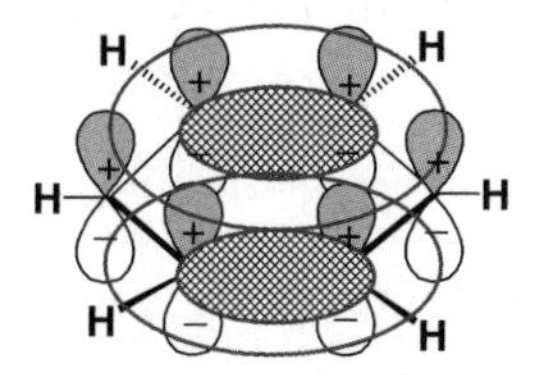

Doughnut-shaped cloud of π electrons

Six *p* atomic orbitals, one from each carbon of the benzene ring, combine to form six *p* molecular orbitals. Three of the molecular orbitals have energies lower than that of an isolated *p* orbital, and are known as *bonding molecular orbitals*. Another three of the molecular orbitals have energies higher than that of an isolated *p* orbital and are called *antibonding molecular orbitals*. Two of the bonding orbitals have the same energy, as do the antibonding orbitals. Such orbitals are said to be *degenerate*.

Stability of benzene

Benzene has a closed bonding shell of delocalized π electrons. This closed bonding shell partly accounts for the stability of benzene. Benzene is more stable than the Kekulé structure suggests. The stability of benzene can be shown as follows.

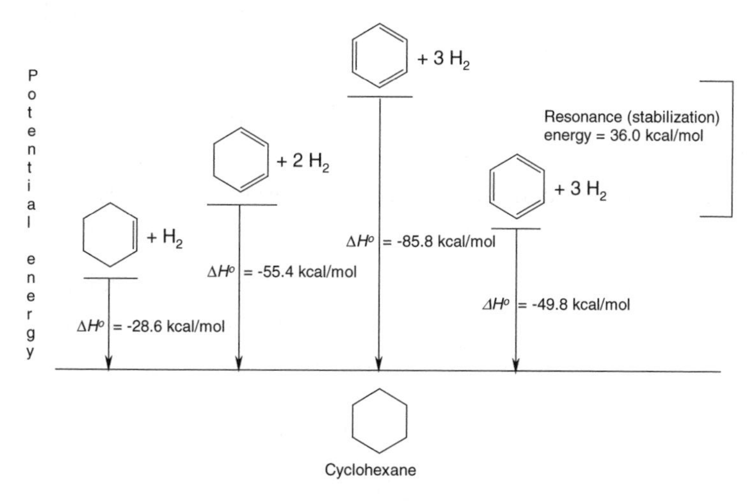

The energy required for the hydrogenation of cyclohexene to cyclohexane is −28.6 kcal/mol. Therefore, in the case of cyclohexadiene, where there are two double bonds, the energy required for the hydrogenation can be calculated as $2 \times -28.6 = -57.2$ kcal/mol. In practice, the experimental value is quite close to this calculated value, and is −55.4 kcal/mol. In this way, if benzene were really a cyclohexatriene as proposed by Kekulé, the calculated required energy would be three times than that of cyclohexene, i.e. $3 \times -28.6 = -85.8$ kcal/mol. In practice, it was found that the required energy for benzene is -49.8 kcal/mol, which means that there is a clear 36 kcal/mol difference between the calculated value and the observed value, and this 36 kcal/mol is known as the *stabilization energy* or *resonance energy*. This explains the stability of benzene. Due to this *stabilization energy*, benzene does not undergo similar reactions to a cycloalkene. This can be depicted with the example as follows.

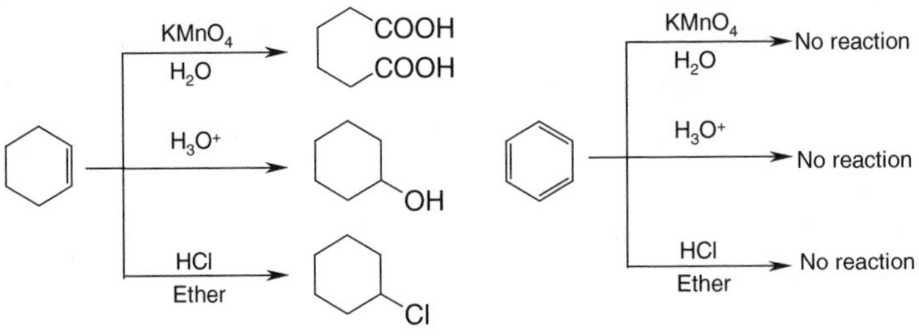

Cyclohexene vs benzene

4.6.7 Nomenclature of benzene derivatives

Benzene derivatives are named by prefixing the name of the substituent group to the word *benzene*, e.g. chlorobenzene and nitrobenzene. Many benzene derivatives have trivial names, which may show no resemblance to the name of the attached substituent group, e.g. phenol, toluene and aniline.

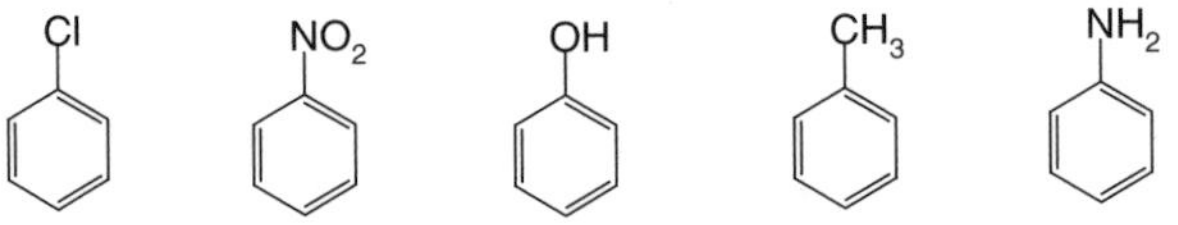

Chlorobenzene Nitrobenzene Phenol (hydroxybenzene) Toluene (methylbenzene) Aniline (aminobenzene)

When two groups are attached to the benzene ring, their relative positions have to be identified. The three possible isomers of a disubstituted benzene are differentiated by the use of the names *ortho*, *meta* and *para*, abbreviated as *o*-, *m*- and *p*-, respectively.

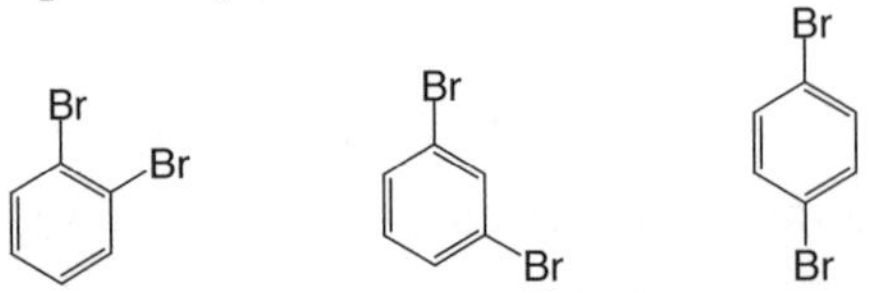

ortho-Dibromobenzene *meta*-Dibromobenzene *para*-Dibromobenzene

If the two groups are different, and neither is a group that gives a trivial name to the molecule, the two groups are named successively, and the word *benzene* is added. If one of the two groups is the kind that gives a trivial name to the molecule, then the compound is named as a derivative of this compound. In both cases, relative position should also be designated.

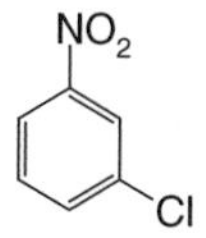

NO_2 Cl OH Br

meta-Chloronitrobenzene  *para*-Bromophenol

When more than two groups are attached to the benzene ring, numbers are used to indicate their relative positions. If the groups are the same, each is given a number, the sequence being the one that gives the lowest combination of numbers; if the groups are different, then the last-named group is understood to be in position 1, and the other numbers conform to this. If one of the groups that give a trivial name is present, then the compound is named as having the special group in position 1.

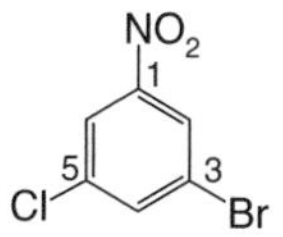

NO_2 1 3 5 Cl Br Br 1 Br 2 4 Br

3-Bromo-5-chloronitrobenzene 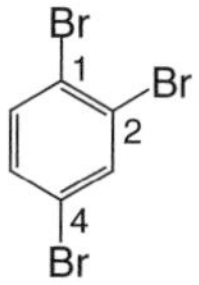 1,2,4-Tribromobenzene

4.6.8 Electrophilic substitution of benzene

Benzene is susceptible to electrophilic attack, and unlike any alkene it undergoes substitution reactions rather than addition reactions. Before we go into any details of such reactions, let us try to understand the following terms.

Arenes. Aromatic hydrocarbons, as a class, are called arenes.
Aryl group. An aromatic hydrocarbon with a hydrogen atom removed is called an aryl group, designated by Ar—.
Phenyl group. The benzene ring with one hydrogen atom removed (C_6H_5—) is called the phenyl group, designated by Ph—.
Electrophile. Electron loving. Cations, E^+, or electron-deficient species. For example, Cl^+ or Br^+ (halonium ion) and $^+NO_2$ (nitronium ion).

An electrophile (E^+) reacts with the benzene ring and substitutes for one of its six hydrogen atoms. A cloud of π electrons exists above and below the plane of the benzene ring. These π electrons are available to electrophiles. Benzene's closed shell of six π electrons gives it a special stability.

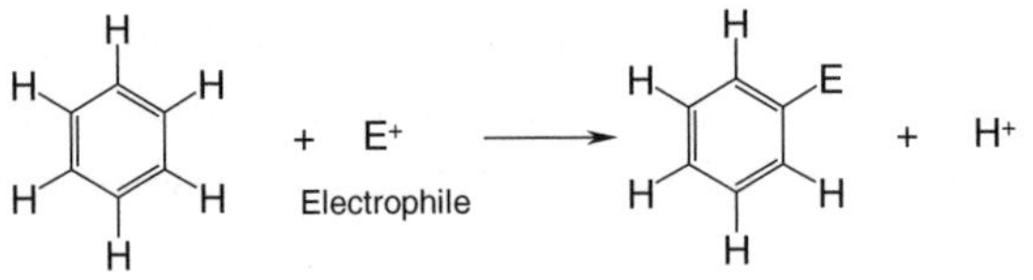

Substitution reactions allow the aromatic sextet of π electrons to be regenerated after attack by the electrophile has occurred. Electrophiles attack the π system of benzene to form a delocalized nonaromatic carbocation (*arenium ion* or σ complex). Some specific examples of electrophilic substitution reactions of benzene are summarized below (see Chapter 5).

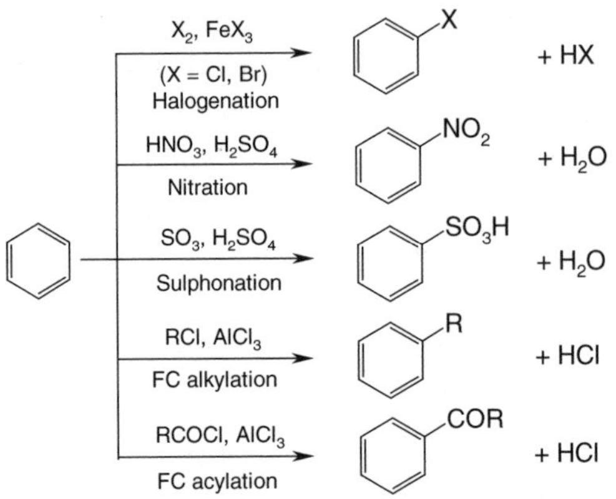

Reactivity and orientation in electrophilic substitution of substitued benzene

When substituted benzene undergoes electrophilic attack, groups already on the ring affect the reactivity of the benzene ring as well as the orientation of the reaction. A summary of these effects of substituents on reactivity and orientation of electrophilic substitution of substituted benzene is presented below.

Substituent	Reactivity	Orientation	Inductive effect	Resonance effect
$-CH_3$	Activating	*ortho*, *para*	Weak electron donating	None
$-OH$, $-NH_2$	Activating	*ortho*, *para*	Weak electron withdrawing	Strong electron donating
$-F$, $-Cl$, $-Br$, $-I$	Deactivating	*ortho*, *para*	Strong electron withdrawing	Weak electron donating
$-N^+(CH_3)_3$	Deactivating	*meta*	Strong electron withdrawing	None
$-NO_2$, $-CN$, $-CHO$, $-COOCH_3$, $COCH_3$, $-COOH$	Deactivating	*meta*	Strong electron withdrawing	Strong electron withdrawing

Reactivity Groups already present on the benzene ring may activate the ring (*activating groups*), making it more reactive towards electrophilic substitution than benzene, e.g. the —OH substituent makes the ring 1000 times more reactive than benzene, or may deactivate the ring (*deactivating groups*), making it less reactive than benzene, e.g. the $-NO_2$ substituent makes the ring more than 10 million times less reactive. The relative rate of reaction depends on whether the substituent group (—S) withdraws or releases electrons relative to hydrogen. When —S is an electron-releasing group the reaction is faster, whereas when this group is an electron-withdrawing group a slower rate of reaction is observed.

–S releases electron + E^+ → Transition state is stabilized → Reaction is faster → Arenium ion is stabilized

–S withdraws electron + E^+ → Transition state is destabilized → Reaction is slower → Arenium ion is destabilized

Orientation Similarly, groups already present on the benzene ring direct the orientation of the new substituent to *ortho*, *para* or *meta* positions. For example, nitration of chlorobenzene yields *ortho*-nitrochlorobenzene (30%) and *para*-nitrochlorobenzene (70%).

ortho-Nitrochlorobenzene (30%)

para-Nitrochlorobenzene (70%)

All activating groups are *ortho* and *para* directing, and all deactivating groups other than halogens are *meta* directing. The halogens are unique in being deactivating but *ortho* and *para* directing. A summary of various groups and their effects on the benzene ring in relation to reactivity and orientation is presented below.

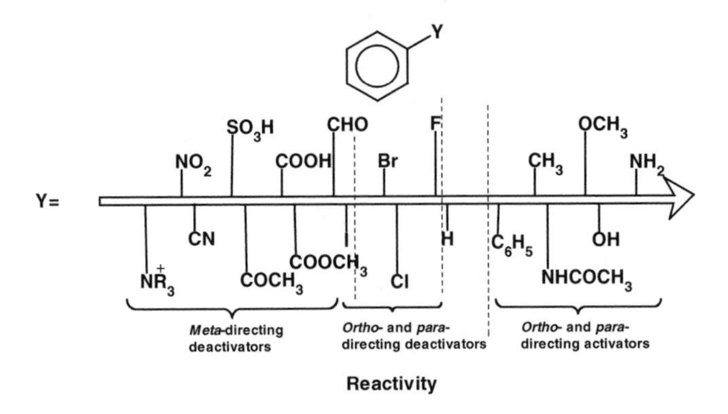

Inductive effect of substituent present on the benzene ring

An *inductive effect* is the withdrawal or donation of electrons through a σ bond due to electronegativity and the polarity of bonds in functional groups (*electrostatic interaction*). When the substituent ($-S$) bonded to a benzene ring is a more electronegative atom (or group) than carbon; e.g. F, Cl or Br, the benzene ring will be at the positive end of the dipole. These substituents will withdraw electron from the ring. As a consequence, an electrophilic attack will be less favoured because of an additional full positive charge on the ring.

S = F, Cl or Br   S = CH_3

If a substituent ($-S$) bonded to a benzene ring is less electron withdrawing than a hydrogen, the electrons in the σ bond that attaches the substituent to the benzene ring will move toward the ring more readily than will those in the s bond that attaches a hydrogen to the ring. Such a substituent (e.g. CH_3), compared with a hydrogen atom, donates electrons inductively into the ring. Inductive electron donation makes the ring more reactive towards electrophilic substitution because of the increased availability of electrons.

Resonance effect of substituent present on the benzene ring

A resonance effect is the withdrawal (e.g. by $-CO$, $-CN$ or $-NO_2$) or donation (e.g. by $-X$, $-OH$ or $-OR$) of electrons through a π bond due to the overlap of a *p* orbital on the substituent with a *p* orbital on the aromatic ring. The presence of a substituent may increase or decrease the resonance stabilization of the intermediate arenium ion complex.

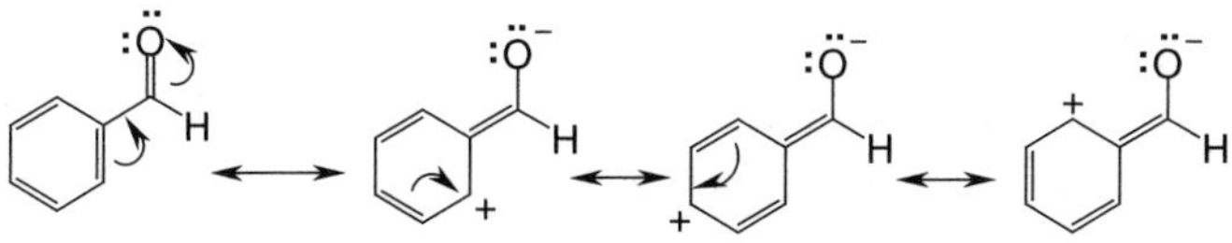

Electron withdrawing effect by an aldehyde (CHO) group

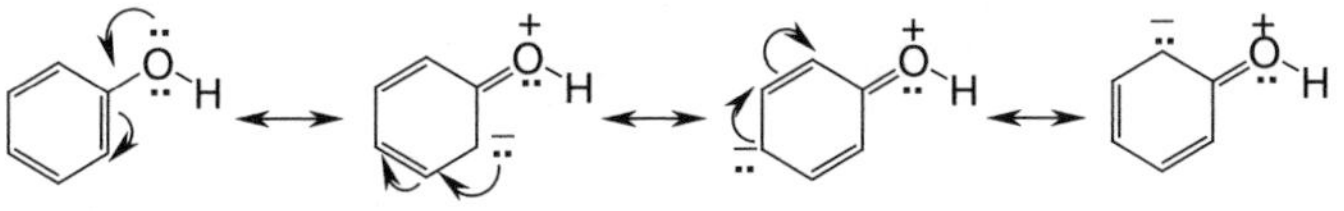

Electron donating phenolic hydroxyl group (OH)

The electron-donating resonance effect applies with decreasing strength in the following order:

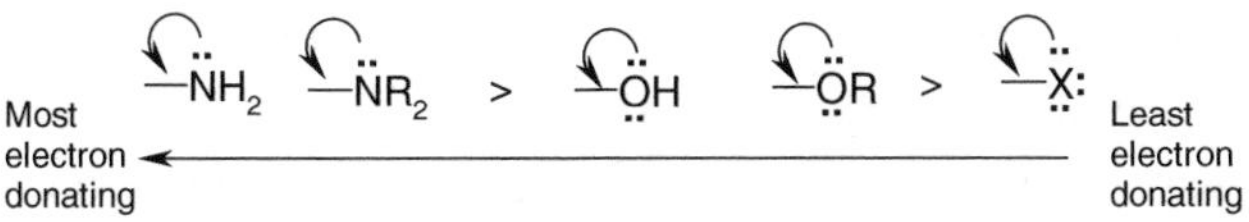

Most electron donating $-NH_2$ $-NR_2$ > $-OH$ $-OR$ > $-X$: Least electron donating

Why the $-CF_3$ group is meta directing

All *meta*-directing groups have either a partial positive charge or a full positive charge on the atom directly linked to the benzene ring. In the trifluoromethyl group (CF_3), there are three electronegative fluorine atoms, which make this group strongly electron withdrawing. As a result $-CF_3$ deactivates the benzene

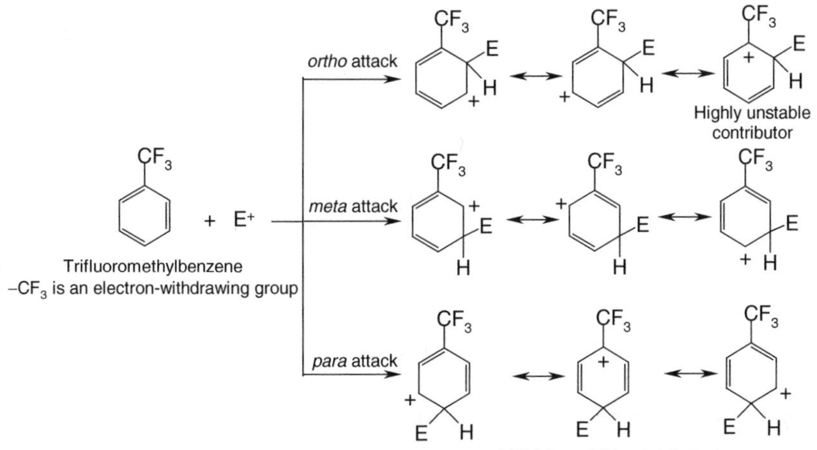

Trifluoromethylbenzene
$-CF_3$ is an electron-withdrawing group

ortho attack — Highly unstable contributor

meta attack

para attack — Highly unstable contributor

ring, and directs further substitutions to *meta* positions. The *ortho* and *para* electrophilic attacks in trifluoromethyl benzene result in one highly unstable contributing resonance structure of the arenium ion, but no such highly unstable resonance structure is formed from *meta* attack. In the case of *ortho* and *para* attacks, the positive charge in one of the resulting contributing resonance structures is located on the ring carbon that bears the electron-withdrawing group. The arenium ion formed from *meta* attack is the most stable among the three, and thus the substitution in the *meta*-position is favoured. Therefore, the trifluoromethyl group is a *meta*-directing group.

Why the $-CH_3$ group is ortho–para directing

The stability of the carbocation intermediate formed in the rate-determining step is actually the underlying factor for a substituent to direct an incoming electrophile to a particular position, *ortho*, *meta* or *para*. The methyl group ($-CH_3$) donates electrons inductively, and in the presence of this electron-donating group the resonance contributors formed from *ortho*, *meta* and *para* attacks are shown below. In the most stable contributors, arising from *ortho* and *para* attacks, the methyl group is attached directly to the positively charged carbon, which can be stabilized by donation of electrons through the inductive effect. From *meta* attack no such stable contributor is formed. Thus, the substitutions in *ortho* and *para* positions are favoured. Therefore, the methyl group is an *ortho* and *para* directing group.

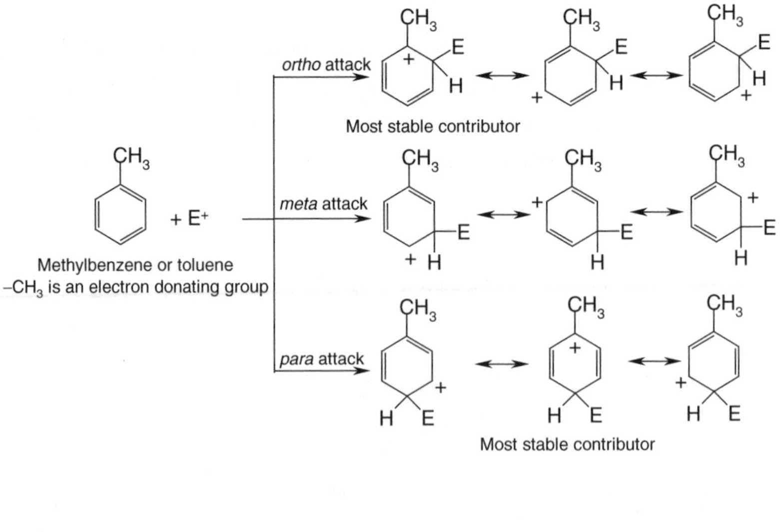

ortho attack
Most stable contributor
$+ E^+$
Methylbenzene or toluene
$-CH_3$ is an electron donating group
meta attack
para attack
Most stable contributor

Why halogens are ortho–para directing

Halogens are the only deactivating substituents that are *ortho–para* directors. However, they are the weakest of the deactivators. Halogens withdraw electrons from the ring through the inductive effect more strongly than they donate electrons by resonance. It is the resonance-aided electron-donating effect that causes halogens to be *ortho–para*-directing groups. Halogens can stabilize the transition states leading to reaction at the *ortho* and *para* positions. On the other hand, the electron-withdrawing inductive effect of halogens influences the reactivity of halobenzenes. A halogen atom, e.g. Cl, donates an unshared pair of electrons, which give rise to relatively stable resonance structures contributing to the hybrids for the *ortho*- and *para*-substituted arenium ions. Thus, despite being deactivators, halogens are *ortho*- and *para*-directors. The resonance contributors formed from *ortho*, *meta* and *para* attacks on the chlorobenzene are shown below.

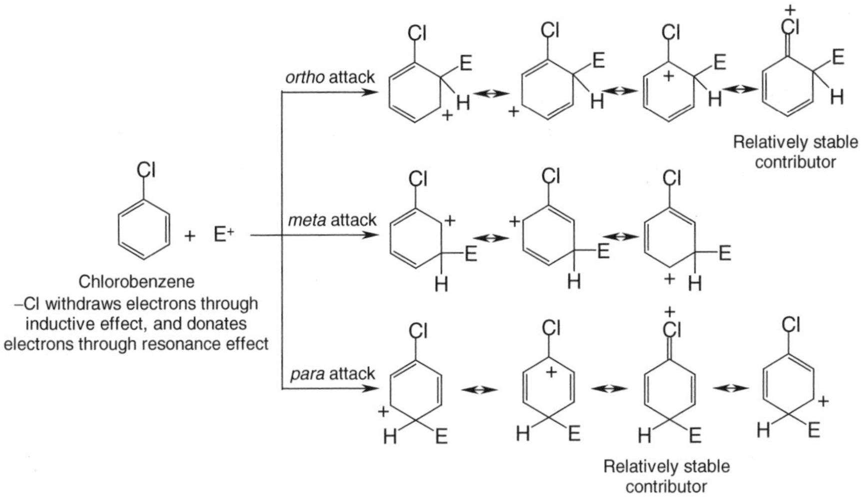

4.6.9 Alkylbenzene: toluene

Toluene, also known as methylbenzene, is the simplest member of the series known as *alkylbenzenes*, where an alkyl group, e.g. CH_3, is directly attached to the benzene ring. As the use of benzene as a nonpolar solvent has long been prohibited because of its adverse effect on the central nervous system (CNS) and on bone marrow, as well as its carcinogenic property, toluene has replaced benzene as a nonpolar solvent. Although it has a CNS depressant property like benzene, it does not cause leukaemia or aplastic anaemia.

Toluene, like benzene, undergoes electrophilic substitutions, where the substitutions take place in *ortho* and *para* positions. As the $-CH_3$ group is an activating group, the reaction rate is much faster than usually observed with benzene. For example, the nitration of toluene produces *ortho*-nitrotoluene (61%) and *para*-nitrotoluene (39%).

Toluene + HNO_3 $\xrightarrow{H_2SO_4}$ *ortho*-Nitrotoluene (61%) + *para*-Nitrotoluene (39%)

Apart from the usual electrophilic aromatic substitution reactions, other reactions can be carried out involving the methyl group in toluene, e.g. oxidation and halogenation of the alkyl group.

Oxidation of toluene

Regardless of the length of the alkyl substituent in any alkylbenzene, it can be oxidized to a carboxylic acid provided that it has a hydrogen atom, bonded to the benzylic carbon. So, reaction can occur with 1° and 2°, but not 3°, alkyl side chains. Toluene is oxidized to benzoic acid.

Toluene $\xrightarrow[H_2O]{KMnO}$ Benzoic acid

Benzylic bromination of toluene

Bromine selectively substitutes for a benzylic hydrogen in toluene in a radical substitution reaction to produce bromomethylbenzene or benzylbromide. *N*-bromosuccinimide is used to carry out benzylic bromination of toluene.

Toluene $\xrightarrow[CCl_4]{\text{N-bromosuccinimide}}$ Bromomethylbenzene

Bromomethylbenzene or benzylbromide can be subjected to further nucleophilic reactions. Bromine can be replaced by a variety of nucleophiles by means of an S_N2 and S_N1 reaction, resulting in various monosubstituted benzenes.

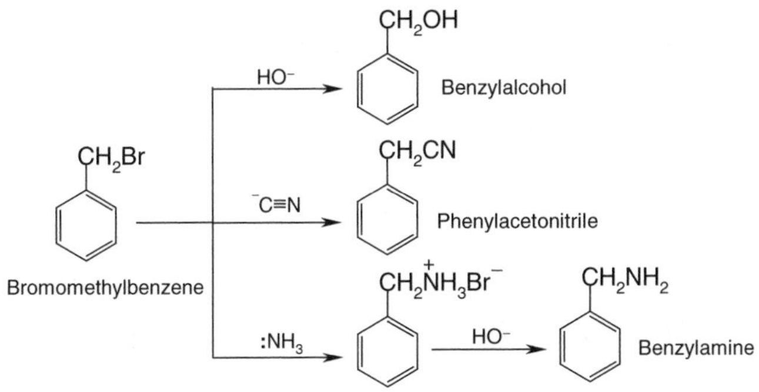

CH_2Br Bromomethylbenzene → (HO^-) CH_2OH Benzylalcohol

→ ($^-C{\equiv}N$) CH_2CN Phenylacetonitrile

→ ($:NH_3$) $CH_2\overset{+}{N}H_3Br^-$ → (HO^-) CH_2NH_2 Benzylamine

4.6.10 Phenols

Phenols are compounds of the general formula ArOH, where Ar is a phenyl, a substituted phenyl, or one of the other aryl groups, e.g. naphthyl. Phenols differ from alcohols in having the —OH group attached directly to an aromatic ring. Hydroxybenzene, the simplest member of the phenols, is generally referred to as *phenol.*

OH — Phenol
–OH is directly linked to the aromatic ring carbon

CH_2OH — Benzylalcohol
–OH is not directly linked to the aromatic ring carbon

Many pharmaceutically and pharmacologically important compounds, either of natural or synthetic origin, belong to this class of compounds, e.g. salicylic acid and quercetin.

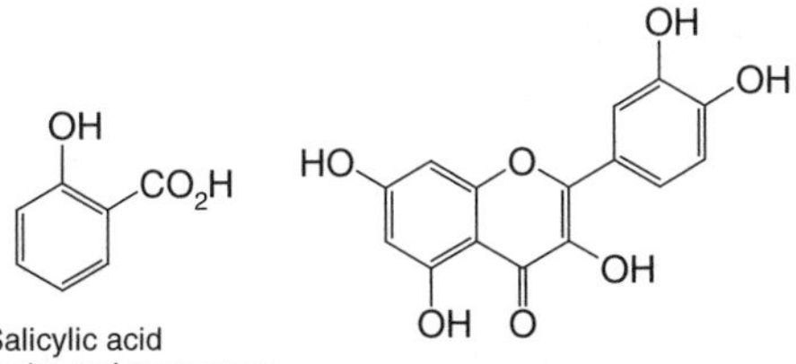

Salicylic acid
An analgesic, and a precusor for aspirin

Quercetin
A natural antioxidant

Nomenclature of phenols

Phenols are generally named as derivatives of the simplest member of the family, phenol, e.g. *o*-chlorophenol. Sometimes trivial or special names are also used, e.g. *m*-cresol. Occasionally, phenols are named as hydroxy-compounds, e.g. *para*-hydroxybenzoic acid. Numbering is often used to denote the position(s) of the substituent(s) on a phenol skeleton, e.g. 2,4-dinitrophenol.

ortho-Chlorophenol *m*-Cresol *para*-Hydroxybenzoic acid 2,4-Dinitrophenol

Physical properties of phenols

The simplest phenols are liquids or low-melting solids. Because of hydrogen bonding, phenols have a quite high boiling point (e.g. the boiling point of *m*-cresol is 201 °C). Phenol itself is somewhat soluble in water (9 g per 100 g of water) because of hydrogen bonding with water. Most other phenols are insoluble in water.

Generally, phenols themselves are colourless. However, they are easily oxidized to form coloured substances. Phenols are acidic in nature and most phenols have K_a values of $\sim 10^{-10}$.

Physical properties of nitrophenols Physical properties of *o*-, *m*- and *p*-nitrophenols differ considerably.

ortho-Nitrophenol *meta*-Nitrophenol *para*-Nitrophenol

Nitrophenols	Boiling point at 70 mm in °C	Solubility in g/100 g H_2O
ortho-nitrophenol	100	0.2
meta-nitrophenol	194	1.35
para-nitrophenol	Decomposes	1.69

Among the nitrophenols, *meta*- and *para*-nitrophenols have high boiling points because of *intermolecular hydrogen bonding*, as shown below.

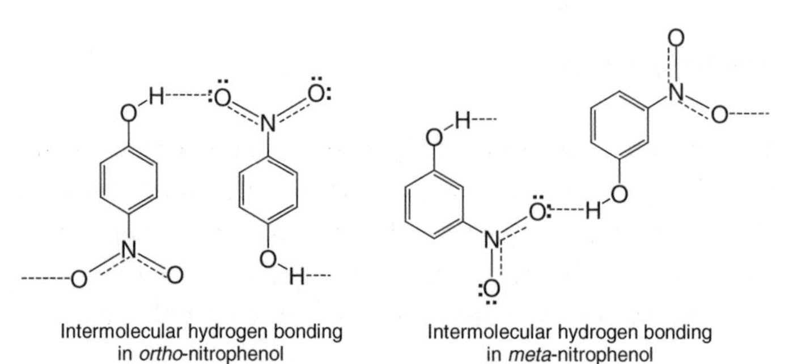

Intermolecular hydrogen bonding in *ortho*-nitrophenol

Intermolecular hydrogen bonding in *meta*-nitrophenol

These two nitrophenols are also soluble in water due to intermolecular hydrogen bonding with water molecules as outlined below.

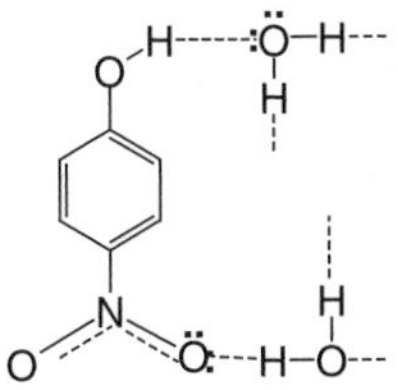

Intermolecular hydrogen bonding of *para*-nitrophenol with water

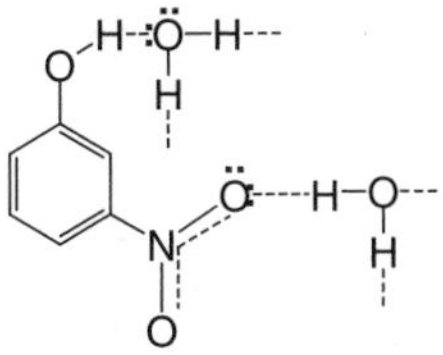

Intermolecular hydrogen bonding of *meta*-nitrophenol with water

However, in the case of *ortho*-nitrophenol, the $-NO_2$ and $-OH$ groups are located exactly right for the formation of a hydrogen bond within a single molecule, i.e. *intramolecular hydrogen bonding*, as shown below. This intramolecular hydrogen bonding takes the place of intermolecular hydrogen bonding with other phenol molecules or water molecules.

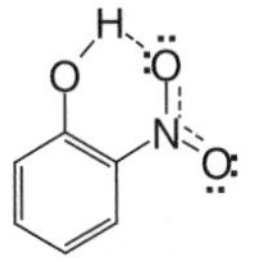

Intramolecular hydrogen bonding in *ortho*-nitrophenol

As a consequence, *o*-nitrophenol has lower boiling point than *m*- and *p*-nitrophenols, and extremely poor solubility in water compared with that of *m*- and *p*-nitrophenols.

Acidity of phenols Phenols are fairly acidic compounds. Aqueous hydroxides, e.g. NaOH, convert phenols into their salts (not by aqueous bicarbonates).

$$ArOH + HO^- \longrightarrow ArO^- + H_2O$$

Aqueous mineral acids, carboxylic acids or carbonic acid convert the salts back to free phenols.

$$ArO^- + H_2CO_3 \longrightarrow ArOH + HCO_3^-$$

Most phenols (K_a values of $\sim 10^{-10}$) are considerably weaker acids than carboxylic acids (K_a values of $\sim 10^{-5}$). Although weaker than carboxylic acids, phenols are more acidic than alcohols (K_a values around 10^{-16} to 10^{-18}). The benzene ring of a phenol acts as if it were an electron-withdrawing group. It withdraws electrons from the $-OH$ group and makes the oxygen positive.

The acidity of phenols is mainly due to an electrical charge distribution in phenols that causes the $-OH$ oxygen to be more positive. As a result, the proton is held less strongly, and phenols can easily give this loosely held proton away to form a *phenoxide ion* as outlined below.

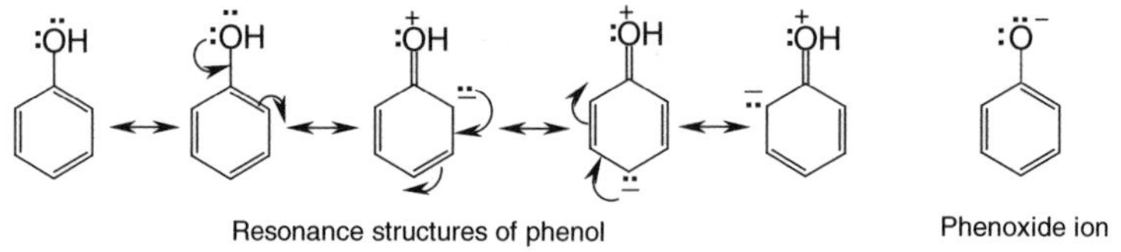

Resonance structures of phenol

Phenoxide ion

Preparation of phenols

In the laboratory, phenols are predominantly prepared by either hydrolysis of diazonium salts or alkali fusion of sulphonates.

Hydrolysis of diazonium salts Diazonium salts react with water in the presence of mineral acids to yield phenols.

$$Ar\text{-}N_2^+ + H_2O \xrightarrow{H^+} Ar\text{-}OH + N_2$$

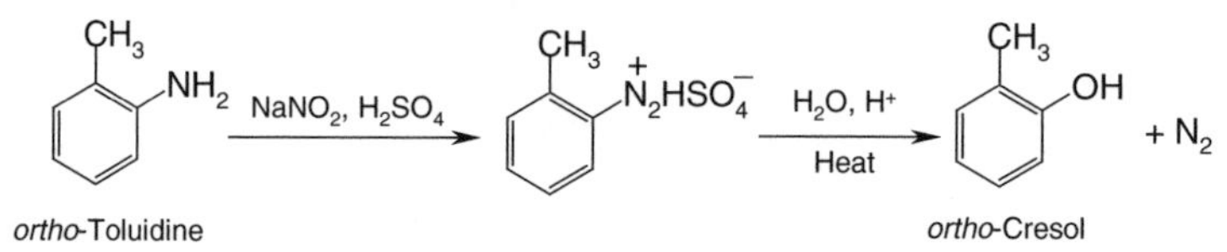

ortho-Toluidine $\xrightarrow{NaNO_2,\ H_2SO_4}$ $\xrightarrow[Heat]{H_2O,\ H^+}$ *ortho*-Cresol $+ N_2$

Alkali fusion of sulphonates Phenols can be prepared from the corresponding sulphonic acids by fusion with alkali.

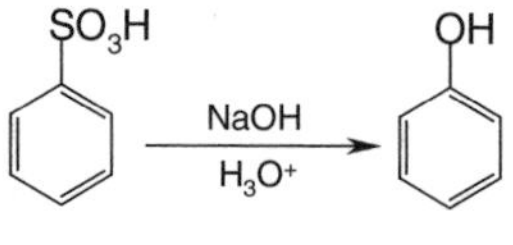

Reactions of phenols

Phenols undergo electrophilic substitutions. In phenol, the substitutions take place in *ortho* and *para* positions. As the —OH group is an activating group, the reaction rate is much faster than usually observed with benzene. For example, the bromination of phenol produces *ortho*-bromophenol (12%) and *para*-bromophenol (88%).

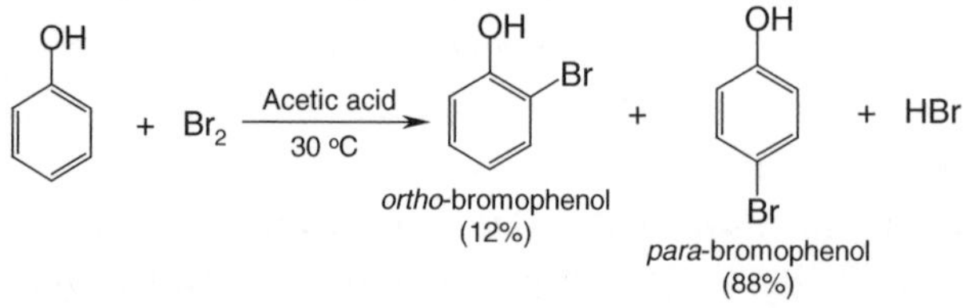

$+ Br_2 \xrightarrow[30\ ^{\circ}C]{Acetic\ acid}$ *ortho*-bromophenol (12%) + *para*-bromophenol (88%) + HBr

A number of other reactions can also be carried out with phenols as follows.

Salt formation Phenol is acidic in nature, and can form a salt with alkali, e.g. NaOH.

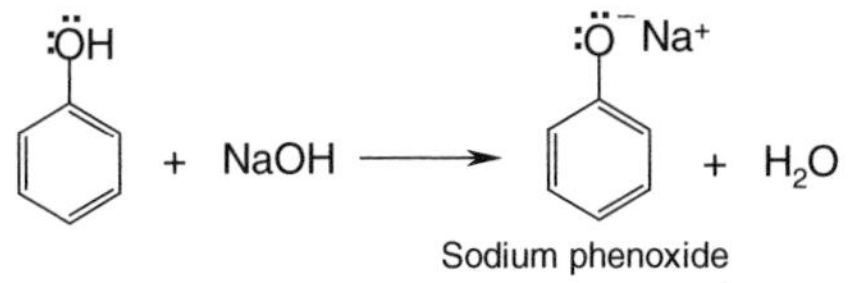

:ÖH + NaOH → :Ö⁻ Na+ + H_2O

Sodium phenoxide

Ether formation Phenol reacts with ethyliodide (C_2H_5I), in the presence of aqueous NaOH, to produce ethylphenylether, also known as *phenetole.*

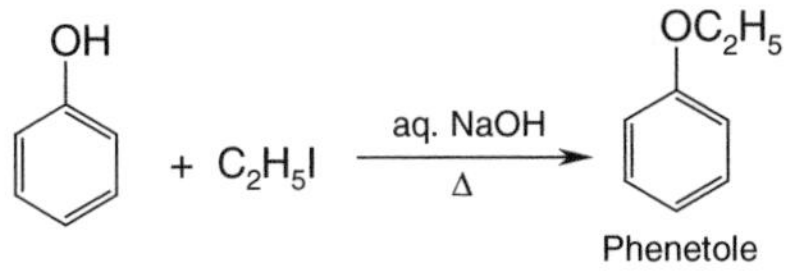

OH + C_2H_5I —aq. NaOH, Δ→ OC_2H_5

Phenetole

Ester formation Phenols can undergo esterification, and produce corresponding esters. For example, phenol reacts with benzoylchloride to yield phenylbenzoate, and bromophenol reacts with toluenethionyl chloride to produce bromophenyltoluene sulphonate.

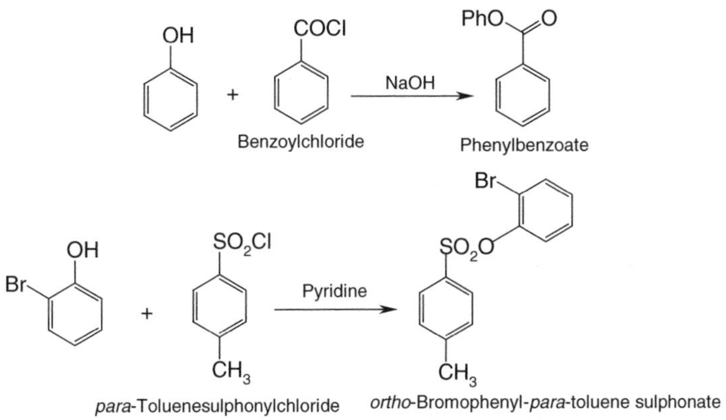

OH + COCl —NaOH→ PhO O

Benzoylchloride

Phenylbenzoate

Br OH + SO_2Cl CH_3 —Pyridine→ Br SO_2O CH_3

para-Toluenesulphonylchloride

ortho-Bromophenyl-*para*-toluene sulphonate

Carbonation: Kolbe reaction Treatment of a salt of a phenol with CO_2 replaces a ring hydrogen with a carboxyl group. This reaction is applied in the conversion of phenol itself into *ortho*-hydroxybenzoic acid, known as salicylic acid. Acetylation of salicylic acid produces acetylsalicylic acid (aspirin), which is the most popular painkiller in use today.

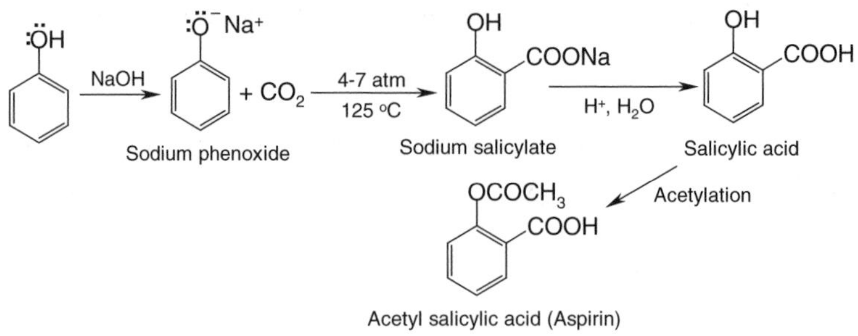

NaOH
Sodium phenoxide
$+ CO_2$ 4-7 atm, 125 °C
Sodium salicylate
H^+, H_2O
Salicylic acid
Acetylation
Acetyl salicylic acid (Aspirin)

Aldehyde formation: Reimer–Tiemann reaction Treatment of a phenol with chloroform ($CHCl_3$) and aqueous hydroxide introduces an aldehyde group (—CHO) onto the aromatic ring, generally *ortho* to the —OH group. A substituted benzalchloride is initially formed, but is hydrolysed by the alkaline medium. Salicylaldehyde can be produced from phenol by this reaction. Again, salicylaldehyde could be oxidized to salicylic acid, which could be acetylated to aspirin.

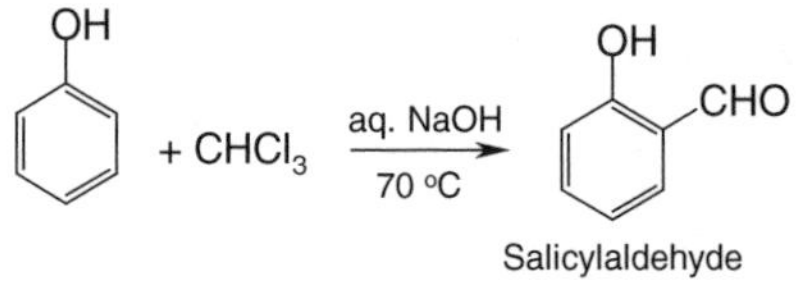

$+ CHCl_3$ aq. NaOH, 70 °C
Salicylaldehyde

Reaction with formaldehyde (formation of phenol–formaldehyde resins) Phenol reacts with formaldehyde (HCHO) to produce *ortho*-hydroxymethylphenol, which reacts with phenol to produce *ortho*-(*para*-hydroxybenzyl)-phenol. This reaction continues to form polymer.

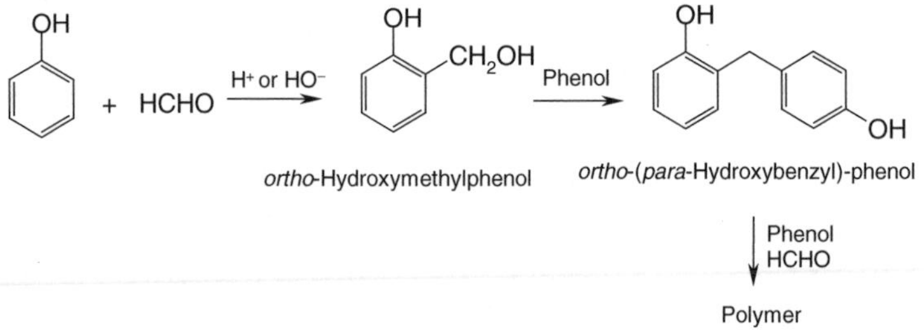

+ HCHO H^+ or HO^-
ortho-Hydroxymethylphenol
Phenol
ortho-(*para*-Hydroxybenzyl)-phenol
Phenol, HCHO
Polymer

4.6.11 Aromatic amines: aniline

An amine has the general formula RNH_2 (1° amine), R_2NH (2° amine) or R_3N (3° amine), where R = alkyl or aryl group, e.g. methylamine CH_3NH_2, dimethylamine $(CH_3)_2NH$ and trimethylamine $(CH_3)_3N$.

When an amino group ($-NH_2$) is directly attached to the benzene ring, the compound is known as *aniline*.

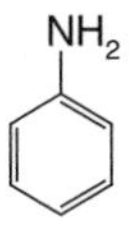

Aniline
$-NH_2$ group is attached
directly to the benzene ring

Physical properties of aniline

Aniline is a polar compound, and can form *intermolecular hydrogen bonding* between two aniline molecules. Aniline has higher b.p. (184 °C) than nonpolar compounds of the same molecular weight. It also forms hydrogen bonds with water. This hydrogen bonding accounts for the solubility of aniline in water (3.7 g/100 g water).

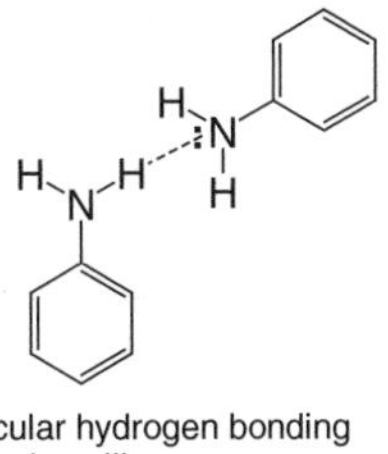

Intermolecular hydrogen bonding in aniline

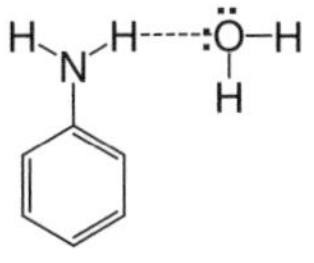

Intermolecular hydrogen bonding of aniline with water

Basicity of aniline Aniline, like all other amines, is a basic compound ($K_b = 4.2 \times 10^{-10}$). Anilinium ion has a p$K_a = 4.63$, whereas methylammonium ion has a p$K_a = 10.66$. Arylamines, e.g. aniline, are less basic than alkylamines, because the nitrogen lone pair electrons are *delocalized* by interaction with the aromatic ring π electron system and are less available for bonding to H^+. Arylamines are stabilized relative to alkylamines because of the five resonance structures as shown below. Resonance stabilization is lost on protonation, because only two resonance structures are possible for the arylammonium ion.

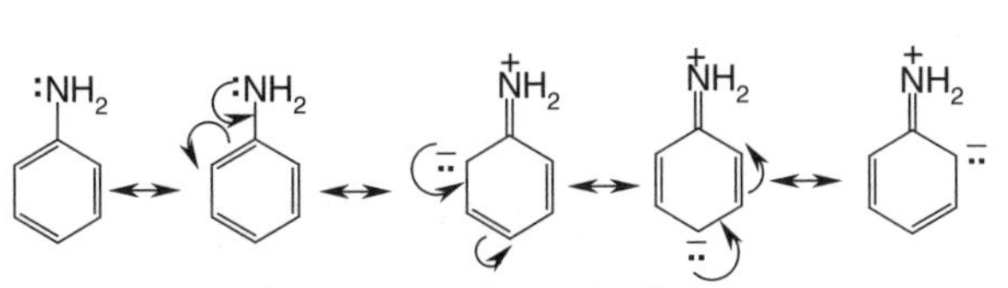

Resonance contributors of aniline

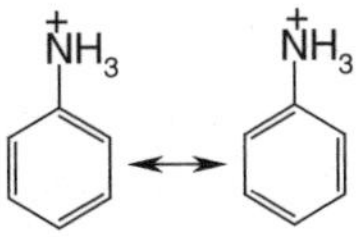

Arylammonium ion

The energy difference ΔG° between protonated and nonprotonated forms, as shown in the following diagram, is higher for arylamines than it is for alkylamines. This is why arylamines are less basic.

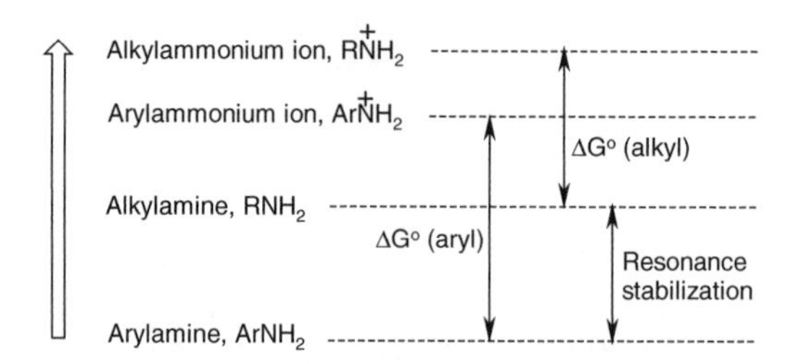

Effect of substituents on the basicity of aniline The effect of substituent(s) on the basicity of aniline is summarized below. Electron-donating substituents (Y = $-CH_3$, $-NH_2$, $-OCH_3$) activate the ring, and increase the basicity of aniline, whereas electron-withdrawing substituents (Y = $-Cl$, $-NO_2$, $-CN$) deactivate the ring, and decrease the basicity.

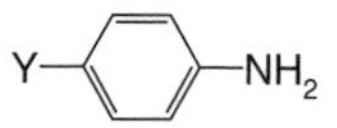

Y–C_6H_4–NH_2

	Substituent Y	pK_a	Effect on reactivity
Stronger base	$-NH_2$	6.15	Activating
	$-OCH_3$	5.34	Activating
	$-CH_3$	5.08	Activating
	$-H$	4.63	
	$-Cl$	3.98	Deactivating
	$-CN$	1.74	Deactivating
Weaker base	$-NO_2$	1.00	Deactivating

Preparation of aniline

Reduction of nitrobenzene Aniline can be prepared from nitrobenzene by either chemical reduction using acid and metal or catalytic hydrogenation using molecular hydrogen.

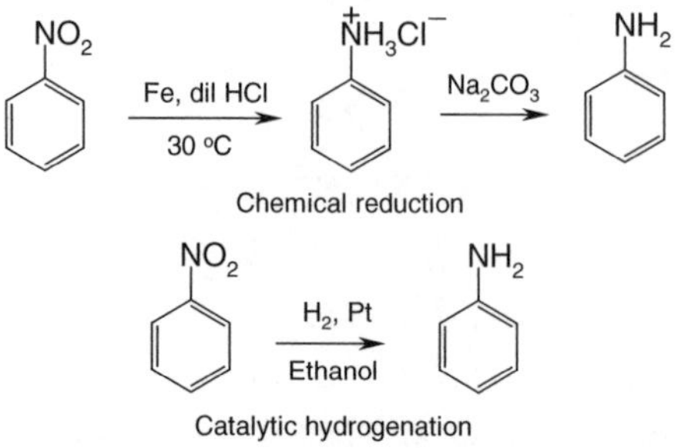

$C_6H_5NO_2$ —(Fe, dil HCl, 30 °C)→ $C_6H_5\overset{+}{N}H_3Cl^-$ —(Na_2CO_3)→ $C_6H_5NH_2$

Chemical reduction

$C_6H_5NO_2$ —(H_2, Pt, Ethanol)→ $C_6H_5NH_2$

Catalytic hydrogenation

From chlorobenzene Treatment of chlorobenzene with ammonia (NH_3) at high temperature and high pressure in the presence of a catalyst yields aniline.

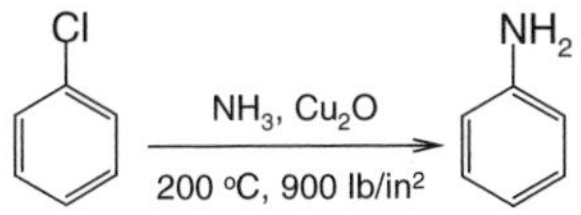

Cl → NH$_2$ (NH_3, Cu_2O; 200 °C, 900 lb/in^2)

Hofmann degradation of benzamide This reaction produces aniline, which contains one less carbon than the starting material (benzamide). The group (phenyl) attached to the carbonyl carbon in the amide (benzamide) is found joined to nitrogen in the product (aniline). This is an example of *molecular rearrangement.*

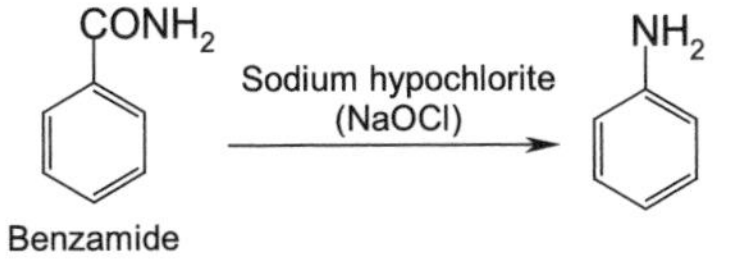

$CONH_2$ → NH_2 (Sodium hypochlorite (NaOCl))

Benzamide

Substituted benzamides produce substituted aniline, and show the following order of reactivity: $Y = -OCH_3 > -CH_3 > -H > -Cl > -NO_2$.

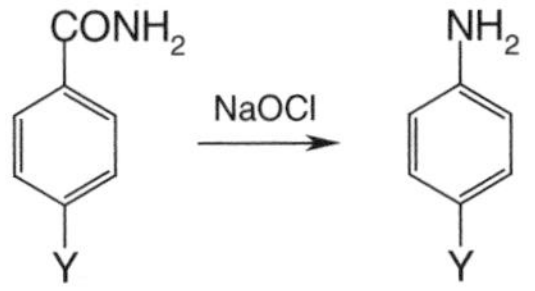

$CONH_2$ (Y) → NH_2 (Y) (NaOCl)

Reactions of aniline

Aniline undergoes electrophilic substitutions. In aniline, the substitutions take place in *ortho* and *para* positions. As the $-NH_2$ group is a strong activating group, the reaction rate is much faster than usually observed with benzene. A number of other types of reaction can also be carried out with aniline. Some of these reactions are discussed here.

Salt formation As aniline is a base, it forms salt with mineral acids.

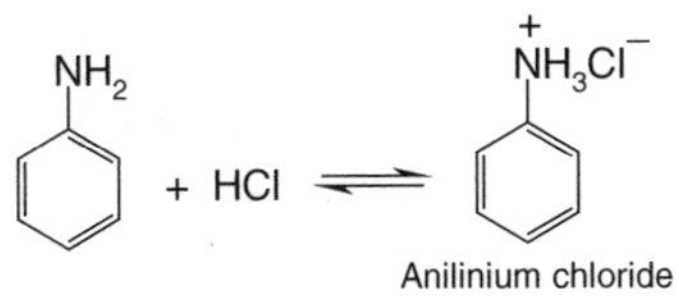

NH_2 + HCl ⇌ $\overset{+}{N}H_3Cl^-$

Anilinium chloride

***N*-alkylation** The hydrogen atoms of the amino group in aniline can be replaced by alkyl substituent to produce *N*-alkylated aniline.

Aniline reacts with CH_3Cl to produce *N*-methylaniline, which again reacts with CH_3Cl to produce *N,N*-dimethylaniline and finally the quaternary salt.

NH2 →(CH3Cl) NHCH3 →(CH3Cl) N(CH3)2 →(CH3Cl) +N(CH3)3Cl−

N-Methylaniline *N.N*-Dimethylaniline

The alkyl halide, CH_3Cl, undergoes nucleophilic substitution (see Section 5.5.1) with the basic aniline serving as the nucleophilic reagent. One of the hydrogen atoms attached to the nitrogen is replaced by an alkyl group. The final stage of this reaction involves the formation of a quaternary ammonium salt where four organic groups are covalently bonded to nitrogen, and the positive charge of this ion is balanced by the negative chloride (Cl^-) ion.

Formation of amide Aniline reacts with acid chloride to form corresponding amide. For example, when aniline is treated with benzoylchloride in the presence of pyridine, it produces benzanilide.

NH2 + COCl →(Pyridine) Ph–NH–C(=O)–Ph

Benzoyl chloriode Benzanilide

Formation of sulphonamide Aniline reacts with sulphonylchloride to form the corresponding *sulphonamide*. For example, when aniline is treated with benzenesulphonylchloride in the presence of a base, it produces the sulphonamide *N*-phenylbenzenesulphonamide.

NH2 + SO2Cl →(Base) HN–Ph, O=S=O

Benzenesulphonylchloride *N*-phenylbenzenesulphonamide

Application in reductive aminition Aniline can be used in reductive aminition reactions. For example, aniline can be converted to *N*-isopropylaniline by the reaction with acetone (CH_3COCH_3) in the presence of the reducing agent sodium borohydride ($NaBH_4$).

NH2 + H3C–C(=O)–CH3 →(NaBH4; CH3CO2− Na+, CH3COOH) N-isopropylaniline

Acetone *N*-isopropylaniline

Diazonium salt formation Primary arylamines react with nitrous acid (HNO_2) to yield stable arenediazonium salts, $Ar{-}N^+{\equiv}NX^-$. Alkylamines also react with nitrous acid, but the alkanediazonium salts are so reactive that they cannot be isolated.

Aniline is a primary arylamine, and it reacts with nitrous HNO_2 to yield stable benzene diazonium salt ($Ph{-}N^+{\equiv}NX^-$).

NH_2 $\xrightarrow[H_2SO_4]{HNO_2}$ $\overset{+}{N}{\equiv}NHSO_4^-$ $+ 2\,H_2O$

Benzene diazonium salt

The drive to form a molecule of stable nitrogen gas causes the leaving group of a diazonium ion to be easily displaced by a wide variety of nucleophiles ($Nu{:}^-$).

$\overset{+}{N}{\equiv}NHSO_4^-$ + $Nu{:}^-$ $\longrightarrow$ Nu $+ N_2\uparrow$

Benzene diazonium salt

Nucleophile

The mechanism by which a nucleophile displaces the diazonium group depends on the nucleophile. While some displacements involve phenyl cations, others involve radicals. Nucleophiles, e.g. CN^-, Cl^- and Br^-, replace the diazonium group if the appropriate cuprous salt is added to the solution containing the arene diazonium salt. The reaction of an arene diazonium salt with cuprous salt is known as a *Sandmeyer reaction.*

$\overset{+}{N}{\equiv}NBr^-$ $\xrightarrow{CuBr}$ Br $+ N_2\uparrow$

Benzene diazonium bromide

Bromobenzene

This diazotization reaction is compatible with the presence of a wide variety of substituents on the benzene ring. Arenediazonium salts are extremely important in synthetic chemistry, because the diazonio group ($N{\equiv}N$) can be replaced by a nucleophile in a radical substitution reaction, e.g. preparation of phenol, chlorobenzene and bromobenzene. Under proper conditions, arenediazonium salts react with certain aromatic compounds to yield products of the general formula Ar–N=N–Ar′, called *azo compounds.* In this coupling reaction, the nitrogen of the diazonium group is retained in the product.

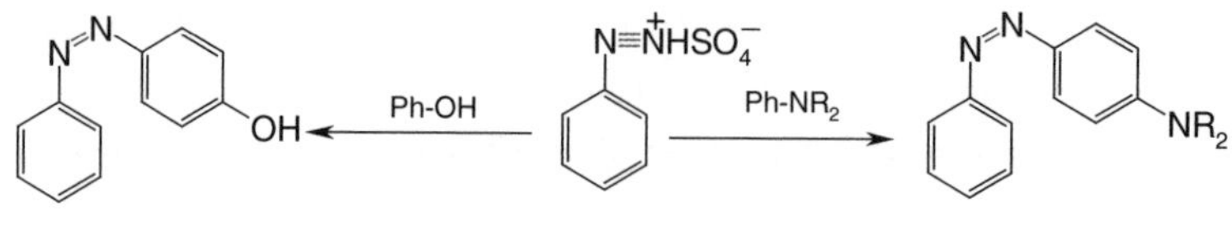

Ph-OH Ph-NR$_2$

Synthesis of sulpha drugs from aniline

Antimicrobial sulpha drugs, e.g. sulphanilamide, are the amide of sulphanilic acid, and certain related substituted amides. Sulphanilamide, the first of the sulpha drugs, acts by inhibiting the bacterial enzyme that incorporates *para*-aminobenzoic acid into folic acid. Sulphanilamide is a bacteriostatic drug, i.e. inhibits the further growth of the bacteria.

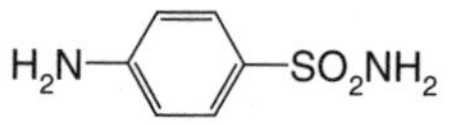

H_2N–C_6H_4–SO_2NH_2

Sulphanilamide
(*para*-aminobenzenesulphonamide)

Multistep synthesis, starting from aniline, as depicted in the following scheme, can achieve the product, sulpha drug.

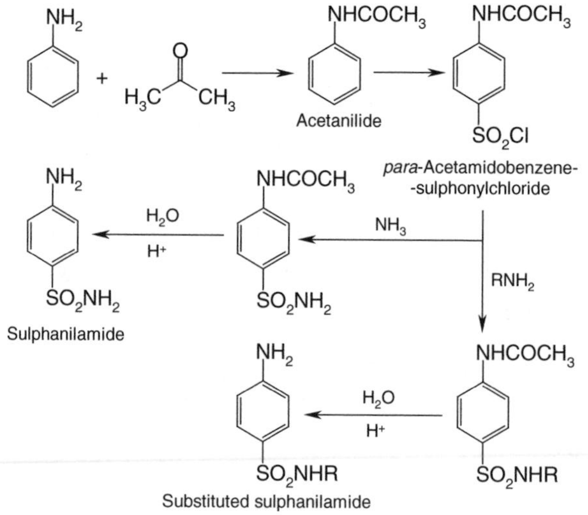

Acetanilide

para-Acetamidobenzene-
-sulphonylchloride

Sulphanilamide

Substituted sulphanilamide

Separation of aniline and a neutral compound from a mixture by solvent extraction

If a mixture contains aniline and a neutral compound, both the constituents can easily be separated and purified by the solvent extraction method. To purify these compounds, the mixture is dissolved in diethylether, HCl and water are added and the solution is shaken in a separating funnel. Once two layers, aqueous and ether layers, are formed, they are separated. The lower layer (aqueous) contains the salt of aniline, and the ether layer has the neutral compound. Ether is evaporated from the ether layer using a rotary evaporator to obtain purified neutral compound. To the aqueous layer, sodium hydroxide and ether are added, and the resulting solution is shaken in a separating funnel. Two layers are separated. The ether layer (top layer) contains free aniline, and the aqueous layer (bottom layer) has the salt, sodium chloride. Ether is evaporated from the ether layer using a rotary evaporator to obtain purified aniline.

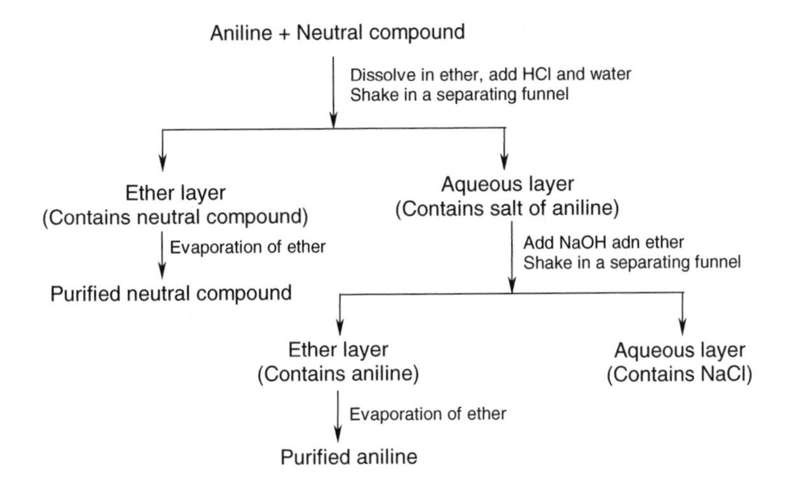

4.6.12 Polycyclic benzenoids

Two or more benzene rings fused together form a number of polycyclic benzenoid aromatic compounds, naphthalene, anthracene and phenanthrene, and their derivatives. All these hydrocarbons are obtained from coal tar. Naphthalene is the most abundant (5%) of all constituents of coal tar.

Napthalene Anthracene Phenanthrene

Synthesis of naphthalene from benzene: Haworth synthesis

Naphthalene can be synthesized from benzene through multi-step synthesis involving, notably, Friedel–Crafts (FC) acylation, Clemmensen reduction and aromatization reactions as outlined in the following scheme.

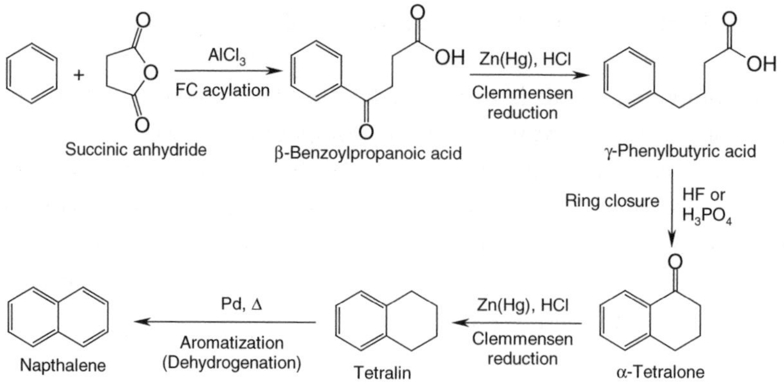

$AlCl_3$ FC acylation; Zn(Hg), HCl Clemmensen reduction

Succinic anhydride β-Benzoylpropanoic acid γ-Phenylbutyric acid

Ring closure HF or H_3PO_4

Pd, Δ Aromatization (Dehydrogenation); Zn(Hg), HCl Clemmensen reduction

Napthalene Tetralin α-Tetralone

Reactions of naphthalene

Naphthalene undergoes electrophilic substitutions on the ring, resulting in its various derivatives. In addition to the usual electrophilic substitutions, naphthalene can also undergo oxidation and reduction reactions under specific conditions as outlined below.

Oxidation Oxidation of naphthalene by oxygen in the presence of vanadium pentoxide (V_2O_5) destroys one ring and yields phthalic anhydride (an important industrial process). However, oxidation in the presence of CrO_3 and acetic acid (AcOH) destroys the aromatic character of one ring and yields naphthoquinone (a diketo compound).

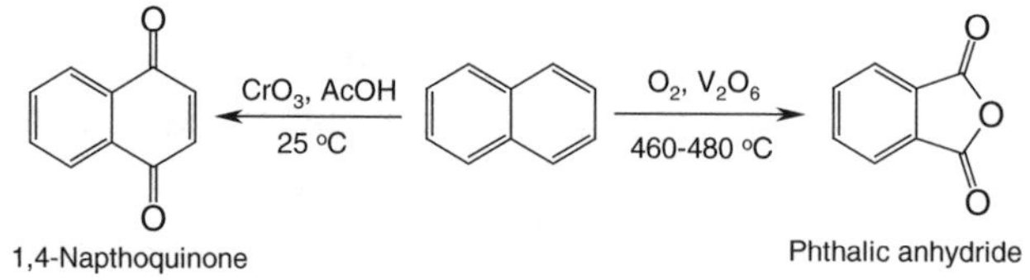

CrO_3, AcOH 25 °C; O_2, V_2O_6 460-480 °C

1,4-Napthoquinone Phthalic anhydride

Reduction One or both rings of naphthalene can be reduced partially or completely, depending upon the reagents and reaction conditions.

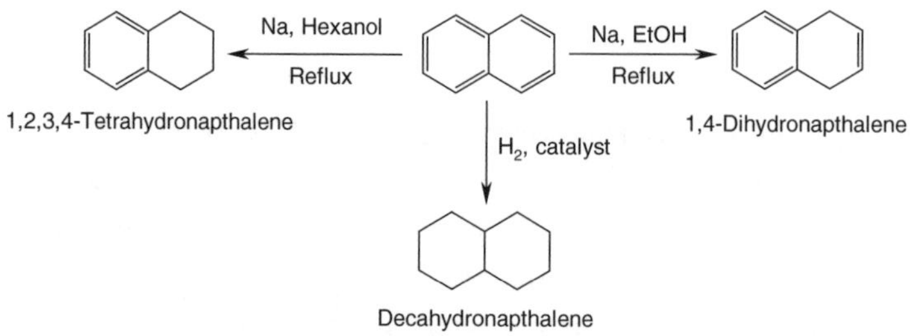

Na, Hexanol Reflux; Na, EtOH Reflux; H_2, catalyst

1,2,3,4-Tetrahydronapthalene 1,4-Dihydronapthalene

Decahydronapthalene

4.7 Heterocyclic compounds and their derivatives

Cyclic compounds that have one or more of atoms other than carbon, e.g. N, O or S (hetero-atoms), in their rings are called *heterocyclic compounds* or *heterocycles*, e.g. pyridine, tetrahydrofuran, thiophene and so on.

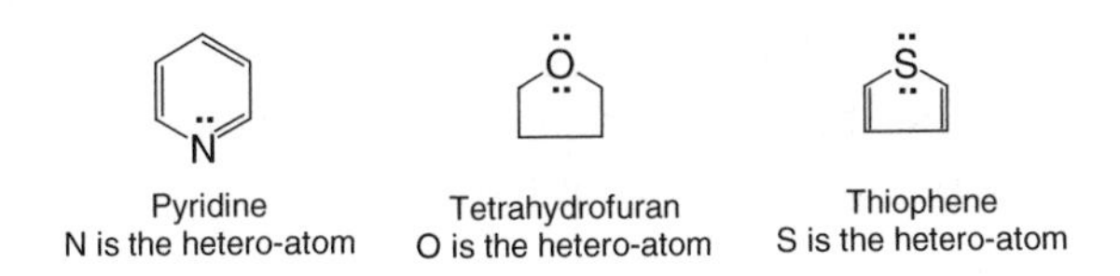

Among the heterocyclic compounds, there are aromatic, e.g. pyridine, as well as nonaromatic, e.g. tetrahydrofuran, compounds. Similarly, there are saturated (e.g. tetrahydrofuran) and unsaturated (e.g. pyridine) heterocyclic compounds. Heterocycles also differ in their ring sizes, e.g. pyridine has a six-membered ring, whereas tetrahydrofuran is a five-membered oxygen-containing heterocyclic compound.

4.7.1 Medicinal importance of heterocyclic compounds

More than 50% of all known organic compounds are *heterocyclic compounds*. They play important roles in medicine and biological systems. A great majority of important drugs and natural products, e.g. caffeine, nicotine, morphine, penicillins and cephalosporins, are heterocyclic compounds. The purine and pyrimidine bases, two nitrogenous heterocyclic compounds, are structural units of RNA and DNA. Serotonin, a neurotransmitter found in our body, is responsible for various bodily functions.

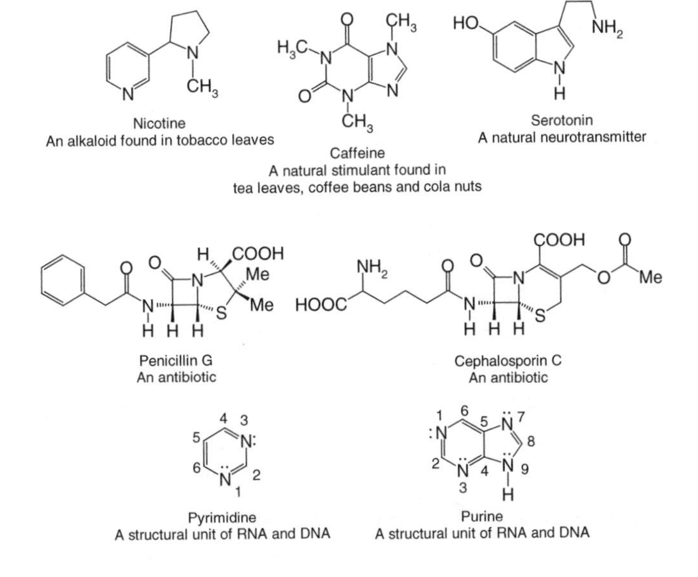

4.7.2 Nomenclature of heterocyclic compounds

Most of the heterocycles are known by their trivial names, e.g. pyridine, indole, quinoline, thiophene and so on. However, there are some general rules to be followed in a heterocycle, especially in the use of suffixes to indicate the ring size, saturation or unsaturation as shown in the following table. For example, from the name, pyrid*ine*, where the suffix is *-ine*, one can understand that this heterocyclic compound contains nitrogen, has a six-membered ring system and is unsaturated.

	Ring with nitrogen		Ring without nitrogen	
Ring size	*Maximum unsaturation*	*Saturation*	*Maximum unsaturation*	*Saturation*
3	irine	iridine	irene	irane
4	ete	etidine	ete	etane
5	ole	olidine	ole	olane
6	ine	–	ine	ane
7	epine	–	epine	epane
8	ocine	–	ocine	ocane
9	onine	–	onine	onane
10	ecine	–	ecine	ecane

Monocyclic heterocycles containing three to ten members, and one or more hetero-atoms, are named systematically by using a prefix or prefixes to indicate the nature of the hetero-atoms as presented in the following table. For example, *thia*cyclobutane contains the hetero-atom sulphur (S).

Element	Prefix	Element	Prefix	Element	Prefix
O	oxa	P	phospha	Ge	germa
S	thia	As	arsa	Sn	stanna
Se	selena	Sb	stiba	Pb	plumba
Te	tellura	Bi	bisma	B	bora
N	aza	Si	sila	Hg	mercura

Two or more identical hetero-atoms are indicated by use of the multiplying prefixes *di-*, *tri-* or *tetra-*. When more than one distinct hetero-atom is present, the appropriate prefixes are cited in the name in descending order of group number in the periodic table, e.g. *oxa-* takes precedence over *aza-*. If both lie within the same group of the periodic table, then the order is determined by increasing atomic number, e.g. *oxa-* precedes *thia-*.

In unsaturated heterocycles, if the double bonds can be arranged in more than one way, their positions are defined by indicating the N or C atoms that are not multiply bonded, and consequently carry an 'extra' hydrogen atom, by 1H-, 2H- and so on, for example 1H-azepine and 2H-azepine.

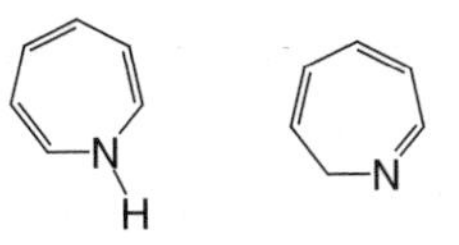

1H-Azepine 2H-Azepine

Important aromatic heterocycles that contain a single hetero-atom include pyridine, quinoline, isoquinoline, pyrrole, thiophene, furan and indole.

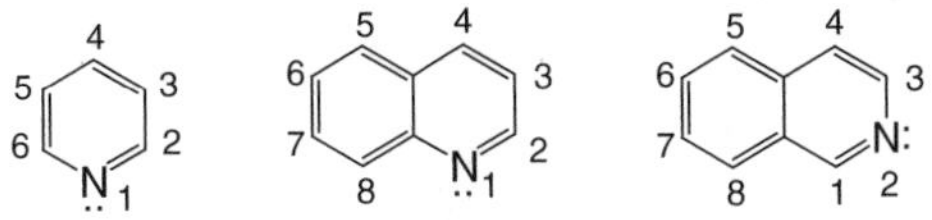

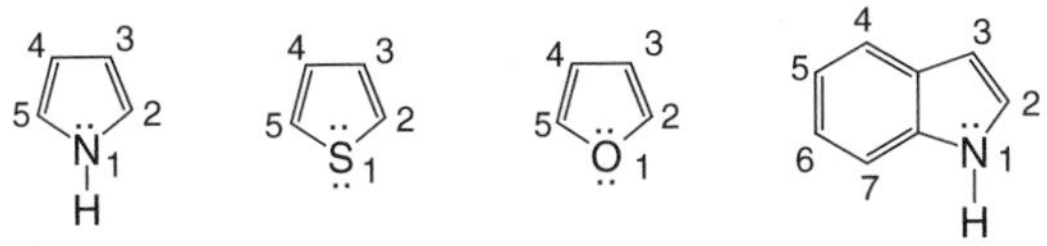

Pyridine Quinoline Isoquinoline

Pyrrole Thiophene Furan Indole

Derivatives of these heterocyclic compounds are named in the same way as other compounds, by adding the name of the substituent, in most cases as a prefix to the name of the heterocycle, and a number to indicate its position on the ring system, e.g. 2-methylpyridine, 5-methylindole and 3-phenylthiophene.

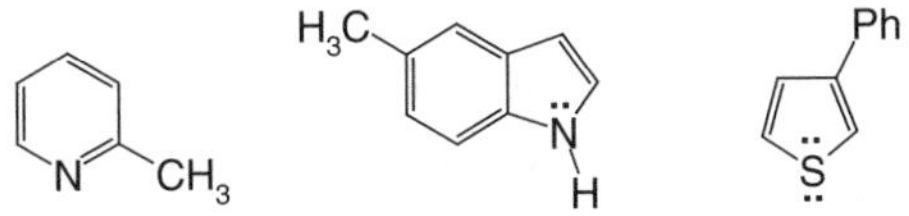

2-Methylpyridine 5-Methylindole 3-Phenylthiophene

Heterocyclic aromatic compounds can also have two or more hetero-atoms. If one of the hetero-atoms is a nitrogen atom, and the compound has a five-membered system, their names all end in *-azole*, and the rest of the name indicates other hetero-atoms. For example, pyrazole and imidazole are two isomeric heterocycles that contain two nitrogen atoms in the ring, thiazole has a sulphur atom and a nitrogen atom in the ring, and oxazole contains an oxygen atom and a nitrogen atom. In imidazole and oxazole, two hetero-atoms are separated by a carbon atom, whereas in their isomers, pyrazole and isoxazole, the hetero-atoms are directly linked to each other. The six-membered aromatic heterocycles with two nitrogens can exist in three isomeric forms, the most important being pyrimidine.

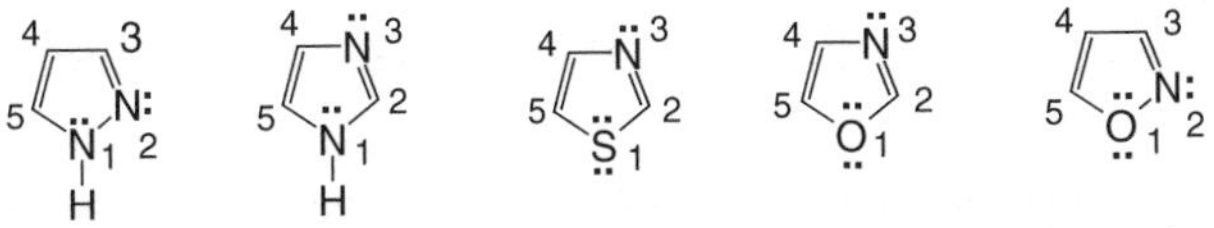

Pyrazole Imidazole Thiazole Oxazole Isoxazole

There are a number of fully saturated nonaromatic heterocycles. For example, pyrrolidine, tetrahydrofuran, isoxazolidine and piperidine are fully saturated derivatives of pyrrole, furan, isoxazole and pyridine, respectively. Partially saturated derivatives, e.g. 2-pyroline, 2-isoxazoline and 1,4-dihydropyridine, are also known.

Unsaturated	Partially saturated	Fully saturated
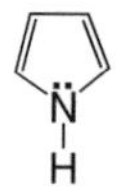 Pyrrole	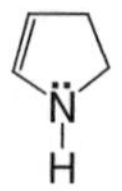 2-Pyrroline	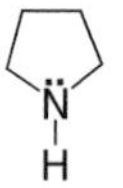 Pyrrolidine
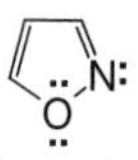 Isoxazole	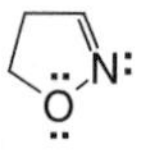 2-isoxazoline	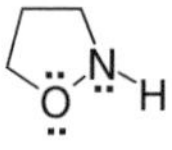 Isoxazolidine
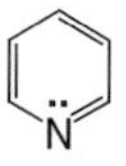 Pyridine	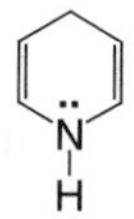 1,4-Dihydropyridine	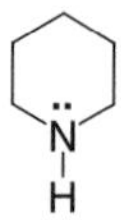 Piperidine

4.7.3 Physical properties of heterocyclic compounds

A large number of structurally diverse compounds belong to the class *heterocycles*. This makes it extremely difficult to generalize the physical properties of these compounds, because they vary significantly depending on the saturation–unsaturation status, aromatic–nonaromatic behaviour, ring sizes and type and number of hetero-atoms present. Saturated heterocycles, known as *alicyclic heterocycles*, containing five or more atoms have physical and chemical properties typical of acyclic compounds that contain the same hetero-atoms. These compounds undergo the same reactions as their open chain analogues. On the other hand, aromatic heterocycles display very characteristic and often complex reactivity. However, aromatic heterocycles show general patterns of reactivity associated with certain 'molecular fragments' such that the reactivity of a given heterocycle can be anticipated. Physical and chemical properties of selected important heterocyclic compounds are discussed under each compound sub-heading.

4.7.4 Pyrrole, furan and thiophene: five-membered unsaturated heterocycles

Pyrrole is a nitrogen-containing unsaturated five-membered heterocyclic aromatic compound. It shows aromaticity by delocalization of a lone pair of electrons from nitrogen. In pyrrole, there are four π electrons, two short of the Hückel criteria for aromaticity. The nitrogen atom is sp^2-hybridized, formally containing a lone pair of electrons in the p orbital at right angles to the ring. However, the system delocalizes and pushes the lone pair of electrons into the ring to complete the sextet required for aromaticity. The nonbonding electrons on the nitrogen atom become a part of the aromatic sextet. A small number of simple pyrroles occur in nature. However, biologically more significant natural pyrroles are rather less simple; they are tetrameric pyrrole derivatives, known as porphyrins, e.g. chlorophyll-a and haem.

Furan, also known as furane and furfuran, is an oxygen-containing five-membered aromatic heterocyclic compound that is usually produced when wood, especially pine wood, is distilled. The highly electronegative oxygen holds on the electron density tightly. Although it has a lone pair of electrons, these electrons cannot delocalize easily, and so the system is generally considered to be almost nonaromatic or weakly aromatic.

Thiophene is a sulphur-containing five-membered unsaturated heterocycle. The lone pair electrons of the sulphur are in the 3s orbital, and are less able to interact with the π electrons of the double bonds. Therefore, thiophene is considered weakly aromatic. Acetylenic thiophene is found in some higher plant species. However, the thiophene ring is present in many important pharmaceutical products.

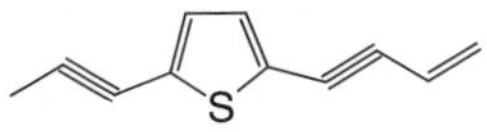

S

Acetylenic thiophene

Physical properties of pyrrole, furan and thiophene

Pyrrole is a weakly basic compound. However, as the nonbonding electrons on the nitrogen atom are part of the aromatic sextet, and no longer available for protonation, it has an extremely low basicity (pK_a = ~ 15). Pyrrole accepts a proton on one of the carbon atoms adjacent to the nitrogen atom, whereas the proton on the nitrogen atom can be removed by hydroxide ion to yield its conjugate base.

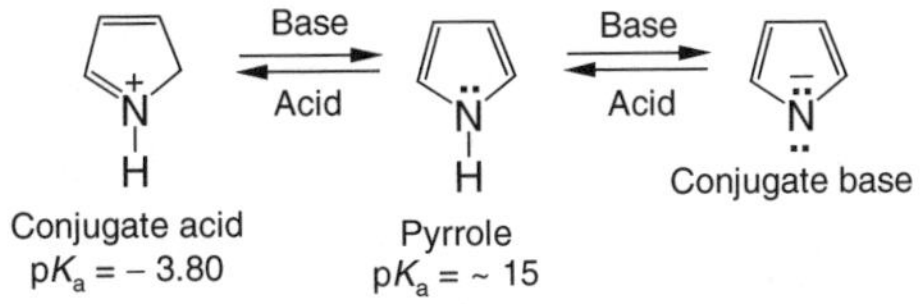

Base
Acid
Base
Acid

Conjugate acid
pK_a = − 3.80

Pyrrole
pK_a = ~ 15

Conjugate base

Salts containing the pyrrole anion can easily be prepared by this way. The pair of nonbonding electrons on N in pyrrole is much less available for protonation than the pair on ammonia. Thus, pyrrole is much less basic than NH_3 (pK_a = 36), i.e. a much stronger acid than NH_3.

Furan and thiophene are both clear and colourless liquids at room temperature. While furan is extremely volatile and highly flammable with a boiling point close to room temperature (31.4 °C), the b.p. of thiophene is 84 °C. Thiophene possesses a mildly pleasant odour.

Preparation of pyrrole, furan and thiophene

A general way of synthesizing heterocyclic compounds is by cyclization of a dicarbonyl or diketo compound using a nucleophilic reagent that introduces the desired hetero-atom.

Paal–Knorr synthesis It is a useful and straightforward method for the synthesis of five-membered heterocyclic compounds, e.g. pyrrole, furan and thiophene. However, necessary precursors, e.g. dicarbonyl compounds, are not readily available. Ammonia, primary amines, hydroxylamines or hydrazines are used as the nitrogen component for the synthesis of pyrrole.

RNH_2

Diketo compound → Substituted pyrrole

Paal–Knorr synthesis can also be used to synthesize furan and thiophene ring systems. A simple dehydration of a 1,4-dicarbonyl compound provides the furan system, whereas thiophene or substituted thiophenes can be prepared by treating 1,4-dicarbonyl compounds with hydrogen sulphide (H_2S) and hydrochloric acid (HCl).

H_3PO_4; $-H_2O$

Substituted dihydrofuran system → Substituted furan system

H_2S/HCl

Diketo compound → Substituted thiphene

Commercial preparation of pyrrole, furan and thiophene Pyrrole is obtained commercially from coal tar or by treating furan with NH_3 over an alumina catalyst at 400 °C.

NH_3, H_2O; Al_2O_3, 400 °C

Furan → Pyrrole

Furan is synthesized by decarbonylation of furfural (furfuraldehyde), which itself can be prepared by acidic dehydration of the pentose sugars found in oat hulls, corncobs and rice hulls.

$C_5H_{10}O_5$ (Pentose mixture) $\xrightarrow{H_3O^+}$ Furfural $\xrightarrow[280\ °C]{\text{Ni catalyst}}$ Furan + CO

Thiophene is found in small amounts in coal tar, and commercially it is prepared from the cyclization of butane or butadiene with sulphur at 600 °C.

1,3-Butadiene + S $\xrightarrow{600\ °C}$ Thiophene + H_2S

Hantszch synthesis A reaction of an α-haloketone with a β-ketoester and NH_3 or a primary amine yields substituted pyrrole.

α-Ketoester + α-Haloketone $\xrightarrow{RNH_2}$ Substituted pyrrole

Substituted furan can be prepared by using the *Feist–Bénary synthesis*, which is similar to the *Hantszch synthesis* of the pyrrole ring. In this reaction, α-haloketones react with 1,3-dicarbonyl compounds in the presence of pyridine to yield substituted furan.

1,3-Dicarbonyl compound + α-Haloketone $\xrightarrow{\text{Pyridine}}$ Substituted furan

Reactions of pyrrole, furan and thiophene

Pyrrole, furan and thiophene undergo electrophilic substitution reactions. However, the reactivity of this reaction varies significantly among these heterocycles. The ease of electrophilic substitution is usually furan > pyrrole > thiophene > benzene. Clearly, all three heterocycles are more reactive than benzene towards electrophilic substitution. Electrophilic substitution generally occurs at C-2, i.e. the position next to the hetero-atom.

Vilsmeier reaction Formylation of pyrrole, furan or thiophene is carried out using a combination of phosphorus oxychloride ($POCl_3$) and

N, *N*-dimethylformamide (DMF). This reaction proceeds by formation of the electrophilic Vilsmeier complex, followed by electrophilic substitution of the heterocycle. The formyl group is generated in the hydrolytic workup.

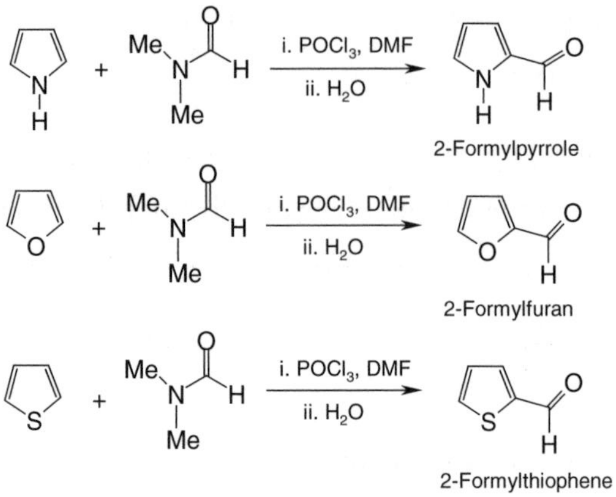

i. $POCl_3$, DMF
ii. H_2O

2-Formylpyrrole

2-Formylfuran

2-Formylthiophene

Mannich reaction Pyrrole and alkyl substituted furan undergo the *Mannich reaction*. Thiophene also undergoes this reaction, but, instead of acetic acid, hydrochloric acid is used.

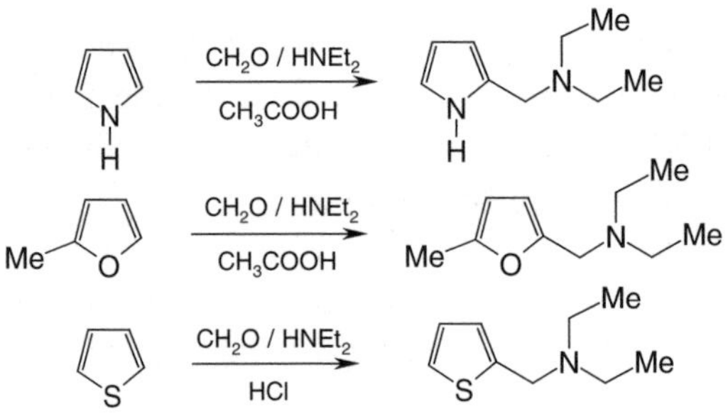

CH_2O / $HNEt_2$
CH_3COOH

CH_2O / $HNEt_2$
HCl

Sulphonation Pyrrole, furan and thiophene undergo sulphonation with the pyridine–sulphur trioxide complex ($C_5H_5N^+SO_3^-$).

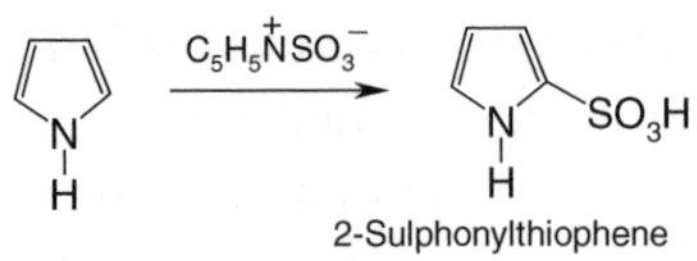

$C_5H_5\overset{+}{N}SO_3^-$

2-Sulphonylthiophene

Nitration Instead of a mixture of nitric acid and sulphuric acid, nitration of these three heterocycles is carried out with acetyl nitrate (formed from nitric acid and acetic anhydride). Nitration is in place mainly at one of the carbon atoms next to the hetero-atom.

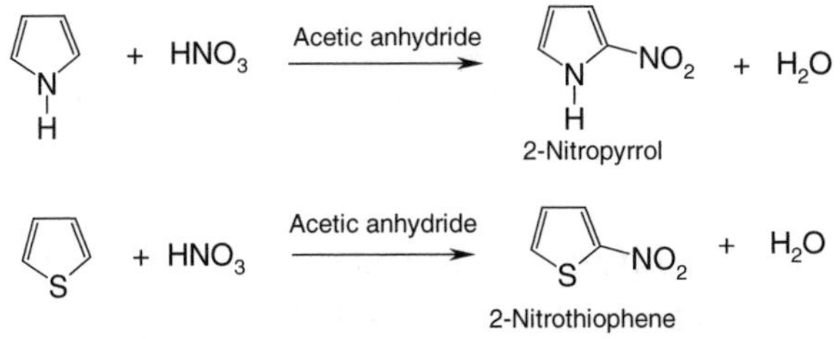

Acetic anhydride

+ HNO_3 → NO_2 + H_2O

2-Nitropyrrol

Acetic anhydride

+ HNO_3 → NO_2 + H_2O

2-Nitrothiophene

Bromination The five-membered aromatic heterocycles are all more reactive toward electrophiles than benzene is, and the reactivity is similar to that of phenol. These compounds undergo electrophilic bromination. However, reaction rates vary considerably, and for pyrrole, furan and thiophene the rates are 5.6×10^8, 1.2×10^2 and 1.00, respectively. While unsubstituted five-membered aromatic heterocycles produce a mixture of bromo-derivatives, e.g. bromothiphenes, substituted heterocycles produce a single product.

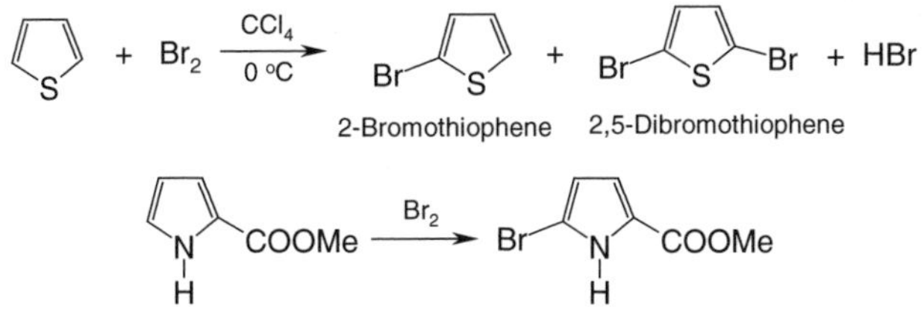

+ Br_2 (CCl_4, 0 °C) → + + HBr

2-Bromothiophene 2,5-Dibromothiophene

COOMe → (Br_2) Br COOMe

FC acylation and alkylation As pyrroles and furans are not stable in the presence of Lewis acids, which are necessary for FC alkylations and acylations, only thiophene, which is stable in Lewis acids, can undergo these reactions. Thiophene reacts with benzoyl chloride in the presence of aluminium chloride to produce phenyl 2-thienyl ketone.

COCl + → ($AlCl_3$, CS_2, 25 °C) Ph + HCl

Phenyl 2-thienyl ketone

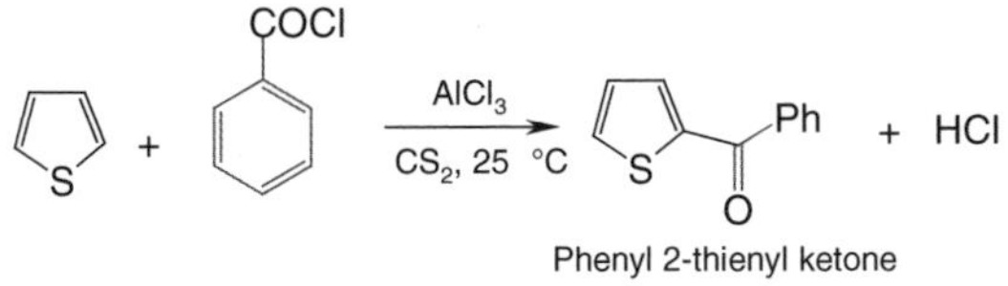

Alkylthiophene reacts with bromothane in the presence of a Lewis acid to bring in 3-ethyl substituent on the ring.

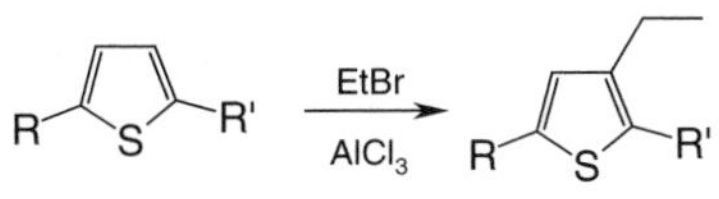

R R' → (EtBr, $AlCl_3$) R R'

Ring opening of substituted furan Furan may be regarded as a cyclic hemi-acetal that has been dehydrated, and is hydrolysed back to a dicarbonyl compound when heated with dilute mineral acid.

2,5-Dimethylfuran + H_2O $\xrightarrow[\Delta]{H_2SO_4,\ CH_3CO_2H}$ 2,5-Hexanedione (86%)

Addition reaction of furan Furan reacts with bromine by 1,4-addition reactions, not electrophilic substitution. When this reaction is carried out in methanol (MeOH), the isolated product is formed by solvolysis of the intermediate dibromide.

Furan + Br_2 $\xrightarrow[\text{Benzene, } -5\ ^{\circ}C]{Na_2CO_3,\ MeOH}$ [dibromide] $\xrightarrow{MeOH}$ 2,5-Dimethoxy-2,5-dihydrofuran

Catalytic hydrogenation of furan Catalytic hydrogenation of furan with a palladium catalyst gives tetrahydrofuran, which is a clear, low-viscosity liquid with a diethyl-ether-like smell.

Furan $\xrightarrow[Pd\text{-}C]{H_2}$ Tetrahydrofuran

4.7.5 Pyridine

Pyridine (C_5H_5N) is a nitrogen-containing unsaturated six-membered heterocyclic aromatic compound. It is similar to benzene, and conforms to Hückel's rule for aromaticity. Pyridine, a tertiary amine, has a lone pair of electrons instead of a hydrogen atom, but the six π electrons are essentially the same as benzene. A number of drug molecules possess pyridine or a modified pyridine skeleton in their structures, e.g. the antihypertensive drug amlodipine and the antifungal drug pyridotriazine.

Amlodipine
An antihypertensive agent

Pyridotriazine
An antifungal drug

Physical properties of pyridine

Pyridine is a liquid (b.p. 115 °C) with an unpleasant smell. It is a polar aprotic solvent and is miscible with both water and organic solvents. The dipole moment of pyridine is 1.57 D. Pyridine is an excellent donor ligand in metal complexes. It is highly aromatic and moderately basic in nature, with a pK_a 5.23, i.e. a stronger base than pyrrole but weaker than alkylamines. The lone pair of electrons on the nitrogen atom in pyridine is available for bonding without interfering with its aromaticity. Protonation of pyridine results in a pyridinium ion ($pK_a = 5.16$), which is a stronger acid than a typical ammonium ion (e.g. piperinium ion, $pK_a = 11.12$), because the acidic hydrogen of a pyridinium ion is attached to an sp^2-hybridized nitrogen that is more electronegative than an sp^3-hybridized nitrogen.

Pyridinium ion ⇌ Pyridine + H^+ Piperidinium ion ⇌ Piperidine + H^+

Preparation of pyridine

Among the methods available for the synthesis of the pyridine system, *Hantzsch synthesis* is probably the most important and widely used synthetic route. However, the pyridine ring can be synthesized from the reaction between pentan-2,4-dione and ammonium acetate. Cyclization of 1,5-diketones is also considered as a convenient method for the synthesis of corresponding pyridine derivatives. Commercially, pyridine is obtained from distillation of coal tar.

Hantzsch synthesis The reaction of 1,3-dicarbonyl compounds with aldehydes and NH_3 provides a 1,4-dihydropyridine, which can be aromatized by oxidation with nitric acid or nitric oxide. Instead of NH_3, primary amine can be used to give 1-substituted 1,4-dihydropyridines.

Methylacetoacetate —(Benzaldehyde, NH_3)→ 1,4-dihydropyridine derivative —(HNO_3)→ Substituted pyridine

Cyclization of 1,5-diketones The reaction between 1,5-diketones and NH_3 produces dihydropyridine systems, which can easily be oxidized to pyridines.

1,5-Diketone + NH_3 $\xrightarrow{-H_2O}$ Dihydropyridine system $\xrightarrow[-2H]{[O]}$ Pyridine system

Reactions of pyridine

Electrophilic substitutions Pyridine's electron-withdrawing nitrogen causes the ring carbons to have significantly less electron density than the ring carbons of benzene. Thus, pyridine is less reactive than benzene towards electrophilic aromatic substitution. However, pyridine undergoes some electrophilic substitution reactions under drastic conditions, e.g. high temperature, and the yields of these reactions are usually quite low. The main substitution takes place at C-3.

Pyridine $\xrightarrow[300\ ^\circ C]{Br_2,\ FeBr_3}$ 3-Bromopyridine (30%)

Pyridine $\xrightarrow[230\ ^\circ C]{H_2SO_4}$ Pyridine-3-sulphonic acid (71%)

Pyridine $\xrightarrow[300\ ^\circ C]{HNO_3,\ H_2SO_4}$ 3-Nitropyridine (22%)

Nucleophilic aromatic substitutions Pyridine is more reactive than benzene towards nucleophilic aromatic substitutions because of the presence of electron-withdrawing nitrogen in the ring. Nucleophilic aromatic substitutions of pyridine occur at C-2 (or C-6) and C-4 positions.

Pyridine + $NaNH_2$ $\xrightarrow[\Delta]{\text{Toluene}}$ 2-Aminopyridine + $H_2\uparrow$

These nucleophilic substitution reactions are rather facile when better leaving groups, e.g. halide ions, are present. Reaction occurs by addition of the nucleophile to the C=N bond, followed by loss of halide ion from the anion intermediate.

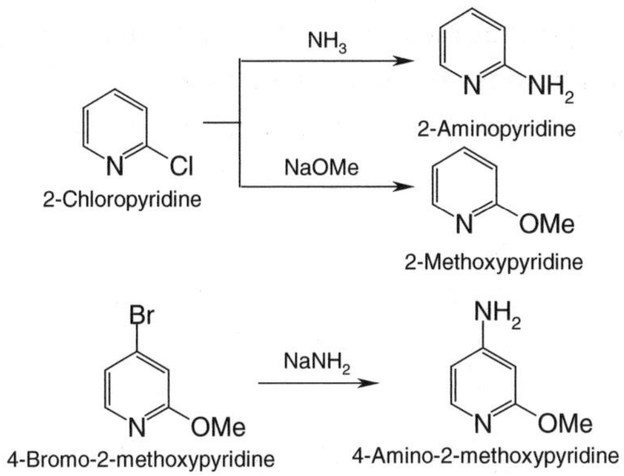

NH_3

2-Aminopyridine

NaOMe

2-Chloropyridine

2-Methoxypyridine

$NaNH_2$

4-Bromo-2-methoxypyridine

4-Amino-2-methoxypyridine

Reactions as an amine Pyridine is a tertiary amine, and undergoes reactions characteristic to tertiary amines. For example, pyridine undergoes S_N2 reactions with alkyl halides, and it reacts with hydrogen peroxide to form an *N*-oxide.

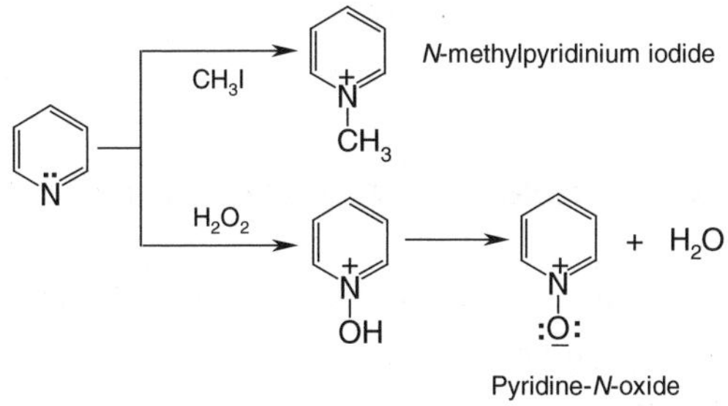

CH_3I

N-methylpyridinium iodide

H_2O_2

$+ H_2O$

Pyridine-*N*-oxide

4.7.6 Oxazole, imidazole and thiazole

Oxazole, imidazole and thiazole systems contain a five-membered ring and two hetero-atoms, one of which is a nitrogen atom. The hetero-atoms are separated by a carbon atom in the ring. The second hetero-atoms are oxygen, nitrogen and sulphur for oxazole, imidazole and thiazole systems, respectively.

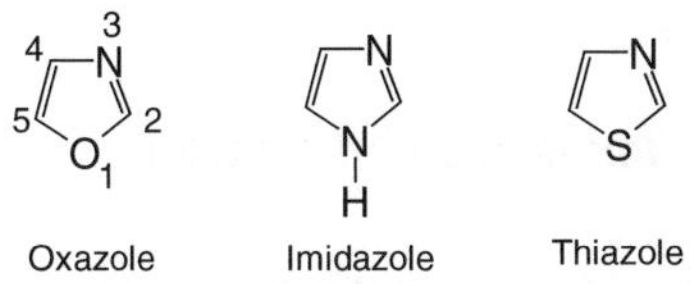

Oxazole Imidazole Thiazole

These compounds are isomeric with the 1,2-azoles, e.g. isoxazole, pyrazole and isothiazole. The aromatic characters of the oxazole, imidazole and thiazole systems arise from delocalization of a lone pair of electrons from the second hetero-atom.

Histamine, an important mediator of inflammation, gastric acid secretion and other allergic manifestations, contain an imidazole ring system. Thiamine, an essential vitamin, possesses a quaternized thiazole ring.

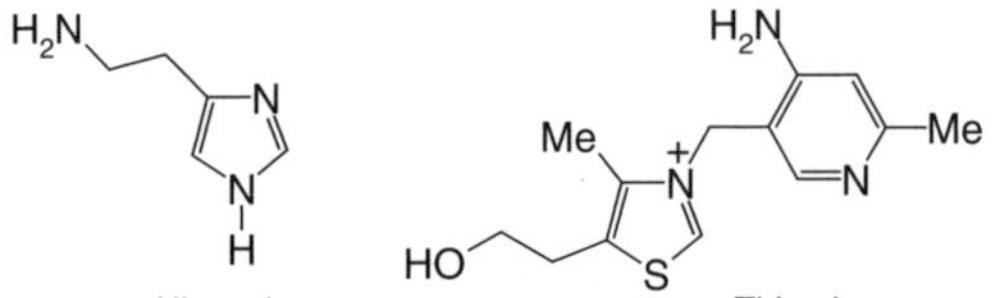

Histamine Thiamine

Apart from some plant and fungal secondary metabolites, the occurrence of oxazole ring system in nature is rather limited. However, the following anti-inflammatory drug contains an oxazole ring system.

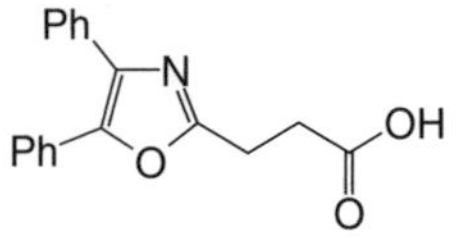

An anti-inflammatory drug

Physical properties of oxazole, imidazole and thiazole

Among these 1,3-azoles, imidazole is the most basic compound. The increased basicity of imidazole can be accounted for from the greater electron-releasing ability of two nitrogen atoms relative to a nitrogen atom and a hetero-atom of higher electronegativity. Some of the physical properties of these compounds are presented below.

1,3-azoles	pK_a	b.p. (°C)	Water solubility	Physical state
Oxazole	0.8	69–70	Sparingly soluble	Clear to pale yellow liquid
Imidazole	7.0	255–256	Soluble	Clear to pale yellow crystalline flake
Thiazole	2.5	116–118	Sparingly soluble	Clear to pale yellow liquid

Preparation of oxazole, imidazole and thiazole

Preparation of oxazole Cyclocondensation of amides, through dehydration, leads to the formation of corresponding oxazoles. This synthesis is known as *Robinson-Gabriel synthesis*. A number of acids or acid anhydrides, e.g. phosphoric acid, phosphorus oxychloride, phosgene and thionyl chloride, can bring about this dehydration.

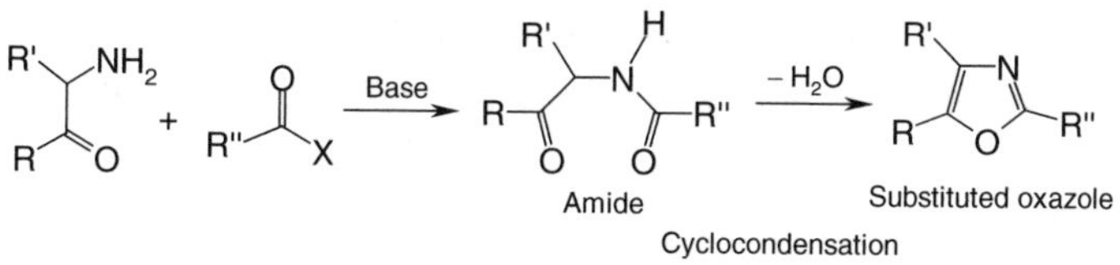

Amide

Substituted oxazole

Cyclocondensation

Preparation of imidazole The condensation of a 1,2-dicarbonyl compound with ammonium acetate and an aldehyde results in the formation of an imidazole skeleton.

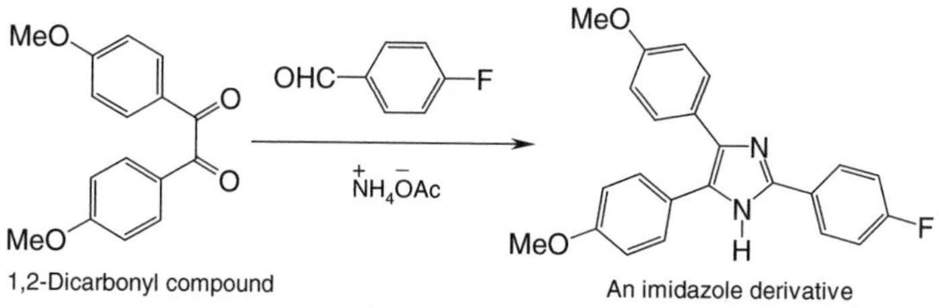

1,2-Dicarbonyl compound

An imidazole derivative

Preparation of thiazole *Hantzsch synthesis* can be applied to synthesize the thiazole system from thioamides. The reaction involves initial nucleophilic attack by sulphur followed by a cyclocondensation.

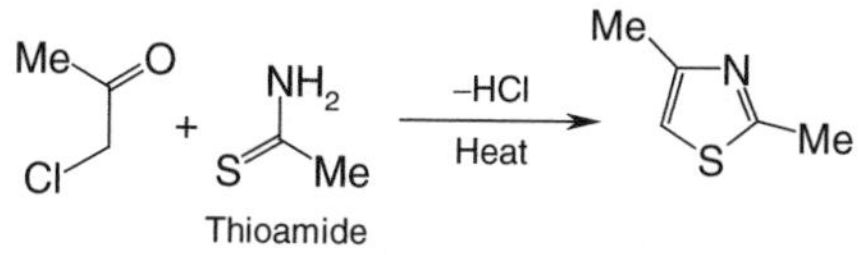

Thioamide

A modification of the above method involves the use of thiourea instead of a thioamide.

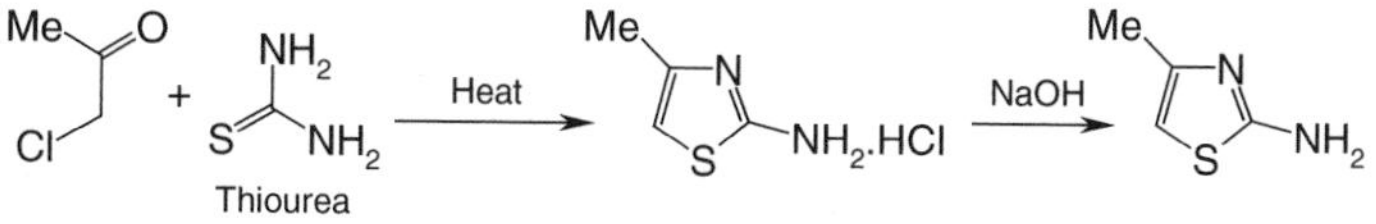

Thiourea

Reactions of oxazole, imidazole and thiazole

The presence of the pyridine-like nitrogen deactivates the 1,3-azoles toward electrophilic attack, and increases their affinity towards nucleophilic attack.

Electrophilic substitutions Although oxazole, imidazole and thiazoles are not very reactive towards aromatic electrophilic substitution reactions, the presence of any electron-donating group on the ring can facilitate electrophilic substitution. For example, 2-methoxythiazole is more reactive than thiazole itself. Some examples of electrophilic substitutions of oxazole, imidazole and thiazoles and their derivatives are presented below.

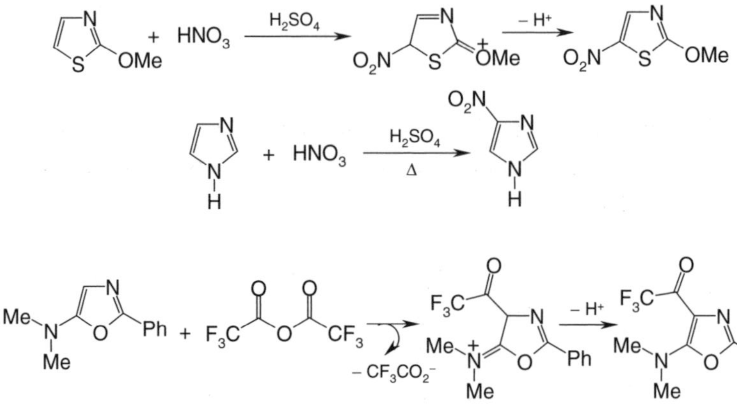

Nucleophilic aromatic substitutions 1,3-azoles are more reactive than pyrrole, furan or thiaphene towards nucleophilic attack. Some examples of nucleophilic aromatic substitutions of oxazole, imidazole and thiazoles and their derivatives are given below. In the reaction with imidazole, the presence of a nitro-group in the reactant can activate the reaction because the nitro-group can act as an electron acceptor.

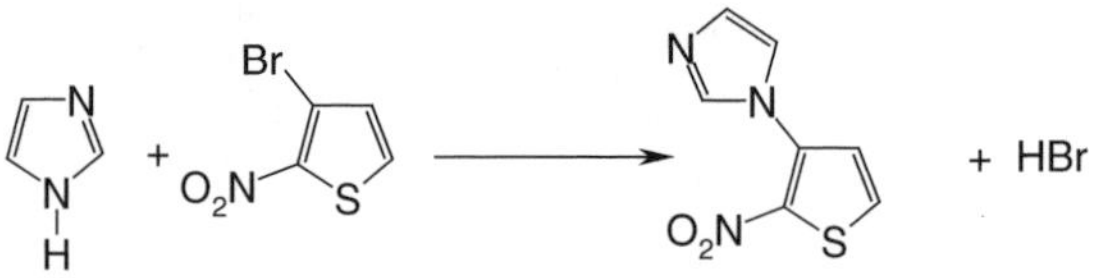

No activation is required for 2-halo-1,3-azoles, which can undergo nucleophilic aromatic substitutions quite easily.

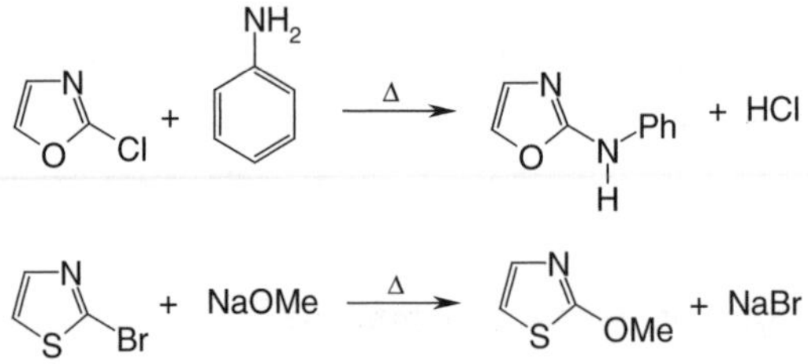

4.7.7 Isoxazole, pyrazole and isothiazole

Isoxazole, pyrazole and isothiazole constitute the 1,2-azole family of heterocycles that contain two hetero-atoms, one of which is a nitrogen atom. The second hetero-atom is oxygen, nitrogen or sulphur, respectively, for isoxazole, pyrazole and isothiazole. The aromaticity of these compounds is due to the delocalization of a lone pair of electrons from the second hetero-atom to complete the aromatic sextet.

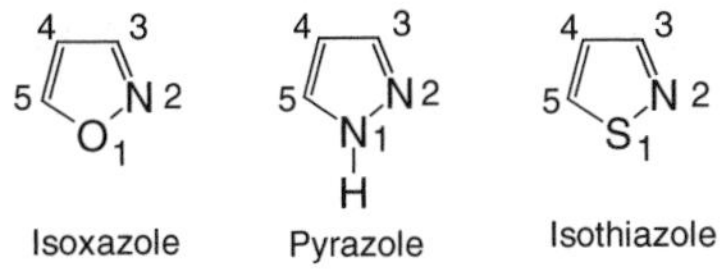

Isoxazole Pyrazole Isothiazole

The 1,2-azole family of heterocycles is important in medicine. For example, the following drug used in the treatment of bronchial asthma possesses a substituted isoxazole system.

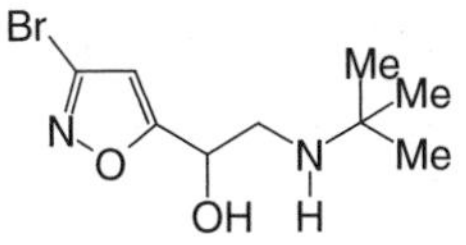

Physical properties of isoxazole, pyrazole and isothiazole

The 1,2-azoles are basic compounds because of the lone pair of electrons on the nitrogen atom, which is available for protonation. However, these compounds are much less basic than their isomers, 1,3-azoles, owing to the electron-withdrawing effect of the adjacent hetero-atom. Some of the physical properties of these compounds are as follows.

1,2-azoles	pK_a	b.p. (°C)	m.p. (°C)	Physical state
Isoxazole	−2.97	95	–	Liquid
Pyrazole	2.52	186–188	60–70	Solid
Isothiazole	–	114	–	Liquid

Preparation of isoxazole, pyrazole and isothiazole

Isoxazole and pyrazole synthesis While 1,3-diketones undergo condensation with hydroxylamine to produce isoxazoles, with hydrazine they yield corresponding pyrazoles.

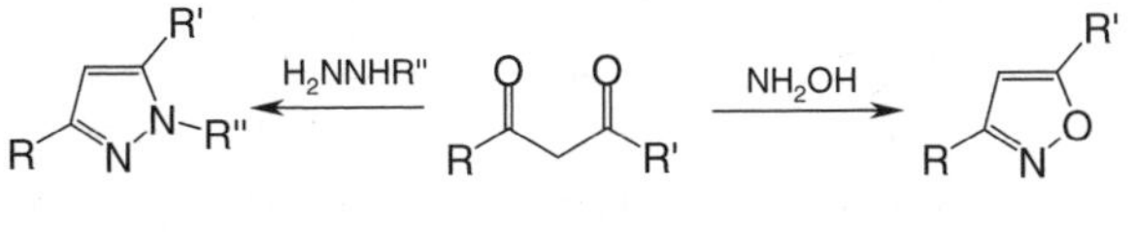

H_2NNHR'' NH_2OH

Isothiazole synthesis Isothiazole can be prepared from thioamide in the following way.

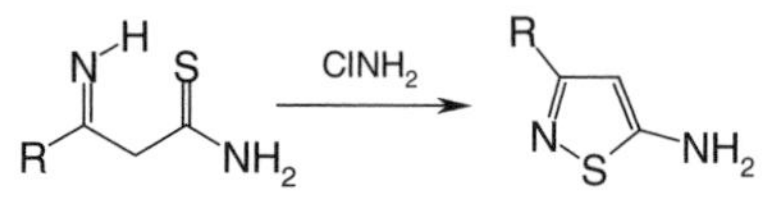

Reactions of isoxazole, pyrazole and isothiazole

Like 1,3-azoles, due to the presence of a pyridine-like nitrogen atom in the ring, 1,2-azoles are also much less reactive towards electrophilic substitutions than furan, pyrrole or thiophene. However, 1,2-azoles undergo electrophilic substitutions under appropriate reaction conditions, and the main substitution takes place at the C-4 position, for example bromination of 1,2-azoles. Nitration and sulphonation of 1,2-azoles can also be carried out, but only under vigorous reaction conditions.

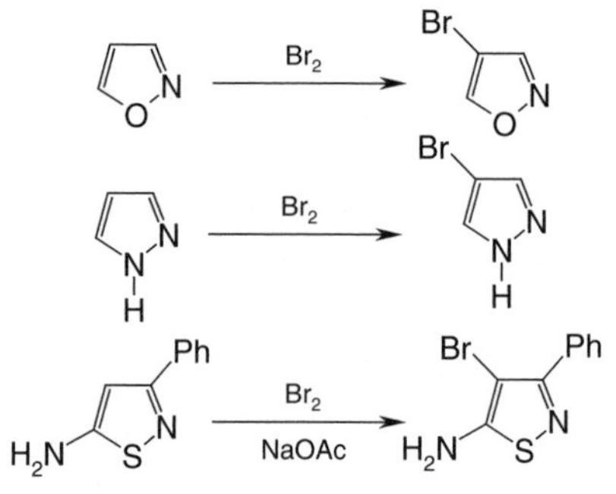

4.7.8 Pyrimidine

Pyrimidine is a six-membered aromatic heterocyclic compound that contains two nitrogen atoms, separated by a carbon atom, in the ring. Nucleic acids, DNA and RNA, contain substituted purines and pyrimidines. Cytosine, uracil, thymine and alloxan are just a few of the biologically significant modified pyrimidine compounds, the first three being the components of the nucleic acids.

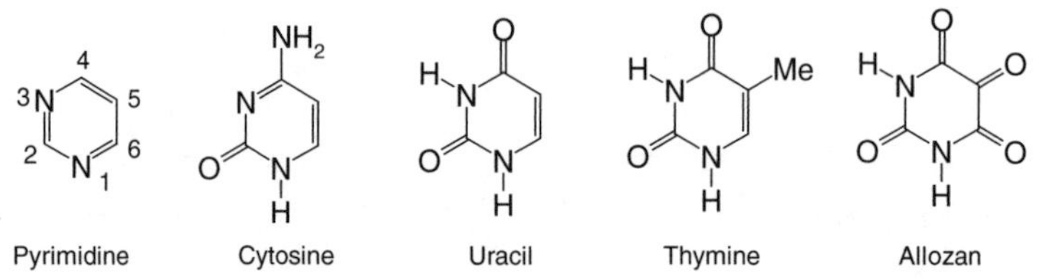

Pyrimidine Cytosine Uracil Thymine Allozan

A number of drug molecules contain a modified pyrimidine skeleton, the best known examples being the anticancer drug 5-fluorouracil, which is structurally similar to thymine, the antiviral drug AZT, currently being used in the treatment of AIDS, and phenobarbital, a well known sedative.

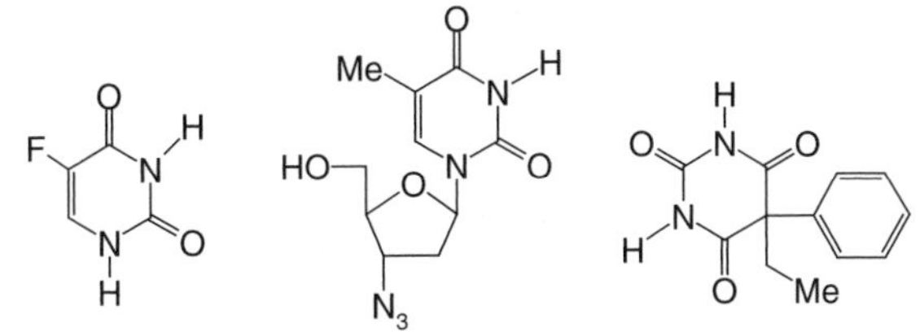

5-Fluorouracil
An anticancer drug

AZT
An antiviral drug

Phenobarbital
A sedative

Two positional isomers of pyrimidine are pyridiazine and pyrazine, which only differ structurally from pyrimidine in terms of the position of the nitrogen atoms in the ring. These three heterocycles together with their derivatives are known as *diazines*.

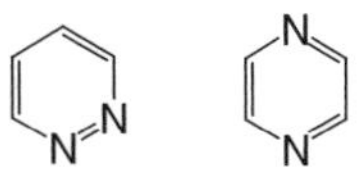

Pyridiazine   Pyrazine

Physical properties of pyrimidine

Pyrimidine is a weaker base than pyridine because of the presence of the second nitrogen. Its conjugate acid is a much stronger acid ($pK_a = 1.0$). The pK_a values of the *N*-1 hydrogen in uracil, thymine and cytosine are 9.5, 9.8 and 12.1, respectively. Pyrimidine is a hygroscopic solid (b.p. 123–124 °C, m.p. 20–22 °C) and soluble in water.

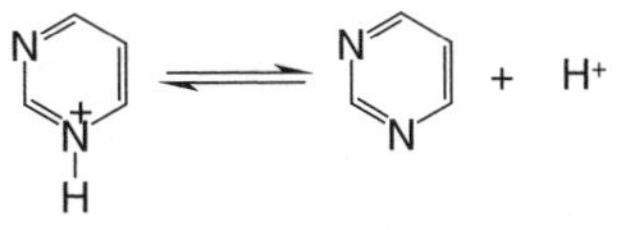

Conjugate acid of pyrimidine

Preparation of pyrimidine

The combination of bis-electrophilic and bis-nucleophilic components is the basis of general pyrimidine synthesis. A reaction between an amidine (urea or thiourea or guanidine) and a 1,3-diketo compound produces corresponding pyrimidine systems. These reactions are usually facilitated by acid or base catalysis.

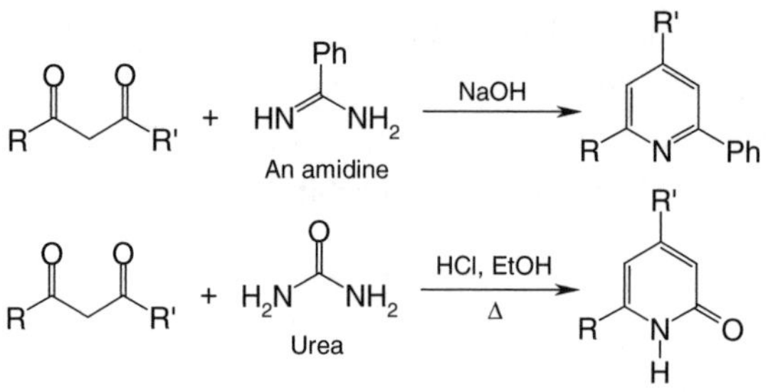

NaOH

An amidine

HCl, EtOH

Δ

Urea

Reactions of pyrimidine

Electrophilic aromatic substitutions The chemistry of pyrimidine is similar to that of pyridine with the notable exception that the second nitrogen in the aromatic ring makes it less reactive towards electrophilic substitutions. For example, nitration can only be carried out when there are two ring-activating substituents present on the pyrimidine ring (e.g. 2,4-dihydroxypyrimidine or uracil). The most activated position towards electrophilic substitution is C-5.

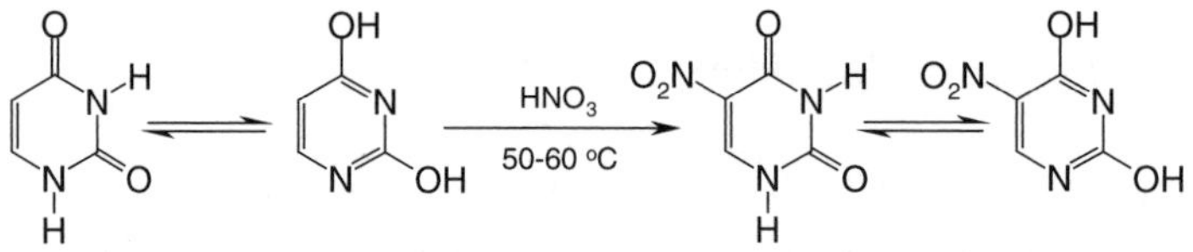

HNO_3

50-60 °C

Keto-enol tautomeric forms of uracil

2,4-Dihydroxy-5-nitropyrimidine (5-Nitrouracil)

Nucleophilic aromatic substitutions Pyrimidine is more reactive than pyridine towards nucleophilic aromatic substitution, again due to the presence of the second electron-withdrawing nitrogen in the pyrimidine ring. Leaving groups at C-2, C-4 or C-6 positions of pyrimidine can be displaced by nucleophiles.

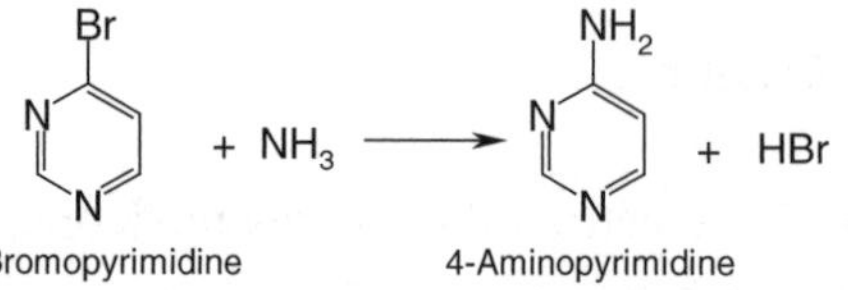

4-Bromopyrimidine + NH_3 → 4-Aminopyrimidine + HBr

2-Chloropyrimidine + NaOMe → 4-Aminopyrimidine + NaCl

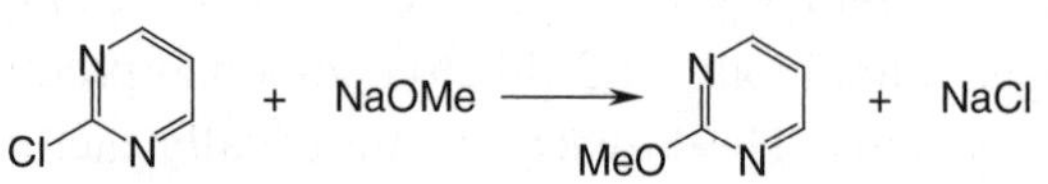

4.7.9 Purine

Purine contains a pyrimidine ring fused with an imidazole nucleus. Guanine and adenine are two purine bases that are found in nucleic acids, DNA and RNA.

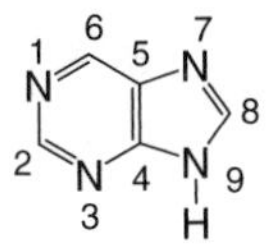

Purine 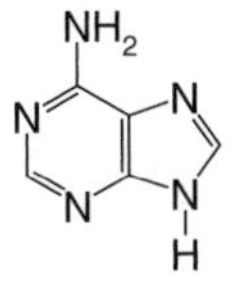 Adenine 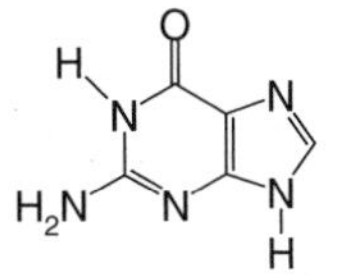 Guanine

Several purine derivatives are found in nature, e.g. xanthine, hypoxanthine and uric acid. The pharmacologically important (CNS-stimulant) xanthine alkaloids, e.g. caffeine, theobromine and theophylline, are found in tea leaves, coffee beans and coco. The actual biosynthesis of purines involves construction of a pyrimidine ring onto a pre-formed imidazole system.

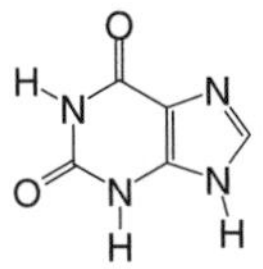

Xanthine 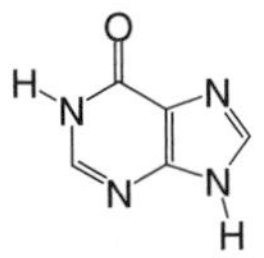 Hypoxanthine 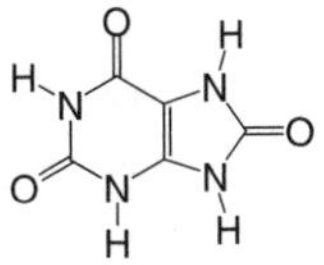 Uric acid

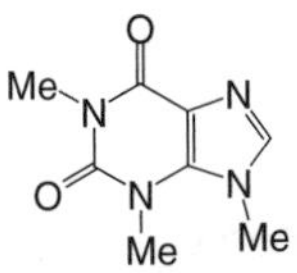

Caffeine 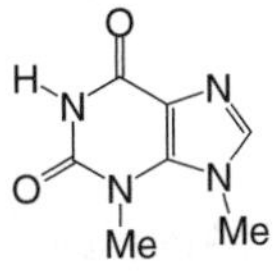 Theobromine 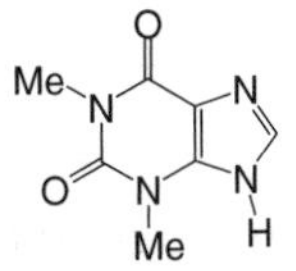 Theophylline

The purine and pyrimidine bases play an important role in the metabolic processes of cells through their involvement in the regulation of protein synthesis. Thus, several synthetic analogues of these compounds are used to interrupt the cancer cell growth. One such example is an adenine mimic, 6-mercaptopurine, which is a well known anticancer drug.

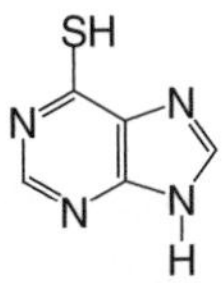

6-Mercaptopurine
An anticancer drug

Physical properties of purine

Purine is a basic crystalline solid (m.p. 214 °C). As it consists of a pyrimidine ring fused to an imidazole ring, it possesses the properties of both rings. The electron-donating imidazole ring makes the protonated pyrimidine part less acidic (p$K_a = 2.5$) than unsubstituted protonated pyrimidine (p$K_a = 1.0$). On the other hand, the electron-withdrawing pyrimidine ring makes hydrogen on *N*-9 (p$K_a = 8.9$) more acidic than the corresponding *N*-1 hydrogen of imidazole (p$K_a = 14.4$).

$$\text{H-N}^+\text{(protonated purine)} \rightleftharpoons \text{purine} + H^+ \rightleftharpoons \text{purine anion} + H^+$$

Reactions of purine

Nucleophilic substitutions Aminopurines react with dilute nitrous acid to yield the corresponding hydroxy compounds.

$$\text{Adenine} \xrightarrow[\text{AcOH, } H_2O,\ \text{AcONa},\ 0\,^{\circ}C]{NaNO_2} \text{6-hydroxypurine}$$

Deamination of aminopurines Adenine undergoes deamination to produce hypoxanthine, and guanine is deaminated to xanthine.

$$\text{Adenine} \xrightarrow[H_2O]{H^+} \text{Hypoxanthine} \qquad \text{Guanine} \xrightarrow[H_2O]{H^+} \text{Xanthine}$$

Oxidation of xanthine and hypoxanthine Xanthine and hypoxanthine can be oxidized enzymatically with xanthine oxidase to produce uric acid.

$$\text{Hypoxanthine} \xrightarrow{\text{Xanthine oxidase}} \text{Uric acid} \xleftarrow{\text{Xanthine oxidase}} \text{Xanthine}$$

4.7.10 Quinoline and isoquinoline

Quinoline and isoquinoline, known as benzopyridines, are two isomeric heterocyclic compounds that have two rings, a benzene and a pyridine ring, fused together. In quinoline this fusion is at C2/C3, whereas in isoquinoline this is at C3/C4 of the pyridine ring. Like benzene and pyridine, these benzopyridines are also aromatic in nature.

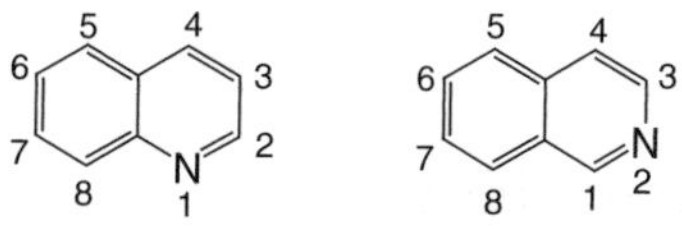

Quinoline

Isoquinoline

A number of naturally occurring pharmacologically active alkaloids possess quinoline and isoquinoline skeleton. For examples, papaverine from *Papaver somniferum* is an isoquinoline alkaloid and quinine from *Cinchona* barks is a quinoline alkaloid that has antimalarial properties.

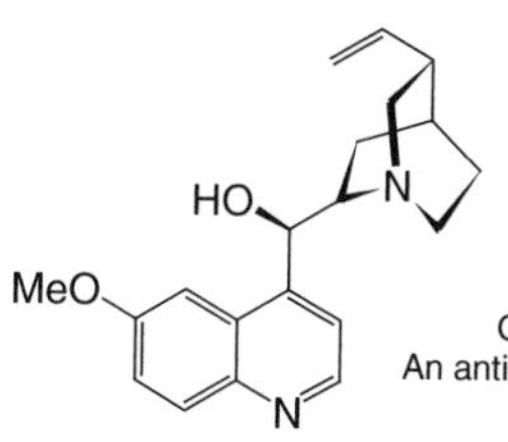

Quinine
An antimalarial drug

Physical properties of quinoline and isoquinoline

Quinoline and isoquinoline are basic in nature. Like pyridine, the nitrogen atom of quinoline and isoquinoline is protonated under the usual acidic conditions. The conjugate acids of quinoline and isoquinoline have similar pK_a values (4.85 and 5.14, respectively) to that of the conjugate acid of pyridine.

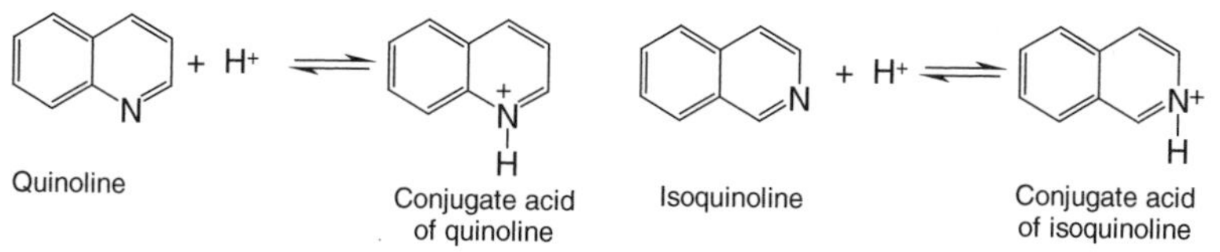

Quinoline + H^+ ⇌ Conjugate acid of quinoline

Isoquinoline + H^+ ⇌ Conjugate acid of isoquinoline

Quinoline, when exposed to light, forms first a yellow liquid, and slowly a brown liquid. It is only slightly soluble in water but dissolves readily in many organic solvents. Isoquinoline crystallizes to platelets and is sparingly soluble in water but dissolves well in ethanol, acetone, diethyl ether, carbon disulphide and other common organic solvents. It is also soluble in dilute acids as the protonated derivative. Some other physical properties of these compounds are shown below.

	b.p. (°C)	m.p. (°C)	Physical state
Quinoline	238	−15.0	A colourless hygroscopic liquid with a strong odour
Isoquinoline	242	26–28	A colourless hygroscopic liquid at room temperature with a penetrating, unpleasant odour.

Preparation of quinoline and isoquinoline

Quinoline synthesis *Skraup synthesis* is used to synthesize the quinoline skeleton by heating aniline with glycerol, using sulphuric acid as a catalyst and dehydrating agent. Ferrous sulphate is often added as a moderator, as the reaction can be violently exothermic. The most likely mechanism of this synthesis is that glycerol is dehydrated to acrolein, which undergoes conjugate addition to the aniline. This intermediate is then cyclized, oxidized and dehydrated to give the quinoline system.

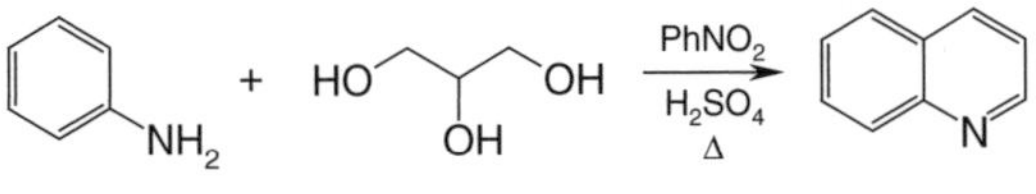

NH_2 + HO OH OH $\xrightarrow[H_2SO_4,\ \Delta]{PhNO_2}$ N

A modified version of *Friedlnder synthesis* utilizing a 2-nitroaryl carbonyl compound is sometimes used to synthesize quinoline skeleton. *Friedlnder synthesis* itself is somewhat complicated because of the difficulty in preparing the necessary 2-aminoaryl carbonyl compounds.

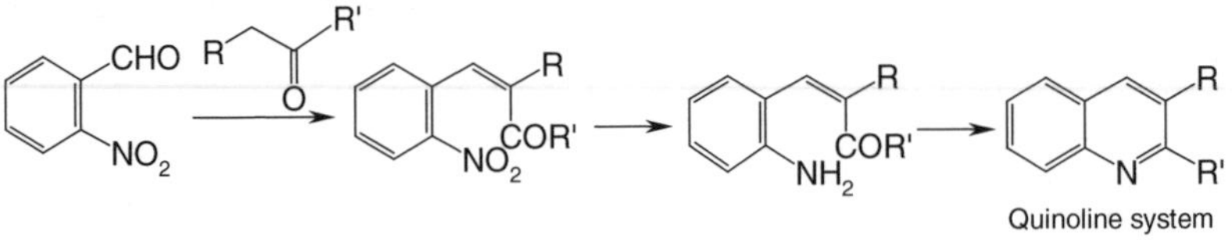

CHO NO_2 R R' O ⟶ R COR' NO_2 ⟶ R COR' NH_2 ⟶ R N R'

Quinoline system

Isoquinoline synthesis *Bischler–Napieralski synthesis* is used to synthesize isoquinolines. β-phenylethylamine is acylated, and then cyclodehydrated using phosphoryl chloride, phosphorus pentoxide or other Lewis acids to yield dihydroisoquinoline, which can be aromatized by dehydrogenation with palladium, for example in the synthesis of papaverine, a pharmacologically active isoquinoline alkaloid.

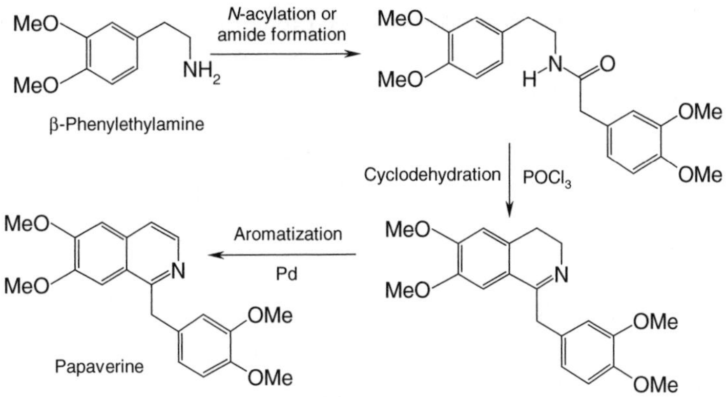

Pictet–Spengler synthesis is another method of preparing isoquinolines. β-phenylethylamine reacts with an aldehyde to produce an imine, which undergoes acid-catalysed cyclization, resulting in the synthesis of the tetrahydroisoquinoline system. Again, tetrahydroisoquinoline can be aromatized by palladium dehydrogenation to produce an isoquinoline system.

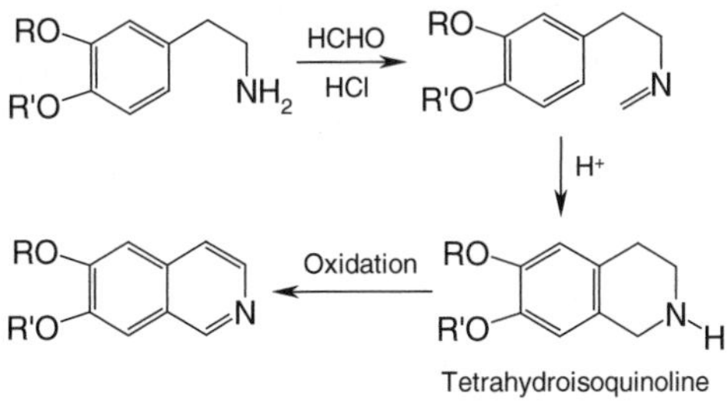

Reactions of quinoline and isoquinoline

Electrophilic aromatic substitutions Quinoline and isoquinoline undergo electrophilic aromatic substitution on the benzene ring, because a benzene ring is more reactive than a pyridine ring towards such reaction. Substitution generally occurs at C-5 and C-8, e.g. bromination of quinoline and isoquinoline.

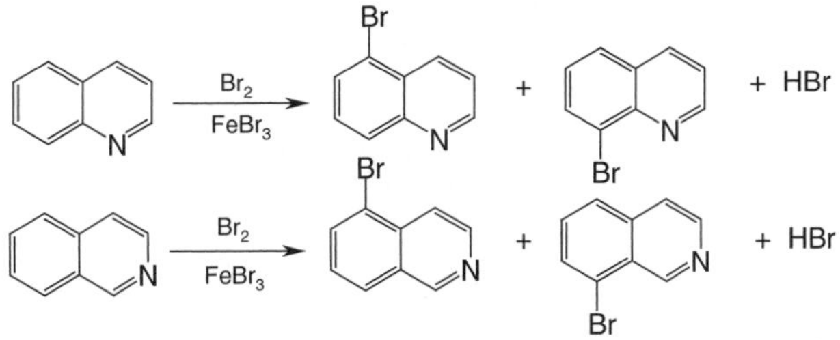

Nucleophilic substitutions Nucleophilic substitutions in quinoline and isoquinoline occur on the pyridine ring because a pyridine ring is more reactive than a benzene ring towards such reaction. While this substitution takes place at C-2 and C-4 in quinoline, isoquinoline undergoes nucleophilic substitution only at C-1.

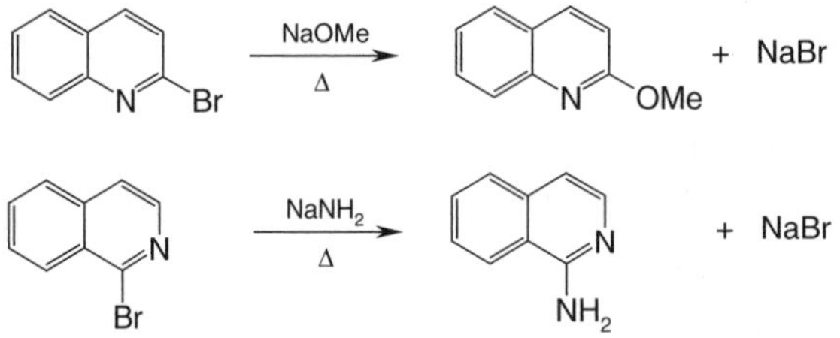

2-Bromoquinoline $\xrightarrow[\Delta]{NaOMe}$ 2-methoxyquinoline (OMe) + NaBr

1-Bromoisoquinoline $\xrightarrow[\Delta]{NaNH_2}$ 1-aminoisoquinoline (NH_2) + NaBr

4.7.11 Indole

Indole contains a benzene ring fused with a pyrrole ring at C-2/C-3, and can be described as benzopyrrole. Indole is a ten π electron aromatic system achieved from the delocalization of the lone pair of electrons on the nitrogen atom. Benzofuran and benzothiaphene are very similar to benzopyrrole (indole), with different hetero-atoms, oxygen and sulphur respectively.

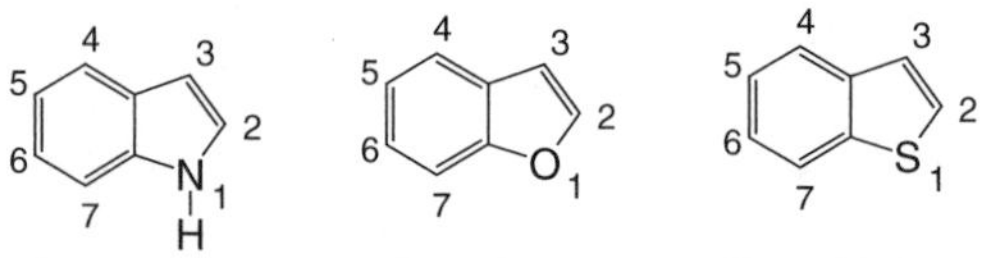

Benzopyrrole
Indole

Benzofuran

Benzothiophene

The indole group of compounds is one of the most prevalent groups of alkaloids found in nature. A number of important pharmacologically active medicinal products and potential drug candidates contain the indole system. For example, serotonin, a well known neurotransmitter, has a substituted indole system.

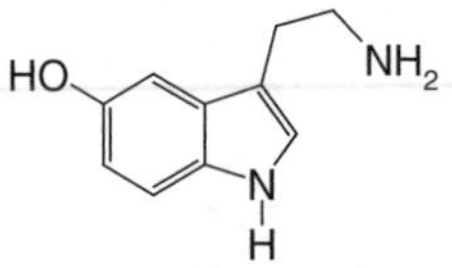

Serotonin
(5-Hydroxytryptamine)
A neurotransmitter

Physical properties of indole

Indole is a weakly basic compound. The conjugate acid of indole is a strong acid ($pK_a = -2.4$). Indole is a white solid (b.p. 253–254 °C, m.p. 52–54 °C)

at room temperature, and posseses an intense faecal smell. However, at low concentrations it has a flowery smell. Indole is slightly soluble in water, but readily soluble in organic solvents, e.g. ethanol, ether and benzene.

Preparation of indole

Fischer indole synthesis Cyclization of arylhydrazones by heating with an acid or Lewis acid catalyst yields an indole system. The most commonly used catalyst is $ZnCl_2$. The disadvantage of this reaction is that unsymmetrical ketones give mixtures of indoles if R′ also has an α-methylene group.

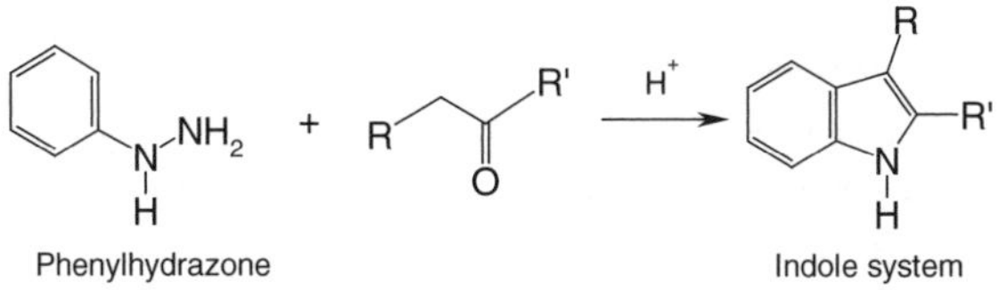

NH_2 N H + R R' O H^+ R R' N H

Phenylhydrazone

Indole system

Leimgruber synthesis Aminomethylenation of nitrotoluene followed by hydrogenation yield indole.

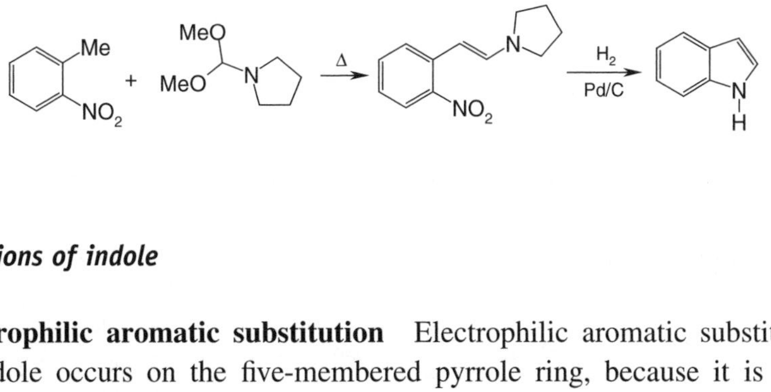

Me NO_2 + MeO MeO N Δ N NO_2 H_2 Pd/C N H

Reactions of indole

Electrophilic aromatic substitution Electrophilic aromatic substitution of indole occurs on the five-membered pyrrole ring, because it is more reactive towards such reaction than a benzene ring. As an electron-rich heterocycle, indole undergoes electrophilic aromatic substitution primarily at C-3, for example bromination of indole.

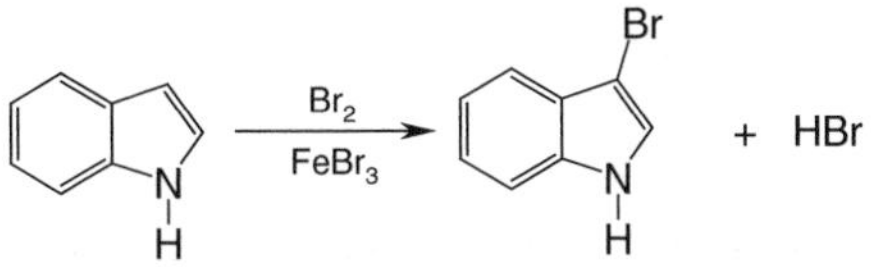

N H Br_2 $FeBr_3$ Br N H + HBr

The *Mannich reaction* is another example of electrophilic aromatic substitution where indole can produce an aminomethyl derivative.

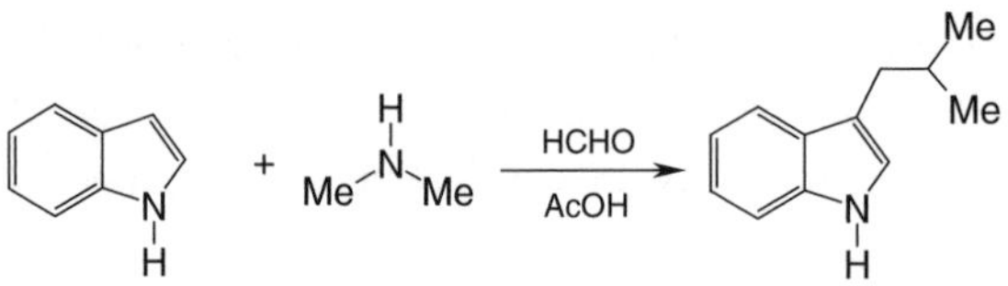

Similarly, using the *Vilsmeier reaction* an aldehyde group can be brought in at C-3 of indole.

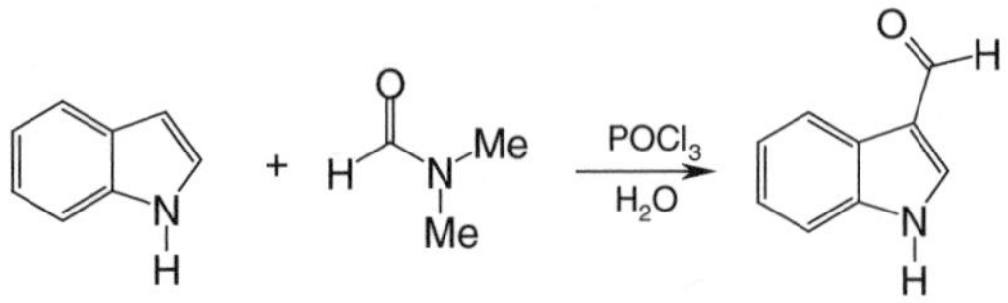

Test for indole Indole is a component of the amino acid tryptophan, which can be broken down by the bacterial enzyme tryptophanase. When tryptophan is broken down, the presence of indole can be detected through the use of *Kovacs' reagent*. Kovacs' reagent, which is yellow, reacts with indole and produces a red colour on the surface of the test tube. Kovacs' reagent is prepared by dissolving 10 g of *p*-aminobenzaldehyde in 150 mL of isoamylalcohol and then slowly adding 50 mL of concentrated HCl.

4.8 Nucleic acids

The nucleic acids, deoxyribonucleic acid (DNA) and ribonucleic acid (RNA), are the chemical carriers of a cell's genetic information. Nucleic acids are biopolymers made of *nucleotides* joined together to form a long chain. These biopolymers are often found associated with proteins, and in this form they are called *nucleoproteins*. Each nucleotide comprises a *nucleoside* bonded to a phosphate group, and each nucleoside is composed of an aldopentose sugar, ribose or 2-deoxyribose, linked to a heterocyclic purine or pyrimidine base (see Section 4.7).

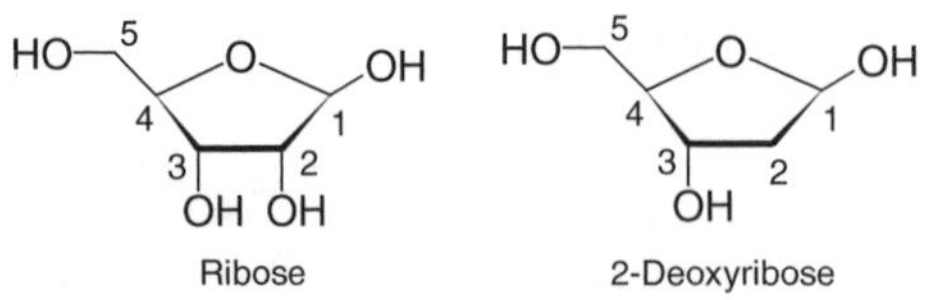

Ribose

2-Deoxyribose

The sugar component in RNA is *ribose*, whereas in DNA it is *2-dexoyribose*. In deoxyribonucleotides, the heterocyclic bases are purine bases, adenine and guanine, and pyrimidine bases, cytosine and thymine. In ribonucleotides, adenine, guanine and cytosine are present, but not thymine, which is replaced by uracil, another pyrimidine base.

In the nucleotides, while the heterocyclic base is linked to C-1 of the sugar through an *N*-glycosidic β-linkage, the phosphoric acid is bonded by a phosphate ester linkage to C-5. When the sugar is a part of a nucleoside, the numbering of sugars starts with 1′, i.e. C-1 becomes C-1′, for example 2′-deoxyadenosine 5′-phosphate and uridine 5′-phosphate.

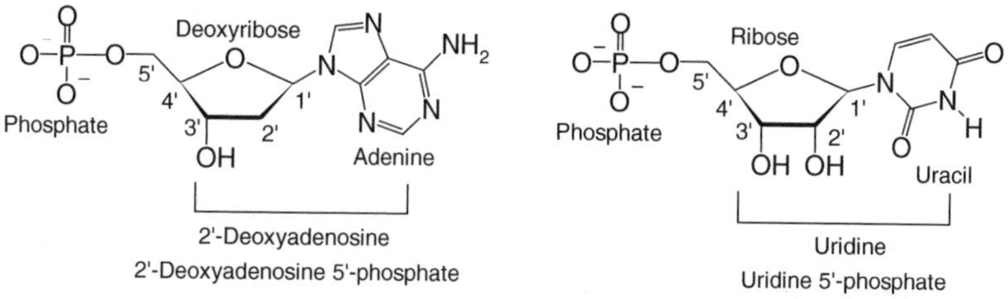

2'-Deoxyadenosine
2'-Deoxyadenosine 5'-phosphate

Uridine
Uridine 5'-phosphate

Despite being structurally similar, DNA and RNA differ in size and in their functions within a cell. The molecular weights of DNA, found in the nucleus of cells, can be up to 150 billion and lengths up to 12 cm, whereas the molecular weight of RNA, found outside the cell nucleus, can only be up to 35 000.

Deoxyribonucleic acid (DNA)	
Name of the nucleotide	*Composition*
2′-Deoxyadenosine 5′-phosphate	Adenine + deoxyribose + phosphate Nucleoside is 2′-deoxyadenosine, composed of adenine and deoxyribose
2′-Deoxyguanosine 5′-phosphate	Guanine + deoxyribose + phosphate Nucleoside is 2′-deoxyguanosine, composed of guanine and deoxyribose
2′-Deoxycytidine 5′-phosphate	Cytosine + deoxyribose + phosphate Nucleoside is 2′-deoxycytidine, composed of cytosine and deoxyribose
2′-Deoxythymidine 5′-phosphate	Thymine + deoxyribose + phosphate Nucleoside is 2′-deoxythymidine, composed of thymine and deoxyribose
Ribonucleic acid (RNA)	
Adenosine 5′-phosphate	Adenine + ribose + phosphate Nucleoside is adenosine, composed of adenine and ribose
Guanosine 5′-phosphate	Guanine + ribose + phosphate Nucleoside is guanosine, composed of guanine and ribose
Cytidine 5′-phosphate	Cytosine + ribose + phosphate Nucleoside is cytidine, composed of cytosine and ribose
Uridine 5′-phosphate	Uracil + ribose + phosphate Nucleoside is uridine, composed of uracil and ribose

4.8.1 Synthesis of nucleosides and nucleotides

A reaction between a suitably protected ribose or 2-deoxyribose and an appropriate purine or pyrimidine base yields a nucleoside. For example, guanosine can be synthesized from a protected ribofuranosyl chloride and a chloromercurieguanine.

Nucleosides can also be prepared through the formation of the heterocyclic base on a protected ribosylamine derivative.

Phosphorylation of nucleosides produces corresponding nucleotides. Phosphorylating agents, e.g. dibenzylphosphochloridate, are used in this reaction. To carry out phosphorylation at C-5′, the other two hydroxyl functionalities at C-2′ and C-3′ have to be protected, usually with an isopropylidine group. At the final step, this protecting group can be removed by mild acid-catalysed hydrolysis, and a hydrogenolysis cleaves the benzylphosphate bonds.

4.8.2 Structure of nucleic acids

Primary structure

Nucleotides join together in DNA and RNA by forming a phosphate ester bond between the 5′-phosphate group on one nucleotide and the 3′-hydroxyl group on the sugar (ribose or 2′-deoxyribose) of another nucleotide. In the nucleic acids, these phosphate ester links provide the nucleic acids with a long unbranched chain with a 'backbone' of sugar and phosphate units with heterocyclic bases sticking out from the chain at regular intervals. One end of the nucleic acid polymer has a free hydroxyl at C-3′ (the *3′-end*), and the other end has a phosphate at C-5′ (the *5′-end*).

The structure of nucleic acids depends on the sequence of individual nucleotides. The actual base sequences for many nucleic acids from various species are available to date. Instead of writing the full name of each nucleotide, abbreviations are used, e.g. A for adenine, T for thymidine, G for guanosine and C for cytidine. Thus, a typical DNA sequence might be presented as TAGGCT.

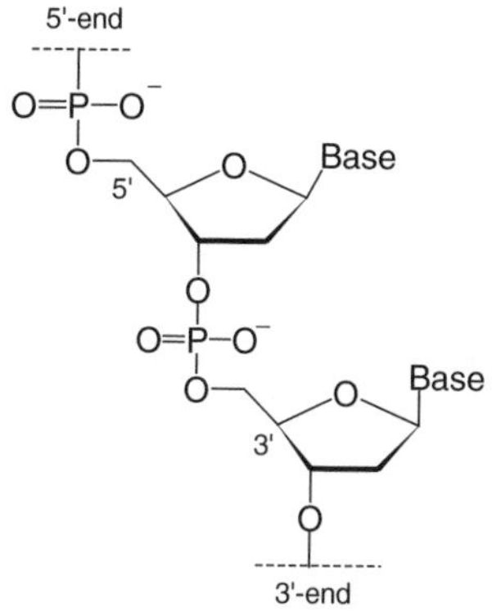

Generalized structure of DNA

Secondary structure: base pairing

The base sequence along the chain of a DNA contains the genetic information. Samples of DNA isolated from different tissues of the same species have the same proportions of heterocyclic bases, but the samples from different species often have different proportions of bases. For example, human thymus DNA comprises 30.9% adenine, 29.4% thymine, 19.9% guanine and 19.8% cytosine, while the bacterium *Staphylococcus aureus* contains 30.8% adenine, 29.2% thymine, 21% guanine and 19% cytosine. In these examples, it is clear that the bases in DNA occur in pairs. Adenine and thymine are usually present in equal amounts; so are cytosine and guanine. In the late 1940s, E. Chargaff pointed out these regularities and summarized as follows.

(a) The total mole percentage of purines is approximately equal to that of the pyrimidines; i.e., (%G + %A)/(%C + %T) ≅ 1.

(b) The mole percentage of adenine is nearly equal to that of thymine, i.e %A/%T ≅ 1, and same is true for guanine and cytosine, i.e. %G/%C ≅ 1.

To provide explanations for some of these earlier findings, the secondary structure of DNA was first proposed by James Watson and Francis Crick in 1953, and was verified shortly thereafter through X-ray crystallographic analysis by Wilkins. According to the Watson–Crick model, DNA consists of two polynucleotide strands coiled around each other in a *double helix* like the handrails on a spiral staircase. The two strands run in opposite directions and are held together by hydrogen bonds between specific pairs of bases. Adenine and thymine form strong hydrogen bonds to each other, but not to cytosine or guanine. Similarly, cytosine and guanine form strong hydrogen bonds to each other, but not to adenine or thymine.

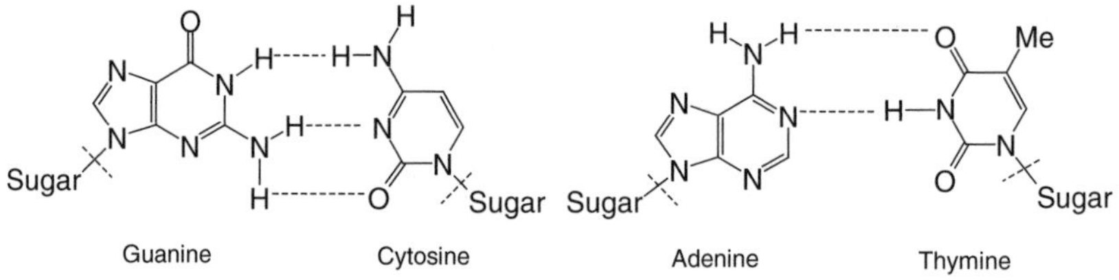

Guanine Cytosine Adenine Thymine

Hydrogen bonding between base pairs of the DNA double helix

The base pairs are on the inside of the helix, and the sugar–phosphate backbone is on the outside. The pitch of the helix is such that ten successive nucleotide pairs form one complete turn in 34 Å (the repeat distance). The exterior width of the spiral is about 20 Å, and the internal distance between 1′-positions of ribose units on opposite chains is about 11 Å.

The two strands of DNA double helix are not identical, but complementary to each other in such a way that whenever cytosine occurs in one strand a guanine occurs opposite in the other strand, and the same situation is true for adenine and thymine. This complementary pairing of bases explains why A and T are always found in equal amounts, as are C and G. It is this complementary behaviour of the two strands that explains how a DNA molecule replicates itself at the time of cell division and thereby passes on the genetic information to each of the two daughter cells.

The two strands of the double helix coil in such a way that two types of 'groove' are formed, a major groove 1.2 nm wide and a minor groove 600 pm wide. A number of flat, polycyclic molecules fit sideways into the groove between the strands and intercalate, or insert themselves, between the stacked base pairs. Many cancer causing and cancer preventing agents exert their actions through intercalating with DNA.

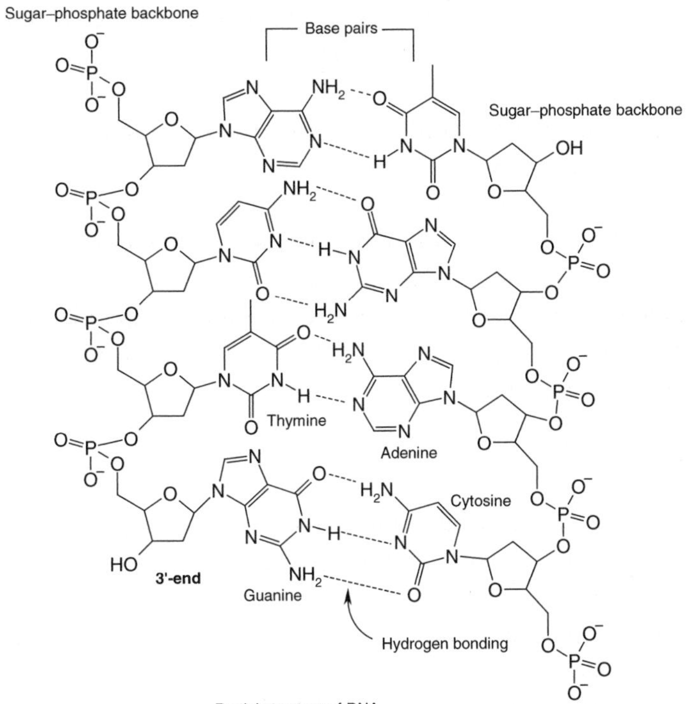

Sugar–phosphate backbone
Base pairs
Sugar–phosphate backbone
Thymine
Adenine
Cytosine
Guanine
HO
3'-end
Hydrogen bonding
5'-end

Partial structure of DNA

While the sugar–phosphate backbone of DNA is completely regular, the sequence of heterocyclic base pairs along the backbone can be of different permutations. It is the precise sequence of base pairs that carries the genetic information.

Adenine
Thymine
Guanine
Cytosine
DNA backbone

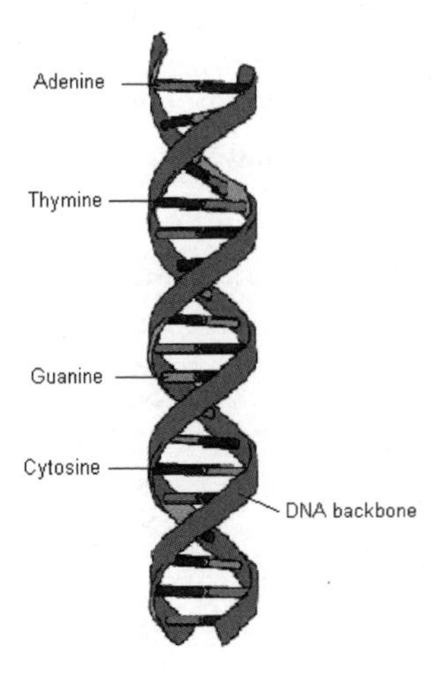

4.8.3 Nucleic acids and heredity

The genetic information of an organism is stored as a sequence of deoxyribonucleotides strung together in the DNA chain. Three fundamental processes are involved in the transfer of this stored genetic information.

- *Replication.* This process creates the identical copies of DNA, so that information can be preserved and handed down to offspring.
- *Transcription.* This process reads the stored genetic information and brings it out of the nucleus to ribosomes, where protein synthesis occurs.
- *Translation.* In this process, the genetic messages are decoded and used to build proteins.

Replication of DNA

Replication of DNA is an enzymatic process that starts with the partial unwinding of the double helix. Just before the cell division, the double strand begins to unwind. As the strands separate and bases are exposed, new nucleotides line up on each strand in a complementary fashion, A to T, and C to G. Two new strands now begin to grow, which are complementary to their old template strands. Two new identical DNA double helices are produced in this way, and these two new molecules can then be passed on, one to each daughter cell. As each of the new DNA molecules contains one strand of old DNA, and one new, the process is called *semiconservative replication.*

Addition of new nucleotide units to the growing chain occurs in the 5′ to C′ direction, and is catalysed by the enzyme DNA polymerase. The most important step is the addition of a 5′-mononucleoside triphosphate to the free 3′-hydroxyl group of the growing chain as the 3′-hydroxyl attacks the triphosphate and expels a diphosphate leaving group.

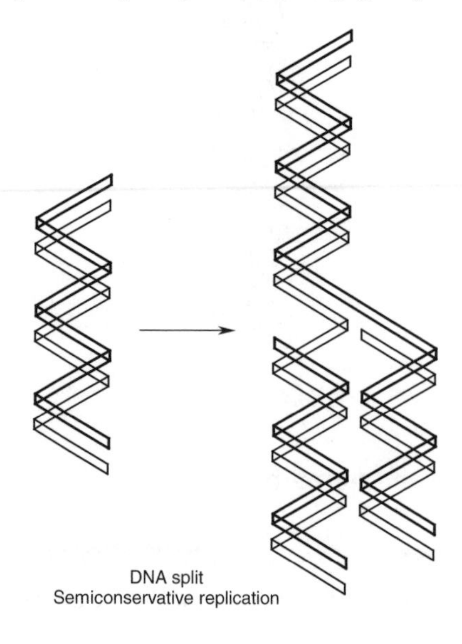

DNA split
Semiconservative replication

Transcription: synthesis of RNA

Transcription starts with the process by which the genetic information is transcribed onto a form of RNA, called mRNA. Ribonucleic acid, RNA, is structurally similar to DNA with the exceptions that its nucleotides contain ribose, instead of a 2′-deoxyribose, and the base thymine is replaced by uracil. There are three major types of RNA depending on their specific functions. However, all three types of RNA are much smaller than DNA and they are single stranded, rather than double stranded.

(a) *Messenger RNA (mRNA)* carries genetic information from DNA to ribosomes where protein synthesis occurs.

(b) *Ribosomal RNA (rRNA),* complexed with proteins (nucleoproteins), provides the physical make up of ribosomes.

(c) *Transfer RNA (tRNA)* transports amino acids to the ribosomes for protein synthesis.

Protein synthesis takes place in the cell nucleus with the synthesis of mRNA. Part of the DNA double helix unwinds adequately to expose on a single chain a portion corresponding to at least one *gene*. Ribonucleotides, present in the cell nucleus, assemble along the exposed DNA chain by pairing with the bases of DNA in a similar fashion that is observed in DNA base pairing. However, in RNA uracil replaces thymine. The ribonucleotide units of mRNA are joined into a chain by the enzyme *RNA polymerase*. Once the mRNA is synthesized, it moves into the cytoplasm, where it acts as a template for protein synthesis. Unlike what is seen in DNA replication, where both strands are copied, only one of the two DNA strands is transcribed into mRNA. The strand that contains the gene is called the *coding strand* or *sense strand*. The strand that gets transcribed is known as the *template strand* or *antisense strand*. As the template strand and the coding strand are complementary, and as the template strand and the RNA molecule are also complementary, the RNA molecule produced during transcription is a copy of the coding strand, with the only exception that the RNA molecule contains a U everywhere the DNA coding strand has a T.

Ribosomes are small granular bodies scattered throughout the cytoplasm, and this is the place where protein synthesis starts. rRNA itself does not directly govern protein synthesis. A number of ribosomes get attached to a chain of mRNA and form a *polysome*, along which, with mRNA acting as the template, protein synthesis occurs. One of the major functions of rRNA is to bind the ribosomes to the mRNA chain.

tRNA is the smallest of all three types of RNA mentioned above, and consequently much more soluble than mRNA and rRNA. This is why tRNA is also sometimes called *soluble RNA*. tRNA transports amino acids, building blocks of protein synthesis, to specific areas of the mRNA of the polysome. tRNAs are composed of a small number of nucleotide units (70–90 units) folded into several loops or arms through base pairing along the chain.

Translation: RNA and protein biosynthesis

Translation is the process by which mRNA directs protein synthesis. In this process, the message carried by mRNA is read by tRNA. Each mRNA is divided into codons, ribonucleotide triplets that are recognized by small amino-acid-carrying molecules of tRNA, which deliver the appropriate amino acids needed for protein synthesis.

RNA directs biosynthesis of various peptides and proteins essential for any living organisms. Protein biosynthesis seems to be catalysed by mRNA rather than protein-based enzymes and occur on the ribosome. On the ribosome, the mRNA acts as a template to pass on the genetic information that it has transcribed from the DNA. The specific ribonucleotide sequence in mRNA forms an '*instruction*' or *codon* that determines the order in which different amino acid residues are to be joined. Each 'instruction' or codon along the mRNA chain comprises a sequence of three ribonucleotides that is specific for a given amino acid. For example, the codon U–U–C on mRNA directs incorporation of the amino acid phenylalanine into the growing protein.

4.8.4 DNA fingerprinting

DNA fingerprinting, also known as DNA typing, is a method of identification that compares fragments of DNA. This technique was first developed in 1985, originally used to detect the presence of genetic diseases. With the exception of identical twins, the complete DNA of each individual is unique.

In 1984, it was discovered that human genes contain short, repeating sequence of noncoding DNA, called *short tandem repeats* (STRs). The STR loci are slightly different for every individual except identical twins. By sequencing these loci, a unique pattern for each individual can be obtained. On the basis of this fundamental discovery, the technique of DNA fingerprinting was developed.

The DNA fingerprinting technique has now been applied almost routinely in all modern forensic laboratories to solve various crimes. When a DNA sample is obtained from a crime scene, e.g. from blood, hair, skin or semen, the sample is subjected to cleavage with restriction endonucleases to cut out fragments containing the STR loci. The fragments are then amplified using the polymerase chain reaction (PCR), and the sequence of the fragments is determined. If the DNA profile from a known individual and that obtained from the DNA from the crime scene matches, the probability is approximately 82 billion to 1 that the DNA is from the same person.

The DNA of father and offspring are related, but not completely identical. Thus, in paternity cases the DNA fingerprinting technique comes very handy, and the identity of the father can be established with a probability of 100000 to 1.

4.9 Amino acids and peptides

Amino acids, as the name implies, contain both an amino and a carboxylic acid group, and are the building blocks of proteins. Twenty different amino acids are used to synthesize proteins, and these are alanine (Ala, A), arginine (Arg, R), asparagine (Asn, N), aspartic acid (Asp, D), cysteine (Cys, C), glutamine (Gln, Q), glutamic acid (Glu, E), glycine (Gly, G), histidine (His, H), isoleucine (Ile, I), leucine (Leu, L), lysine (Lys, K), methionine (Met, M), phenylalanine (Phe, F), proline (Pro, P), serine (Ser, S), threonine (Thr, T), tryptophan (Trp, W), tyrosine (Tyr, Y) and valine (Val, V). The shape and other properties of each protein are dictated by the precise sequence of amino acids in it. Most amino acids are optically active, and almost all the 20 naturally occurring amino acids that comprise proteins are of the L-form. While the (*R*) and (*S*)-system can be used to describe the absolute stereochemistry of amino acids, conventionally the D and L-system is more popular for amino acids.

Aliphatic aminoacids

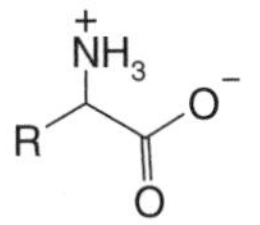

Alanine R = Me
Glycine R = H
Leucine R = $CH_2CH(CH_3)_2$
Valine R = $CH(CH_3)_2$

Isoleucine R =

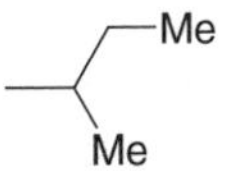

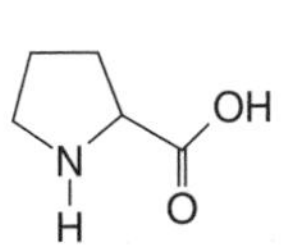

Proline
A cyclic amino acid

Aromatic amino acids

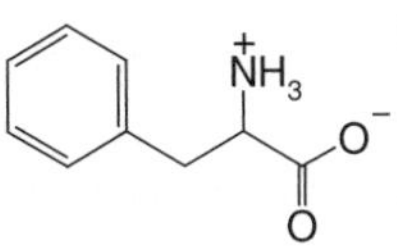

Phenyl alanine

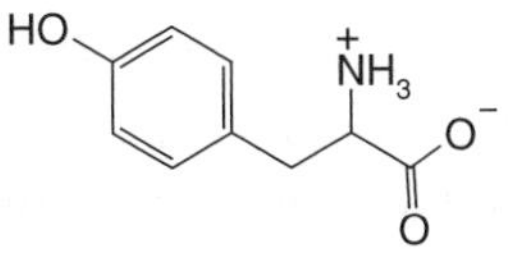

Tyrosine

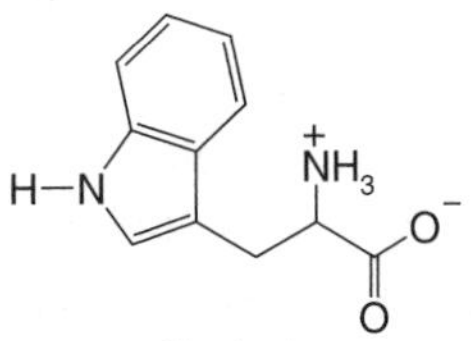

Tryptophan

Acidic amino acids

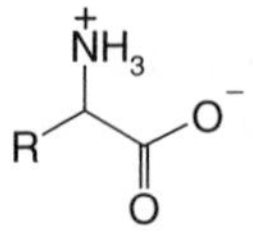

Aspartic acid R = CH_2COOH
Glutamic acid R = CH_2CH_2COOH

Basic amino acids

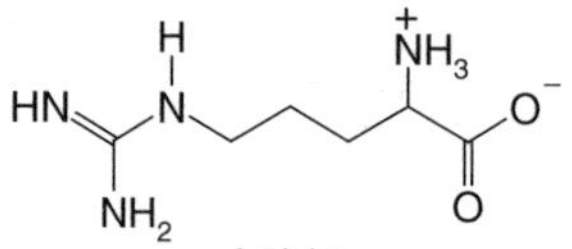

Arginine 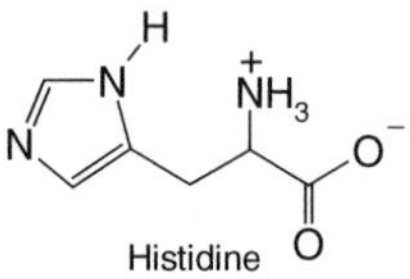 Histidine 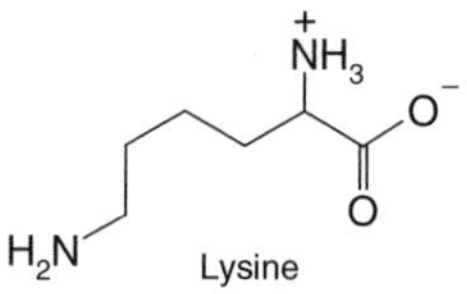 Lysine

Hydroxylic amino acids

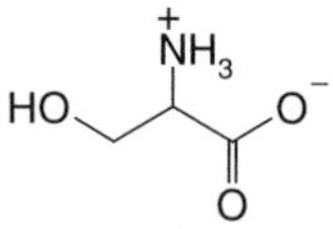

Serine 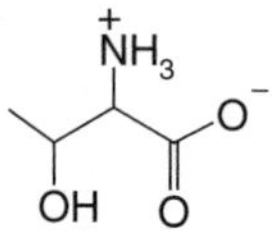 Threonine

Sulphur-containing amino acids

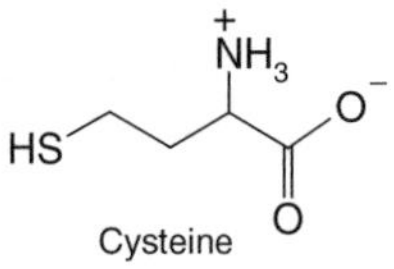

Cysteine 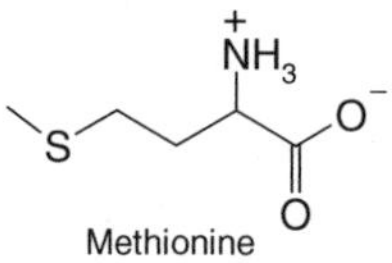 Methionine

Amidic amino acids

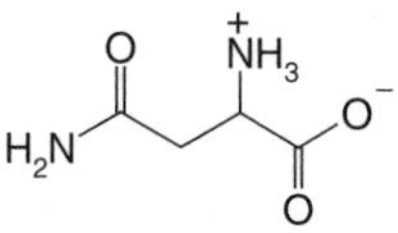

Asparagine 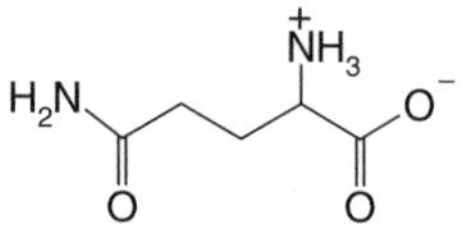 Glutamine

Peptides are biologically important polymers in which α-amino acids are joined into chains through amide linkages, called *peptide bonds*. A peptide bond is formed from the amino group ($-NH_2$) of one amino acid and the carboxylic acid group ($-$COOH) of another. The term *peptide bond* implies the existence of the peptide group, which is commonly written in text as $-$CONH$-$. Two molecules (amino acids) linked by a peptide bond form a *dipeptide*. A chain of molecules linked by peptide bonds is called a *polypeptide*. Proteins are large peptides. A protein is made up of one or more polypeptide chains, each of which consists of amino acids. Instead of writing out complex formulae, sequences of amino acids are commonly written using the three- or one-letter codes e.g. Ala–Val–Lys (three letter) or

AVK (one letter). The ends of a peptide are labelled as the amino end or amino terminus, and the carboxy end or carboxy terminus.

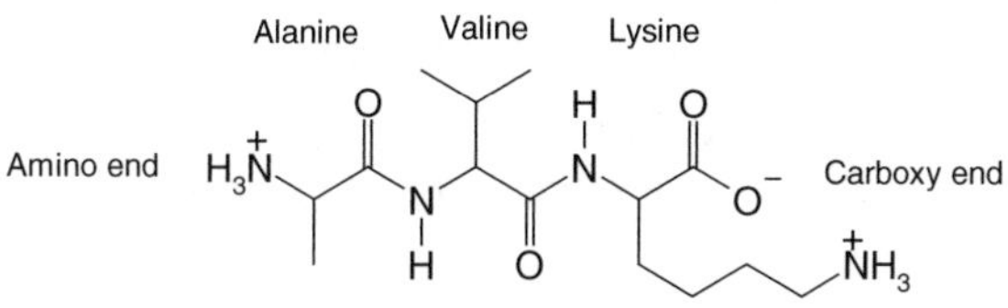

Alanylvalyllysine
(Ala-Val-Lys or AVK)

Large peptides of biological significance are known by their trivial names; e.g., insulin is an important peptide composed of 51 amino acid residues.

4.9.1 Fundamental structural features of amino acids

Each amino acid consists of a carbon atom to which is attached a hydrogen atom, an amino group ($-NH_2$), a carboxyl group ($-COOH$) and one of 20 different 'R' groups. It is the structure of the R group (side chain) that determines the identity of an amino acid and its special properties. The side chain (R group), depending on the functional groups, can be aliphatic, aromatic, acidic, basic, hydroxylic, sulphur containing or amidic (containing amide group). However, proline has an unusual ring structure, where the side chain is bonded at its terminus to the main chain nitrogen.

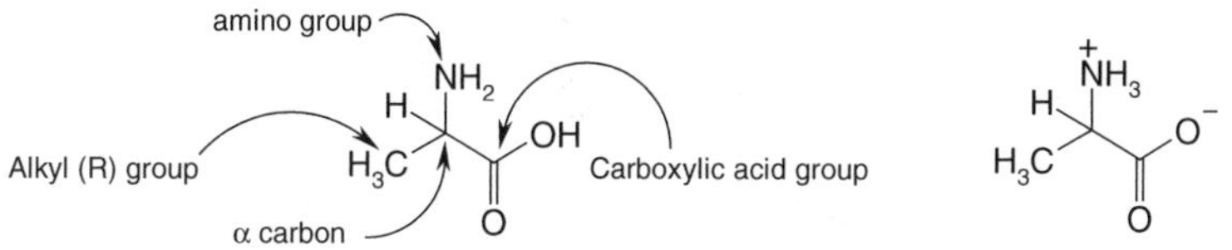

Alanine
An amino acid

The zwitterionic
structure of alanine

An amino acid, with an overall charge of zero, can contain within the same molecule two groups of opposite charge. Molecules containing oppositely charged groups are known as *zwitterions*. For amino acids, a *zwitterionic* structure is possible because the basic amino group can accept a proton and the acidic carboxylic group can donate a proton.

4.9.2 Essential amino acids

All living organisms can synthesize amino acids. However, many higher animals are deficient in their ability to synthesize all of the amino acids they need for their proteins. Thus, these higher animals require certain amino acids as a part of their diet. Human beings also must include in their diet adequate amounts of eight different amino acids, which they cannot synthesize in their body. These are known as *essential* amino acids. The eight essential amino acids are valine, leucine, isoleucine, phenylalanine, tryptophan, threonine, methionine and lysine. Sometimes, arginine and histidine are also included in the category of essential amino acids.

4.9.3 Glucogenic and ketogenic amino acids

The carbon skeletons of the amino acids can be used to produce metabolic energy. Several amino acids can be classified as *glucogenic* and *ketogenic* because of their degradation products.

Amino acids that are converted to glucose or glycogen are called glucogenic amino acids. Alanine, arginine, asparagine, cysteine, glutamine, glycine, histidine, hydroxyproline, methionine, proline, serine and valine are *glucogenic amino acids*. Glucogenic amino acids give rise to a net production of pyruvate or TCA cycle, such as α-ketoglutarate or oxaloacetate, all of which are precursors to glucose via *gluconeogenesis*.

Amino acids that give rise to ketone bodies (acetylCoA or acetoacetyl-CoA, neither of which can bring about net glucose production) are called ketogenic amino acids. Leucine and lysine are *ketogenic amino acids*. Some amino acids, e.g. threonine, isoleucine, phenylalanine, tyrosine and tryptophan, can be both ketogenic and glycogenic.

4.9.4 Amino acids in human body

All human tissues are capable of synthesizing the nonessential amino acids, amino acid remodelling and conversion of non-amino-acid carbon skeletons into amino acids and other derivatives that contain nitrogen. However, the liver is the major site of metabolism of nitrogenous compounds in the body. Dietary proteins are the primary source of essential amino acids (or nitrogen). Digestion of dietary proteins produces amino acids, which are absorbed through epithelial cells and enter the blood. Various cells take up these amino acids that enter the cellular pools.

In our bodies, amino acids are used for the synthesis of proteins and other nitrogen-containing compounds, or they are oxidized to produce energy. Cellular proteins, hormones (thyroxine, epinephrine and insulin), neurotransmitters, creatine phosphate, the haem of haemoglobin, cytochrome, melanin (skin pigment) and nucleic acid bases (purine and pyrimidine) are examples of amino-acid-derived nitrogen-containing biologically important group of compounds found in humans.

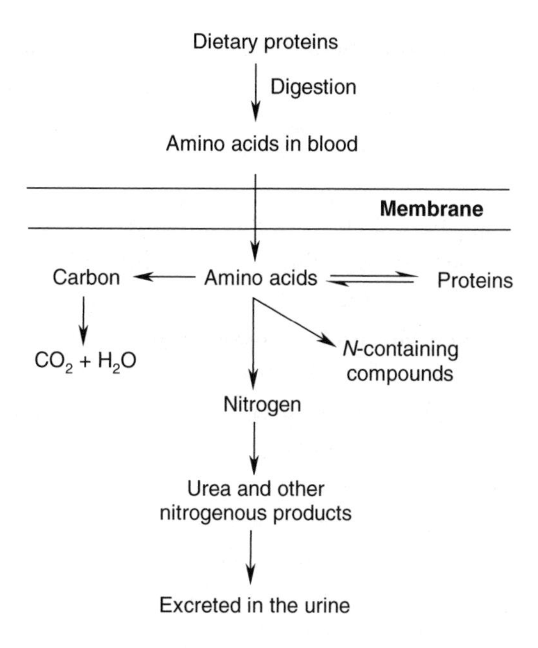

4.9.5 Acid–base properties of amino acids

The neutral forms of amino acids are *zwitterions*. This is why amino acids are insoluble in apolar aprotic solvents, e.g. ether, but most nonprotonated amines and unionized carboxylic acids dissolve in ether. For the same reason, amino acids usually have high melting points, e.g. the m.p. of glycine is 262 °C, and large dipole moments. The high melting points and greater water solubility than in ether are saltlike characteristics, not the characteristics of uncharged organic molecules. This saltlike characteristic is found in all zwitterionic compounds. Water is the best solvent for most amino acids because it solvates ionic groups much as it solvates the ions of a salt. A large dipole moment is characteristic of zwitterionic compounds that contain great deal of separated charge. The pK_a values for amino acids are also typical of zwitterionic forms of neutral molecules. Peptides can also exist as zwitterions; i.e., at pH values near 7, amino groups are protonated and carboxylic acid groups are ionized.

4.9.6 Isoelectric points of amino acids and peptides

Isoelectric point (*pI*) or *isoelectric pH* is the pH at which a molecule carries no net electrical charge, i.e. zero charge. It is an important measure of the acidity or basicity of an amino acid. To have a sharp *isoelectric point*, a molecule must be amphoteric, i.e. it must have both acidic and basic functional groups, as found in amino acids. For an amino acid with only one amino and one carboxylic acid group, the pI can be calculated from the pK_a values of this molecule.

$$pI = \frac{pK_{a1} + pK_{a2}}{2}$$

For amino acids with more than two ionizable groups, e.g. lysine, the same formula is used but the two pK_a values used are those of the two groups that lose and gain a charge from the neutral form of the amino acid.

The process that separates proteins according to their isoelectric point is called *isoelectric focusing*. At a pH below the pI proteins carry a net positive charge, whereas above the pI they carry a net negative charge. Appling this principle, *gel electrophoretic methods* have been developed to separate proteins. The pH of an electrophoretic gel is determined by the buffer used for that gel. If the pH of the buffer is above the pI of the protein being run, the protein will migrate to the positive pole (negative charge is attracted to a positive pole). Similarly, if the pH of the buffer is below the pI of the protein being run, the protein will migrate to the negative pole of the gel (positive charge is attracted to the negative pole). If the protein is run with a buffer pH that is equal to the pI, it will not migrate at all. This also applies for individual amino acids.

4.10 Importance of functional groups in determining drug actions and toxicity

In Chapter 2, you have already learned that most drugs bind to the appropriate receptor molecules to exhibit their pharmacological actions, and also toxicity, which in fact is the adverse pharmacological action. A drug's pharmacological activity is inherently related to its chemical structure. Various functional groups present in the drug molecules are involved in the drug–receptor binding or interaction. For example, drugs containing hydroxyl or amino groups tend to be involved in hydrogen bonding with the receptor.

Any changes in the functional groups in a drug molecule can render significant changes in the activity and toxicity, and this is the basis of any structure–activity-relationship (SAR) study of drug molecules. The SAR study is the study for understanding the relationship of chemical structure to activity. The activity can be a pharmacological response, binding, toxicity or any other quantifiable event. In SAR studies, essential functional groups or structural features of a drug molecule, which are responsible for the optimum pharmacological actions with minimum toxicity index, are identified or optimized. These essential functional groups for the pharmacological activities are called *pharmacophores*.

By changing the functional groups of any drug, several analogues are usually synthesized in an effort to improve its binding to the receptor, facilitate absorption by the human body, increase specificity for different organs/tissue types, broaden the spectrum of activity or reduce the toxicity/side-effects. Toxicity prevents many compounds from being developed and approved. A number of approved drugs have also been forced to be withdrawn from the market beacuse of toxicities. For example, in 2004, Merck's arthritis drug Vioxx was withdrawn owing to severe cardiovascular side-effects, and the Parke-Davis and Warner-Lambert antidiabetic drug troglitazone (Rezulin) was withdrawn from the market in 2000 after it was found to cause severe liver toxicity. The drug industries expend considerable time and effort trying to avoid or minimize toxic effects by identifying and altering the functional groups responsible for toxic effects. A change in functional groups leading to toxicty can be demonstrated by paracetamol toxicity.

The sulpha drugs and the penicillin group of antibacterial agents can be the ideal examples for demonstrating the importance of functional groups in drug actions and effectiveness. In Chapter 6, you will also see how a small change in the functional group(s) of steroidal molecules can render remarkable changes in their pharmacological and hormonal functions.

4.10.1 Structure–activity relationships of sulpha drugs

To date, over 10 000 structural analogues of sulphanilamide, the parent of all sulpha drugs, have been synthesized and used in the SAR studies. However, only about 40 of them have ever been used as prescribed drugs. Sulpha drugs are *bactereostatic,* i.e. they inhibit bacterial growth but do not actively kill bacteria. These drugs act on the biosynthetic pathway of tetrahydrofolic acid, inhibit dihydropteroate synthetase and mimic the shape of PABA (*para*-aminobenzoic acid).

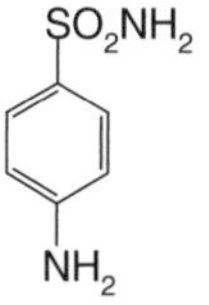

SO_2NH_2 / NH_2
Sulphanilamide
The first sulpha drug

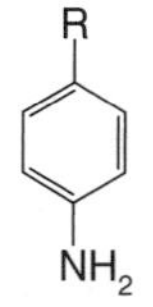

R / NH_2
General structure of sulphonamides
R = SO_2NHR' or SO_3H

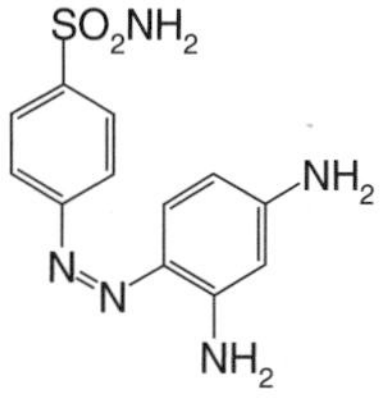

SO_2NH_2 / N=N / NH_2 / NH_2
Prontosil

From numerous studies, it has now been established that the amino functional groups ($-NH_2$) is essential for the activity. In addition, the following structural features have to be present in sulpha drugs for the optimum antibacterial activity.

(a) The amino and the sulphonyl groups have to be *para* to each other, i.e. a *para*-disubstituted benzene ring is essential.

(b) The anilino (Ph-NH_2) amino group may be substituted, but optimum activity is observed with the unsubstituted form.

(c) Replacement of the central benzene ring (aromatic) or additional functional groups on the benzene ring diminishes activity.

(d) *N′*-monosubstitution on SO_2NH_2 increases potency, especially with heteroaromatic groups.

(e) *N′*-disubstitution on SO_2NH_2 leads to inactive compounds.

The structure of Prontosil, an azo dye, is quite similar to the structure of sulphanilamide with the modification that the $-NH_2$ is substituted. As result, it does not have any *in vitro* antibacterial activity, but *in vivo* Prontosil is converted via reduction of the $-N{=}N-$ linkage to its active metabolite sulphanilamide.

N-heterocyclic derivatives of sulphanilamide, e.g. sulphadiazine, sulphathiazole and sulphoxazole, have broad-spectrum antimicrobial activity. They are generally more water soluble, and thus better absorbed and retained better, i.e. excreted slowly.

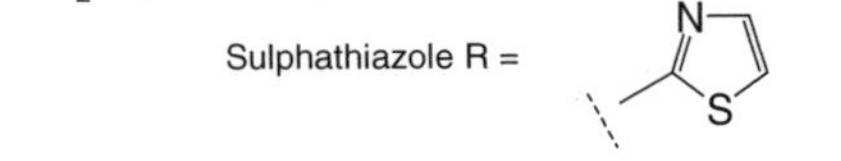

4.10.2 Structure–activity relationships of penicillins

The penicillin group of antibiotics, also known as β-lactam antibiotics, has revolutionized the history of modern medicine, by their effectiveness against several pathogenic bacterial species that cause various forms of infections. Penicillin G, the parent of all these antibiotics, was first isolated from a fungal species, *Penicillium notatum*. Since the discovery of this antibiotic, several modifications have been introduced to the parent structure in order to enhance the activity, increase the acid resistance, facilitate bioavailability and reduce toxicity. Penicillin G is rather a complex molecule, and possesses various types of functional group, e.g. phenyl, alkyl, amide, carboxylic acid and β-lactam.

All penicillins are susceptible to attack in acidic solution via intramolecular attack of the amide carbonyl oxygen on the β-lactam carbonyl, leading to the complete destruction of the β-lactam ring, and thus the antibacterial activity. Similarly, penicillins are unstable in basic solution because of β-lactam ring opening by free basic nucleophiles. Thus, for the antibacterial activity, the stability of the β-lactam functional group in penicillins is of paramount importance.

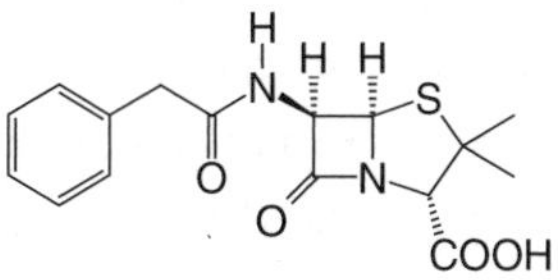

Penicillin G
The first penicillin of the penicillin group of antibiotics

The degree of instability of the β-lactam ring depends on the availability of the electrons for attack, so modification of penicillins with the addition of electron withdrawing groups near the amide carbonyl decreases the availability of these electrons and significantly improves acid stability. For example, the amino group of amoxicillin and ampicillin makes these molecules acid stable.

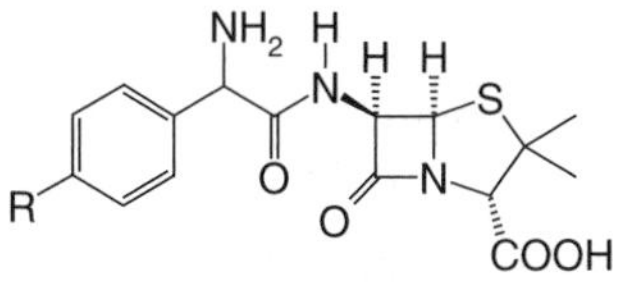

Amoxicillin R = OH
Ampicillin R = H
Stable in acidic condition

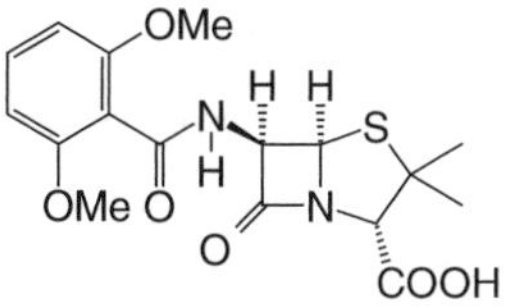

Methicillin

From numerous studies with semisynthetic penicillins, it has been established that the penicillins that contain more polar groups are able to cross easily the Gram-negative cell wall and will have a greater spectrum of antibacterial activity. For example, the amino group in amoxicillin gives the molecule polarity, and makes it effective against both Gram-positive and Gram-negative bacteria. The SAR of penicillin can be summarized as follows.

(a) Oxidation of the sulphur to a sulphone or sulphoxide decreases the activity of penicillins but provides better acid stability.

(b) The β-lactam carbonyl and nitrogen are absolutely necessary for activity.

(c) The amide carbonyl is essential for activity.

(d) The group attached to the amide carbonyl (the R group) is the basis for the changes in activity, acid stability and susceptibility to resistance.

(e) Any other changes generally decrease activity.

(f) A bulky group directly adjacent to the amide carbonyl usually offers a β-lactamase resistant property.

A bulky group directly adjacent to the amide carbonyl will prevent the penicillin from entering the active site of penicillin-destroying enzymes, e.g. β-lactamases, but still allow them to enter the active site of penicillin binding proteins. For example, methicillin has a bulky group directly adjacent to the amide carbonyl, and is β-lactamase resistance.

The addition of polar groups to the R group, i.e. the group directly linked to the amide carbonyl, generally allows the penicillin molecule, e.g. amoxicillin, to more easily pass through the Gram-negative cell wall, and thus increases antibacterial activity.

4.10.3 Paracetamol toxicity

Bioactivation is a classic toxicity mechanism where the functional group or the chemical structure of the drug molecule is altered by enzymatic reactions. For example, the enzymatic breakdown of the analgesic acetaminophen (paracetamol), where the aromatic nature and the hydroxyl functionality in paracetamol are lost, yields *N*-acetyl-*p*-benzoquinone imine, a hepatotoxic agent. Paracetamol can cause liver damage and even liver failure, especially when combined with alcohol.

Bioactivation

Paracetamol
(Acetaminophen)

N-Acetyl-*p*-benzoquinone imine
The hepatotoxic metabolite

4.11 Importance of functional groups in determining stability of drugs

In Section 4.10, you have already seen that, just by introducing a new functional group on a penicillin molecule, the acid stability of penicillins can be improved remarkably, and similarly the introduction of bulky functional groups in penicillin offers stability against β-lactamases. Thus, functional groups play a vital role in the stability of drugs.

Certain functional groups in drug molecules are prone to chemical degradation. Drugs with an ester functional group are easily hydrolysed; e.g., aspirin is easily hydrolysed to salicylic acid. Similarly, many drug molecules are susceptible to oxidation because of certain oxidizable functional groups, e.g. alcohol.

$OCOCH_3$ CO_2H Hydrolysis OH CO_2H

Aspirin Salicylic acid

Insulin is a protein and contains amide linkages, which makes this compound unstable in acidic medium, and unsuitable for oral administration. Like any other proteins in the gastrointestinal tract, insulin is reduced to its amino acid components, and the activity is totally lost. Many drugs having olefinic double bonds exhibit *trans–cis* isomerism in the presence of light. Similarly, because of the presence of certain functional groups or the chemical structure, a drug can be sensitive to heat. Many monoterpenyl or sesquiterpenyl drugs are unstable at high temperatures.

Recommended further reading

Clayden, J., Greeves, N., Warren, S. and Wothers, P. *Organic Chemistry*, Oxford University Press, Oxford, 2001.

5

Organic reactions

Learning objectives

After completing this chapter the student should be able to –

- recognize various types of organic reaction;
- describe the mechanisms and examples of addition, substitution, radical, oxidation–reduction, elimination and pericyclic reactions.

5.1 Types of organic reaction

Reaction types	Brief definition	Where they occur
Radical reactions	New bond is formed using radical from each reactant.	Alkanes and alkenes.
Addition reactions	Addition means two systems combine to a single entity.	Alkenes, alkynes, aldehydes and ketones.
Elimination reactions	Elimination refers to the loss of water, hydrogen halide or halogens from a molecule.	Alcohols, alkyl halides and alkyl dihalides.
Substitution reactions	Substitution implies that one group replaces the other.	Alkyl halides, alcohols, epoxides, carboxylic acid and its derivatives, and benzene and its derivatives.
Oxidation–reduction reactions	Oxidation = loss of electrons.	Alkenes, alkynes, 1° and 2° alcohols, aldehydes and ketones.

(*Continued*)

Chemistry for Pharmacy Students Satyajit D Sarker and Lutfun Nahar

Reaction types	Brief definition	Where they occur
	Reduction = gain of electrons.	Alkene, alkyne, aldehydes, ketones, alkyl halides, nitriles, carboxylic acid and its derivatives, and benzene and its derivatives.
Pericyclic reactions	Concerted reaction that takes place as a result of a cyclic rearrangement of electrons.	Conjugated dienes and α,β-unsaturated carbonyl compounds.

5.2 Radical reactions: free radical chain reactions

A *radical*, often called a *free radical*, is a highly reactive and short lived species with an unpaired electron. Free radicals are electron-deficient species, but usually uncharged. So their chemistry is very different from the chemistry of even-electron and electron-deficient species, e.g. carbocations and carbenes. A radical behaves like an electrophile, as it requires only a single electron to complete its octet.

Radical reactions are often called *chain reactions*. All chain reactions have three steps: chain initiation, chain propagation and chain termination. For example, the halogenation of alkane is a free radical chain reaction.

5.2.1 Preparation of alkyl halides

Chlorine or bromine reacts with alkanes in the presence of light (hν) or high temperatures to give alkyl halides. Usually, this method gives mixtures of halogenated compounds containing mono-, di-, tri- and tetra-halides. However, this reaction is an important reaction of alkanes as it is the only way to convert inert alkanes to reactive alkyl halides. The simplest example is the reaction of methane with Cl_2 to yield a mixture of chlorinated methane derivatives.

$$\underset{\text{Methane}}{CH_4} + Cl_2 \xrightarrow{h\nu} \underset{\text{Methyl chloride}}{CH_3Cl} + \underset{\text{Dichloro methane}}{CH_2Cl_2} + \underset{\text{Chloroform}}{CHCl_3} + \underset{\text{Carbon tetrachloride}}{CCl_4}$$

To maximize the formation of monohalogenated product, a radical substitution reaction must be carried out in the presence of excess alkane. For example, when a large excess of methane is used, the product is almost completely methyl chloride (chloromethane).

$$\underset{\text{Methane (Large excess)}}{CH_4} + Cl_2 \xrightarrow{h\nu} \underset{\text{Methyl chloride}}{CH_3Cl} + HCl$$

Similarly, when a large excess of cyclopentane is heated with chlorine at 250°C, the major product is chlorocyclopentane (95%), along with small amounts of dichlorocyclopentanes.

+ Cl_2 $\xrightarrow{250\ ^{\circ}C}$ Cl + Cl Cl + Cl Cl

Chloro cyclopentane

1,2-Dichloro cyclopentane

1,3-Dichloro cyclopentane

A free radical chain reaction is also called a *radical substitution* reaction, because radicals are involved as intermediates, and the end result is the substitution of a halogen atom for one of the hydrogen atoms of alkane.

H H + Cl_2 $\xrightarrow{h\nu}$ H Cl + HCl

Chlorocyclohexane (50%)

$H_3C-C(CH_3)_2-H + Br_2 \xrightarrow{h\nu} H_3C-C(CH_3)_2-Br + HBr$

t-Butane

t-Butylbromide (90%)

The high temperature or light supplies the energy to break the chlorine–chlorine bond homolytically. In the *homolytic bond cleavage*, one electron of the covalent bond goes to each atom. A single headed arrow indicates the movement of one electron. The chlorine molecule (Cl_2) dissociates into two chlorine radicals in the first step, known as the *initiation step*, leading to the substitution reaction of chlorine atoms for hydrogen atoms in methane.

- *Initiation.* Initiation generates a reactive intermediate. A chlorine atom is highly reactive because of the presence of an unpaired electron in its valence shell. It is electrophilic, seeking a single electron to complete the octet. It acquires this electron by abstracting a hydrogen atom from methane.

$Cl-Cl \xrightarrow{h\nu} 2\ Cl\cdot$

Chlorine — Chlorine radicals — Unpaired electron

- *Propagation.* In this step, the intermediate reacts with a stable molecule to produce another reactive intermediate and a product molecule. The propagation step yields a new electrophilic species, the methyl radical, which has an unpaired electron. In a second propagation step, the methyl radical abstracts a chlorine atom from a chloromethane molecule, and generates a chlorine radical.

$$H_3C{-}H + \cdot Cl \longrightarrow \cdot CH_3 + HCl$$

Methane + Chlorine radical → Methyl radical + Hydrogen chloride

$$\cdot CH_3 + Cl{-}Cl \xrightarrow{h\nu} CH_3Cl + \cdot Cl$$

Methyl radical + Chlorine → Methyl chloride + Chlorine radical

- *Termination.* Various reactions between the possible pairs of radicals allow for the formation of ethane, Cl_2 or the methyl chloride. In this step, the reactive particles are consumed, but not generated.

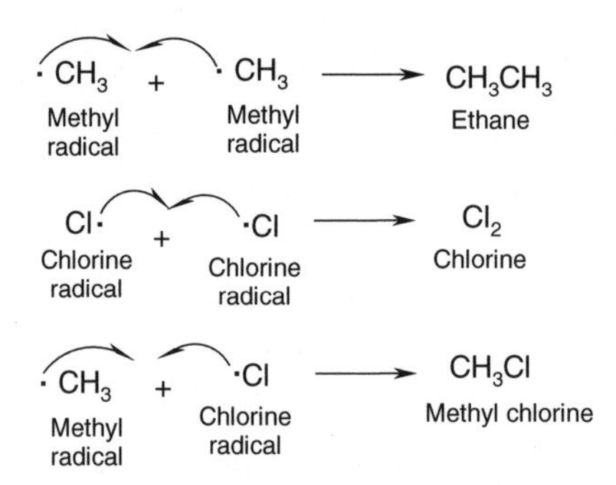

Bromination of alkanes follows the same mechanism as chlorination. The only difference is the reactivity of the radical; i.e., the chlorine radical is much more reactive than the bromine radical. Thus, the chlorine radical is much less selective than the bromine radical, and it is a useful reaction when there is only one kind of hydrogen in the molecule. If a radical substitution reaction yields a product with a chiral centre, the major product is a racemic mixture. For example, radical chlorination of *n*-butane produces a 71% racemic mixture of 2-chlorobutane, and bromination of *n*-butane produces a 98% racemic mixture of 2-bromobutane.

$$CH_3CH_2CH_2CH_3 \xrightarrow[Cl_2]{h\nu} CH_3CH_2CH(Cl)CH_3 + HCl$$

n-Butane → 2-Chlorobutane (Racemic mixture) 71%

$$CH_3CH_2CH_2CH_3 \xrightarrow[Br_2]{h\nu} CH_3CH_2CH(Br)CH_3 + HBr$$

n-Butane → 2-Bromobutane (Racemic mixture) 98%

5.2.2 Relative stabilities of radicals

The formation of different radicals from the same starting compound offers a way to estimate relative radical stabilities, which correspond directly to the stabilities of the corresponding carbocations. Carbocations are classified according to the number of alkyl groups that are bonded to the positively charged carbon. A primary (1°) carbocation has one alkyl group, a secondary (2°) has two and a tertiary (3°) has three alkyl groups.

Alkyl groups are able to decrease the concentration of positive charge on the carbocation by donating electrons inductively, thus increasing the stability of the carbocation. The greater the number of alkyl groups bonded to the positively charged carbon, the more stable is the carbocation. Therefore, a 3° carbocation is more stable than a 2° carbocation, and a 2° carbocation is more stable than a 1° carbocation, which in turn is more stable than a methyl cation.

In molecular orbital terms, alkyl groups can stabilize a carbocation by *hyperconjugation*. This is the overlap of the filled σ orbitals of the C—H or C—C bonds adjacent to the carbocation with an empty *p* orbital on the positively charged carbon atom. As a result, the positive charge is delocalized onto more than one atom, and thus increases the stability of the system. The more alkyl groups there are attached to the carbocation, the more σ bonds there are for hyperconjugation, and the more stable is the carbocation.

R_3C^+ > R_2HC^+ > RH_2C^+ > H_3C^+

3° carbocation 2° carbocation 1° carbocation Methyl cation

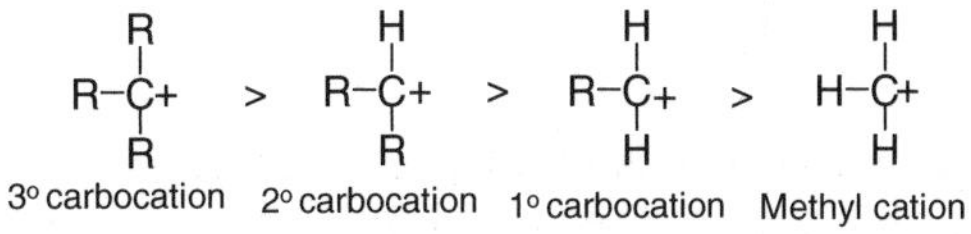

The relative stabilities of radicals follow the same trend as for carbocations. Like carbocations, radicals are electron deficient, and are stabilized by hyperconjugation. Therefore, the most substituted radical is most stable. For example, a 3° alkyl radical is more stable than a 2° alkyl radical, which in turn is more stable than a 1° alkyl radical. Allyl and benzyl radicals are more stable than alkyl radicals, because their unpaired electrons are delocalized. Electron delocalization increases the stability of a molecule. The more stable a radical, the faster it can be formed. Therefore, a hydrogen atom, bonded to either an allylic carbon or a benzylic carbon, is substituted more selectively in the halogenation reaction. The percentage substitution at allylic and benzylic carbons is greater in the case of bromination than in the case of chlorination, because a bromine radical is more selective.

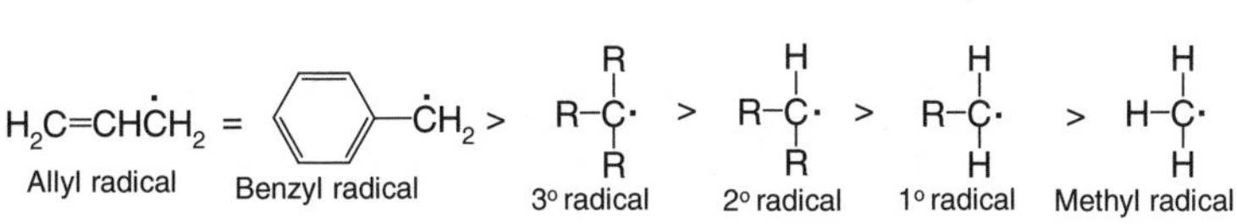

$H_2C{=}CH\dot{C}H_2$ = $C_6H_5\dot{C}H_2$ > $R_3C\cdot$ > $R_2HC\cdot$ > $RH_2C\cdot$ > $H_3C\cdot$

Allyl radical Benzyl radical 3° radical 2° radical 1° radical Methyl radical

5.2.3 Allylic bromination: preparation of alkene halides

Under high temperature or UV light and in the gas phase, cyclohexene can undergo free radical substitution by halogens. A common reagent for allylic bromination is *N*-bromosuccinimide (NBS), because it continually generates small amounts of Br_2 through the reaction with HBr. The bromination of cyclohexene produces 3-bromocyclohexene. An allylic hydrogen atom is substituted for a bromine atom. Allylic means the substituent is adjacent to a carbon–carbon double bond.

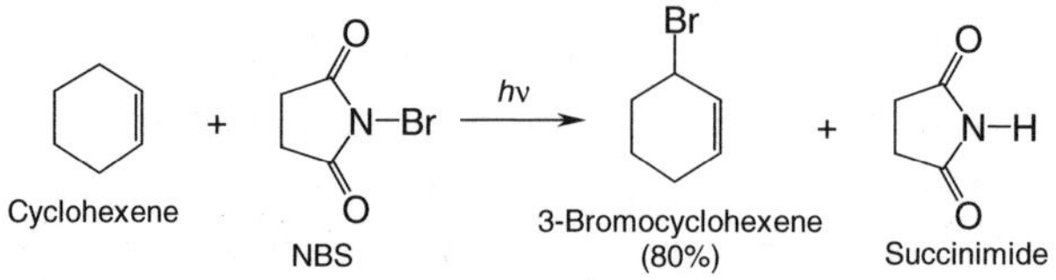

Cyclohexene + NBS —hν→ 3-Bromocyclohexene (80%) + Succinimide

Mechanism

Homolytic cleavage of the N—Br bond of NBS generates radicals.

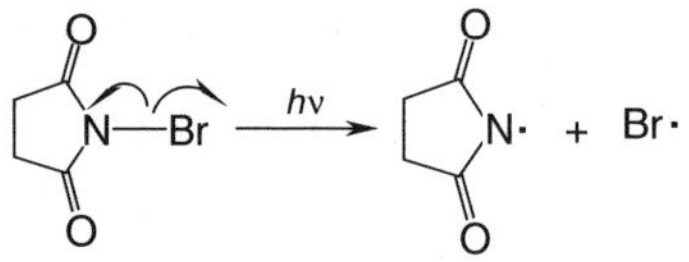

N—Br —hν→ N· + Br·

The bromine radical abstracts an allylic hydrogen atom of the cyclohexene, and forms a resonance stabilized allylic radical and hydrogen bromide.

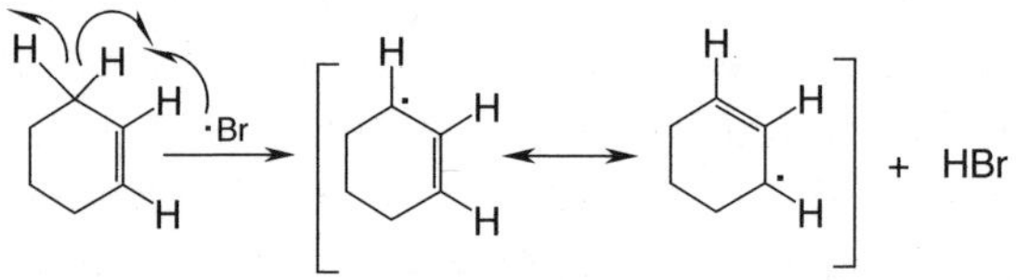

·Br → [⟷] + HBr

Hydrogen bromide reacts with NBS to produce a Br_2 molecule, which reacts with the allylic radical to form 3-bromocyclohexene, and a bromine radical is produced to continue the chain.

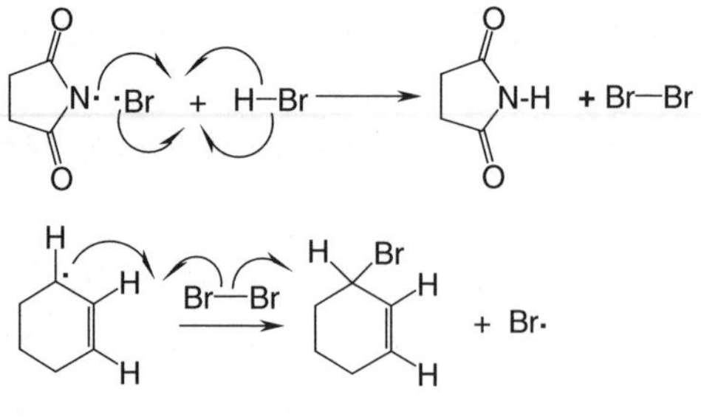

N· ·Br + H—Br → N-H + Br—Br

Br—Br → + Br·

5.2.4 Radical inhibitors

Radical inhibitors are used as antioxidants or preservatives. They preserve food by preventing unwanted radical reactions. Butylated hydroxyanisol

(BHA) and butylated hydroxytoluene (BHT) are synthetic preservatives that are added to many packaged foods.

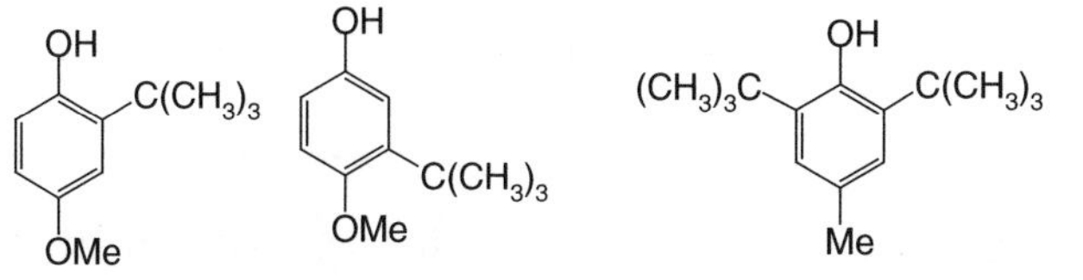

Butylated hydroxyanisole (BHA) Butylated hydroxytoluene (BHT)

Vitamin C, also known as ascorbic acid, and vitamin E, also known as α-tocopherol, are the two most common examples of radical inhibitors that are present in biological systems.

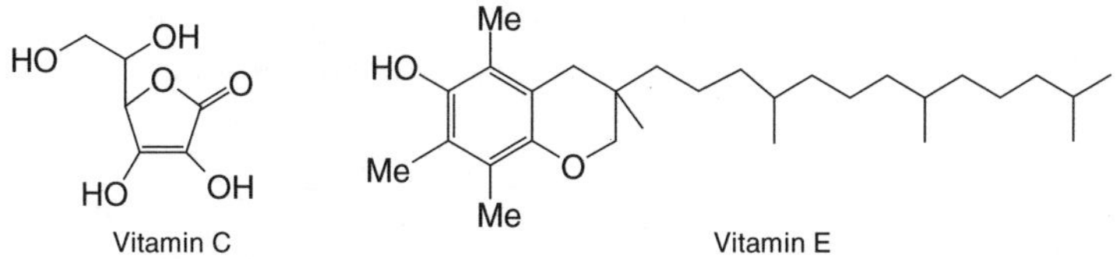

Vitamin C
Ascorbic acid

Vitamin E
α-Tocopherol

5.3 Addition reactions

Addition reactions occur in compounds having π electrons in carbon–carbon double (alkenes) or triple bonds (alkynes) or carbon–oxygen double bonds (aldehydes and ketones). Addition reactions are of two types: *electrophilic addition* to alkenes and alkynes, and *nucleophilic addition* to aldehydes and ketones. In an addition reaction, the product contains all of the elements of the two reacting species.

5.3.1 Electrophilic addition

Alkenes and alkynes readily undergo electrophilic addition reactions. They are nucleophilic and commonly react with electrophiles. The π bonds of alkenes and alkynes are involved in the reaction, and reagents are added to the double or triple bonds. In the case of alkynes, two molecules of reagent are needed for each triple bond for the total addition.

An alkyne is less reactive than an alkene. A vinyl cation is less able to accommodate a positive charge, as the hyperconjugation is less effective in stabilizing the positive charge on a vinyl cation than on an alkyl cation. The vinyl cation is more stable with positive charge on the more substituted carbon. Electrophilic addition reactions allow the conversion of alkenes and alkynes into a variety of other functional groups.

General reaction and mechanism

$$\text{>C=C<} \;(\text{Alkene}) + \overset{\delta^+}{E}\text{—}\overset{\delta^-}{Nu} \;(\text{Reagent}) \xrightarrow{\text{Addition}} \text{—}\underset{Nu}{\overset{|}{C}}\text{—}\underset{E}{\overset{|}{C}}\text{—} \;(\text{Product})$$

The π electrons attack the electrophile, the positive part of the reagent, usually the H^+, and form a carbocation intermediate.

$$\text{>C=C<} + \overset{\delta^+}{E}\text{—}\overset{\delta^-}{Nu} \xrightarrow{\text{Slow}} \text{—}\underset{+}{\overset{|}{C}}\text{—}\underset{E}{\overset{|}{C}}\text{—} + Nu:^-$$

The nucleophile ($Nu:^-$), the negative part of the reagent, usually X^-, HO^- and so on, attacks the carbocation to form the product.

$$\text{—}\underset{+}{\overset{|}{C}}\text{—}\underset{E}{\overset{|}{C}}\text{—} \;\; Nu:^- \xrightarrow{\text{Fast}} \text{—}\underset{Nu}{\overset{|}{C}}\text{—}\underset{E}{\overset{|}{C}}\text{—} \;(\text{Product})$$

Addition of hydrogen atoms to alkenes and alkynes: catalytic hydrogenation Preparation of alkanes

Addition of hydrogen atoms in the presence of a metal catalyst to double or triple bonds is known as *hydrogenation* or *catalytic hydrogenation*. Alkenes and alkynes are reduced to alkanes by the treatment with H_2 over a finely divided metal catalyst such as platinum (Pt—C), palladium (Pd—C) or Raney nickel (Ni). The platinum catalyst is also frequently used in the form of PtO_2, which is known as *Adams's catalyst*. The catalytic hydrogenation reaction is a reduction reaction.

In the catalytic hydrogenation, two new C—H σ bonds are formed simultaneously from H atoms absorbed into the metal surface. Thus, catalytic hydrogenation is stereospecific, giving only the *syn* addition product. If the atoms are added on the same side of the molecule, the addition is known as *syn* addition. If the atoms are added on opposite sides of the molecule, the addition is called an *anti* addition. For example, 2-butene reacts with H_2 in the presence of a metal catalyst to give *n*-butane.

$$\underset{\text{2-Butene}}{CH_3CH{=}CHCH_3} + H_2 \xrightarrow{\text{Pt/C}} \underset{n\text{-Butane}}{CH_3CH_2CH_2CH_3}$$

Similarly, 2-methyl-1-butene and 3-methylcyclohexene react with H_2 in the presence of a metal catalyst to give 2-methylbutane and methylcyclohexane, respectively.

$$CH_3CH_2C(CH_3){=}CH_2 + H_2 \xrightarrow[\text{Pd-C}]{\text{Pt-C or}} CH_3CH_2CH(CH_3)CH_3$$

2-Methyl-1-butene → 2-Methylbutane

3-Methylcyclohexene + H_2 $\xrightarrow{\text{Raney Ni}}$ Methylcyclohexane

Hydrogen adds twice to alkynes in the presence of a catalyst to generate alkanes. For example, acetylene reacts with hydrogen in the presence of a metal catalyst to give ethane.

$$HC{\equiv}CH + 2\,H_2 \xrightarrow[25\,^{\circ}C]{\text{Pt-C or Pd-C}} CH_3CH_3$$

Acetylene → Ethane

The reduction of alkynes occurs in two steps: addition of one mole of hydrogen atoms to form alkenes, and then addition of the second mole of hydrogen to alkenes to form alkanes. This reaction proceeds through a *cis*-alkene intermediate, but cannot be stopped at this stage except with the use of a special catalyst.

Selective hydrogenation of alkynes

Preparation of *cis*-alkenes *Lindlar's catalyst*, which is also known as *poisoned catalyst*, consists of barium sulphate, palladium and quinoline, and is used in selective and partial hydrogenation of alkynes to produce *cis*-alkenes. Hydrogen atoms are delivered simultaneously to the same side of the alkyne, resulting in *syn* addition (*cis*-alkenes). Thus, the *syn* addition of alkyne follows same procedure as the catalytic hydrogenation of alkyne.

$$R{-}C{\equiv}C{-}R \xrightarrow[\text{Quinoline, } CH_3OH]{H_2,\ Pd/BaSO_4} (R)(H)C{=}C(R)(H)$$

cis-Alkene
syn addition

Preparation of *trans*-alkenes The *anti* addition (*trans*-alkenes) is achieved in the presence of an alkali metal, e.g. sodium or lithium, in ammonia at −78°C.

$$R{-}C{\equiv}C{-}R \xrightarrow[\text{Liq. } NH_3]{\text{Na}} (R)(H)C{=}C(H)(R)$$

trans-Alkene
anti addition

Electrophilic addition to symmetrical and unsymmetrical π bonds

When the same substituents are at each end of the double or triple bond, it is called symmetrical. Unsymmetrical means different substituents are at each end of the double or triple bond. Electrophilic addition of unsymmetrical reagents to unsymmetrical double or triple bonds follows *Markovnikov's rule*. According to Markovnikov's rule, addition of unsymmetrical reagents, e.g. HX, H_2O or ROH, to an unsymmetrical alkene proceeds in a way that the hydrogen atom adds to the carbon that already has the most hydrogen atoms. The reaction is not stereoselective since it proceeds via a planar carbocation intermediate. However, when reaction proceeds via a cyclic carbocation intermediate, it produces regiospecific and stereospecific product (see below). A *regioselective reaction* is a reaction that can potentially yield two or more constitutional isomers, but actually produces only one isomer. A reaction in which one stereoisomer is formed predominantly is called a *stereoselective reaction*.

$$R_2C{=}CHR + HX \longrightarrow R_2CX{-}CHRH$$

Unsymmetrical alkene → Markovnikov addition

The modern Markovnikov rule states that, in the ionic addition of an unsymmetrical reagent to a double bond, the positive portion of the adding reagent adds to a carbon atom of the double bond to yield the more stable carbocation as an intermediate. Thus, Markovnikov addition to unsymmetrical π bonds produces regioselective product.

Addition of hydrogen halides to alkenes: preparation of alkyl halides

Alkenes are converted to alkyl halides by the addition of HX (HCl, HBr or HI). Addition of HX to unsymmetrical alkenes follows Markovnikov's rule. The reaction is regioselective, and occurs via the most stable carbocation intermediate. For example, addition of hydrogen bromide (HBr) to propene yields 2-bromopropane as the major product.

$$H_2C{=}CH(CH_3) + HBr \longrightarrow CH_3CH(Br)CH_3 + CH_3CH_2CH_2Br$$

Propene; 2-Bromopropane (Major product); 1-Bromopropane (Minor product)

Mechanism. The double bond π electrons attack the electrophile. Protonation of the double bond yields a secondary carbocation inter mediate. The bromine nucleophile attacks the carbocation to form 2-bromopropane.

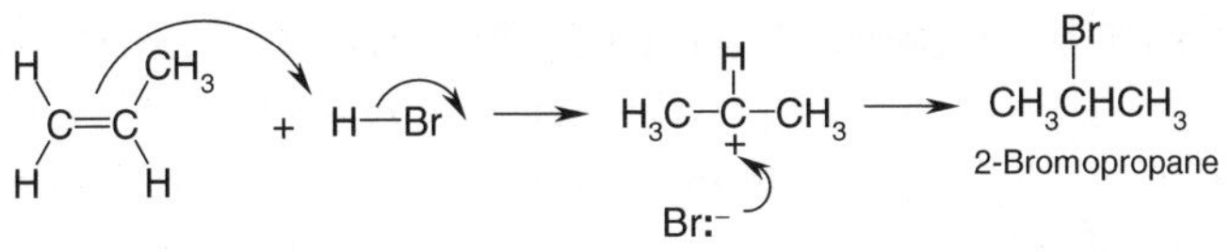

$$H_2C{=}CHCH_3 + H{-}Br \longrightarrow H_3C{-}\overset{+}{C}H{-}CH_3 \;(+\ Br:^-) \longrightarrow CH_3CHBrCH_3$$

2-Bromopropane

Addition of HBr to 2-methylpropene gives mainly *tert*-butyl bromide, because the product with the more stable carbocation intermediate always predominates in this type of reaction.

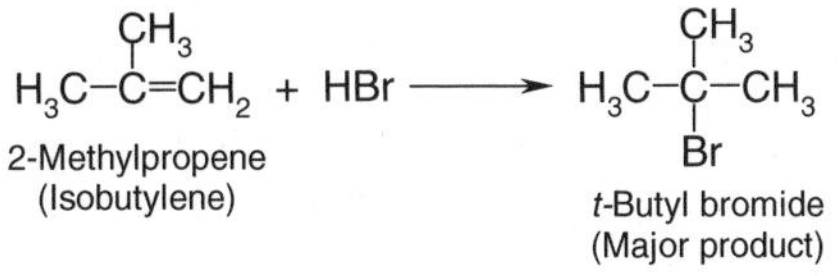

$$H_3C{-}C(CH_3){=}CH_2 + HBr \longrightarrow H_3C{-}C(CH_3)(Br){-}CH_3$$

2-Methylpropene (Isobutylene) → *t*-Butyl bromide (Major product)

Mechanism.

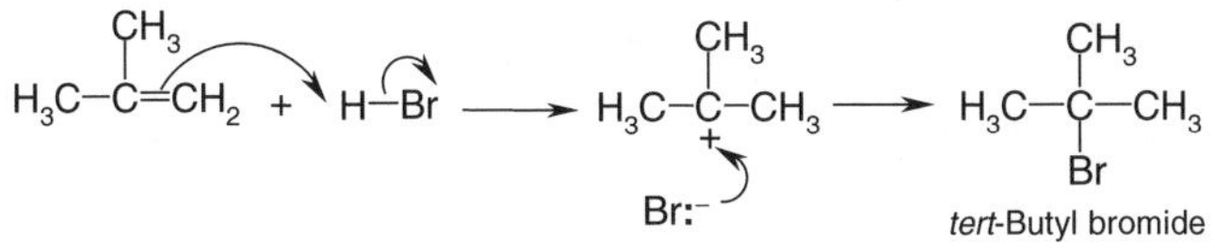

$$H_3C{-}C(CH_3){=}CH_2 + H{-}Br \longrightarrow H_3C{-}\overset{+}{C}(CH_3){-}CH_3 \;(+\ Br:^-) \longrightarrow H_3C{-}C(CH_3)(Br){-}CH_3$$

tert-Butyl bromide

Addition of HBr to 1-butene yields a chiral molecule. The reaction is *regioselective* and a racemic mixture is formed.

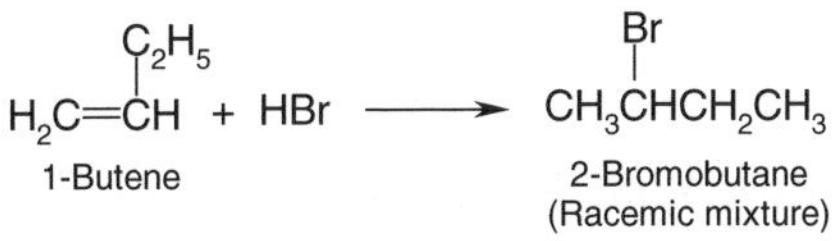

$$H_2C{=}CH(C_2H_5) + HBr \longrightarrow CH_3CHBrCH_2CH_3$$

1-Butene → 2-Bromobutane (Racemic mixture)

Mechanism.

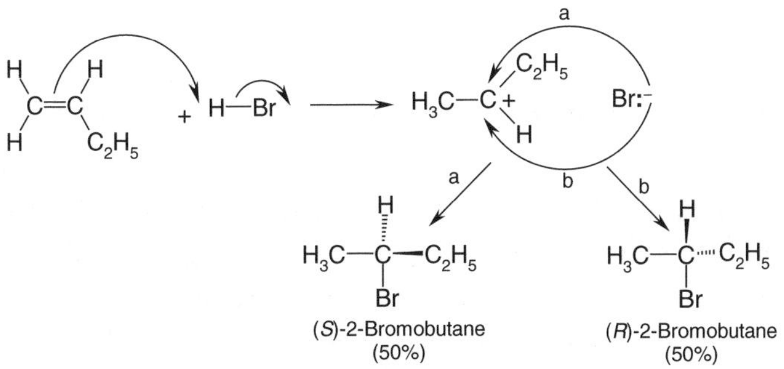

$$H_2C{=}CH(C_2H_5) + H{-}Br \longrightarrow H_3C{-}\overset{+}{C}H(C_2H_5) \;(+\ Br:^-)$$

a → (*S*)-2-Bromobutane (50%)

b → (*R*)-2-Bromobutane (50%)

Addition of hydrogen halides to alkynes: preparation of alkyl dihalides and tetrahalides

Electrophilic addition to terminal alkynes (unsymmetrical) is regioselective and follows Markovnikov's rule. Hydrogen halides can be added to alkynes just like alkenes, to form first the vinyl halide, and then the *geminal* alkyl dihalide. The addition of HX to an alkyne can be stopped after the first addition of HX. A second addition takes place when excess HX is present. For example, 1-propyne reacts with one equivalent of HCl to produce 2-chloropropene; a second addition of HCl gives 2,2-dichloropropane, a *geminal*-dihalide.

$$H_3C{-}C{\equiv}CH + 2\,HCl \longrightarrow H_3C{-}CCl_2{-}CH_3$$

1-Propyne → 2,2-Dichloropropane, A *geminal*-dihalide

Mechanism. The vinyl cation is more stable with positive charge on the more substituted carbon, because a secondary vinylic cation is more stable than a primary vinylic cation.

$$CH_3C{\equiv}CH + H{-}Cl \longrightarrow (H_3C)\overset{+}{C}{=}CH_2 \;(Cl:^-) \longrightarrow (H_3C)(Cl)C{=}CH_2 \xrightarrow{H-Cl} H_3C{-}\overset{+}{C}Cl{-}CH_3 \;(Cl:^-) \longrightarrow H_3C{-}CCl_2{-}CH_3$$

Markovnikov addition

Markovnikov addition

2,2-Dichloropropane, A *geminal*-dihalide

Addition of hydrogen halides to an internal alkyne is not regioselective. When the internal alkyne has identical groups attached to the *sp* carbons, only one *geminal*-dihalide is produced.

$$CH_3C{\equiv}CCH_3 + HCl \longrightarrow H_3C{-}CCl_2{-}C_2H_5$$

2-Butyne + HCl (Excess) → 2,2-Dichlorobutane

When the internal alkyne has different groups attached to the *sp* carbons, two *geminal*-dihalides are formed, since both intermediate cations are substituted. For example, 2-pentyne reacts with excess HBr to yield 3,3-dibromopentane and 2,2-dibromopentane.

$$C_2H_5C{\equiv}CCH_3 \xrightarrow{HBr} (C_2H_5)(H)C{=}C(Br)(CH_3) + (C_2H_5)(Br)C{=}C(H)(CH_3) \xrightarrow{HBr} C_2H_5{-}CH_2{-}CBr_2{-}CH_3 + C_2H_5{-}CBr_2{-}C_2H_5$$

2-Pentyne → 2,2-Dibromopentane + 3,3-Dibromopentane

Free radical addition of HBr to alkenes: peroxide effect. Preparation of alkyl halides

It is possible to obtain *anti*-Markovnikov products when HBr is added to alkenes in the presence of *free radical initiators*, e.g. hydrogen peroxide (HOOH) or alkyl peroxide (ROOR). The free radical initiators change the mechanism of addition from an electrophilic addition to a free radical addition. This change of mechanism gives rise to the *anti*-Markovnikov regiochemistry. For example, 2-methyl propene reacts with HBr in the presence of peroxide (ROOR) to form 1-bromo-2-methyl propane, which is an *anti*-Markovnikov product. Radical additions do not proceed with HCl or HI.

$$H_3C{-}C(CH_3){=}CH_2 + HBr \xrightarrow{ROOR} CH_3CH(CH_3){-}CH_2Br$$

2-Methyl propene; 1-Bromo-2-methyl propane, *anti*-Markovnikov addition

Initiation The oxygen–oxygen bond is weak, and is easily homolytically cleaved to generate two alkoxy radicals, which in turn abstract hydrogen to generate bromine radicals.

$$RO{-}OR \xrightarrow{h\nu} RO^{\cdot} + RO^{\cdot} \qquad RO^{\cdot} + H{-}Br \longrightarrow RO{-}H + Br^{\cdot}$$

Alkoxy radicals; Alcohol; Bromine radical

Propagation The bromine radical is electron deficient and electrophilic. The radical adds to the double bond, generating a carbon-centred radical. This radical abstracts hydrogen from HBr, giving the product and another bromine radical. The orientation of this reaction is *anti*-Markovnikov. The reversal of regiochemistry through the use of peroxides is called the *peroxide effect.*

$$(H_3C)_2C{=}CH_2 + Br^{\cdot} \longrightarrow H_3C{-}\dot{C}(CH_3){-}CH_2Br$$

Tertiary alkyl radical

$$H_3C{-}\dot{C}(CH_3){-}CH_2Br + H{-}Br \longrightarrow CH_3CH(CH_3){-}CH_2Br + Br^{\cdot}$$

Tertiary alkyl radical; 1-Bromo-2-methyl propane, *anti*-Markovnikov addition

Termination Any two radicals present in the reaction mixture can combine in a termination step, and end the radical chain reaction. Thus, radical reactions produce a mixture of products.

$$Br\cdot + Br\cdot \longrightarrow Br_2$$

Bromine radical, Bromine radical, Bromine molecule

$$H_3C-\dot{C}(CH_3)-CH_2Br + Br\cdot \longrightarrow H_3C-C(CH_3)(Br)-CH_2Br$$

Tertiary alkyl radical, Bromine radical, 1,2-Dibromo-2-methyl propane

$$H_3C-\dot{C}(CH_3)-CH_2Br \quad H_3C-\dot{C}(CH_3)-CH_2Br \longrightarrow BrCH_2-C(CH_3)_2-C(CH_3)_2-CH_2Br$$

Tertiary alkyl radical, Tertiary alkyl radical, 1,5-Dibromo-3,3,4,4-tetramethyl butane

Free radical addition of HBr to alkynes: peroxide effect. Preparation of bromoalkenes

The peroxide effect is also observed with the addition of HBr to alkynes. Peroxides (ROOR) generate *anti*-Markovnikov products, e.g. 1-butyne reacts with HBr in the presence of peroxide to form 1-bromobutene.

$$C_2H_5C{\equiv}CH + HBr \xrightarrow{ROOR} C_2H_5(H)C{=}C(H)Br$$

1-Butyne → 1-bromobutene, *anti*-Markovnikov addition

Addition of water to alkenes: preparation of alcohols

Addition of water is known as a *hydration reaction.* The hydration reaction occurs when alkenes are treated with aqueous acids, most commonly H_2SO_4, to form alcohols. This is called *acid-catalysed hydration* of alkenes, which is the reverse of the *acid-catalysed dehydration* of an alcohol.

Addition of water to an unsymmetrical alkene follows Markovnikov's rule. The reaction is highly regiospecific. According to Markovnikov's rule, in the addition of water (H—OH) to alkene, the hydrogen atom adds to the least substituted carbon of the double bond. For example, 2-methylpropene reacts with H_2O in the presence of dilute H_2SO_4 to form *t*-butyl alcohol. The reaction proceeds via protonation to give the more stable tertiary carbocation intermediate. The mechanism is the reverse of that for dehydration of an alcohol.

$$H_3C-C(CH_3){=}CH_2 + H_2O \xrightleftharpoons{H_2SO_4} H_3C-C(CH_3)(OH)-CH_3$$

2-Methylpropene (isobutylene) → *t*-Butyl alcohol (Major product)

Mechanism.

$$H_3C-\overset{CH_3}{\overset{|}{C}}=CH_2 + H-OSO_3H \xrightleftharpoons{H_2O} H_3C-\overset{CH_3}{\overset{|}{\underset{+}{C}}}-CH_3 + HSO_4^-$$

$$\Big\Updownarrow H_2\ddot{O}:$$

$$H_2SO_4 + H_3C-\overset{CH_3}{\overset{|}{\underset{OH}{\underset{|}{C}}}}-CH_3 \rightleftharpoons H_3C-\overset{CH_3}{\overset{|}{\underset{H-\overset{+}{\ddot{O}}-H}{\underset{|}{C}}}}-CH_3 \quad HSO_4^-$$

Hydration of alkenes can also be achieved either by oxymercuration–reduction (Markovnikov addition of water) or hydroboration–oxidation (*anti*-Markovnikov addition of water). Addition of water by oxymercuration–reduction or hydroboration–oxidation has two advantages over the acid-catalysed addition of water. These procedures do not require acidic condition, and carbocation rearrangements never occur. Thus, they give high yields of alcohols.

Oxymercuration–reduction of alkenes: preparation of alcohols Addition of water to alkenes by oxymercuration–reduction produces alcohols via Markovnikov addition. This addition is similar to the acid-catalysed addition of water. Oxymercuration is regiospecific and *anti*-stereospecific. In the addition reaction, Hg(OAc) bonds to the less substituted carbon, and the OH to the more substituted carbon of the double bond. For example, propene reacts with mercuric acetate in the presence of an aqueous THF to give a hydroxy-mercurial compound, followed by reduction with sodium borohydride ($NaBH_4$) to yield 2-propanol.

$$\underset{\text{Propene}}{CH_3CH{=}CH_2} \xrightarrow[\text{ii. } NaBH_4,\ NaOH]{\text{i. } Hg(OAc)_2,\ H_2O,\ THF} \underset{\substack{\text{2-Propanol} \\ \text{Markovnikov addition}}}{CH_3\overset{OH}{\overset{|}{C}}H-CH_3}$$

$$AcO = CH_3\overset{O}{\overset{\|}{C}}\text{-}O$$
Acetate

Mechanism. The reaction is analogous to the addition of bromine molecules to an alkene. The electrophilic mercury of mercuric acetate adds to the double bond, and forms a cyclic mercurinium ion intermediate rather than a planer carbocation. In the next step, water attacks the most substituted carbon of the mercurinium ion to yield the addition product. The hydroxymercurial compound is reduced *in situ* using $NaBH_4$ to give alcohol. The removal of Hg(OAc) in the second step is called *demercuration*. Therefore, the reaction is also known as *oxymercuration–demercuration*.

$$CH_3CH{=}CH_2 + Hg(OAc)OAc \longrightarrow H_3C{-}CH(^{+}HgOAc){-}CH_2 + AcO^-$$

$$\xrightarrow{H_2\ddot{O}:} H_3C{-}CH(H{-}\overset{+}{O}{-}H){-}CH_2{-}Hg{-}OAc \;(AcO^-)$$

$$\longrightarrow AcOH + H_3C{-}CH(OH){-}CH_2{-}Hg{-}OAc \xrightarrow{NaBH_4,\ NaOH} CH_3CH(OH){-}CH_3$$

2-Propanol
Markovnikov addition

Hydroboration–oxidation of alkenes: preparation of alcohols Addition of water to alkenes by hydroboration–oxidation gives alcohols via *anti*-Markovnikov addition. This addition is opposite to the acid-catalysed addition of water. Hydroboration is regioselective and *syn* stereospecific. In the addition reaction, borane bonds to the less substituted carbon, and hydrogen to the more substituted carbon of the double bond. For example, propene reacts with borane and THF complex, followed by oxidation with basic hydrogen peroxide (H_2O_2), to yield propanol.

$$\underset{\text{Propene}}{CH_3CH{=}CH_2} \xrightarrow[\text{ii. } H_2O_2,\ KOH]{\text{i. } BH_3.THF} \underset{\text{Propanol}}{CH_2CH_2CH_2OH}$$

anti-Markovnikov addition

Mechanism.

$$CH_3CH{=}CH_2 + \overset{\delta^-}{H}{-}\overset{\delta^+}{B}H{-}H \xrightarrow{THF} CH_3CH_2{-}CH_2(BH_2) \xrightarrow[H_2O_2]{^-{:}OH} CH_3CH_2CH_2OH$$

Propanol
anti-Markovnikov addition

Addition of water to alkynes: preparation of aldehydes and ketones

Internal alkynes undergo acid-catalysed addition of water in the same way as alkenes, except that the product is an enol. Enols are unstable, and tautomerize readily to the more stable keto form. Thus, enols are always in equilibrium with their keto forms. This is an example of *keto–enol tautomerism.*

Addition of water to an internal alkyne is not regioselective. When the internal alkyne has identical groups attached to the sp carbons, only one ketone is obtained. For example, 2-butyne reacts with water in the presence of acid catalyst to yield 2-butanone.

$$\underset{\text{2-Butyne}}{CH_3C{\equiv}CCH_3} \xrightarrow[H_2O]{H_2SO_4} \underset{\text{Enol}}{H_3C(HO)C{=}C(H)CH_3} \rightleftharpoons \underset{\text{2-Butanone}}{H_3C{-}C({=}O){-}C_2H_5}$$

When the internal alkyne has different groups attached to the *sp* carbons, two ketones are formed, since both intermediate cations are substituted. For example, 2-pentyne reacts with water in the presence of acid catalyst to yield 3-pentanone and 2-pentanone.

$$\underset{\text{2-Pentyne}}{C_2H_5C{\equiv}CCH_3} \xrightarrow[H_2O]{H_2SO_4} C_2H_5(HO)C{=}C(H)CH_3 + C_2H_5(H)C{=}C(OH)CH_3$$

$$\rightleftharpoons \underset{\text{3-Pentanone}}{C_2H_5{-}C({=}O){-}C_2H_5} + \underset{\text{2-Pentanone}}{C_3H_7{-}C({=}O){-}CH_3}$$

Terminal alkynes are less reactive than internal alkynes towards the acid-catalysed addition of water. Therefore, terminal alkynes require Hg salt ($HgSO_4$) catalyst for the addition of water to yield aldehydes and ketones. Addition of water to acetylene gives acetaldehyde, and all other terminal alkynes give ketones. The reaction is regioselective and follows Markovnikov addition. For example, 1-butyne reacts with water in the presence of H_2SO_4 and $HgSO_4$ to yield 2-butanone.

$$\underset{\text{1-Butyne}}{C_2H_5C{\equiv}CH} \xrightarrow[HgSO_4]{H_2O,\ H_2SO_4} \underset{\text{Enol (Markovnikov addition)}}{C_2H_5(HO)C{=}CH_2} \rightleftharpoons \underset{\text{2-Butanone}}{C_2H_5{-}C({=}O){-}CH_3}$$

Mechanism. Addition of $HgSO_4$ generates a cyclic mercurinium ion, which is attacked by a nucleophilic water molecule on the more substituted carbon. Oxygen loses a proton to form a mercuric enol, which under work-up produces enol (vinyl alcohol). The enol is rapidly converted to 2- butanone.

$$C_2H_5C{\equiv}CH \xrightarrow[H_2\ddot{O}:]{Hg(OAc)_2} C_2H_5C{=}CH\ (\text{cyclic } \overset{+}{H}gSO_4) \longrightarrow C_2H_5(H{-}\overset{+}{O}H)C{=}C(H)HgSO_4 \xrightarrow{H_2\ddot{O}:} \underset{\text{Mercuric enol}}{C_2H_5(HO)C{=}C(H)HgSO_4}$$

$$\xrightarrow{H_2O} \underset{\text{Enol (Markovnikov addition)}}{C_2H_5(HO)C{=}CH_2} \rightleftharpoons \underset{\text{2-Butanone}}{C_2H_5{-}C({=}O){-}CH_3}$$

Hydroboration–oxidation of alkynes: preparation of aldehydes and ketones Hydroboration–oxidation of terminal alkynes gives *syn* addition of water across the triple bond. The reaction is regioselective and follows *anti*-Markovnikov addition. Terminal alkynes are converted to aldehydes, and all other alkynes are converted to ketones. A sterically hindered dialkylborane must be used to prevent the addition of two borane molecules. A vinyl borane is produced with *anti*-Markovnikov orientation, which is oxidized by basic hydrogen peroxide to an enol. This enol tautomerizes readily to the more stable keto form.

RC≡CH →[(CH3)2BH] H H / R B(CH3)2 →[H2O2, KOH] H OH / R H ⇌ R H / O
Aldehyde
anti-Markovnikov addition

Mechanism.

RC≡CH + H—B(CH3)2 (δ−, δ+) →[THF] H, R C=CH B(CH3)2 :ÖH⁻ →[H2O2] H OH / R H ⇌ R H / O
anti-Markovnikov addition

Addition of sulphuric acid to alkenes: preparation of alcohols

Addition of concentrated H_2SO_4 to alkenes yields acid-soluble alkyl hydrogen sulphates. The addition follows Markovnikov's rule. The sulphate is hydrolysed to obtain the alcohol. The net result is Markovnikov addition of acid-catalysed hydration to an alkene. The reaction mechanism of H_2SO_4 addition is similar to that of acid-catalysed hydration.

$H_2C{=}CHCH_3$ (Propene) →[Conc H_2SO_4, No heat] $CH_3CH(OSO_3H)CH_3$ (Isopropyl hyrogen sulphate) →[H_2O, Heat] $CH_3CH(OH)CH_3$ (Propanol) + H_2SO_4

Addition of alcohols to alkenes: acid catalysed. Preparation of ethers

Alcohols react with alkenes in the same way as water does. The addition of alcohols in the presence of an acid catalyst, most commonly aqueous

H_2SO_4, produces ethers. Addition of alcohol to an unsymmetrical alkene follows Markovnikov's rule. The reaction proceeds via protonation to give the more stable carbocation intermediate. The mechanism is the reverse of that for dehydration of ethers.

For example, 2-methylpropane reacts with methanol (CH_3OH) in the presence of aqueous H_2SO_4 to form methyl *t*-butyl ether.

$$(H_3C)_2C{=}CH_2 + CH_3OH \xrightleftharpoons{H_2SO_4} H_3C{-}C(CH_3)(OMe){-}CH_3$$

2-Methylpropene Methyl *t*-butyl ether

Alkoxymercuration–reduction of alkenes

Addition of alcohol to alkenes by alkoxymercuration–reduction produces ethers via Markovnikov addition. This addition is similar to the acid-catalysed addition of an alcohol. For example, propene reacts with mercuric acetate in aqueous THF, followed by reduction with $NaBH_4$, to yield methyl propyl ether. The second step is known as *demercuration*, where Hg(OAc) is removed by $NaBH_4$. Therefore, this reaction is also called *alkoxymercuration–demercuration*. The reaction mechanism is exactly the same as the oxymercuration–reduction of alkenes.

$$\underset{\text{Propene}}{CH_3CH{=}CH_2} \xrightarrow[\text{ii. } NaBH_4,\ NaOH]{\text{i. } Hg(OAc)_2,\ CH_3OH,\ THF} CH_3CH(OMe)CH_3$$

Methyl propyl ether
Markovnikov addition

Addition of halides to alkenes: preparation of alkyl dihalides

Addition of X_2 (Br_2 and Cl_2) to alkenes gives *vicinal*-dihalides. This reaction is used as a test for unsaturation (π bonds), because the red colour of the bromine reagent disappears when an alkene or alkyne is present. For example, when ethylene is treated with Br_2 in CCl_4 in the dark at room temperatures, the red colour of Br_2 disappears rapidly, forming 1,2-dibromoethane, a colourless product.

$$\underset{\text{(Colourless)}}{H_2C{=}CH_2} + \underset{\text{(Red colour)}}{Br_2} \xrightarrow[\text{Dark, r.t.}]{CCl_4} H_2C(Br){-}CH_2(Br)$$

1,2-Dibromoethane
(colourless)

Mechanism. When Br_2 approaches to the double bond it becomes polarized. The positive part of the bromine molecule is attacked by the electron rich π bond, and forms a cyclic bromonium ion. The negative part of bromine is the nucleophile, which attacks the less substituted carbon to open up the cyclic bromonium ion and forms 1,2-dibromoethane (*vicinal*-dihalide).

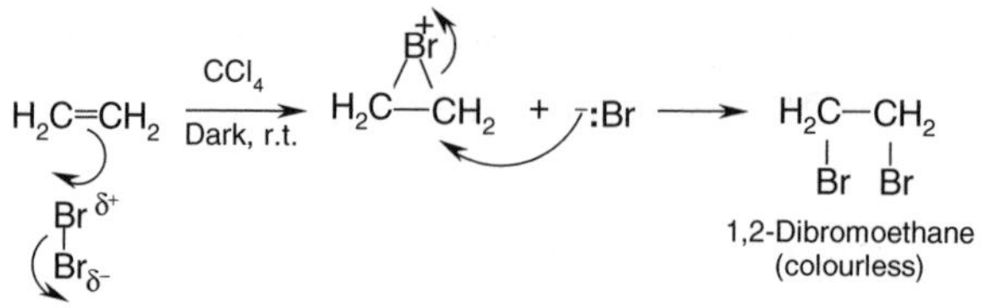

$H_2C{=}CH_2 + Br^{\delta+}{-}Br^{\delta-} \xrightarrow[\text{Dark, r.t.}]{CCl_4}$ cyclic bromonium ion $+ \; {}^-{:}Br \longrightarrow H_2C(Br){-}CH_2(Br)$

1,2-Dibromoethane (colourless)

Halogenation of double bonds is *stereospecific*. A reaction is stereospecific when a particular stereoisomeric form of the starting material gives a specific stereoisomeric form of the product. For example, the halogenation of *cis*- and *trans*-2-butene produces a racemic mixture of 2,3-dibromobutane and *meso*-2,3-dibromobutane, respectively.

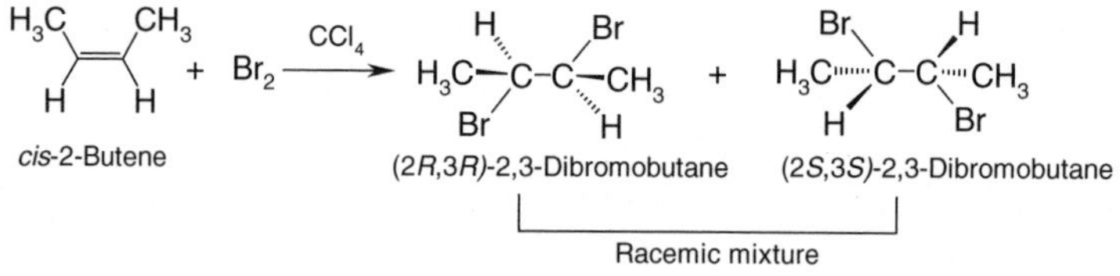

cis-2-Butene $+ Br_2 \xrightarrow{CCl_4}$ (2*R*,3*R*)-2,3-Dibromobutane + (2*S*,3*S*)-2,3-Dibromobutane

Racemic mixture

trans-2-Butene $+ Br_2 \xrightarrow{CCl_4}$ 2*R*,3*S*-2,3-Dibromobutane (Meso compound)

When cyclopentene reacts with Br_2, the product is a racemic mixture of *trans*-1,2-dibromocyclopentane. Addition of Br_2 to cycloalkenes gives a cyclic bromonium ion intermediate instead of the planar carbocation. The reaction is stereospecific, and gives only *anti* addition of dihalides.

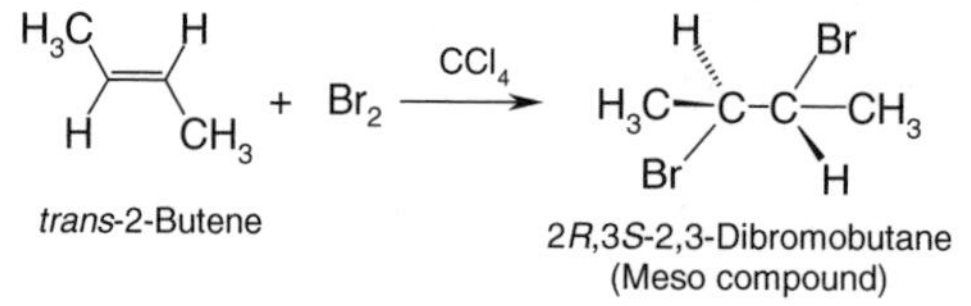

Cyclopentene $+ Br_2 \xrightarrow[\text{Dark, r.t}]{CCl_4}$ *trans*-1,2-Dibromocyclopentane (two enantiomers)

Mechanism.

Cyclopentene $+ Br_2 \rightleftharpoons$ Bromonium ion $\xrightarrow{a}$ / $\xrightarrow{b}$ *trans*-1,2-Dibromocyclopentane

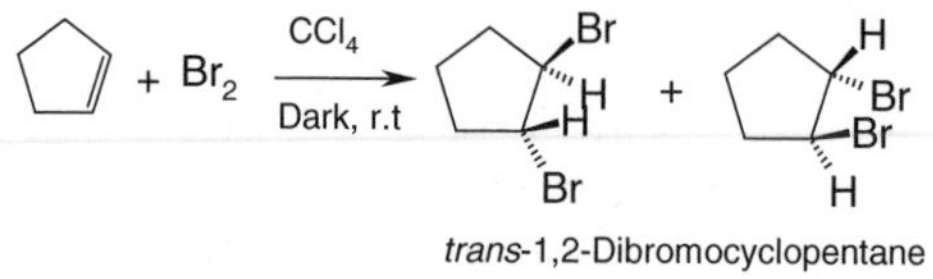

Addition of halides to alkynes: preparation of alkyl dihalides and tetrahalides

Halides (Cl_2 or Br_2) add to alkynes in an analogous fashion as for alkenes. When one mole of halogen is added, a dihaloalkene is produced, and a mixture of *syn* and *anti* addition is observed.

$$R{-}C{\equiv}C{-}R' + X_2 \longrightarrow (X)(R)C{=}C(X)(R') + (X)(R)C{=}C(R')(X)$$

cis-Alkene *trans*-Alkene

It is usually hard to control the addition of just one equivalent of halogen, and it is more common to add two equivalents to generate tetrahalides.

$$R{-}C{\equiv}C{-}R' + 2\,X_2 \longrightarrow R{-}CX_2{-}CX_2{-}R'$$

Acetylene undergoes electrophilic addition reaction with bromine in the dark. Bromine adds successively to each of the two π bonds of the alkyne. In the first stage of the reaction, acetylene is converted to an alkene, 1,2-dibromoethene. In the final stage, another molecule of bromine is added to the π bond of this alkene, and produces 1,1,2,2-tetrabromoethane.

$$HC{\equiv}CH + Br_2 \xrightarrow[25\ ^{o}C]{CCl_4,\ dark} H{-}C(Br){=}C(Br){-}H + Br_2 \xrightarrow[25\ ^{o}C]{CCl_4,\ dark} H{-}C(Br)_2{-}C(Br)_2{-}H$$

Acetylene (Alkyne); Bromine (Red); 1,2-Dibromoethene; 1,1, 2, 2-Dibromoethane (Colourless)

Addition of halides and water to alkenes: preparation of halohydrins

When halogenation of alkenes is carried out in aqueous solvent, a *vicinal* halohydrin is obtained. The reaction is regioselective, and follows the Markovnikov rule. The halide adds to the less substituted carbon atom via a *bridged halonium ion* intermediate, and the hydroxyl adds to the more substituted carbon atom. The reaction mechanism is similar to the halogenation of alkenes, except that instead of the halide nucleophile, the water attacks as a nucleophile.

$$R_2C{=}CHR + X_2 \xrightarrow[X\ =\ Cl_2\ or\ Br_2]{H_2O} R{-}C(R)(OH){-}C(R)(X){-}H + HX$$

Halohydrin
Markovinkov addition

Mechanism.

Slow $H_2\ddot{O}$: R–C–C–H + X:⁻ Fast → R–C–C–H with H–O^+–H and X; $H_2\ddot{O}$: Fast → R–C(R)(OH)–C(R)(X)–H

Addition of carbenes to alkenes: preparation of cyclopropanes

Carbenes are divalent carbon compounds, also known as methylene. They have neutral carbons with a lone pair of electrons, and are highly reactive. Methylene can be prepared by heat or light initiated decomposition of diazomethane (explosive and toxic gas).

$$:\bar{C}H_2\text{—}\overset{+}{N}{\equiv}\bar{N}: \xrightarrow[\text{or heat}]{h\nu} :\bar{C}H_2 + :N{\equiv}N:$$

Diazomethane → Methylene (Carbene) + Nitrogen

Addition of methylene (CH_2) to alkenes gives substituted cyclopropanes. For example, methylene reacts with ethylene to form cyclopropane.

$$H_2C{=}CH_2 + :CH_2 \longrightarrow \text{cyclo-}(CH_2)_3$$

Ethylene + Methylene (Carbene) → Cyclopropane

5.3.2 Nucleophilic addition to carbonyl groups

The most common reaction of aldehyde and ketone is *nucleophilic addition*. Aldehyde generally undergoes nucleophilic addition more readily than ketone. In the nucleophilic addition reaction, carbonyl compound can behave as both Lewis acid and Lewis base, depending on the reagents. The carbonyl group is strongly polarized, with the oxygen bearing partial negative charge (δ^-) and the carbon bearing partial positive charge (δ^+). So the carbon is electrophilic, and therefore readily attacked by the nucleophile. The attacking nucleophile can be either negatively charged (Nu:−) or a neutral (Nu:) molecule. Aldehydes and ketones react with nucleophiles to form addition products followed by protonation.

If the nucleophile is a negatively charged anion (good nucleophile, e.g. HO^-, RO^- and H^-), it readily attacks the carbonyl carbon, and forms an alkoxide tetrahedral intermediate, which is usually protonated in a subsequent step either by the solvent or by added aqueous acid.

$$R{-}\overset{\delta^-}{C}({=}O){-}Y \xrightarrow{Nu:^-} R{-}C(O^-)(Nu){-}Y \underset{}{\overset{H_3O^+}{\rightleftharpoons}} R{-}C(OH)(Nu){-}Y$$

Alkoxide
Tetrahedral intermediate

Alcohol
Y = H or R

If the nucleophile is a neutral molecule with a lone pair of electrons (weaker nucleophile, e.g. water or alcohol), it requires an acid catalyst. The carbonyl oxygen is protonated by acid, which increases the susceptibility of the carbonyl carbon to nucleophilic attack.

$$R{-}C({=}O){-}Y \overset{H^+}{\rightleftharpoons} R{-}C({=}\overset{+}{O}H){-}Y \xrightarrow{Nu:^-} R{-}C(OH)(Nu){-}Y$$

Y = H or R

Alcohol

If the attacking nucleophile has a pair of nonbonding electrons available in the addition product, water is eliminated in the presence of anhydrous acid from the addition product. This is known as *nucleophilic addition–elimination* reaction.

$$R{-}C(OH)(Nu){-}Y \overset{\text{Dry } H^+}{\rightleftharpoons} R{-}C(\overset{+}{O}H_2)(Nu){-}Y \rightleftharpoons R{-}C({=}\overset{+}{N}u){-}Y + H_2O$$

Nu: = ROH, RNH_2

Y = H or R

Addition of organometallic reagents to carbonyl compounds

Aldehydes and ketones react with organometallic reagents to yield different classes of alcohols depending on the starting carbonyl compound. Nucleophilic addition of Grignard reagent (RMgX) or organolithium (RLi) to the carbonyl (C=O) group is a versatile and useful synthetic reaction. These reagents add to the carbonyl, and protonated in a separate step by the solvent or by added acid.

General mechanism.

$$R{-}C({=}O){-}Y + \overset{\delta^-}{R'}{-}\overset{\delta^+}{MgBr} \xrightarrow{\text{Dry ether}} R{-}C(\overset{-}{O}\overset{+}{Mg}Br)(R'){-}Y \xrightarrow{H_3O^+} R{-}C(OH)(R'){-}Y$$

Y = H or R

Addition of organometallic reagent to formaldehyde: preparation of primary alcohols

Formaldehyde reacts with a Grignard or organolithium reagent to generate a primary alcohol, which contains one more carbon atom than the original Grignard reagent. For example, formaldehyde reacts with methyl magnesium bromide to yield ethanol.

$$\underset{\text{Formaldehyde}}{H-\overset{:O:}{\overset{\|}{C}}-H} \xrightarrow[\text{Dry ether}]{CH_3MgBr} H-\overset{:\ddot{O}^{-}\overset{+}{M}gBr}{\overset{|}{\underset{CH_3}{\underset{|}{C}}}}-H \xrightarrow{H_3O^+} \underset{\substack{\text{Ethanol} \\ (1^\circ\ \text{Alcohol})}}{CH_3CH_2OH}$$

Addition of organometallic reagent to aldehyde: preparation of secondary alcohols

Reaction of an aldehyde with a Grignard or organolithium reagent generates a secondary alcohol. For example, acetaldehyde reacts with methyl magnesium bromide to give 2-propanol.

$$\underset{\text{Acetaldehyde}}{H_3C-\overset{:O:}{\overset{\|}{C}}-H} \xrightarrow[\text{Dry ether}]{CH_3MgBr} H_3C-\overset{:\ddot{O}^{-}\overset{+}{M}gBr}{\overset{|}{\underset{CH_3}{\underset{|}{C}}}}-H \xrightarrow{H_3O^+} \underset{\substack{\text{2-Propanol} \\ (2^\circ\ \text{Alcohol})}}{CH_3\overset{CH_3}{\overset{|}{C}}HOH}$$

Addition of organometallic reagent to ketone: preparation of tertiary alcohols

Addition of Grignard or organolithium reagents to a ketone gives tertiary alcohol. For example, acetone reacts with methyl magnesium bromide to yield *t*-butanol.

$$\underset{\text{Acetone}}{H_3C-\overset{:O:}{\overset{\|}{C}}-CH_3} \xrightarrow[\text{Dry ether}]{CH_3MgBr} H_3C-\overset{:\ddot{O}^{-}\overset{+}{M}gBr}{\overset{|}{\underset{CH_3}{\underset{|}{C}}}}-CH_3 \xrightarrow{H_3O^+} \underset{\textit{tert}\text{-Butanol}\ (3^\circ\ \text{alcohol})}{H_3C-\overset{CH_3}{\overset{|}{\underset{CH_3}{\underset{|}{C}}}}-OH}$$

Carbonation of Grignard reagent: preparation of carboxylic acids

Grignard reagent reacts with CO_2 to give magnesium salts of carboxylic acid. Addition of aqueous acid produces carboxylic acid.

$$RCH_2{-}MgX \xrightarrow{CO_2} RCH_2{-}\overset{O}{\overset{\|}{C}}{-}\overset{-\ +}{OMgX} \xrightarrow{H_3O^+} RCH_2{-}\overset{O}{\overset{\|}{C}}{-}OH$$

Carboxylic acid

Addition of organometallic reagent to nitrile: preparation of ketones

Grignard or organolithium reagent attacks nitrile to generate the magnesium or lithium salt of imine. Acid hydrolysis of this salt generates a ketone. Since the ketone is not formed until the work-up, the organometallic reagent does not have the opportunity to react with the ketone.

$$\underset{\text{Nitrile}}{RC{\equiv}N} + R'MgBr \longrightarrow \left[\begin{matrix} R' \\ R \end{matrix} {>}C{=}N{-}MgBr \right] \xrightarrow{H_3O^+} \underset{\text{Ketone}}{\begin{matrix} R' \\ R \end{matrix} {>}C{=}O} + NH_4^+$$

Addition of acetylides and alkynides to carbonyl compounds

Acetylide (RC≡CNa) and alkynide (RC≡CMgX and RC≡CLi) are good nucleophiles. They react with carbonyl group to from alkoxide, which under acidic work-up gives alcohol. The addition of acelylides and alkynides produces similar alcohols to organometallic reagents.

$$R{-}\overset{:O:}{\overset{\|}{C}}{-}H \xrightarrow{R'C{\equiv}\overset{-\ +}{CNa}} R{-}\underset{H}{\overset{:\ddot{O}^-}{C}}{-}C{\equiv}C{-}R' \xrightarrow{H_3O^+} R{-}\underset{H}{\overset{:\ddot{O}H}{C}}{-}C{\equiv}C{-}R'$$

$$R{-}\overset{:O:}{\overset{\|}{C}}{-}R \xrightarrow[\text{or } R'C{\equiv}\overset{-\ +}{CLi}]{R'C{\equiv}\overset{-\ +}{CMgX}} R{-}\underset{R}{\overset{:\ddot{O}^-}{C}}{-}C{\equiv}C{-}R' \xrightarrow{H_3O^+} R{-}\underset{R}{\overset{:\ddot{O}H}{C}}{-}C{\equiv}C{-}R'$$

Addition of phosphorus ylide to carbonyl compounds: Wittig reaction

Georg Wittig (1954) discovered that the addition of a phosphorus ylide (stabilized anion) to an aldehyde or a ketone generates an alkene, not an alcohol. This reaction is known as *Wittig reaction.*

$$\underset{Y\,=\,H\text{ or }R}{\begin{matrix} R \\ Y \end{matrix}{>}{=}O} + \underset{\text{Phosphorus ylide}}{\begin{matrix} R' \\ R' \end{matrix}{>}\overset{-}{C}{-}\overset{+}{P}(Ph)_3} \longrightarrow \underset{\text{Alkene}}{\begin{matrix} R \\ Y \end{matrix}{>}{=}{<}\begin{matrix} R' \\ R' \end{matrix}} + \underset{\text{Triphenylphosphine oxide}}{Ph_3P{=}O}$$

Preparation of phosphorus ylide Phosphorus ylides are produced from the reaction of triphenylphosphine and alkyl halides. *Phosphorus ylide* is a molecule that is overall neutral, but exists as a carbanion bonded to a positively charged phosphorus. The ylide can also be written in the double-bonded form, because phosphorus can have more than eight valence electrons.

$$\underset{\text{Alkyl halide}}{RCH_2{-}X} \xrightarrow[\text{ii. BuLi, THF}]{\text{i. }(Ph)_3P:} \underset{\text{Phosphorus ylide}}{(Ph)_3\overset{+}{P}{-}\overset{-}{C}HR \longleftrightarrow (Ph)_3P{=}CHR}$$

Mechanism. In the first step of the reaction, the nucleophilic attack of the phosphorus on the primary alkyl halide generates an alkyl triphenylphosphonium salt. Treatment of this salt with a strong base, e.g. butyllithium, removes a proton to generate the ylide. The carbanionic character of the ylide makes it a powerful nucleophile.

$$RCH_2{-}X + (Ph)_3P: \longrightarrow (Ph)_3\overset{+}{P}{-}CH(R){-}H + X^- \xrightarrow[\text{THF}]{\overset{-}{Bu}\overset{+}{Li}} \underset{\text{Phosphorus ylide}}{(Ph)_3\overset{+}{P}{-}\overset{-}{C}HR \longleftrightarrow (Ph)_3P{=}CHR}$$

Preparation of alkenes Ketone reacts with phosphorus ylide to give alkene. By dividing a target molecule at the double bond, one can decide which of the two components should best come from the carbonyl, and which from the ylide. In general, the ylide should come from an unhindered alkyl halide since triphenyl phosphine is bulky.

$$\underset{\text{Phosphorus ylide}}{R_2\overset{-}{C}{-}\overset{+}{P}(Ph)_3} + \underset{\text{Ketone}}{R'_2C{=}O} \longrightarrow \underset{\text{Alkene}}{R_2C{=}CR'_2} + \underset{\text{Triphenylphosphine oxide}}{Ph_3P{=}O}$$

Mechanism.

$$(Ph)_3\overset{+}{P}{-}\overset{-}{C}R_2 + R'_2C{=}O \longrightarrow \underset{\text{Betaine}}{Ph_3\overset{+}{P}{-}CR_2{-}CR'_2{-}O^-} \longrightarrow Ph_3P{-}O \text{ (four-membered ring with } R_2C{-}CR'_2) \longrightarrow Ph_3P{=}O + \underset{\text{Alkene}}{R_2C{=}CR'_2}$$

Phosphorus ylide reacts rapidly with aldehydes and ketones to produce an intermediate called a *betaine*. Betaines are unusual since they contain negatively charged oxygen and positively charged phosphorus. Phosphorus and oxygen always form strong bonds, and these groups therefore combine to generate a four-membered ring, an oxaphosphetane ring. This four-membered ring quickly decomposes to generate an alkene and a stable triphenyl phosphine oxide ($Ph_3P{=}O$). The net result is replacement of the carbonyl oxygen atom by the $R_2C{=}$ group, which was originally bonded to the phosphorus atom. This is a good synthetic route to make alkenes from aldehydes and ketones.

Addition of hydrogen cyanide to carbonyl compounds: preparation of cyanohydrins

Addition of hydrogen cyanide to aldehyde and ketone forms cyanohydrin. The reaction is usually carried out using sodium or potassium cyanide with HCl. Hydrogen cyanide is a toxic volatile liquid, and a weak acid. Therefore, the best way to carry out this reaction is to generate it *in situ* by adding HCl to a mixture of aldehydes or ketones and excess sodium or potassium cyanide. Cyanohydrins are useful in organic reaction, because the cyano group can be converted easily to an amine, amide or carboxylic acid.

$$R{-}C({=}O){-}R \underset{H_2O}{\overset{KCN,\ HCl}{\rightleftharpoons}} R{-}C(O^-)(CN){-}R \xrightarrow{HCN} R{-}C(OH)(CN){-}R$$

Cyanohydrin

Addition of ammonia and its derivatives to carbonyl compounds: preparation of oximes and imine derivatives (Schiff's bases)

Ammonia and its derivatives, e.g. primary amines (RNH_2), hydroxylamine (NH_2OH), hydrazine (NH_2NH_2) and semicarbazide ($NH_2NHCONH_2$), react with aldehydes and ketones in the presence of an acid catalyst to generate imines or substituted imines. An imine is a nitrogen analogue of an aldehyde or a ketone with a C=N nitrogen double bond instead of a C=O. Imines are nucleophilic and basic. Imines obtained from ammonia do not have a substituent other than a hydrogen atom bonded to the nitrogen. They are relatively unstable to be isolated, but can be reduced *in situ* to primary amines.

$$R{-}C({=}O){-}Y + NH_3 \underset{H_3O^+}{\overset{\text{Dry } H^+}{\rightleftharpoons}} \left[R{-}C({=}NH){-}Y\right] + H_2O$$

Y = H or R; Unstable imine

Imines obtained from hydroxylamines are known as *oximes*, and imines obtained from primary amines are called *Schiff's bases*. An imine is formed in the presence of an anhydrous acid catalyst.

$$\mathrm{R{-}C({=}O){-}Y}\ (Y = H\ or\ R) \xrightarrow[\text{Dry } H^+]{NH_2{-}OH} \mathrm{R{-}C({=}N{-}OH){-}Y}\ \text{(Oxime)}$$

$$\xrightarrow[\text{Dry } H^+]{R'NH_2} \mathrm{R{-}C({=}N{-}R'){-}Y}\ \text{(Imine (Schiff's base))}$$

$$\xrightarrow[\text{Dry } H^+]{R'_2NH} \mathrm{R{-}C({=}N^+R'_2){-}Y}\ \text{(Iminium salt)}$$

The reaction is reversible, and the formation of all imines (Schiff's base, oxime, hydrazone and semicarbazide) follows the same mechanism. In aqueous acidic solution, imines are hydrolysed back to the parent aldehydes or ketones, and amines.

Mechanism. The neutral amine nucleophile attacks the carbonyl carbon to form a dipolar tetrahedral intermediate. The intramolecular proton transfer from nitrogen and oxygen yields a neutral carbinolamine tetrahedral intermediate. The hydroxyl group is protonated, and the dehydration of the protonated carbinolamine produces an iminium ion and water. Loss of proton to water yields the imine and regenerates the acid catalyst.

$$\mathrm{R{-}C({=}\ddot{O}){-}Y} + \mathrm{R'\ddot{N}H_2} \underset{H_3O^+}{\rightleftharpoons} \mathrm{R{-}C(O^-)(^+NHR'H){-}Y} \underset{}{\overset{\pm H^+}{\rightleftharpoons}} \mathrm{R{-}C(\ddot{O}H)(\ddot{N}HR'){-}Y} \overset{\text{Dry } H^+}{\rightleftharpoons} \mathrm{R{-}C(\overset{+}{O}H_2)(\ddot{N}HR'){-}Y}$$

Neutral carbinolamine tetrahedral intermediate

$$\overset{H_3O^+}{\rightleftharpoons} \mathrm{H_2\ddot{O}} + \mathrm{R{-}C({=}\overset{+}{N}H{-}R'){-}Y} \rightleftharpoons \mathrm{H_3\ddot{O}^+} + \mathrm{R{-}C({=}NR'){-}Y}$$

Preparation of hydrazone and semicarbazones Imines obtained from hydrazines are known as *hydrazones*, and imines obtained from semicarbazides are called *semicarbazones*.

$$\underset{\text{Hydrazone}}{\mathrm{R{-}C({=}N{-}NH_2){-}Y}} \xleftarrow[\text{Dry } H^+]{NH_2NH_2} \underset{Y = H\ or\ R}{\mathrm{R{-}C({=}O){-}Y}} \xrightarrow[\text{Dry } H^+]{NH_2NHCONH_2} \underset{\text{Semicarbazone}}{\mathrm{R{-}C({=}N{-}NHCONH_2){-}Y}}$$

Mechanism. The hydrazine nucleophile attacks the carbonyl carbon, and forms a dipolar tetrahedral intermediate. Intramolecular proton transfer produces a neutral tetrahedral intermediate. The hydroxyl group is protonated, and the dehydration yields an ionic hydrazone and water. Loss of a proton to water produces the hydrazone and regenerates the acid catalyst.

$$R{-}C(=O){-}Y + NH_2NH_2 \underset{H_3O^+}{\rightleftharpoons} R{-}C(O^-)(Y){-}{}^{+}NH_2NH_2 \overset{\pm H^+}{\rightleftharpoons} R{-}C(OH)(Y){-}NHNH_2 \overset{\text{Dry } H^+}{\rightleftharpoons} R{-}C({}^{+}OH_2)(Y){-}NHNH_2$$

Neutral tetrahedral intermediate

$$\rightleftharpoons R{-}C(Y){=}{}^{+}NH{-}NH_2 + H_2O \rightleftharpoons H_3O^+ + R{-}C(Y){=}NNH_2$$

Hydrazone

Addition of secondary amine to carbonyl compounds: preparation of enamines

Secondary amine reacts with aldehyde and ketone to produce enamine. An *enamine* is an α,β-unsaturated tertiary amine. Enamine formation is a reversible reaction, and the mechanism is exactly the same as the mechanism for imine formation, except the last step of the reaction.

$$RCH_2{-}C(=O){-}Y + R'_2NH \underset{H_3O^+}{\overset{\text{Dry } H^+}{\rightleftharpoons}} RCH{=}C(NR'_2){-}Y + H_2O$$

Y = H or R; product: Enamine

Mechanism.

$$RCH_2{-}C(=O){-}Y + R'_2NH \rightleftharpoons RCH_2{-}C(O^-)(Y){-}{}^{+}NHR'_2 \overset{\pm H^+}{\rightleftharpoons} RCH_2{-}C(OH)(Y){-}NR'_2$$

Neutral tetrahedral intermediate

$$\overset{\text{Dry } H^+}{\rightleftharpoons} RCH_2{-}C({}^{+}OH_2)(Y){-}NR'_2 \rightleftharpoons RCH_2{-}C(Y){=}{}^{+}NR'_2 + H_2O \rightleftharpoons RCH{=}C(NR'_2){-}Y$$

Enamine

Addition of water to carbonyl compounds: acid-catalysed hydration. Preparation of diols

Aldehyde and ketone reacts with water in the presence of aqueous acid or base to form hydrate. A *hydrate* is a molecule with two hydroxyl groups on the same carbon. It is also called *gem*-diol. Hydration proceeds through the two classic nucleophilic addition mechanisms with water in acid conditions or hydroxide in basic conditions. Hydrates of aldehydes or ketones are generally unstable to isolate.

$$\text{R-C(=O)-Y} + H_2O \xrightleftharpoons{H_3O^+ \text{ or NaOH}} \text{R-C(OH)}_2\text{-Y}$$

Y = H or R; Hydrate (*gem*-diol)

Acid conditions

$$\text{R-C(=O)-Y} \underset{}{\overset{H\text{-}\overset{+}{O}H_2}{\rightleftharpoons}} \text{R-C(=}\overset{+}{O}\text{H)-Y} \rightleftharpoons \text{R-C(OH)(}\overset{+}{O}H_2\text{)-Y} \rightleftharpoons \text{R-C(OH)}_2\text{-Y} + H_3O^+$$

Y = H or R

Basic conditions

$$\text{R-C(=O)-Y} \underset{H_2O}{\overset{HO^-}{\rightleftharpoons}} \text{R-C(O}^-\text{)(OH)-Y} \overset{H\text{-}OH}{\rightleftharpoons} \text{R-C(OH)}_2\text{-Y} + HO^-$$

Y = H or R

Addition of alcohol to carbonyl compounds: preparation of acetal and ketal

In a similar fashion to the formation of hydrate with water, aldehyde and ketone react with alcohol to form acetal and ketal, respectively. In the formation of an acetal, two molecules of alcohol add to the aldehyde, and one mole of water is eliminated. An alcohol, like water, is a poor nucleophile. Therefore, the acetal formation only occurs in the presence of anhydrous acid catalyst. Acetal or ketal formation is a reversible reaction, and the formation follows the same mechanism. The equilibrium lies towards the formation of acetal when an excess of alcohol is used. In hot aqueous acidic solution, acetals or ketals are hydrolysed back to the carbonyl compounds and alcohols.

$$\text{R-C(=O)-H} \underset{H_3O^+}{\overset{ROH/\,H^+}{\rightleftharpoons}} \text{R-C(OH)(OR')-H} \underset{H_3O^+}{\overset{ROH/\,H^+}{\rightleftharpoons}} \text{R-C(OR')}_2\text{-H} + H_2O$$

Aldehyde; Hemiacetal; Acetal

Mechanism. The first step is the typical acid-catalysed addition to the carbonyl group. Then the alcohol nucleophile attacks the carbonyl carbon, and forms a tetrahedral intermediate. Intramolecular proton transfer from nitrogen and oxygen yields a hemiacetal tetrahedral intermediate. The hydroxyl group is protonated, followed by its leaving as water to form hemi-acetal, which reacts further to produce the more stable acetal.

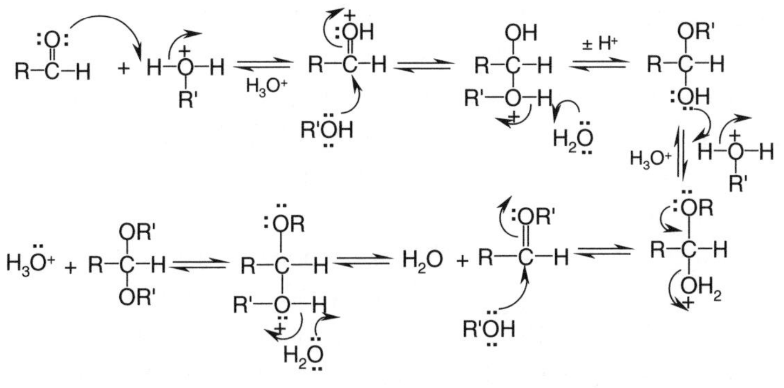

Instead of two molecules of alcohols, a diol is often used. This produces cyclic acetals. 1,2-ethanediol (ethylene glycol) is usually the diol of choice, and the products are called ethylene acetals.

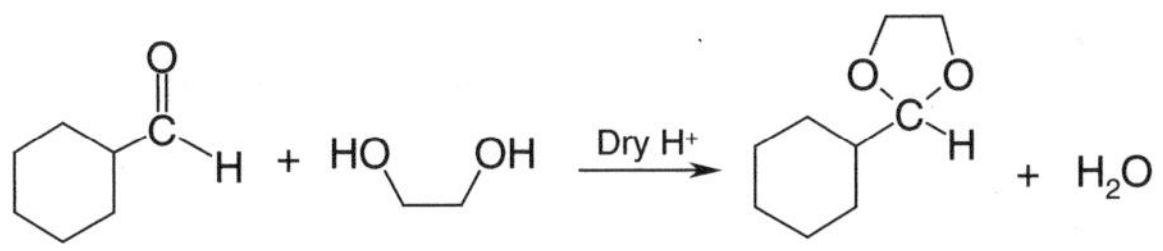

Acetal as protecting group

A protecting group converts a reactive functional group into a different group that is inert to the reaction conditions in which the reaction is carried out. Later, the protecting group is removed. Acetals are hydrolysable under acidic conditions, but are stable to strong bases and nucleophiles. These characteristics make acetals ideal protecting groups for aldehydes and ketones. They are also easily formed from aldehydes and ketones, and easily converted back to the parent carbonyl compounds. They can be used to protect aldehydes and ketones from reacting with strong bases and nucleophiles, e.g. Grignard reagents and metal hydrides.

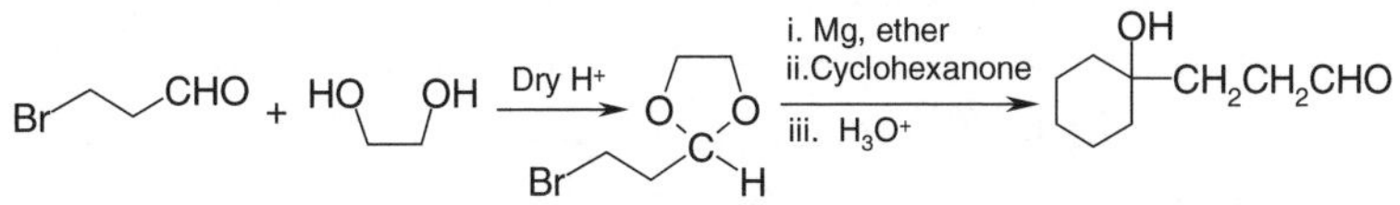

Aldehydes are more reactive than ketones. Therefore, aldehydes react with ethylene glycol to form acetals preferentially over ketones. Thus, aldehydes can be protected selectively. This is a useful way to perform reactions on ketone functionalities in molecules that contain both aldehyde and ketone groups.

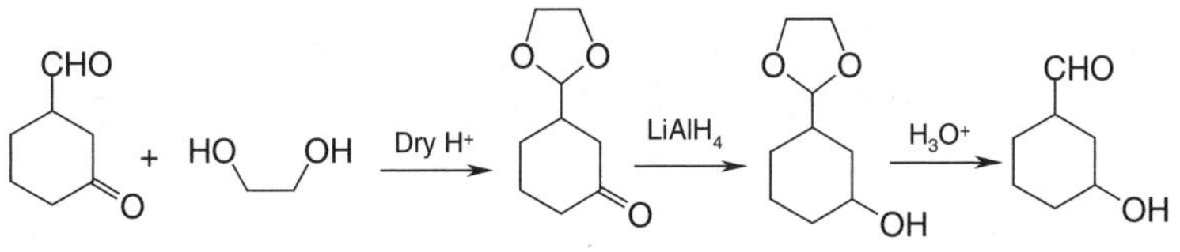

Aldol condensation

In aldol condensation, the enolate anion of one carbonyl compound reacts as a nucleophile, and attacks the electrophilic carbonyl group of another one to form a larger molecule. Thus, the aldol condensation is a *nucleophilic addition* reaction.

The hydrogen (known as an α-hydrogen) bonded to a carbon adjacent to a carbonyl carbon (called an α-carbon) is acidic enough to be removed by a strong base, usually NaOH, to form an enolate anion. The enolate anion adds to the carbonyl carbon of a second molecule of aldehyde or ketone via nucleophilic addition reaction.

$$R{-}CH(H){-}C(=O){-}Y \xrightarrow[\;]{NaOH,\ H_2O} R{-}\overset{-}{C}H{-}C(=O){-}Y \longleftrightarrow R{-}CH{=}C(O^-){-}Y$$

Y = H or R

Resonance stabilized enolate anion

Aldol condensation reaction may be either acid or base catalysed. However, base catalysis is more common. The product of this reaction is called an *aldol*, i.e. *ald* from aldehyde and *ol* from alcohol. The product is either a β-hydroxyaldehyde or β-hydroxyketone, depending on the starting material. For example, two acetaldehyde (ethanal) molecules condense together in the presence of an aqueous base (NaOH), to produce 3-hydroxybutanal (a β-hydroxyaldehyde).

$$H_3C{-}C(=O){-}H + H_3C{-}C(=O){-}H \xrightarrow{NaOH,\ H_2O} H_3C{-}CH(OH){-}CH_2{-}C(=O){-}H$$

Acetaldehyde Acetaldehyde 3-Hydroxybutanal (A β-Hydroxyaldehyde)

Mechanism. Removal of an α-hydrogen from the acetaldehyde by NaOH produces a resonance-stabilized enolate anion. Nucleophilic addition of the enolate to the carbonyl carbon of another acetaldehyde gives an alkoxide tetrahedral intermediate. The resulting alkoxide is protonated by the solvent, water, to give 3-hydroxybutanal and regenerate the hydroxide ion.

$$H{-}CH(H){-}C(=O){-}H \rightleftharpoons \overset{-}{C}H_2{-}C(=O){-}H \longleftrightarrow H_2C{=}C(O^-){-}H$$

Resonance stabilized enolate anion

$$H_3C{-}C(=O){-}H + \overset{-}{C}H_2{-}C(=O){-}H \rightleftharpoons H_3C{-}CH(O^-){-}CH_2{-}C(=O){-}CH_3 \overset{H_2O}{\rightleftharpoons} H_3C{-}CH(OH){-}CH_2{-}C(=O){-}CH_3 + HO^-$$

3-Hydroxybutanal

5.4 Elimination reactions: 1,2-elimination or *β*-elimination

The term *elimination* can be defined as the electronegative atom or a leaving group being removed along with a hydrogen atom from adjacent carbons in the presence of strong acids or strong bases and high temperatures. Alkenes can be prepared from alcohols or alkyl halides by elimination reactions. The two most important methods for the preparation of alkenes are *dehydration* ($-H_2O$) of alcohols, and *dehydrohalogenation* ($-HX$) of alkyl halides. These reactions are the reverse of the *electrophilic addition* of water and hydrogen halides to alkenes.

β Carbon α Carbon

$$CH_3CH_2OH \xrightarrow[\text{Heat}]{\text{Conc. } H_2SO_4} CH_2{=}CH_2 + H_2O$$

Ethyl alcohol → Ethylene

$$CH_3CH_2Cl \xrightarrow[\text{Heat}]{\text{Alcoholic KOH}} CH_2{=}CH_2 + HCl$$

Ethyl chloride → Ethylene

In 1,2-elimination, e.g. dehydrohalogenation of alkyl halide, the atoms are removed from adjacent carbons. This is also called β-*elimination*, because a proton is removed from a β-carbon. The carbon to which the functional group is attached is called the α-carbon. A carbon adjacent to the α-carbon is called a β-carbon.

Depending on the relative timing of the bond breaking and bond formation, different pathways are possible: *E1 reaction* or unimolecular elimination and *E2 reaction* or bimolecular elimination.

$$\text{H–C–C–X} + B{:}^- \xrightarrow{\text{Heat}} \text{C=C} + \text{B-H} + X{:}^-$$

Alkyl halide + Base → Alkene

5.4.1 E1 reaction or first order elimination

E1 reaction or *first order elimination* results from the loss of a leaving group to form a carbocation intermediate, followed by the removal of a proton to form the C=C bond. This reaction is most common with good leaving groups, stable carbocations and weak bases (strong acids). For example, 3-bromo-3-methyl pentane reacts with methanol to give 3-methyl-2-pentene. This reaction is unimolecular, i.e. the rate-determining step involves one molecule, and it is the slow ionization to generate a carbocation. The second step is the fast removal of a proton by the base (solvent) to form the C=C bond. In fact, any base in the reaction mixture (ROH, H_2O, HSO_4^-) can remove the proton in the elimination reaction. The E1 is not particularly useful from a synthetic point of view, and occurs in competition with S_N1 reaction of tertiary alkyl halides. Primary and secondary alkyl halides do not usually react with this mechanism.

$$H{-}C(CH_3)(H){-}C(CH_3)(Br){-}C_2H_5 \xrightarrow[\text{Heat}]{CH_3OH} (H_3C)(H)C{=}C(CH_3)(C_2H_5) + HBr + CH_3OH$$

3-Bromo-3-methyl pentane → 3-Methyl-2-pentene

Mechanism.

$$H{-}C(CH_3)(H){-}C(CH_3)(Br){-}C_2H_5 \underset{}{\overset{\text{Slow}}{\rightleftharpoons}} H{-}C(CH_3)(H){-}C^{+}(CH_3){-}C_2H_5 + Br:^-$$

$$H_3C{-}\ddot{O}H \quad \xrightarrow{\text{Fast}} \quad (H_3C)(H)C{=}C(CH_3)(C_2H_5) + CH_3\overset{+}{O}H_2 + Br:^-$$

3-Methyl-2-pentene

5.4.2 E2 reaction or second order elimination

E2 elimination or *second order elimination* takes place through the removal of a proton and simultaneous loss of a leaving group to form the C=C bond. This reaction is most common with high concentration of strong bases (weak acids), poor leaving groups and less stable carbocations. For example, 3-chloro-3-methyl pentane reacts with sodium methoxide to give 3-methyl-2-pentene. The bromide and the proton are lost simultaneously to form the alkene. The E2 reaction is the most effective for the synthesis of alkenes from primary alkyl halides.

$$H{-}C(CH_3)(H){-}C(CH_3)(Cl){-}C_2H_5 \xrightarrow[CH_3OH,\ \text{heat}]{CH_3ONa} (H_3C)(H)C{=}C(CH_3)(C_2H_5) + CH_3OH + NaCl$$

3-Chloro-3-methyl pentane → 3-Methyl-2-pentene

Mechanism.

$$H{-}C(CH_3)(H){-}C(CH_3)(Cl){-}C_2H_5 + CH_3\ddot{O}^{-}Na^{+} \xrightarrow{\text{Fast}} (H_3C)(H)C{=}C(CH_3)(C_2H_5) + CH_3OH + NaCl$$

3-Methyl-2-pentene

5.4.3 Dehydration of alcohols: preparation of alkenes

The dehydration of alcohols is a useful synthetic route to alkenes. Alcohols typically undergo elimination reactions when heated with strong acid catalysts, e.g. H_2SO_4 or phosphoric acid (H_3PO_4), to generate an alkene and water. The hydroxyl group is not a good leaving group, but under acidic conditions it can be protonated. The ionization generates a molecule of water and a cation, which then easily deprotonates to give alkene. For example, the dehydration of 2-butanol gives predominately (*E*)-2-butene. The reaction is reversible, and the following equilibrium exists.

$$CH_3CH_2CH(OH)CH_3 \underset{H_2O}{\overset{H_2SO_4,\ heat}{\rightleftharpoons}} CH_3CH{=}CHCH_3 + CH_2CH_2CH{=}CH_2$$

2-Butanol; (*E*)-2-Butene (Major product); (*Z*)-1-Butene (Minor product)

Mechanism.

$$CH_3CH_2CH(\ddot{O}H)CH_3 + H\text{-}O\text{-}SO_3H \overset{Heat}{\rightleftharpoons} CH_3CH_2CH(\overset{+}{O}H_2)CH_3 + HSO_4^-$$

$$H_2SO_4 + CH_3CH_2CH{=}CH_2 \rightleftharpoons CH_3CH_2\overset{+}{C}H\text{-}CH_2\text{-}H + H_2O + HSO_4^-$$

Similarly, the dehydration of 2,3-dimethylbut-2-ol gives predominantly 2,3-dimethylbutene via E1 reaction.

$$H_3C\text{-}CH(CH_3)\text{-}C(OH)(CH_3)\text{-}CH_3 \underset{Heat}{\xrightarrow{H_2SO_4}} (H_3C)_2C{=}C(CH_3)_2 + H_2O + H_2SO_4$$

2,3-Dimethylbut-2-ol; 2,3-Dimethylbutene

Mechanism.

$$H_3C\text{-}CH(CH_3)\text{-}C(\ddot{O}H)(CH_3)\text{-}CH_3 + H\text{-}OSO_3H \longrightarrow H_3C\text{-}CH(CH_3)\text{-}C(\overset{+}{O}H_2)(CH_3)\text{-}CH_3 + HSO_4^-$$

$$H_3C\text{-}CH(CH_3)\text{-}\overset{+}{C}(CH_3)\text{-}CH_3 + H_2O \xrightarrow{HSO_4^-} H_2SO_4 + (H_3C)_2C{=}C(CH_3)_2$$

2,3-Dimethylbutene

While dehydration of 2° and 3° alcohols is an E1 reaction, dehydration of 1° alcohols is an E2 reaction. Dehydration of 2° and 3° alcohols involves the formation of a carbocation intermediate, but formation of a primary carbocation is rather difficult and unstable. For example, dehydration of propanol gives propene via E2.

$$\underset{\text{Propanol}}{CH_3CH_2CH_2OH} \underset{H_2O}{\overset{H_2SO_4,\ \text{heat}}{\rightleftharpoons}} \underset{\text{Propene}}{CH_3CH{=}CH_2}$$

Mechanism.

$$CH_3CH_2CH_2\ddot{O}H + H{-}OSO_3H \rightleftharpoons CH_3CH(H){-}CH_2{-}\overset{+}{O}H_2 + HSO_4^- \rightleftharpoons CH_3CH{=}CH_2 + H_2O + H_2SO_4$$

An E2 reaction occurs in one step: first the acid protonates the oxygen of the alcohol; a proton is removed by a base (HSO_4^-) and simultaneously carbon–carbon double bond is formed via the departure of the water molecule.

Use of concentrated acid and high temperature favours alkene formation, but use of dilute aqueous acid favours alcohol formation. To prevent the alcohol formation, alkene can be removed by distillation as it is formed, because it has a much lower boiling point than the alcohol. When two elimination products are formed, the major product is generally the more substituted alkene.

5.4.4 Dehydration of diols: pinacol rearrangement. Preparation of pinacolone

Pinacol rearrangement is a dehydration of a 1,2-diol to form a ketone. 2,3-dimethyl-2,3-butanediol has the common name *pinacol* (a symmetrical diol). When it is treated with strong acid, e.g. H_2SO_4, it gives 3,3-dimethyl-2-butanone (methyl *t*-butyl ketone), also commonly known as *pinacolone*. The product results from the loss of water and molecular rearrangement. In the rearrangement of pinacol equivalent carbocations are formed no matter which hydroxyl group is protonated and leaves.

$$\underset{\text{Pinacol}}{H_3C{-}C(CH_3)(OH){-}C(CH_3)(OH){-}CH_3} \xrightarrow[\text{Heat}]{H_2SO_4} \underset{\text{Pinacolone}}{H_3C{-}C(CH_3)_2{-}C({=}O){-}CH_3} + H_2O$$

Mechanism. The protonation of OH, followed by the loss of H_2O from the protonated diol, yields a tertiary carbocation, which rearranges with a 1,

2-methyl shift to form a protonated pinacolone. The rearranged product is deprotonated by the base to give pinacolone.

1,2-Methyl shift

Pinacolone

5.4.5 Dehydrohalogenation of alkyl halides

Alkyl halides typically undergo elimination reactions when heated with strong bases, typically hydroxides and alkoxides, to generate alkenes. Removal of a proton and a halide ion is called *dehydrohalogenation.* Any base in the reaction mixture (H_2O, HSO_4^-) can remove the proton in the elimination reaction.

E1 elimination of HX: preparation of alkenes

The E1 reaction involves the formation of a planar carbocation intermediate. Therefore, both *syn* and *anti* elimination can occur. If an elimination reaction removes two substituents from the same side of the C—C bond, the reaction is called a *syn* elimination. When the substituents are removed from opposite sides of the C—C bond, the reaction is called an *anti* elimination. Thus, depending on the substrates E1 reaction forms a mixture of *cis* (*Z*) and *trans* (*E*) products. For example, *tert*-butyl bromide (3° alkyl halide) reacts with water to form 2-methylpropene, following an E1 mechanism. The reaction requires a good ionizing solvent and a weak base. When the carbocation is formed, S_N1 and E1 processes compete with each other, and often mixtures of elimination and substitution products occur. The reaction of *t*-butyl bromide and ethanol gives major product via E1 and minor product via S_N1.

C_2H_5OH, Heat

t-Butyl bromide → 2-Methylpropene, E1 Major product + Ethyl *t*-butyl ether, S_N1 Minor product

Mechanism.

$$(CH_3)_3C\text{—}Br \xrightleftharpoons{\text{Slow}} (CH_3)_3C^+ + Br:^-$$

$$(CH_3)_2\overset{+}{C}\text{—}CH_2\text{—}H + CH_3\ddot{O}H \xrightarrow{\text{Fast}} H_3C\text{—}C(CH_3)\text{=}CH_2 + CH_3\overset{+}{O}H_2$$

E2 elimination of HX: preparation of alkenes

Dehydrohalogenation of 2° and 3° alkyl halides undergo both E1 and E2 reactions. However, 1° halides undergo only E2 reactions. They cannot undergo E1 reaction because of the difficulty of forming primary carbocations. E2 elimination is stereospecific, and it requires an *antiperiplanar* (180°) arrangement of the groups being eliminated. Since only *anti* elimination can take place, E2 reaction predominantly forms one product. The elimination reaction may proceed to alkenes that are constitutional isomers with one formed in excess of the other, described as *regioselectivity*. Similarly, eliminations often favour the more stable *trans*-product over the *cis*-product, described as *stereoselectivity*. For example, bromopropane reacts with sodium ethoxide (EtONa) to give only propene.

$$CH_3CH_2CH_2\text{—}Br \xrightarrow[\text{EtOH, heat}]{C_2H_5ONa} CH_3CH\text{=}CH_2 + CH_3CH_2OH + NaBr$$

Mechanism.

$$H\text{—}CH(CH_3)\text{—}CH_2\text{—}Br + C_2H_5\ddot{O}^-\overset{+}{Na} \xrightarrow[\text{E2}]{\text{EtOH}} CH_3CH\text{=}CH_2 + C_2H_5OH + NaBr$$

The E2 elimination can be an excellent synthetic method for the preparation of alkene when 3° alkyl halide and a strong base, e.g. alcoholic KOH, is used. This method is not suitable for S_N2 reaction.

$$\underset{t\text{-Butyl bromide}}{H_3C\text{—}C(CH_3)_2\text{—}Br} \xrightarrow[\text{Heat}]{\text{KOH}} \underset{\substack{\text{2-Methylpropene} \\ (>90\%)}}{H_3C\text{—}C(CH_3)\text{=}CH_2} + H_2O + KBr$$

A bulky base (a good base, but poor nucleophile) can further discourage undesired substitution reactions. The most common bulky bases are potassium-*t*-butoxide (*t*-BuOK), diisopropylamine and 2,6-dimethylpyridine.

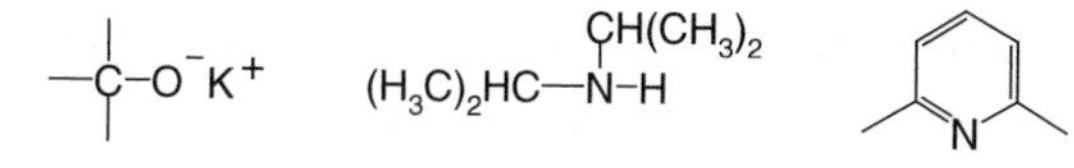

Potassium-t-butoxide   Diisopropylamine   2,6-dimethylpyridine

Cyclohexene can be synthesized from bromocyclohexane in a high yield using diisopropylamine.

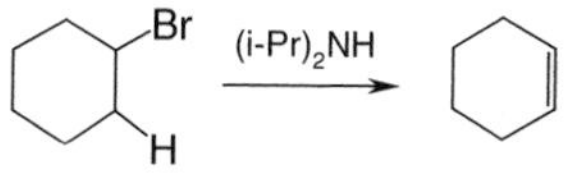

Bromocyclohexane $\xrightarrow{(i\text{-Pr})_2NH}$ Cyclohexene (93%)

Generally, E2 reactions occur with a strong base, which eliminates a proton quicker than the substrate can ionize. Normally, the S_N2 reaction does not compete with E2 since there is steric hindrance around the C—X bond, which retards the S_N2 process.

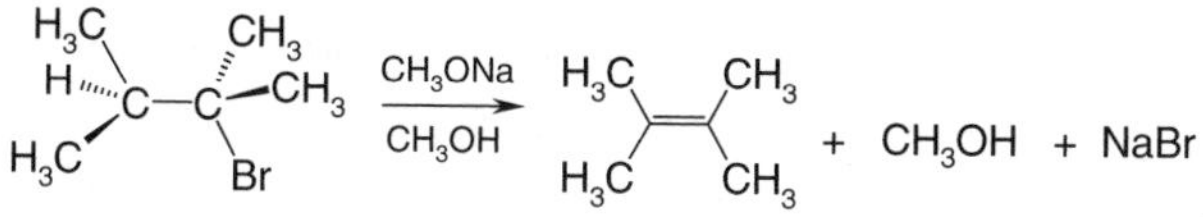

$$(CH_3)_2CH\text{—}C(CH_3)_2Br \xrightarrow[CH_3OH]{CH_3ONa} (CH_3)_2C{=}C(CH_3)_2 + CH_3OH + NaBr$$

Mechanism.

Transition state
Rate=k_2[R–X][B$^-$]

$+ CH_3OH + NaBr$

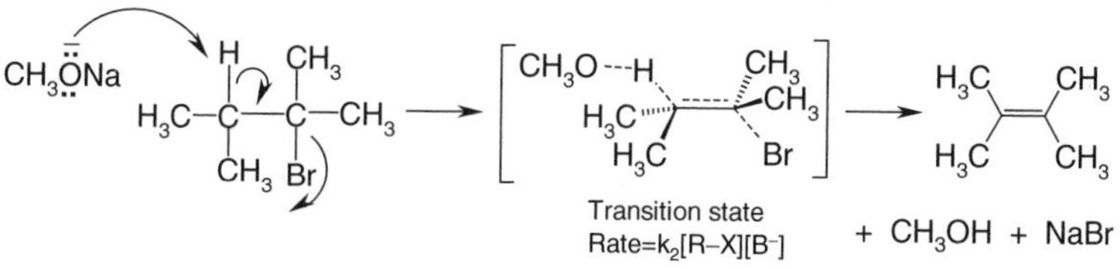

The methoxide (CH_3O^-) is acting as a base rather than a nucleophile. The reaction takes place in one concerted step, with the C—H and C—Br bonds breaking as the CH_3O—H and C=C bonds are forming. The rate is related to the concentrations of the substrate and the base, giving a second order rate equation. The elimination requires a hydrogen atom adjacent to the leaving group. If there are two or more possibilities of adjacent hydrogen atoms, mixtures of products are formed as shown in the following example.

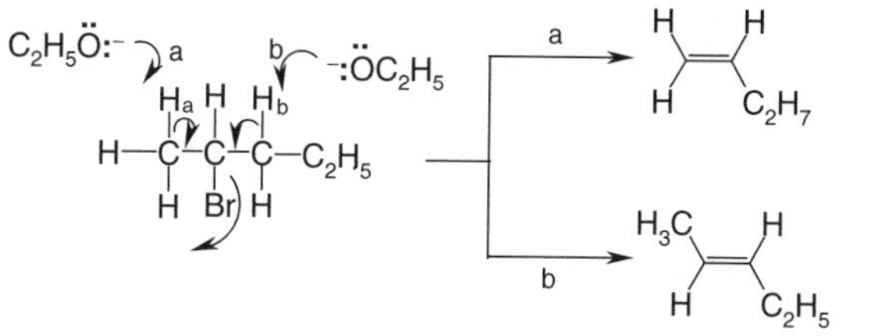

1-Pentene (Minor product)

2-Pentene (Major product)

The major product of elimination is the one with the most highly substituted double bond, and follows the following order.

$R_2C{=}CR_2 > R_2C{=}CRH > RHC{=}CHR$ and $R_2C{=}CH_2 > RCH{=}CH_2$

Stereochemical considerations in the E2 reactions The E2 follows a concerted mechanism, where removal of the proton and formation of the double bond occur at the same time. The partial π bond in the transition state requires the parallel alignment or coplanar arrangement of the *p* orbitals. When the hydrogen and leaving group eclipse each other (0°), this is known as the *syn*-coplanar conformation.

syn-Coplanar (0°) *syn*-Elimination *anti*-Coplanar (180°) *anti*-Elimination

When the leaving group and hydrogen atom are *anti* to each other (180°), this is called the *anti*-coplanar conformation. The *anti*-coplanar conformation is of lower energy, and is by far the most common. In the *anti*-coplanar conformation, the base and leaving group are well separated, thus removing electron repulsions. The *syn*-coplanar conformation requires the base to approach much closer to the leaving group, which is energetically unfavourable.

The E2 reaction is a stereospecific reaction, i.e. a particular stereoisomer reacts to give one specific stereoisomer. It is stereospecific, since it prefers the *anti*-coplanar transition state for elimination. The (*R*,*R*) diastereomer gives a *cis*-alkene, and the (*S*,*R*) diastereomer gives a *trans*-alkene.

(*R*,*R*) → *cis*-configuration + B-H + Br:⁻

(*S*,*R*) → *trans*-configuration + B-H + Br:⁻

E2 elimination of HX in the cyclohexane system

Almost all cyclohexane systems are most stable in the chair conformations. In a chair, adjacent axial positions are in an *anti*-coplanar arrangement, ideal for E2 eliminations. Adjacent axial positions are said to be in a *trans*-diaxial arrangement. E2 reactions only proceed in chair conformations from *trans*-diaxial positions, and chair–chair interconversions allow the hydrogen and the leaving group to attain the *trans*-diaxial arrangement. The elimination of HBr from bromocyclohexane gives cyclohexene. The bromine must be in an axial position before it can leave.

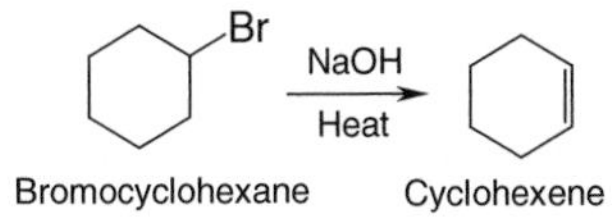

NaOH, Heat

Bromocyclohexane Cyclohexene

Mechanism.

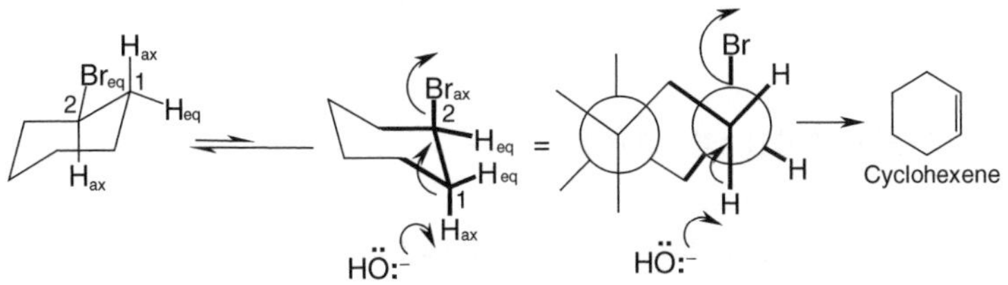

Cyclohexene

E2 elimination of X_2

Preparation of alkenes Dehalogenation of *vicinal*-dihalides with NaI in acetone produces alkene via E2 reactions.

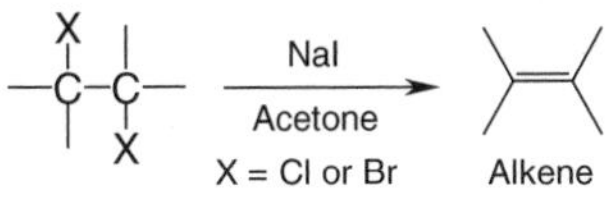

NaI, Acetone

X = Cl or Br Alkene

Preparation of alkynes Alkynes can be produced by elimination of two moles of HX from a *geminal* (halides on the same carbon)-or *vicinal* (halides on the adjacent carbons)-dihalide at high temperatures. Stronger bases (KOH or $NaNH_2$) are used for the formation of alkyne via two consecutive E2 dehydrohalogenations. Under mild conditions, dehydrohalogenation stops at the vinylic halide stage. For example, 2-butyne is obtained from *geminal*- or *vicinal*-dibromobutane.

geminal- or *vicinal*-Dibromobutane

$NaNH_2$

Vinylic bromide

$NaNH_2$

$CH_3C{\equiv}CCH_3$

2-Butyne

Dimethyl acetylene

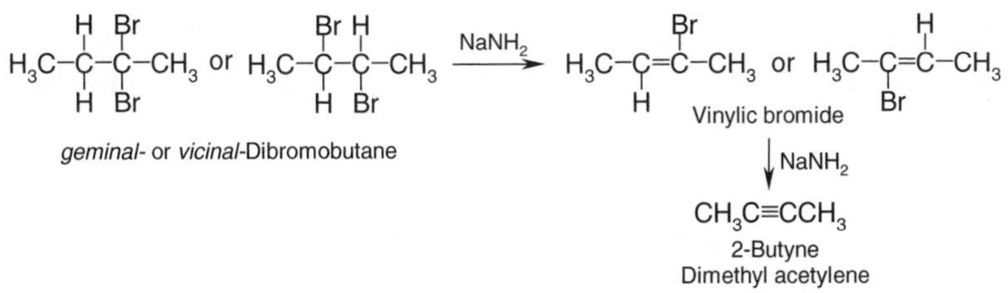

E1 versus E2 mechanism

Criteria	E1	E2
Substrate	Tertiary > secondary > primary	Primary > secondary > tertiary
Rate of reaction	Depends only on the substrate	Depends on both substrate and base
Carbocation	More stable carbocation	Less stable carbocation
Rearrangement	Rearrangements are common	No rearrangements
Geometry	No special geometry required	*anti*-coplanarity required
Leaving group	Good leaving group	Poor leaving group
Base strength	Weak base	Strong and more concentrated base

5.5 Substitution reactions

The word *substitution* implies the replacement of one atom or group by another. Two types of substitution reaction can occur: *nucleophilic substitution* and *electrophilic substitution.*

A *nucleophile* is an electron rich species that reacts with an electrophile. The term *electrophile* literally means 'electron-loving', and is an electron-deficient species that can accept an electron pair. A number of nucleophilic substitution reactions can occur with alkyl halides, alcohols and epoxides. However, it can also take place with carboxylic acid derivatives, and is called *nucleophilic acyl substitution.*

Electrophilic substitution reactions are those where an electrophile displaces another group, usually a hydrogen. Electrophilic substitution occurs in aromatic compounds.

5.5.1 Nucleophilic substitutions

Alkyl halides (RX) are good substrates for substitution reactions. The nucleophile ($Nu:^-$) displaces the leaving group ($X:^-$) from the carbon atom by using its electron pair or lone pair to form a new σ bond to the carbon atom. Two different mechanisms for nucleophilic substitution are S_N1 and S_N2 mechanisms. In fact, the preference between S_N1 and S_N2 mechanisms depends on the structure of the alkyl halide, the reactivity and structure of the nucleophile, the concentration of the nucleophile and the solvent in which reaction is carried out.

$$\text{H X: } -\text{C}-\text{C}- + \text{Nu:}^- \longrightarrow \text{H Nu: } -\text{C}-\text{C}- + \text{X:}^-$$

First order nucleophilic substitution: S_N1 reaction

S_N1 reaction means substitution nucleophilic unimolecular. The S_N1 reaction occurs in two steps, with the first being a slow ionization reaction generating a carbocation. Thus, the rate of an S_N1 reaction depends only on the concentration of the alkyl halide. First, the C—X bond breaks without any help from the nucleophile, and then there is quick nucleophilic attack by the nucleophile on the carbocation. When water or alcohol is the nucleophile, a quick loss of a proton by the solvent gives the final product. For example, the reaction of *t*-butylbromide and methanol gives *t*-butyl methyl ether.

$$(CH_3)_3C\text{—}Br + CH_3OH \longrightarrow (CH_3)_3C\text{—}O\text{—}CH_3 + HBr$$

t-Butyl bromide → *t*-Butyl methyl ether

Mechanism.

$$(CH_3)_3C\text{—}Br \xrightarrow{\text{Slow}} (CH_3)_3C^+ + Br:^- \xrightarrow[\text{CH}_3\ddot{\text{O}}\text{H}]{\text{Fast}} (CH_3)_3C\text{—}\overset{+}{O}(H)\text{—}CH_3 \xrightarrow{CH_3\ddot{O}H} CH_3\overset{+}{O}H_2 + (CH_3)_3C\text{—}\ddot{O}\text{—}CH_3$$

The rate of reaction depends only on the concentration of *t*-butylbromide. Therefore, the rate is first order or unimolecular overall.

$$\text{Rate} = k_1[(CH_3)_3C\text{—}Br]$$

Substituent effects Carbocations are formed in the S_N1 reactions. The more stable the carbocation, the faster it is formed. Thus, the rate depends on carbocation stability, since alkyl groups are known to stabilize carbocations through inductive effects and hyperconjugation (see Section 5.2.1). The reactivities of S_N1 reactions decrease in the order of 3° carbocation > 2° carbocation > 1° carbocation > methyl cation. Primary carbocation and methyl cation are so unstable that primary alkyl halide and methyl halide do not undergo S_N1 reactions. This is the opposite of S_N2 reactivity.

Strength of nucleophiles The rate of the S_N1 reaction does not depend on the nature of the nucleophiles, since the nucleophiles come into play after the rate-determining steps. Therefore, the reactivity of the nucleophiles has no effect on the rate of the S_N1 reaction. Sometimes in S_N1 reaction the solvent is the nucleophile, e.g. water and alcohol. When the solvent is the nucleophile, the reaction is called *solvolysis*.

Leaving group effects Good leaving groups are essential for S_N1 reactions. In the S_N1 reaction, a highly polarizable leaving group helps to stabilize the negative charge through partial bonding as it leaves. The leaving group should be stable after it has left with the bonding electrons, and also be a weak base. The leaving group starts to take on partial negative charge as the cation starts to form. The most common leaving groups are

- anions,: Cl^-, Br^-, I^-, RSO_3^- (sulphonate), RSO_4^- (sulphate), RPO_4^- (phosphate)
- neutral species: H_2O, ROH, R_3N, R_3P.

Solvent effects Protic solvents are especially useful since the hydrogen bonding stabilizes the anionic leaving group after ionization. Ionization requires the stabilization of both positive and negative charges. Solvents with higher dielectric constant (ε), which is a measure of a solvent's polarity, have faster rates for S_N1 reactions.

Stereochemistry of the S_N1 reactions The S_N1 reaction is not stereospecific. The carbocation produced is planar and sp^2-hybridized. For example, the reaction of (*S*)-2-bromobutane and ethanol gives a racemic mixture, (*S*)-2-butanol and (*R*)-2-butanol.

(*S*)-2-Bromobutane
Attack from top face
$Br:^-$
$C_2H_5\ddot{O}H$
(*S*)-2-Butanol (Retention)
Attack from bottom face
$C_2H_5\ddot{O}H$
(*R*)-2-Butanol (Inversion)

The nucleophile may attack from either the top or the bottom face. If the nucleophile attacks from the top face, from which the leaving group departed, the product displays retention of configuration. If the nucleophile attacks from the bottom face, the backside of the leaving group, the product displays an inversion of configuration. A combination of inversion and retention is called *racemization.* Often complete racemization is not achieved since the leaving group will partially block one face of the molecule as it ionizes, thus giving a major product of inversion.

Carbocation rearrangements in S_N1 reactions through 1,2-hydride shift Carbocations often undergo rearrangements, producing more stable ions. This rearrangement produces a more stable tertiary cation instead of a less stable secondary cation. Rearrangements occur when a more stable cation can be produced by a *1,2-hydride shift*. For example, the S_N1 reaction of 2-bromo-3-methylbutane and ethanol gives a mixture of structural isomers, the expected product and a rearranged product.

$$CH_3CH(CH_3)CH(Br)CH_3 \xrightarrow{S_N1} CH_3CH(CH_3)\overset{+}{C}HCH_3 \xrightarrow{C_2H_5OH} CH_3CH(CH_3)CH(OC_2H_5)CH_3$$

2-Bromo-3-methylbutane → 2° carbocation → 3-Methyl-2-ethyl butyl ether (Not rearranged)

2° carbocation ↓ 1,2-hydride shift

$$CH_3\overset{+}{C}(CH_3)CH_2CH_3 \xrightarrow{C_2H_5OH} CH_3C(CH_3)(OC_2H_5)CH_2CH_3$$

3° carbocation → 2-Methyl-2-ethyl butyl ether (Rearranged product)

Carbocation rearrangements in S_N1 reactions through 1,2-methyl shift Carbocation rearrangements often occur when a more stable cation can be produced by an alkyl group or methyl shift. For example, 2,2-dimethyl propyl bromide gives exclusively a rearranged product, which results from a *1,2-methyl shift*. This rearrangement produces a more stable tertiary cation instead of an unstable primary cation. Rearrangements do not occur in S_N2 reactions since carbocations are not formed.

$$H_3C{-}C(OC_2H_5)(CH_3){-}CH_2CH_3 \xleftarrow[\text{Fast}]{C_2H_5OH} H_3C{-}\overset{+}{C}(CH_3){-}CH_2CH_3 \xleftarrow{\text{1,2-Methyl shift}} H_3C{-}C(CH_3)_2{-}CH_2{-}Br \not\rightarrow H_3C{-}C(CH_3)_2{-}\overset{+}{C}H_2$$

2-Methyl-2-ethyl butylether | 3° Carbocation | 2,2-Dimethyl propyl bromide | 1° Carbocation

Second order nucleophilic substitution: S_N2 reaction

S_N2 means bimolecular nucleophilic substitution. For example, the reaction of hydroxide ion with methyl iodide yields methanol. The hydroxide ion is a good nucleophile, since the oxygen atom has a negative charge and a pair of unshared electrons. The carbon atom is electrophilic, since it is bonded to a more electronegative halogen. Halogen pulls electron density away from the carbon, thus polarizing the bond, with carbon bearing partial positive charge and the halogen bearing partial negative charge. The nucleophile attacks the electrophilic carbon through donation of two electrons.

Typically, S_N2 reaction requires a backside attack. The C−X bond weakens as nucleophile approaches. All these occur in one step. This is a *concerted reaction*, as it takes place in a single step with the new bond forming as the old bond is breaking. The S_N2 reaction is *stereospecific*, always proceeding with inversion of stereochemistry. The inversion of configuration resembles the way an umbrella turns inside out in the wind. For example, the reaction between ethyl iodide and hydroxide ion produces ethanol in an S_N2 reaction.

$$\underset{\text{Ethyl iodide}}{C_2H_5{-}I} + HO^- \longrightarrow \underset{\text{Ethanol}}{C_2H_5{-}OH} + I{:}^-$$

Mechanism.

$$HO{:}^- + H_2C(CH_3){-}I \longrightarrow \left[HO{\cdots}C(CH_3)(H)_2{\cdots}I\right] \longrightarrow HO{-}C(CH_3)H_2 + I{:}^-$$

Transition state
Rate = k_2[R–X][HO⁻]

Ethanol

The reaction rate is doubled when the concentration of ethyl iodide [C2H5I] is doubled, and also doubled when the concentration of hydroxide ion [HO−] is doubled. The rate is first order with respect to both reactants, and is second order overall.

$$\text{Rate} = k_2[C_2H_5I][HO^-]$$

Strength of nucleophile The rate of the S_N2 reaction strongly depends on the nature of the nucleophile; i.e., a good nucleophile (nucleophile with a negative charge) gives faster rates than a poor nucleophile (neutral molecule with a lone pair of electrons). Generally, negatively charged species are better nucleophiles than analogous neutral species. For example, methanol (CH_3OH) and sodium methoxide (CH_3ONa) react with CH_3I to produce dimethyl ether in both cases. It is found that CH_3ONa reacts about a million times faster than CH_3OH in S_N2 reactions.

Basicity and nucleophilicity Basicity is defined by the equilibrium constant for abstracting a proton. Nucleophilicity is defined by the rate of attack on an electrophilic carbon atom. A base forms a new bond with a proton. On the other hand, a nucleophile forms a new bond with an atom other than a proton. Species with a negative charge are stronger nucleophiles than analogous species without a negative charge. Stronger bases are also stronger nucleophiles than their conjugate acids.

$$HO^- > H_2O \qquad HS^- > H_2S \qquad {}^-NH_2 > NH_3 \qquad CH_3O^- > CH_3OH$$

Nucleophilicity decreases from left to right across the periodic table. The more electronegative elements hold on more tightly to their nonbonding electrons.

$$HO^- > F^- \qquad NH_3 > H_2O \qquad (CH_3CH_2)_3P > (CH_3CH_2)_2S$$

Nucleophilicity increases down the periodic table with the increase in polarizability and size of the elements.

$$I^- > Br^- > Cl^- > F^- \qquad HSe^- > HS^- > HO^- \qquad (C_2H_5)_3P > (C_2H_5)_3N$$

As the size of an atom increases, its outer electrons move further away from the attractive force of the nucleus. The electrons are held less tightly and are said to be more polarizable. Fluoride is a nucleophile having hard or low polarizability, with its electrons held close to the nucleus, and it must approach the carbon nucleus closely before orbital overlap can occur. The outer shell of the soft iodide has loosely held electrons, and these can easily shift and overlap with the carbon atom at a relatively long distance.

Solvent effects Different solvents have different effects on the nucleophilicity of a species. Solvents with acidic protons are called protic solvents, usually O—H or N—H groups. Polar protic solvents, e.g. dimethyl sulphoxide (DMSO), dimethyl formamide (DMF), acetonitrile (CH_3CN) and acetone (CH_3COCH_3) are often used in S_N2 reactions, since the polar reactants (nucleophile and alkyl halide) generally dissolve well in them.

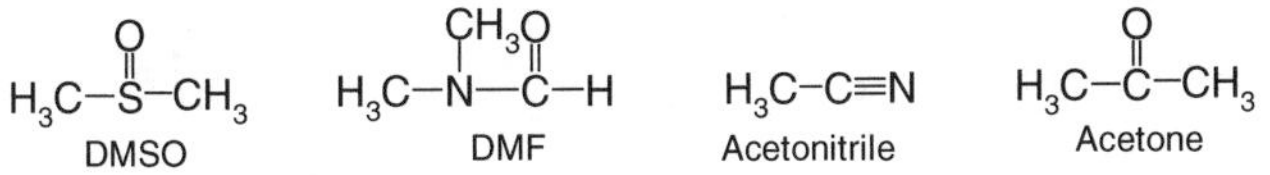

$H_3C-S(=O)-CH_3$ DMSO $H_3C-N(CH_3)-C(=O)-H$ DMF $H_3C-C\equiv N$ Acetonitrile $H_3C-C(=O)-CH_3$ Acetone

Small anions are more strongly solvated than larger anions, and sometimes this can have an adverse effect. Certain anions, e.g. F^-, can be solvated so well in polar protic solvents that their nucleophilicity is reduced by the solvation. For efficient S_N2 reactions with small anions, it is usual to use polar aprotic solvents, which do not have any O—H or N—H bonds to form hydrogen bonds to the small anions.

Steric effects Base strength is relatively unaffected by steric effect, because a base removes a relatively unhindered proton. Thus, the strength of a base depends only on how well the base shares its electrons with a proton. On the other hand, nucleophilicity is affected by the steric effects. A bulky nucleophile has difficulty in getting near the backside of the sp^3 carbon. Therefore, large groups tend to hinder this process.

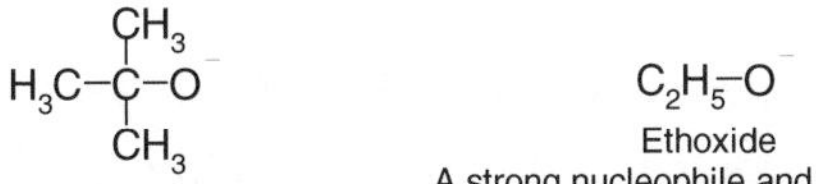

$H_3C-C(CH_3)_2-O^-$
t-Butoxide
A weak nucleophile and strong base

$C_2H_5-O^-$
Ethoxide
A strong nucleophile and weak base

Leaving group effects A good leaving group must be a weak base, and it should be stable after it has left with the bonding electrons. Thus, the weaker the base, the better it is as a leaving group. Good leaving groups are essential for both S_N1 and S_N2 reactions.

Steric effects of the substrate Large groups on the electrophile hinder the approach of the nucleophile. Generally, one alkyl group slows the reaction, two alkyl groups make it difficult and three alkyl groups make it close to impossible.

Relative rates for S_N2 : methyl halides $>$ 1° $>$ 2° $>$ 3° alkyl halides.

Stereochemistry of the S_N2 reaction A nucleophile donates its electron pairs to the C—X bond on the backside of the leaving group, since the leaving group itself blocks attack from any other direction. Inversion of stereochemistry is observed in the product of an S_N2 reaction. The reaction is stereospecific since a certain stereoisomer reacts to give one specific stereoisomer as product.

HO:⁻ + (*S*)-2-Bromobutane → [Transition state] → (*R*)-2-Butanol + Br:⁻

(*S*)-2-Bromobutane Transition state Rate = $k_2[R\text{–}X][OH^-]$ (*R*)-2-Butanol

5.5.2 Nucleophilic substitution reactions of alkyl halides

We have already learnt that alkyl halides react with alcohols and metal hydroxide (NaOH or KOH) to give ethers and alcohols, respectively. Depending on the alkyl halides and the reaction conditions, both S_N1 and S_N2 reactions can occur. Alkyl halides undergo a variety of transformation through S_N2 reactions with a wide range of nucleophiles (alkoxides, cyanides, acetylides, alkynides, amides and carboxylates) to produce other functional groups.

Conversion of alkyl halides

Williamson ether synthesis: preparation of ether The sodium or potassium alkoxides are strong bases and nucleophiles. Alkoxides (RO^-) can react with primary alkyl halides to produce symmetrical or unsymmetrical ethers. This is known as *Williamson ether synthesis*. The reaction is limited to primary alkyl halides. Higher alkyl halides tend to react via elimination. For example, sodium ethoxide reacts with ethyl iodide to produce diethyl ether. The reaction involves S_N2 displacement with backside attack of the alkoxide to form diethyl ether.

$$\underset{\text{Ethyl iodide}}{C_2H_5{-}I} + \underset{\text{Sodium ethoxide}}{C_2H_5{-}ONa} \xrightarrow{\text{EtOH}} \underset{\text{Diethylether}}{C_2H_5{-}OC_2H_5} + NaI$$

Mechanism.

$$C_2H_5\ddot{O}{:}^- \, Na^+ + H\cdots\overset{CH_3}{\underset{H}{C}}{}^{\delta+}{-}I^{\delta-} \longrightarrow C_2H_5{-}OC_2H_5 + NaI$$

Preparation of nitriles Cyanide ion (CN^-) is a good nucleophile, and can displace leaving groups from 1° and 2° alkyl halides. Nitriles are prepared by the treatment of alkyl halides with NaCN or KCN in dimethyl sulphoxide (DMSO). The reaction occurs rapidly at room temperature.

$$\underset{\text{Alkyl halide}}{RCH_2{-}X} + NaCN \xrightarrow{\text{DMSO}} \underset{\text{Nitrile}}{RCH_2{-}CN} + NaX$$

Mechanism.

$$\overset{+}{Na}C{\equiv}N^- \quad H\cdots\overset{R}{\underset{H}{C}}{}^{\delta+}{-}X^{\delta-} \longrightarrow RCH_2C{\equiv}N + NaX$$

Preparation of alkyl azides The azide ion (N_3^-), a good nucleophile, can displace leaving groups from 1° and 2° alkyl halides. Alkyl azides are easily prepared from sodium or potassium azides and alkyl halides. The reaction mechanism resembles the formation of nitrile.

$$\underset{\text{Alkyl halide}}{RCH_2{-}X} \xrightarrow[S_N2]{NaN_3} \underset{\text{Alkyl azide}}{RCH_2{-}N_3}$$

Preparation of primary amines Alkyl halide reacts with sodium amide ($NaNH_2$) to give 1° amine via S_N2 reaction. The reaction mechanism for the formation of 1° amine is similar to the formation of nitrile.

$$\underset{\text{Alkyl halide}}{RCH_2{-}X} \xrightarrow[S_N2]{NaNH_2} \underset{\text{Primary amine}}{RCH_2{-}NH_2}$$

Preparation of alkynes The reaction of primary alkyl halides and metal acetylides or alkynides ($R'C{\equiv}CNa$ or $R'C{\equiv}CMgX$) yields alkynes. The reaction is limited to 1° alkyl halides. Higher alkyl halides tend to react via elimination.

$$\underset{\text{Alkyl halide}}{RCH_2{-}X} \xrightarrow[R'C{\equiv}CMgX]{R'C{\equiv}CNa \text{ or}} \underset{\text{Alkyne}}{RCH_2{-}C{\equiv}CR'}$$

Mechanism.

$$R'C{\equiv}C{:}^- \quad H\text{''''}\overset{R}{\underset{H}{C}}{}^{\delta+}{-}X^{\delta-} \longrightarrow RCH_2C{\equiv}CR' + X{:}^-$$

Preparation of esters Alkyl halide reacts with sodium carboxylate ($R'CO_2Na$) to give ester via S_N2 reaction. The formation of ester follows a similar mechanism to the formation of alkyne.

$$\underset{\text{Alkyl halide}}{RCH_2{-}X} \xrightarrow[DMSO]{R'CO_2Na} \underset{\text{Ester}}{RCH_2{-}CO_2R'}$$

Coupling reaction: Corey–House reaction. Preparation of alkanes

The coupling reaction is a good synthetic way to join two alkyl groups together. Gilman reagent or lithium organocuprate (R'_2CuLi) reacts with alkyl halide (RX) to produce an alkane (R—R′), which has higher carbon number than the starting alkyl halide. The reaction is limited to primary alkyl halide, but the alkyl groups in the Gilman reagents may be 1°, 2° or 3°. This versatile method is also known as the Corey–House reaction.

$$R{-}X + R'_2{-}CuLi \xrightarrow[-78\ ^\circ C]{Ether} R{-}R' + R'{-}Cu + Li{-}X$$

$$\text{(dibromobicyclo compound)}\ Br, Br + (CH_3)_2CuLi \xrightarrow[-78\ ^\circ C]{Ether} CH_3, CH_3$$

5.5.3 Nucleophilic substitution reactions of alcohols

Alcohols are not reactive towards nucleophilic substitution, because the hydroxyl group (—OH) is too basic to be displaced by a nucleophile. The nucleophilic substitution reaction of alcohols only occurs in the presence of an acid. The overall transformation requires the acidic conditions to replace the hydroxyl group (—OH), a poor leaving group, with a good leaving group such as H_2O. Protonation to convert the leaving group to H_2O has limited utility, as not all substrates or nucleophiles can be utilized under acidic conditions without unwanted side reactions. An alternative is to convert the alcohol into alkyl halide or alkyl tosylate (see below), which has a much better leaving group, and reacts with nucleophiles without the need for an acid.

Acid-catalysed condensation of alcohols: preparation of ethers

Bimolecular dehydration is generally used for the synthesis of symmetrical ethers from unhindered 1° alcohols. Industrially, diethyl ether is obtained by heating ethanol at 140 °C in the presence of H_2SO_4. In this reaction, ethanol is protonated in the presence of an acid, which is then attacked by another molecule of ethanol to give diethyl ether. This is an acid-catalysed S_N2 reaction. If the temperature is too high, alkene is formed via elimination.

$$C_2H_5{-}OH + C_2H_5{-}OH \xrightarrow[140\ ^\circ C]{H_2SO_4} \underset{\text{Diethylether}}{C_2H_5{-}O{-}C_2H_5} + H_2O$$

Mechanism.

$$CH_3CH_2{-}\ddot{O}H + H{-}OSO_3H \longrightarrow CH_3CH_2{-}\overset{+}{O}(H){-}H + HSO_4^-$$

$$CH_3CH_2{-}\overset{+}{O}H_2 + C_2H_5{-}\ddot{O}H \longrightarrow C_2H_5{-}\overset{+}{O}(H){-}C_2H_5 + H_2O$$

$$C_2H_5{-}\overset{+}{O}(H){-}C_2H_5 + HSO_4^- \longrightarrow H_2SO_4 + C_2H_5{-}O{-}C_2H_5$$

Conversion of alcohols by hydrogen halides: preparation of alkyl halides

Alcohols react with hydrogen halides (HX) to give alkyl halides. Primary alcohols undergo S_N2 reactions with HX. Primary alcohols with branching on the β-carbon give rearranged products. The temperature must be kept low to avoid the formation of E2 product.

$$\underset{1^\circ\ \text{Alcohol}}{RCH_2{-}OH} \xrightarrow[\text{Heat}]{\text{HX, ether}} \underset{1^\circ\ \text{Alkyl halide}}{RCH_2{-}X} \qquad X = Br, Cl$$

Mechanism.

$$RCH_2{-}\ddot{O}H \overset{H^+}{\rightleftharpoons} RCH_2{-}\overset{+}{O}(H){-}H \xrightarrow{X:^-} RCH_2{-}X + H_2O$$

Secondary and tertiary alcohols undergo S_N1 reactions with hydrogen halides. The reaction of an HX with 3° alcohol proceeds readily at room temperature, whereas the reaction of an HX with a 2° alcohol requires heat.

The reaction occurs via a carbocation intermediate. Therefore, it is possible to form both substitution and elimination products. Secondary alcohols with branching on the β-carbon give rearranged products. The temperature must be kept low to avoid the formation of E1 product.

$$R_3C{-}OH \xrightarrow[\text{Heat}\; X = Br,\, Cl]{\text{HX, ether}} R_3C{-}X$$

3° Alcohol → 3° Alkyl halide

Mechanism.

$$R_3C{-}\ddot{O}H \underset{}{\overset{H^+}{\rightleftharpoons}} R_3C{-}\overset{+}{O}H_2 \;(X:^-) \longrightarrow R_3C{-}X + H_2O$$

Primary alcohol reacts with HCl in the presence of $ZnCl_2$ (a Lewis acid) to produce 1° alkyl chloride. Without the use of $ZnCl_2$, the S_N2 reaction is slow, because chloride is a weaker nucleophile than bromide. The reaction rate is increased when $ZnCl_2$ is used as a catalyst. The $ZnCl_2$ coordinates to the hydroxyl oxygen, and generates a better leaving group. The mixture of HCl and $ZnCl_2$ is known as *Lucas Reagent.*

$$RCH_2{-}OH \xrightarrow[ZnCl_2]{HCl} RCH_2{-}Cl$$

1° Alcohol → 1° Alkyl chloride

Mechanism.

$$RCH_2{-}\ddot{O}H \overset{ZnCl_2}{\rightleftharpoons} RCH_2{-}\overset{+}{O}(H){-}\overset{-}{Z}nCl_2 \;(Cl:^-) \longrightarrow R{-}CH_2{-}Cl$$

1° Alkyl chloride

Secondary and tertiary alcohols react via the S_N1 mechanism with the Lucas reagent. The reaction occurs via a carbocation intermediate. Thus, it is possible to form both S_N1 and E1 products. The temperature must be kept low to avoid the formation of E1 product.

$$R_2CH{-}OH \xrightarrow[ZnCl_2]{HCl} R_2CH{-}Cl$$

2° Alcohol → 2° Alkyl chloride

Mechanism.

$$R_2CH{-}\ddot{O}H \overset{ZnCl_2}{\rightleftharpoons} R_2CH{-}\overset{+}{O}(H){-}\overset{-}{Z}nCl_2 \longrightarrow R_2\overset{+}{C}H \;(Cl:^-) \longrightarrow R_2CH{-}Cl$$

Conversion of alcohols by thionyl chlorides: preparation of alkyl chlorides

Thionyl chloride ($SOCl_2$) is the most widely used reagent for the conversion of 1° and 2° alcohols to corresponding alkyl chlorides. The reaction is often carried out in the presence of a base, e.g. pyridine or triethylamine (Et_3N). The base catalyses the reaction, and also neutralizes the HCl generated during the reaction by forming pyridinium chloride ($C_5H_5NH^+Cl^-$) or triethylammonium chloride ($Et_3NH^+Cl^-$)

$$\underset{\text{1° Alcohol}}{RCH_2{-}OH} + \underset{\text{Thionyl chloride}}{Cl{-}S(=O){-}Cl} \xrightarrow[\text{or } Et_3N]{\text{Pyridine}} \underset{\text{1° Alkyl chloride}}{RCH_2{-}Cl}$$

Mechanism. Thionyl chloride converts the hydroxyl group in an alcohol to a chlorosulphite leaving group that can be displaced by the chloride. Secondary or tertiary alcohols follow S_N1 reactions, whereas primary alcohols proceed via S_N2 reactions.

$$RCH_2{-}\ddot{O}H + Cl{-}S(=O){-}Cl \xrightarrow{\text{Pyridine}} RCH_2{-}\overset{+}{O}(H){-}S(=O){-}Cl + Cl:^-$$

$$RCH_2{-}O{-}S(=O){-}Cl + \underset{\text{Pyridinium chloride}}{C_5H_5\overset{+}{N}HCl^-}$$

$$Cl:^- + SO_2 + \underset{\text{Alkyl chloride}}{RCH_2Cl} \longleftarrow RCH_2{-}O{-}S(=O){-}Cl + Cl:^-$$

Conversion of alcohols by phosphorus halides

Phosphorus halides react with alcohols to yield alkyl halides at low temperature (0 °C). Primary and secondary alcohols undergo S_N2 reactions with PX_3. This type of reaction does not lead to rearranged products, and does not work well with 3° alcohols. PI_3 has to be generated *in situ* via reaction of iodine and phosphorus.

$$\underset{\text{1° Alcohol}}{RCH_2{-}OH} \xrightarrow[\substack{0\,°C \\ X = Br, Cl, I}]{PX_3,\ \text{ether}} \underset{\text{1° Alkyl halide}}{RCH_2{-}X} + HO{-}PX_2$$

Mechanism. The hydroxyl oxygen displaces a halide, a good leaving group, from the phosphorus. The halide attacks the backside of the alkyl group and displaces the positively charged oxygen, which is a good leaving group.

$$RCH_2{-}\ddot{O}H + PX_3 \longrightarrow RCH_2{-}\overset{+}{O}(H){-}PX_2 + X:^- \longrightarrow RCH_2{-}Br + HO{-}PX_2$$

Conversion of alcohols by sulphonyl chlorides: preparation of alkyl tosylates or tosylate esters

Alcohols react with sulphonyl chlorides to yield sulphonate esters via S_N2 reactions. Tosylate esters (alkyl tosylates) are formed from alcohols from the reaction with p-toluenesulphonyl chloride (TsCl). The reaction is most commonly carried out in the presence of a base, e.g. pyridine or triethylamine (Et_3N).

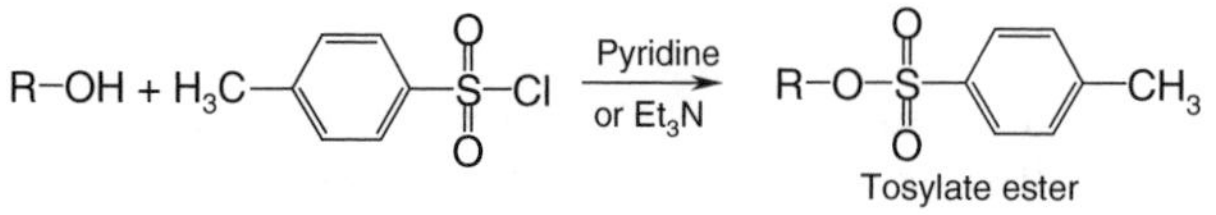

R–OH + H_3C–C_6H_4–SO_2–Cl $\xrightarrow[\text{or } Et_3N]{\text{Pyridine}}$ R–O–SO_2–C_6H_4–CH_3

Tosylate ester

Mechanism.

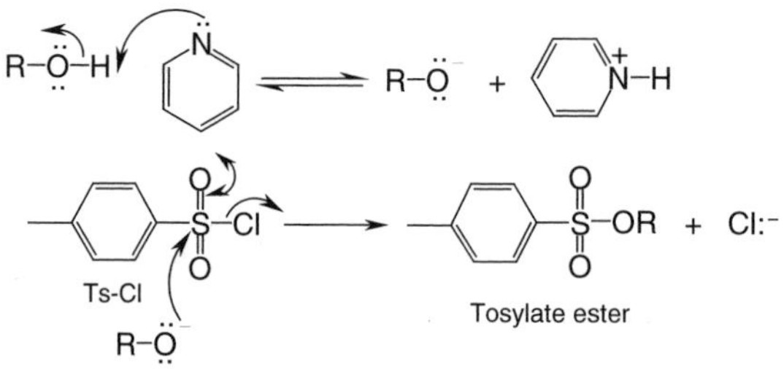

R–Ö–H + pyridine ⇌ R–Ö$^-$ + pyridinium ($\overset{+}{N}$–H)

Ts-Cl + R–Ö$^-$ ⟶ C_6H_4–SO_2–OR + Cl:$^-$

Tosylate ester

Conversion of alkyl tosylates

Tosylates are excellent leaving groups, and can undergo a variety of S_N2 reactions. The reaction is stereospecific, and it occurs with inversion of configuration. For example, (*S*)-2-butanol reacts with TsCl in pyridine to produce (*S*)-2-butane tosylate, which reacts readily with NaI to give (*R*)-2-iodobutane via S_N2 reaction.

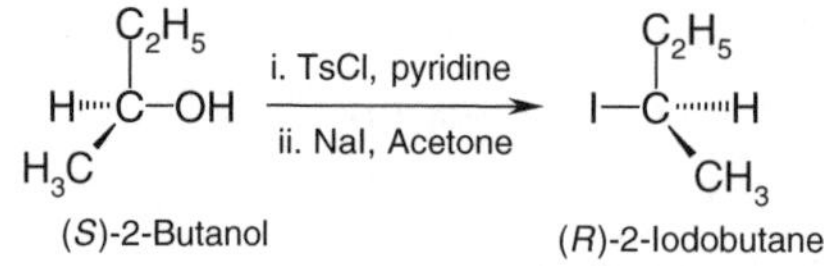

(*S*)-2-Butanol $\xrightarrow[\text{ii. NaI, Acetone}]{\text{i. TsCl, pyridine}}$ (*R*)-2-Iodobutane

Mechanism.

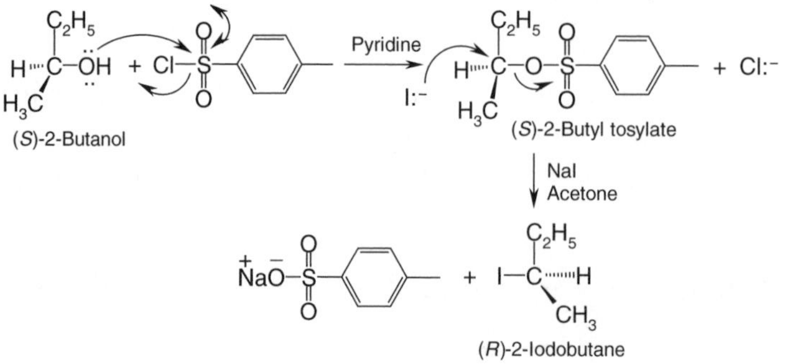

(*S*)-2-Butanol + Cl–SO_2–C_6H_4– $\xrightarrow{\text{Pyridine}}$ (*S*)-2-Butyl tosylate + Cl:$^-$ (attacked by I:$^-$)

(*S*)-2-Butyl tosylate $\xrightarrow{\text{NaI, Acetone}}$ $\overset{+}{Na}\overset{-}{O}$–$SO_2$–$C_6H_4$– + I–C(H)($CH_3$)($C_2H_5$)

(*R*)-2-Iodobutane

Similarly, alkyl tosylate reacts with other nucleophiles, e.g. H^-, X^-, HO^-, $R'O^-$, R'^-, NH_2^- or NH_3, CN^-, N_3^- and $R'CO_2^-$ following the S_N2 reaction mechanism, and produces a number of other functional groups as follows.

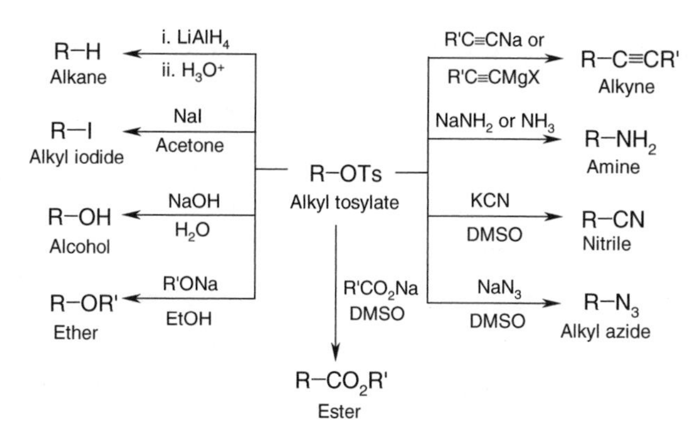

5.5.4 Nucleophilic substitution reactions of ethers and epoxides

Ethers themselves cannot undergo nucleophilic substitution or elimination reactions because the alkoxide anion is not a good leaving group. Thus, acid catalysis is required for the nucleophilic substitution of ethers. Ethers react with HX (usually HBr or HI) at high temperatures to produce alkyl halides. Although an epoxide and an ether have the same leaving group, epoxides are more reactive than ethers due to ring strain in the three membered ring. They undergo ring-opening reactions readily with acids as well as bases. Thus, epoxides are synthetically useful reagents, and they react with a wide variety of nucleophiles. They are easily cleaved by H_2O and ROH in the presence of an acid catalyst via S_N1 reactions, and by strong bases (RMgX, RLi, NaC≡N, NaN_3, RC≡CM, RC≡CMgX, RC≡CLi, $LiAlH_4$ or $NaBH_4$, NaOH or KOH and NaOR or KOR) via S_N2 reactions.

Cleavage of ethers and epoxides by haloacids

Preparation of alkyl halides Ethers can be cleaved at the ether linkage only at high temperatures using haloacids, e.g. HBr or HI at high temperatures. Depending on the structure of the alkyl groups in ether, the reaction can proceed via S_N1 or S_N2. For example, methyl propylether reacts with HBr to give propyl bromide via S_N2 reaction. Protonation of the oxygen in ether creates a good leaving group, a neutral alcohol molecule. Cleavage involves nucleophilic attack by bromide ion on the protonated ether, followed by displacement of the weakly basic CH_3OH to produce propyl bromide.

$$CH_3CH_2CH_2{-}O{-}CH_3 \xrightarrow[\text{Heat}]{\text{HBr}} CH_3CH_2CH_2{-}Br + CH_3OH$$

Methyl propyl ether → Propyl bromide

Mechanism.

$$CH_3CH_2CH_2{-}\ddot{O}{-}CH_3 + H{-}Br \xrightarrow{\text{Heat}} CH_3CH_2CH_2{-}\overset{+}{\ddot{O}}(H){-}CH_3 \;\; Br:^- \xrightarrow{S_N2} CH_3CH_2CH_2Br + CH_3OH$$

Propyl bromide

Preparation of alcohols Ethylene oxide can be easily cleaved by HBr to give bromoethanol. The oxygen is protonated to form a protonated ethylene oxide, which, being attacked by the halide, gives bromoethanol.

$$H_2C{-}CH_2(O) + HBr \longrightarrow BrCH_2CH_2OH$$

Ethylene oxide → Bromoethanol

Mechanism.

$$H_2C{-}CH_2(\ddot{O}) + H{-}Br \rightleftharpoons H_2C{-}CH_2(\overset{+}{\ddot{O}}H) \;\; Br:^- \longrightarrow BrCH_2CH_2OH$$

Bromoethanol

Acid-catalysed cleavage of epoxides

In the case of acid-catalysed unsymmetrical epoxide, the weak nucleophiles (H_2O and ROH) attack the most substituted carbon of the ring, and produce 1-substituted alcohol. This reaction follows S_N1 reaction.

Preparation of diols Acid-catalysed epoxides are easily cleaved by water. Water reacts as the nucleophile, and this is referred to as a *hydrolysis*. For example, hydrolysis of ethylene oxide in the presence of acid-catalyst produces 1,2-ethanediol (ethylene glycol).

$$H_2C{-}CH_2(O) \xrightarrow[H_2O]{H^+} CH_2(OH)CH_2(OH) + H_3O^+$$

Ethylene oxide → 1,2-Ethanediol

Mechanism.

$$H_2C{-}CH_2(\ddot{O}) \overset{H^+}{\rightleftharpoons} H_2C{-}CH_2(:\overset{+}{O}H) \;\; H_2\ddot{O} \longrightarrow H{-}\overset{+}{O}(H){-}CH_2CH_2OH \;\; H_2\ddot{O} \longrightarrow CH_2(OH)CH_2(OH) + H_3\ddot{O}^+$$

Preparation of alkoxy alcohol Acid-catalysed unsymmetrical propylene oxide gives 1-substituted alcohols, resulting from the nucleophilic attack on the most substituted carbon. For example, propylene oxide reacts with alcohol in the presence of acid to give 2-methoxy-1-propanol.

$$H_3C\text{—}CH(O)CH_2 \xrightarrow[CH_3OH]{H^+} CH_3CH(OMe)CH_2OH + CH_3\overset{+}{O}H_2$$

Propylene oxide — 2-Methoxy-1-propanol, Alkoxy alcohol

Mechanism.

$$H_3C\text{—}CH(O)CH_2 \overset{H^+}{\rightleftharpoons} H_3C\text{—}CH(\overset{+}{O}H)CH_2 \xrightarrow{CH_3\ddot{O}H} H_3C\text{—}CH(\overset{+}{O}(H)CH_3)CH_2OH \xrightarrow{CH_3\ddot{O}H} CH_3CH(OMe)CH_2OH + CH_3\overset{+}{O}H_2$$

2-Methoxy-1-propanol

Base-catalysed cleavage of epoxides

Base-catalysed epoxide cleavage follows an S_N2 reaction in which the attack of the nucleophiles (Nu:$^-$) takes place at the least substituted carbon of the ring. Therefore, base-catalysed unsymmetrical epoxide produces 2-substituted alcohol.

$$R\text{—}CH(O)CH_2 \xrightarrow[\text{ii. } H_2O \text{ or } H_3O^+]{\text{i. Nu:}^-} RCH(OH)CH_2Nu$$

Epoxide — 2-Substituted alcohol

Nu:$^-$ = R:$^-$, RC≡C:$^-$, C≡N:$^-$, :N_3^-, H:$^-$, HO:$^-$, RO:$^-$

Mechanism.

$$R\text{—}CH(O)CH_2 + Nu:^- \longrightarrow RCH(\ddot{O}:^-)CH_2Nu \xrightarrow{H_3O^+} RCH(OH)CH_2Nu$$

Nu:$^-$ = R:$^-$, RC≡C:$^-$, C≡N:$^-$, :N_3^-, H:$^-$, HO:$^-$, RO:$^-$

Preparation of alcohols Organometallic reagents (RMgX, RLi) are powerful nucleophiles. They attack epoxides at the least hindered carbon, and generate alcohols. For example, propylene oxide is an unsymmetrical epoxide, which reacts with methyl magnesium bromide to produce 2-butanol, after the acidic work-up.

$$H_3C\text{-}CH(\text{-O-})CH_2 \text{ (Propylene oxide, Epoxide)} \xrightarrow[\text{ii. } H_3O^+]{\text{i. } CH_3MgBr, \text{ ether}} CH_3CH(OH)CH_2CH_3 \text{ (2-Butanol)}$$

Mechanism.

$$H_3C\text{-}CH(\text{-O-})CH_2 + \overset{\delta-}{H_3C}\text{-}\overset{\delta+}{MgBr} \longrightarrow CH_3CH(O^-\,\overset{+}{M}gBr)CH_2CH_3 \xrightarrow{H_3O^+} CH_3CH(OH)CH_2CH_3 \text{ (2-Butanol)}$$

Preparation of alkoxy alcohol Ethylene oxide is a symmetrical epoxide, which reacts with sodium methoxide to produce 2-methoxy-ethanol, after the hydrolytic work-up.

$$H_2C\text{-}CH_2 \text{ (epoxide; Ethylene oxide)} \xrightarrow[\text{ii. } H_2O]{\text{i.} CH_3ONa, CH_3OH} CH_3OCH_2CH_2OH \text{ (2-Methoxy-ethanol)}$$

Mechanism.

$$CH_3\ddot{O}:^- + H_2C\text{-}CH_2 \text{ (epoxide)} \longrightarrow CH_3OCH_2CH_2\ddot{O}:^- \xrightarrow{H_2O} CH_3OCH_2CH_2OH \text{ (2-Methoxy-ethanol)}$$

5.5.5 Nucleophilic acyl substitutions

Carboxylic acid and its derivatives undergo *nucleophilic acyl substitution*, where one nucleophile replaces another on the acyl carbon. Nucleophilic acyl substitution can interconvert all carboxylic acid derivatives, and the reaction mechanism varies depending on acidic or basic conditions. Nucleophiles can either be negatively charged anion ($Nu:^-$) or neutral (Nu:) molecules.

If the nucleophile is a negatively charged anion (R^-, H^-, HO^-, RO^-, CN^-), it will readily attack the carbonyl carbon and form an alkoxide tetrahedral intermediate, which in turn expels the leaving group whilst reforming the carbonyl C=O double bond.

$$R\text{-}\overset{\delta+}{C}(=\overset{\delta-}{O})\text{-}Y + Nu:^- \rightleftharpoons R\text{-}C(O^-)(Nu)\text{-}Y \text{ (Tetrahedral intermediate)} \rightleftharpoons R\text{-}C(=O)\text{-}Nu + Y^-$$

$Y = Cl, RCO_2, OR, NH_2$

If the nucleophile is a neutral molecule with a lone pair of electrons (H_2O, ROH), it requires an acid catalyst for nucleophilic addition reaction to occur. Under acidic conditions, the carbonyl group becomes protonated, and thus is activated towards nucleophilic acyl substitution. Attack by a weak nucleophile generates the tetrahedral intermediate. A simultaneous deprotonation and loss of the leaving group reforms the carbonyl C=O double bond.

Y = OR, NH_2

Tetrahedral intermediate

Fischer esterification

Preparation of esters Esters are obtained by refluxing the parent carboxylic acid and an alcohol with an acid catalyst. The equilibrium can be driven to completion by using an excess of the alcohol, or by removing the water as it forms. This is known as *Fischer esterification.*

$$\mathrm{R{-}C({=}O){-}OH} + \mathrm{R'OH} \xrightleftharpoons{H^+} \mathrm{R{-}C({=}O){-}OR'} + \mathrm{H_3O^+}$$

Mechanism. The carbonyl group of a carboxylic acid is not sufficiently electrophilic to be attacked by the alcohol. The acid catalyst protonates the carbonyl oxygen, and activates it towards nucleophilic attack. The alcohol attacks the protonated carbonyl carbon, and forms a tetrahedral intermediate. Intramolecular proton transfer converts the hydroxyl to a good leaving group as H_2O. A simultaneous deprotonation and loss of H_2O gives an ester.

Transesterification Transesterification occurs when an ester is treated with another alcohol. This reaction can be acid catalysed or base catalysed. This is where the alcohol part of the ester can be replaced with a new alcohol component. The reaction mechanism is very similar to the Fischer esterification.

$$\mathrm{R{-}C({=}O){-}OR} + \mathrm{R'{-}OH} \xrightleftharpoons{H^+ \text{ or } HO^-} \mathrm{R{-}C({=}O){-}OR'} + \mathrm{R{-}OH}$$

Conversion of carboxylic acids

Preparation of acid chlorides The best way to make acid chlorides is the reaction of a carboxylic acid with either thionyl chloride ($SOCl_2$) or oxalyl chloride $(COCl)_2$ in the presence of a base (pyridine). The mechanism of formation of acid chloride is similar to the reaction of alcohol with $SOCl_2$.

$$R{-}\overset{O}{\overset{\|}{C}}{-}OH + Cl{-}\overset{O}{\overset{\|}{S}}{-}Cl \xrightarrow{\text{Pyridine}} R{-}\overset{O}{\overset{\|}{C}}{-}Cl$$

Mechanism.

Pyridine

Cl^- + SO_2 + $R{-}\overset{O}{\overset{\|}{C}}{-}Cl$

Pyridinium chloride

Preparation of acid anhydrides Acid anhydrides are prepared from carboxylic acids by the loss of water. For example, acetic anhydride is prepared industrially by heating acetic acid to 800 °C. Other anhydrides are difficult to prepare directly from the corresponding carboxylic acids. Usually they are prepared from acid chloride and sodium carboxylate salt (see below).

$$\underset{\text{Acetic acids}}{H_3C{-}\overset{O}{\overset{\|}{C}}{-}OH + HO{-}\overset{O}{\overset{\|}{C}}{-}CH_3} \xrightarrow{800\ ^\circ C} \underset{\text{Acetic anhydride}}{H_3C{-}\overset{O}{\overset{\|}{C}}{-}O{-}\overset{O}{\overset{\|}{C}}{-}CH_3} + H_2O$$

Mechanism.

$\pm H^+$

Preparation of amides Ammonia, 1° and 2° amines react with carboxylic acids to produce, respectively, 1°, 2° and 3° amides, through a nucleophilic acyl substitution reaction. The reaction of ammonia and a carboxylic acid initially forms a carboxylate anion and an ammonium cation. Normally the reaction stops at this point, since the carboxylate anion is a poor electrophile. However, by heating the reaction to over 100 °C, the water can be driven off as steam, and amide products are formed. This is an important commercial process for the production of 1° amides.

$$R{-}\overset{O}{\overset{\|}{C}}{-}OH + NH_3 \longrightarrow R{-}\overset{O}{\overset{\|}{C}}{-}O^- + \overset{+}{N}H_4 \xrightarrow{\text{Heat}} R{-}\overset{O}{\overset{\|}{C}}{-}NH_2 + H_2O$$

Conversion of acid chlorides

Preparation of esters Acid chlorides react with alcohols to give esters through a nucleophilic acyl substitution. Because acid chloride is reactive towards weak nucleophile, e.g. alcohol, no catalyst is required for this substitution reaction. The reaction is carried out in base, most commonly in pyridine or triethylamine (Et_3N).

$$R{-}\overset{O}{\overset{\|}{C}}{-}Cl + R'OH \xrightarrow[\text{or } Et_3N]{\text{Pyridine}} R{-}\overset{O}{\overset{\|}{C}}{-}OR'$$

Mechanism. The nucleophilic alcohol attacks the carbonyl carbon of the acid chloride and displaces the chloride ion. The protonated ester loses a proton to the solvent (pyridine or Et_3N) to give the ester.

$$R'{-}\ddot{O}H + R{-}\overset{:\ddot{O}}{\overset{\|}{C}}{-}Cl \longrightarrow R{-}\overset{O}{\overset{\|}{C}}{-}\overset{+}{O}(H){-}R' + Cl^- \xrightarrow{:NEt_3} \underset{\text{Ester}}{R{-}\overset{O}{\overset{\|}{C}}{-}OR'} + Et_3\overset{+}{N}HCl^-$$

Preparation of acid anhydrides Acid chloride reacts with sodium carboxylate to give acid anhydride through nucleophilic acyl substitution reaction. Both symmetrical and unsymmetrical acid anhydrides are prepared in this way.

$$\underset{\text{Acid chloride}}{R{-}\overset{O}{\overset{\|}{C}}{-}Cl} + \underset{\text{Sodium carboxylate}}{R'{-}\overset{O}{\overset{\|}{C}}{-}ONa} \xrightarrow{\text{Ether}} \underset{\text{Acetic anhydride}}{R{-}\overset{O}{\overset{\|}{C}}{-}O{-}\overset{O}{\overset{\|}{C}}{-}R'} + NaCl$$

Mechanism.

$$R'{-}\overset{O}{\overset{\|}{C}}{-}\ddot{\underset{..}{O}}^-\overset{+}{Na} + R{-}\underset{Cl}{C}{=}\ddot{O}: \longrightarrow R'{-}\overset{O}{\overset{\|}{C}}{-}O{-}\overset{O}{\overset{\|}{C}}{-}R' + NaCl$$

Preparation of amides Ammonia and 1° and 2° amines react with acid chlorides and acid anhydrides to give 1°, 2° and 3° amides, respectively, in the presence of excess pyridine (C_5H_5N) or triethylamine (Et_3N). In the case of acid anhydride, two molar equivalents of ammonia or amines are required.

$$R{-}\overset{O}{\overset{\|}{C}}{-}Cl + R'{-}NH_2 \xrightarrow{\text{Pyridine}} R{-}\overset{O}{\overset{\|}{C}}{-}NHR'$$

Mechanism.

$$R{-}\overset{O}{\overset{\|}{C}}{-}NHR' + C_5H_5\overset{+}{N}HCl^-$$

2° Amide Pyridinium chloride

Conversion of acid chlorides and esters by organometallic reagents

Preparation of tertiary alcohols Acid chlorides and esters react with two equivalents of Grignard or organolithium reagents to produce 3° alcohol. A ketone is formed by the first molar equivalent of Grignard reagent, and this immediately reacts with a second equivalent to produce the alcohol. The final product contains two identical alkyl groups at the alcohol carbon that are both derived from the Grignard reagent. This is a good route for the preparation of 3° alcohols with two identical alkyl substituents.

$$R{-}\overset{O}{\overset{\|}{C}}{-}Y \xrightarrow[\text{ii. } H_3O^+]{\text{i. 2 R'MgX or 2 R'Li}} R{-}\overset{OH}{\underset{R'}{C}}{-}R'$$

Y = Cl or OR, Acid chloride or ester → 3° Alcohol

Mechanism.

3° Alcohol

Preparation of ketones Using a weaker organometallic reagent, e.g. Gilman reagent (R_2CuLi, organocuprate), the reaction of acid chlorides can be stopped at the ketonic stage. Gilman reagents do not react with aldehydes, ketones, esters, amides or acid anhydrides. Thus, in the presence of other carbonyl functionalities acid chloride reacts readily with Gilman reagents. The reaction is carried out at −78 °C in ether, and ketone is obtained after the hydrolytic work-up.

$$\underset{\text{Acid chloride}}{R{-}\overset{O}{\overset{\|}{C}}{-}Cl} \xrightarrow[\text{ii. } H_2O]{\text{i. } R'_2CuLi\text{, ether}} \underset{\text{Ketone}}{R{-}\overset{O}{\overset{\|}{C}}{-}R'}$$

Claisen condensation

When two molecules of ester undergo a condensation reaction, the reaction is called a *Claisen condensation.* Claisen condensation, like the aldol condensation, requires a strong base. However, aqueous NaOH cannot be used in Claisen condensation, because the ester can be hydrolysed by aqueous base. Therefore, most commonly used bases are nonaqueous bases, e.g. sodium ethoxide (NaOEt) in EtOH and sodium methoxide (NaOMe) in MeOH. The product of a Claisen condensation is a β-ketoester. As in the aldol condensation, one molecule of carbonyl compound is converted to an enolate anion when an α-proton is removed by a strong base, e.g. NaOEt.

$$R{-}\underset{H}{CH}{-}\overset{:O:}{\overset{\|}{C}}{-}OR' \overset{\text{NaOEt, EtOH}}{\rightleftharpoons} R{-}\overset{-}{C}H{-}\overset{:\ddot{O}}{\overset{\|}{C}}{-}OR' \longleftrightarrow R{-}CH{=}\overset{:\ddot{O}^-}{C}{-}OR'$$

EtÖ:⁻

Resonance stabilized enolate anion

The enolate anion attacks the carbonyl carbon of a second molecule of ester and gives a β-ketoester. Thus, the Claisen condensation is a nucleophilic acyl substitution reaction. For example, two molecules of ethyl acetate condense together to form the enolate of ethyl acetoacetate, which upon addition of an acid produces ethyl acetoacetate (β-ketoester).

$$H_3C{-}\overset{O}{\overset{\|}{C}}{-}OC_2H_5 + H_3C{-}\overset{O}{\overset{\|}{C}}{-}OC_2H_5 \xrightarrow[\text{ii. } H_3O^+]{\text{i. NaOEt, EtOH}} H_3C{-}\overset{O}{\overset{\|}{C}}{-}CH_2{-}\overset{O}{\overset{\|}{C}}{-}OC_2H_5$$

Ethyl acetates — Ethyl acetoacetate (β-Ketoester)

Mechanism. Removal of an α-hydrogen from the ethyl acetate by NaOEt produces a resonance-stabilized enolate anion.

$$H{-}\underset{H}{CH}{-}\overset{:O:}{\overset{\|}{C}}{-}OC_2H_5 \rightleftharpoons \overset{-}{C}H_2{-}\overset{:\ddot{O}}{\overset{\|}{C}}{-}OC_2H_5 \longleftrightarrow H_2C{=}\overset{:\ddot{O}^-}{C}{-}OC_2H_5$$

$C_2H_5\ddot{O}^-$

Resonance-stabilized enolate anion

Nucleophilic attack of the enolate anion to the carbonyl carbon of another ethyl acetate gives an alkoxide tetrahedral intermediate. The resulting alkoxide reforms the carbonyl group by ejecting the ethoxide anion. This ethoxide anion deprotonates the α-hydrogen, and produces a new enolate anion of the resulting condensed product, which is protonated in the next step upon acidification during work-up and yields the ethyl acetoacetate.

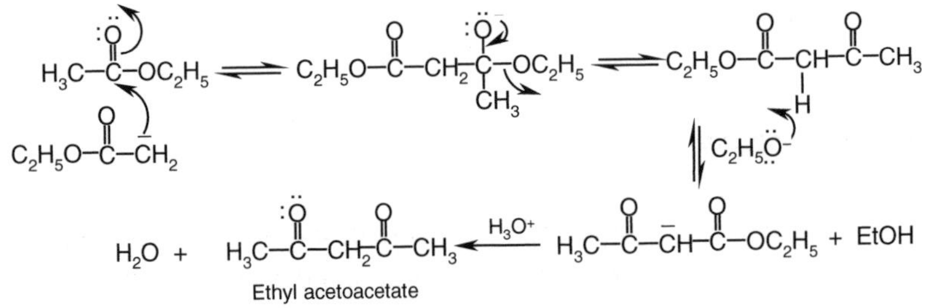

Ethyl acetoacetate

5.5.6 Electrophilic substitutions

Electrophilic aromatic substitution is a reaction where a hydrogen atom in an aromatic system, e.g. benzene, is replaced by an electrophile. Some of the important electrophilic substitution reactions are Friedel–Crafts alkylation and acylation, nitration, halogenation and sulphonation of benzene.

Electrophilic substitution of benzene

Benzene reacts with an electrophile (E^+) (usually in the presence of Lewis acid catalyst) to form the corresponding substituted product.

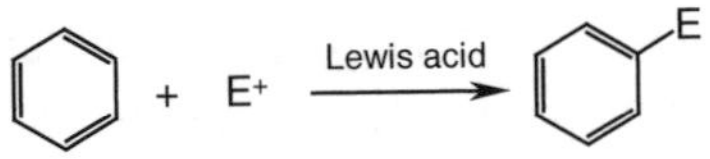

Lewis acid

Mechanism. The electrophile takes two electrons of the six-electron π system to form a σ bond to one of the carbon atoms of the benzene ring. The arenium ion loses a proton from the carbon atom that bears the electrophile to produce the substituted benzene.

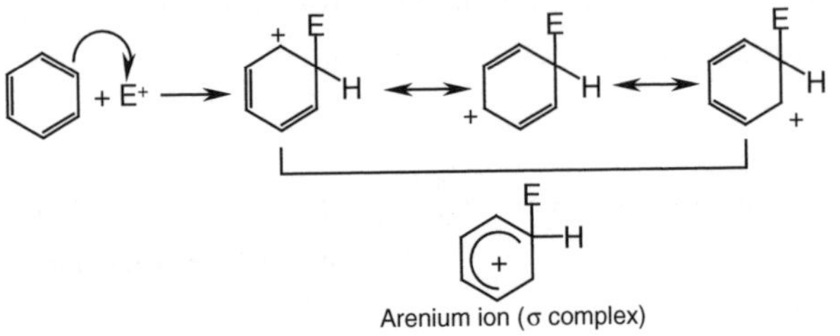

Arenium ion (σ complex)

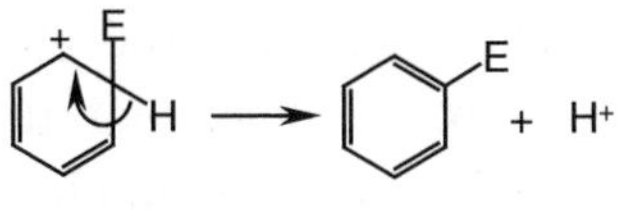

Friedel–Crafts alkylation

First introduced by Charles Friedel and James Crafts in 1877, the FC alkylation is an electrophilic aromatic substitution reaction where the electrophile is a carbocation, R^+. This carbocation is generated by $AlCl_3$-catalysed ionization of alkyl halide. For example, benzene reacts with isopropylchloride in the presence of Lewis acid to produce isopropylbenzene.

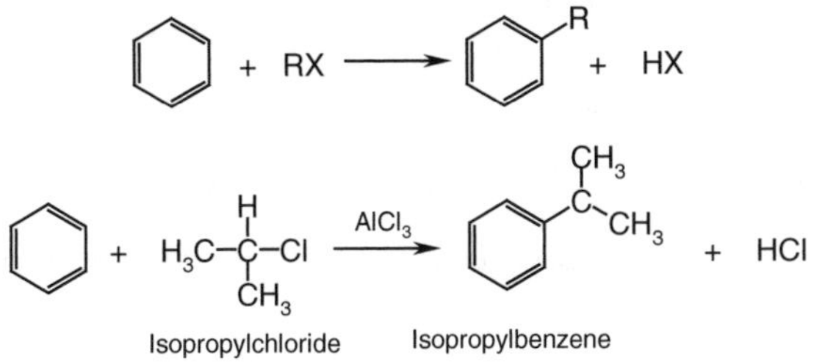

Isopropylchloride Isopropylbenzene

Mechanism.

- *Step 1.* Formation of carbocation

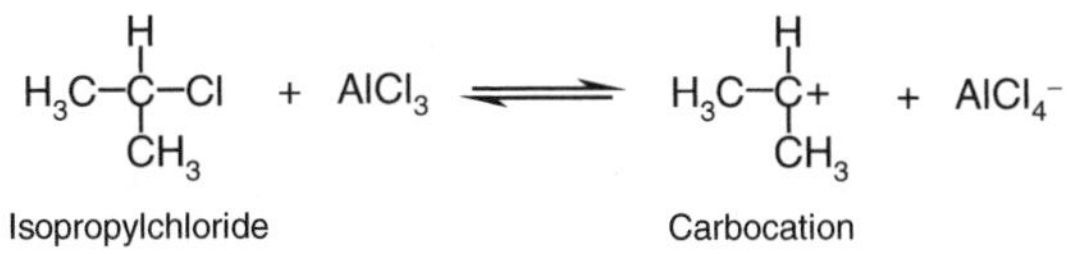

Isopropylchloride Carbocation

- *Step 2.* Formation of arenium ion complex

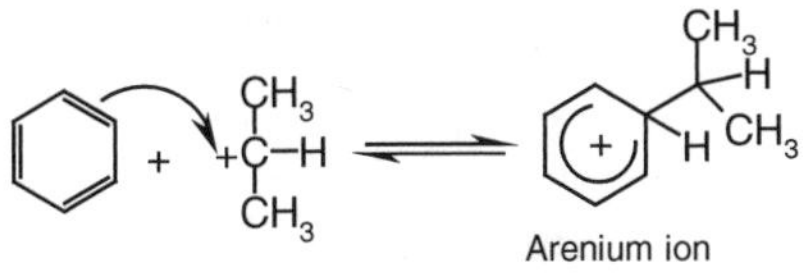

Arenium ion

- *Step 3.* Loss of a proton from the arenium ion

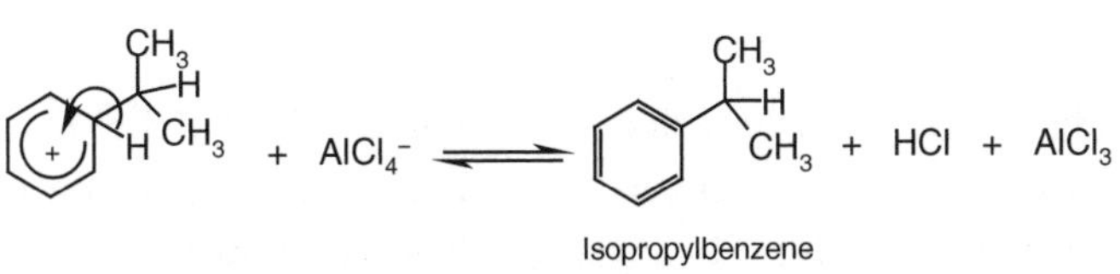

Isopropylbenzene

In the case of 1° alkyl halide, a simple carbocation does not form. $AlCl_3$ forms a complex with 1° alkyl halide, and this complex acts as an electrophile. While this complex is not a simple carbocation, it acts as if it were, and transfers a positive alkyl group to the aromatic ring.

$$\overset{\delta+}{RH_2C}\text{-------------------}\overset{\delta-}{Cl}:AlCl_3$$

FC alkylations are not restricted to the use of RX and $AlCl_3$. Many other pairs of reagents that form carbocations (or carbocation-like species) may be used. For example, an alkene and an acid, or an alcohol and an acid, could be used.

Benzene + $CH_3CH{=}CH_2$ (Propene) + HF (Hydrofluoric acid) ⟶ Isopropylbenzene

Benzene + Cyclohexanol + BF_3 (Boron trifluoride) ⟶ Cyclohexbenzene

Limitations of FC alkylation FC alkylations are limited to alkyl halides. Aryl or vinyl halides do not react. FC alkylation does not occur on aromatic rings containing strong *electron-withdrawing* substituents, e.g. $-NO_2$, $-CN$, $-CHO$, $-COR$, $-NH_2$, $-NHR$ or $-NR_2$ group. Multiple substitutions often take place. Carbocation rearrangements may occur, which result in multiple products.

$CH_3CH_2CH_2CH_2Br$ (Bromobutane) $\xrightarrow{AlCl_3}$ $CH_3CH_2CH(H)\overset{\delta+}{C}H_2\text{..........}\overset{\delta-}{Br}AlCl_3$ ⟶ $CH_3CH_2\overset{+}{C}HCH_3$

$\xrightarrow{AlCl_3,\ HBr,\ \text{benzene}}$ $C_6H_5CH_2CH_2CH_2CH_3$ (Butylbenzene)

$CH_3CH_2\overset{+}{C}HCH_3$ + benzene ⟶ $C_6H_5CH(CH_3)CH_2CH_3$ (*Sec*-butylbenzene)

Friedel–Crafts acylation

First introduced by Charles Friedel and James Crafts, the FC acylation places an acyl group on a benzene ring. Either an acyl halide or an acid anhydride can be used for FC acylation. The acylium ion is the required electrophile, which is formed by the reaction of an acid chloride (acetyl chloride) or an acid anhydride (acetic anhydride) with a Lewis acid ($AlCl_3$).

Benzene + H_3C–CO–Cl (Acetyl chloride) $\xrightarrow[\text{80°C, Excess benzene}]{AlCl_3,\ H_2O}$ Acetophenone (C_6H_5–CO–CH_3) + HCl

Benzene + H_3C–CO–O–CO–CH_3 (Acetic anhydride) $\xrightarrow[\text{80°C, Excess benzene}]{AlCl_3,\ H_2O}$ Acetophenone + HCl

Mechanism.

- *Step 1.* Formation of carbocation (acylium ion)

H_3C–CO–Cl + $AlCl_3$ ⟶ [H_3C–C^+=O ⟷ H_3C–C≡O^+] (Acylium ion) + $AlCl_4^-$

- *Step 2.* Formation of arenium ion complex

Benzene + $^+C(=O)$–CH_3 ⇌ Arenium ion

- *Step 3.* Loss of a proton from the arenium ion complex

Arenium ion + $AlCl_4^-$ ⇌ Acetophenone + HCl + $AlCl_3$

Halogenation of benzene

In the presence of anhydrous Lewis acid (e.g. $FeCl_3$ or $FeBr_3$), benzene reacts readily with halogens (bromine or chlorine) to produce halobenzenes (bromobenzene or chlorobenzene). Fluorine (Fl_2) reacts so rapidly with benzene that it requires special conditions and apparatus to carry out fluorination. On the other hand, iodine (I_2) is so unreactive that an oxidizing agent (e.g. HNO_3) has to be used to carry out iodination.

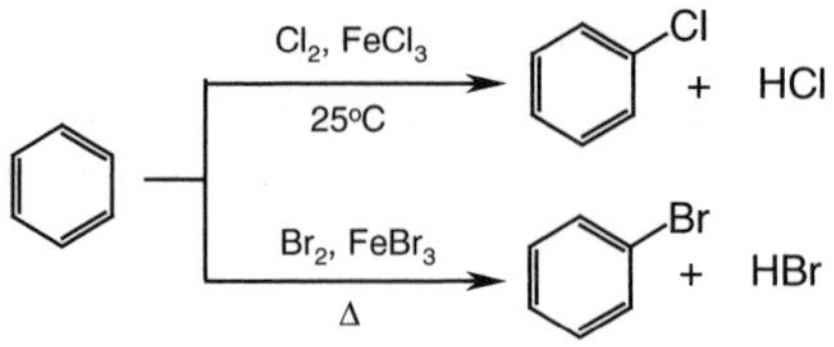

Bromination of benzene follows the same general mechanism of the electrophilic aromatic substitution. The bromine molecule reacts with $FeBr_3$ by donating a pair of its electrons to it, which creates a more polar Br−Br bond.

Mechanism.

- *Step 1.* Formation of carbocation (halonium ion)

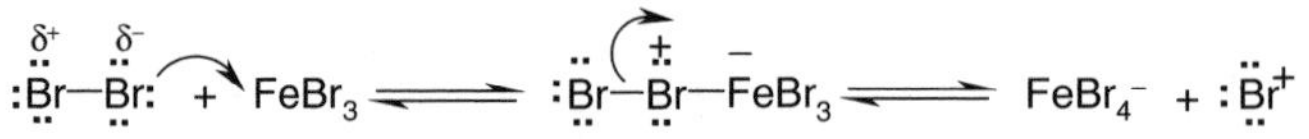

- *Step 2.* Formation of arenium ion complex

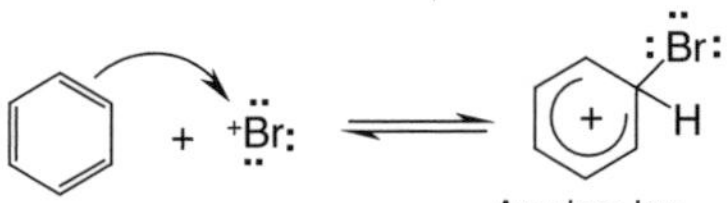

Arenium ion

- *Step 3.* Loss of a proton from the arenium ion complex

Bromobenzene

Nitration of benzene

Benzene reacts slowly with hot concentrated nitric acid (HNO_3) to yield nitrobenzene. The reaction can be faster if a mixture of concentrated HNO_3 and concentrated sulphuric acid (H_2SO_4), which acts as a catalyst, is used. Sulphuric acid protonates HNO_3. Loss of water from protonated HNO_3 forms a nitronium ($^+NO_2$) ion, the electrophile required for nitration. Thus, concentrated H_2SO_4 increases the rate of the reaction by increasing the concentration of electrophile ($^+NO_2$).

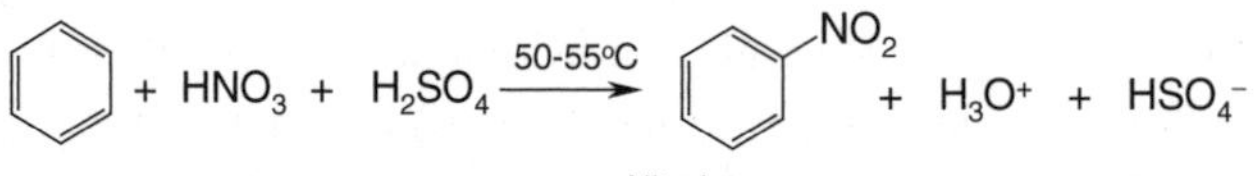

Nitrobenzene

Mechanism.

- *Step 1.* Generation of nitronium ion ($^{+}NO_2$), an electrophile

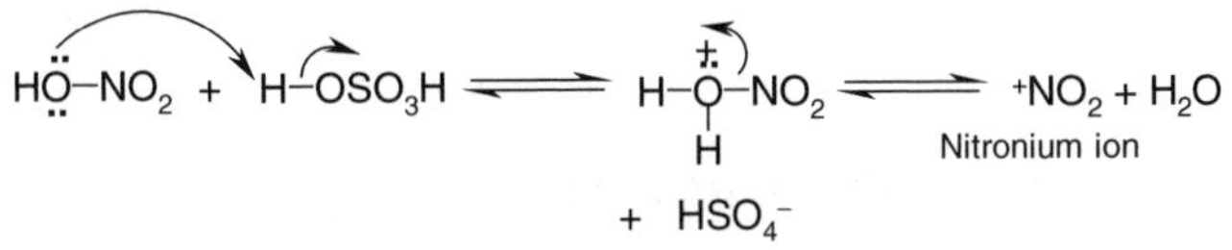

$$H\ddot{O}{-}NO_2 + H{-}OSO_3H \rightleftharpoons H{-}\overset{+}{\ddot{O}}(H){-}NO_2 \rightleftharpoons {}^{+}NO_2 + H_2O$$

$+\ HSO_4^-$

Nitronium ion

- *Step 2.* Formation of arenium ion complex

$$C_6H_6 + \overset{+}{N}O_2 \rightleftharpoons [C_6H_6NO_2]^+$$

Arenium ion

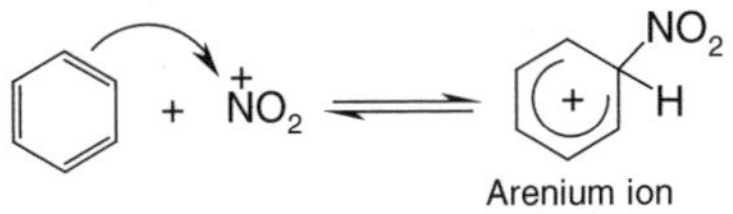

- *Step 3.* Loss of a proton from the arenium ion complex

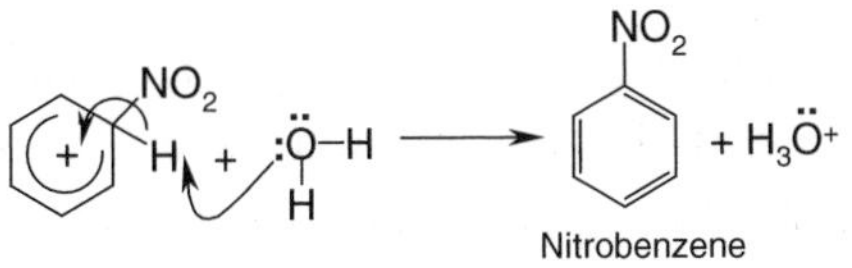

$$[C_6H_6NO_2]^+ + :\ddot{O}{-}H(H) \longrightarrow C_6H_5NO_2 + H_3\ddot{O}^+$$

Nitrobenzene

Sulphonation of benzene

Benzene reacts with fuming sulphuric acid at room temperature to give benzenesulphonic acid. Fuming sulphuric acid contains added sulphur trioxide (SO_3). Sulphonation of benzene can also be carried out with concentrated H_2SO_4, but with slower speed. In both cases, SO_3 acts as an electrophile.

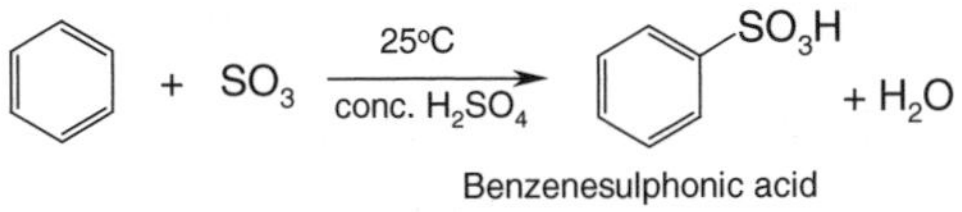

$$C_6H_6 + SO_3 \xrightarrow[\text{conc. } H_2SO_4]{25^{\circ}C} C_6H_5SO_3H + H_2O$$

Benzenesulphonic acid

Mechanism.

- *Step 1.* Generation of electrophile (SO_3)

$$2\,H_2SO_4 \rightleftharpoons SO_3 + H_3O^+ + HSO_4^-$$

- *Step 2.* Formation of arenium ion complex

$$C_6H_6 + SO_3 \rightleftharpoons [C_6H_6SO_3^-]^+$$

Arenium ion

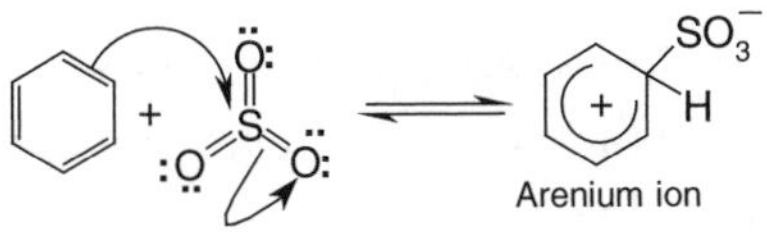

- *Step 3.* Loss of a proton from the arenium ion complex

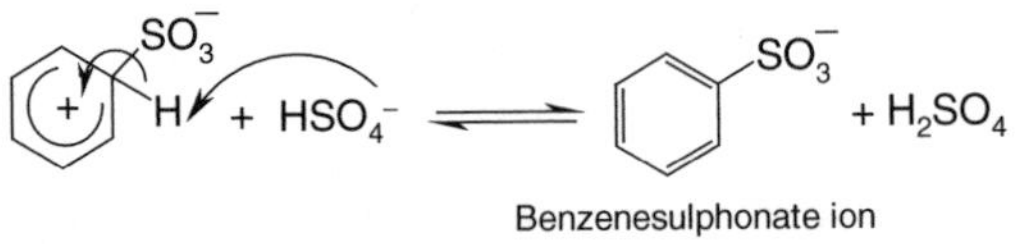

Benzenesulphonate ion

- *Step 4.* Protonation of the benzenesulphonate anion

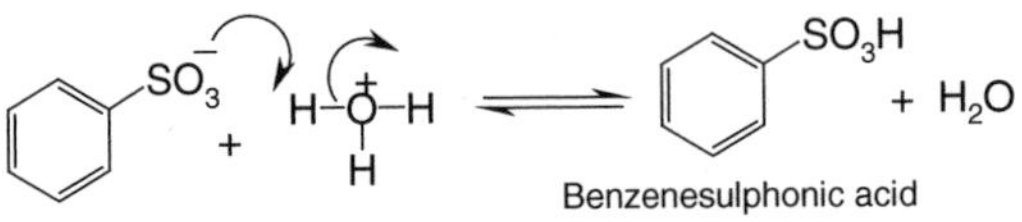

Benzenesulphonic acid

5.6 Hydrolysis

The term *hydrolysis* comes from the word 'hydro' meaning water and 'lysis' meaning breakdown. A hydrolysis reaction is one in which a σ bond is cleaved by the addition of the elements of water to the fragments formed in the cleavage. A hydrolysis reaction is catalysed by acid, base or hydrolysing enzyme. For example, the analgesic drug aspirin (acetyl salicylic acid) is easily hydrolysed in the presence of acid, moisture and heat to form salicylic acid.

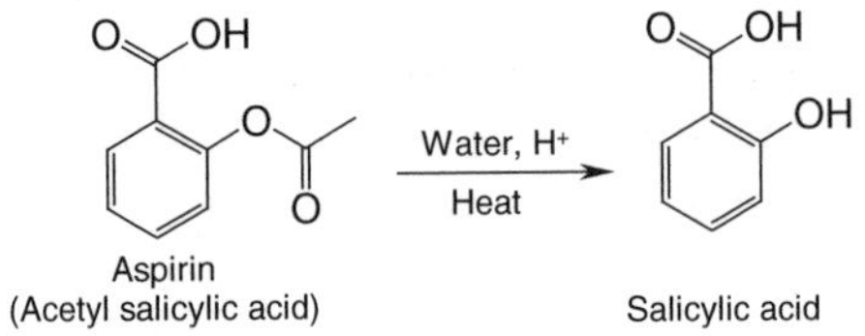

Aspirin (Acetyl salicylic acid) → (Water, H^+, Heat) → Salicylic acid

Glucosidase is a hydrolysing enzyme, and can be used to hydrolyse various glucosides. For example, salicin, found in willow barks, can be hydrolysed to salicyl alcohol by enzyme.

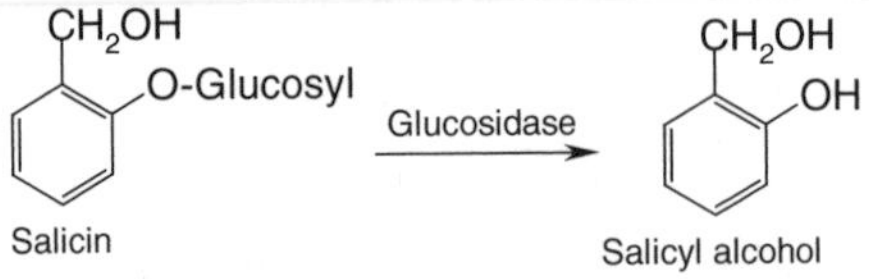

Salicin → (Glucosidase) → Salicyl alcohol

5.6.1 Hydrolysis of carboxylic acid derivatives

All carboxylic acid derivatives yield parent carboxylic acids on hydrolysis, catalysed by either an acid or a base. The reactivity toward hydrolysis varies greatly amongst the derivatives.

Hydrolysis of acid halides and anhydrides

Preparation of carboxylic acids:
Acid halides and anhydrides are so reactive that they react with water under neutral conditions. This can be a potential problem for the storage if these compounds since these compounds can be air (moisture) sensitive. Hydrolysis of these compounds can be avoided by using dry nitrogen atmospheres and anhydrous solvents and reagents.

$$\text{R-C(=O)-Cl} \text{ or } \text{R-C(=O)-O-C(=O)-R} \xrightarrow{H_2O} \text{R-C(=O)-OH}$$

Mechanism.

$$\text{R-C(=O)-Y} + H_2O \rightleftharpoons \text{R-C(O}^-\text{)(Y)-}\overset{+}{O}H_2 \overset{\pm H^+}{\rightleftharpoons} \text{R-C(OH)(OH)-Y} \rightleftharpoons H_3O^+ + Y^- + \text{R-C(=O)-OH}$$

Y = Cl, RCO_2

Hydrolysis of esters: preparation of carboxylic acids

The acid-catalysed hydrolysis of an ester is the reverse reaction of the *Fischer esterification*. Addition of excess water drives the equilibrium towards the acid and alcohol formation. The base-catalysed hydrolysis of esters is also known as *saponification*, and this does not involve the equilibrium process observed for the Fischer esterification.

$$\text{R'OH} + \text{R-C(=O)-OH} \xleftarrow[H_3O^+\text{, heat}]{\text{Acid hydrolysis}} \text{R-C(=O)-OR'} \xrightarrow[\text{NaOH, }H_2O\text{, heat}]{\text{Base hydrolysis}} \text{R-C(=O)-OH} + \text{R'OH}$$

Mechanism.

- *Acid-catalysed hydrolysis.* The carbonyl group of an ester is not sufficiently electrophilic to be attacked by water. The acid catalyst protonates the carbonyl oxygen, and activates it towards nucleophilic attack. The water molecule attacks the protonated carbonyl carbon, and forms a tetrahedral intermediate. Proton transfer from the hydronium ion to a second molecule of water yields an ester hydrate. The intramolecular proton transfer produces a good leaving group as alcohol. A simultaneous deprotonation by the water and loss of alcohol gives a carboxylic acid.

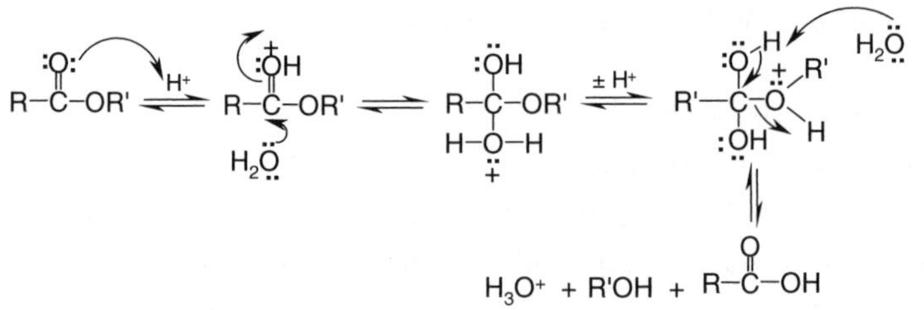

- *Base-catalysed hydrolysis.* Hydroxide ion attacks the carbonyl group to give a tetrahedral intermediate. The negatively charged oxygen can readily expel alkoxide ion, a basic leaving group, and produce a carboxylic acid. The alkoxide ion quickly deprotonates the carboxylic acid, and the resulting carboxylate ion is unable to participate in the reverse reaction. Thus, there is no equilibrium in the base-catalysed hydrolysis, and the reaction goes to completion. Protonation of the carboxylate ion by addition of an aqueous acid in a separate step produces the free carboxylic acid.

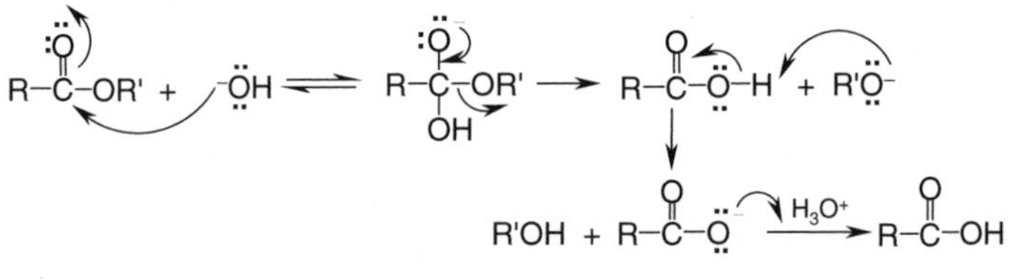

Hydrolysis of amides: preparation of carboxylic acids

Amides are the most reluctant derivatives of carboxylic acids to undergo hydrolysis. However, they can be forced to undergo hydrolysis by the use of vigorous conditions, e.g. heating with 6 M HCl or 40% NaOH for prolonged periods of time.

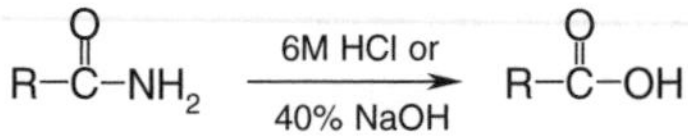

$$\mathrm{R{-}C({=}O){-}NH_2} \xrightarrow[\text{40\% NaOH}]{\text{6M HCl or}} \mathrm{R{-}C({=}O){-}OH}$$

Mechanism.

- *Acid-catalysed hydrolysis.* Under acidic conditions, the hydrolysis of an amide resembles the acid-catalysed hydrolysis of an ester, with protonation of the carbonyl group yielding an activated carbonyl group that undergoes nucleophilic attack by water. The intramolecular proton transfer produces a good leaving group as ammonia. Simultaneous deprotonation by water and loss of ammonia yields a carboxylic acid.

$$R{-}C({=}O){-}NH_2 \underset{}{\overset{H^+}{\rightleftharpoons}} R{-}C({=}\overset{+}{O}H){-}NH_2 \;(+H_2O) \rightleftharpoons R{-}C(OH)(NH_2)({-}\overset{+}{O}H_2) \overset{\pm H^+}{\rightleftharpoons} R{-}C(OH)(OH){-}\overset{+}{N}H_3 \;(H_2O) \longrightarrow H_3O^+ + NH_3 + R{-}C({=}O){-}OH$$

- *Base-catalysed hydrolysis.* Hydroxide ion attacks the carbonyl, and forms a tetrahedral intermediate. The negatively charged oxygen can readily expel amide ion, a basic leaving group, and produce a carboxylic acid. The amide ion quickly deprotonates the carboxylic acid, and the resulting carboxylate ion is unable to participate in the reverse reaction. Thus, there is no equilibrium in the base-catalysed hydrolysis, and the reaction goes to completion. Protonation of the carboxylate by the addition of an aqueous acid in a separate step gives the free carboxylic acid.

$$R{-}C({=}O){-}NH_2 + {}^{-}OH \rightleftharpoons R{-}C(O^-)(OH){-}NH_2 \longrightarrow R{-}C({=}O){-}O{-}H + {}^{-}NH_2 \longrightarrow NH_3 + R{-}C({=}O){-}O^- \xrightarrow{H_3O^+} R{-}C({=}O){-}OH$$

Hydrolysis of nitriles: preparation of primary amides and carboxylic acids

Nitriles are hydrolysed to 1° amides, and then to carboxylic acids either by acid catalysis or base catalysis. It is possible to stop the acid hydrolysis at the amide stage by using H_2SO_4 as an acid catalyst and one mole of water per mole of nitrile. Mild basic conditions (NaOH, H_2O, 50 °C) only take the hydrolysis to the amide stage, and more vigorous basic condition (NaOH, H_2O, 200 °C) is required to convert the amide to a carboxylic acid.

$$R{-}C{\equiv}N \xrightarrow[\text{Heat}]{H_3O^+} R{-}C({=}O){-}NH_2 \xrightarrow[\text{Prolonged period}]{H_3O^+,\ \text{heat}} R{-}C({=}O){-}OH$$

$$R{-}C{\equiv}N \xrightarrow[H_2O,\ 50\ ^\circ C]{NaOH} R{-}C({=}O){-}NH_2 \xrightarrow[H_2O,\ 200\ ^\circ C]{NaOH} R{-}C({=}O){-}OH$$

Mechanism.

- *Acid-catalysed hydrolysis.* The acid-catalysed hydrolysis of nitriles resembles the acid-catalysed hydrolysis of an amide, with protonation of the nitrogen of the cyano group activating the nucleophilic attack by water. The intramolecular proton transfer produces a protonated imidic acid. The imidic acid tautomerizes to the more stable amide via deprotonation on oxygen and protonation on nitrogen. The acid-catalysed amide is converted to carboxylic acid in several steps as discussed earlier for the hydrolysis of amides.

R—C≡N̈ ⇌(H+) R—C≡N̈H (+) ⇌ R—C(=N̈H)—Ö(+)(H)—H ⇌(± H+) R—C(=N̈H₂+)—:ÖH ↕ R—C(=:Ö:)—NH₂ ⇌(H+/H₂O, Several steps) R—C(=O)—OH

- *Base-catalysed hydrolysis.* The hydroxide ion attacks the nitrile carbon, followed by protonation on the unstable nitrogen anion to generate an imidic acid. The imidic acid tautomerizes to the more stable amide via deprotonation on oxygen and protonation on nitrogen. The base-catalysed amide is converted to carboxylic acid in several steps as discussed earlier for the hydrolysis of amides.

R—C≡N: + HÖ⁻ ⇌ R—C(—:ÖH)=N̈:⁻ ⇌(H₂O) R—C(—OH)=N̈H (Imidic acid) + HÖ⁻ ⇌ R—C(=O)—N̈H₂ ⇌(HO⁻/H₂O, Several steps) R—C(=O)—OH

5.7 Oxidation–reduction reactions

Oxidation is a loss of electrons, and reduction is a gain of electrons. However, in the context of organic chemistry, *oxidation* means the loss of hydrogen, the addition of oxygen or the addition of halogen. A general symbol for oxidation is [O]. Thus, oxidation can also be defined as a reaction that increases the content of any element more electronegative than carbon. *Reduction* is the addition of hydrogen, the loss of oxygen or the loss of halogen. A general symbol for reduction is [H]. The conversion of ethanol to acetaldehyde, and that of acetaldehyde to acetic acid, are oxidation reactions, and the reverse reactions are reduction reactions.

H_3C–CH_2–OH (Ethanol) ⇌([O]/[H]) H_3C–CHO (Acetaldehyde) ⇌([O]/[H]) H_3C–COOH (Acetic acid)

5.7.1 Oxidizing and reducing agents

Oxidizing agents are reagents that seek electrons, and are electron-deficient species, e.g. chromic acid (H_2CrO_4), potassium permanganate ($KMnO_4$) and osmium tetroxide (OsO_4). Therefore, oxidizing agents are classified as *electrophiles*. In the process of gaining electrons, oxidizing agents become reduced. Oxidation results in an increase in the number of C—O bonds or a decrease in the number of C—H bonds.

On the other hand, *reducing agents* are reagents that give up electrons, and are electron-rich species, e.g. sodium borohydride ($NaBH_4$), lithium aluminium hydride ($LiAlH_4$). Therefore, reducing agents are classified as *nucleophiles*. In the process of giving up electrons, reducing agents become oxidized. Reduction results in an increase in the number of C—H bonds or a decrease in the number of C—O bonds.

5.7.2 Oxidation of alkenes

Preparation of epoxides

Alkenes undergo a number of oxidation reactions in which the C=C is oxidized. The simplest epoxide, ethylene oxide, is prepared by catalytic oxidation of ethylene with Ag at high temperatures (250 °C).

$$H_2C{=}CH_2 \xrightarrow[250\ ^\circ C]{O_2,\ Ag} \text{Ethylene oxide}$$

Ethylene oxide

Alkenes are also oxidized to epoxides by peracid or peroxyacid (RCO_3H), e.g. peroxybenzoic acid ($C_6H_5CO_3H$). A peroxyacid contains an extra oxygen atom compared with carboxylic acid, and this extra oxygen is added to the double bond of an alkene to give an epoxide. For example, cyclohexene reacts with peroxybenzoic acid to produce cyclohexane oxide.

$$\underset{\text{Alkene}}{RCH{=}CHR} \xrightarrow{RCO_3H} \underset{\text{Epoxide}}{RCH{-}CHR}$$

$$\text{Cyclohexene} \xrightarrow{C_6H_5CO_3H} \text{Cyclohexane oxide}$$

The addition of oxygen to an alkene is stereospecific. Therefore, a *cis*-alkene produces a *cis*-epoxide, and *trans*-alkene gives a *trans*-epoxide.

$$\textit{cis}\text{-Alkene} \xrightarrow{RCO_3H} \textit{cis}\text{-Epoxide} \qquad \textit{trans}\text{-Alkene} \xrightarrow{RCO_3H} \textit{trans}\text{-Epoxide}$$

Preparation of carboxylic acids and ketones

Reaction of an alkene with hot basic potassium permanganate ($KMnO_4$) results in cleavage of the double bond, and formation of highly oxidized carbons. Therefore, unsubstituted carbon atoms become CO_2, mono-substituted carbon atoms become carboxylates, and di-substituted carbon atoms become ketones. This can be used as a chemical test (known as the *Baeyer test*) for alkenes and alkynes, in which the purple colour of the $KMnO_4$ disappears, and a brown MnO_2 residue is formed.

$$H_2C{=}CH_2 \xrightarrow[\text{ii. } H_3O^+]{\text{i. } KMnO_4\text{, NaOH, heat}} 2\,CO_2 + H_2O$$

Ethylene

$$CH_3CH{=}CHCH_3 \xrightarrow[H_2O\text{, heat}]{KMnO_4\text{, NaOH}} 2\,H_3C{-}C({=}O){-}O^- \xrightarrow{H_3O^+} 2\,H_3C{-}C({=}O){-}OH$$

(*cis* or *trans*)-2-Butene; Acetate ion; Acetic acid

$$CH_3CH_2CH_2C(CH_3){=}CH_2 \xrightarrow[\text{ii. } H_3O^+]{\text{i. } KMnO_4\text{, NaOH, heat}} CH_3CH_2CH_2C(CH_3){=}O + CO_2$$

2-Methylpentene; Methyl butanone

5.7.3 *syn-hydroxylation of alkenes: preparation of* **syn-diols**

Hydroxylation of alkenes is the most important method for the synthesis of 1,2-diols (also called glycol). Alkenes react with cold, dilute and basic $KMnO_4$ or osmium tetroxide (OsO_4) and hydrogen peroxide to give *cis*-1,2-diols. The products are always *syn*-diols, since the reaction occurs with *syn* addition.

$$\text{Cyclopentene} \xrightarrow[\text{ii. NaOH, } H_2O]{\text{i. Cold } KMnO_4} \textit{cis}\text{-1,2-Cyclopentane diol}$$

cis-1,2-Cyclopentane diol
A *meso* compound

$$H_2C{=}CH_2 \xrightarrow[\text{ii. } H_2O_2]{\text{i. } OsO_4\text{, Pyridine}} CH_2(OH){-}CH_2(OH)$$

Ethene; *syn*-1,2-Ethanediol
Ethylene glycol

5.7.4 *Anti*-hydroxylation of alkenes: preparation of *anti*-diols

Alkenes react with peroxyacids (RCO_3H) followed by hydrolysis to give *trans*-1,2-diols. The products are always *anti*-diols, since the reaction occurs with *anti* addition.

$$\text{Cyclopentene} \xrightarrow[\text{ii. } H_2O]{\text{i. } RCO_3H} \textit{trans}\text{-1,2-Cyclopentane diol}$$

trans-1,2-Cyclopentane diol
(Racemic mixture)

$$CH_3CH{=}CH_2 \xrightarrow[\text{ii. } H_2O]{\text{i. } RCO_3H} CH_3CH(OH){-}CH_2(OH)$$

Propene; *anti*-1,2-Propanediol
(Propylene glycol)

5.7.5 Oxidative cleavage of *syn*-diols: preparation of ketones and aldehydes or carboxylic acids

The treatment of an alkene by *syn*-hydroxylation, followed by periodic acid (HIO_4) cleavage, is an alternative to the *ozonolysis*, followed by reductive work-up. *Syn*-diols are oxidized to aldehydes and ketones by periodic acid (HIO_4). This oxidation reaction divides the reactant into two pieces, thus it is called an oxidative cleavage.

Alkene $\xrightarrow[H_2O_2]{OsO_4}$ *syn*-diol $\xrightarrow{HIO_4}$ Ketone + Aldehyde

5.7.6 Ozonolysis of alkenes

Alkenes can be cleaved by ozone followed by an oxidative or reductive work-up to generate carbonyl compounds. The products obtained from an ozonolysis reaction depend on the reaction conditions. If ozonolysis is followed by the reductive work-up (Zn/H_2O), the products obtained are aldehydes and/or ketones. Unsubstituted carbon atoms are oxidized to formaldehyde, mono-substituted carbon atoms to aldehydes, and di-substituted carbon atoms to ketones.

When ozonolysis is followed by the oxidative work-up ($H_2O_2/NaOH$), the products obtained are carboxylic acids and/or ketones. Unsubstituted carbon atoms are oxidized to formic acids, mono-substituted carbon atoms to carboxylic acids and di-substituted carbon atoms to ketones.

Preparation of aldehydes and ketones

Alkenes are directly oxidized to aldehydes and/or ketones by ozone (O_3) at low temperatures (−78 °C) in methylene chloride, followed by the reductive work-up. For example, 2-methyl-2-butene reacts with O_3, followed by a reductive work-up to yield acetone and acetaldehyde. This reducing agent prevents aldehyde from oxidation to carboxylic acid.

2-Methyl-2-butene $\xrightarrow[ii.\ Zn,\ H_2O]{i.\ O_3,\ CH_2Cl_2}$ Acetone + Acetaldehyde

Preparation of carboxylic acids and ketones

Alkenes are oxidized to carboxylic acids and/or ketones by ozone (O_3) at low temperatures (−78 °C) in methylene chloride, followed by oxidative work-up. For example, 2-methyl-2-butene reacts with O_3, followed by an oxidative work-up to give acetone and acetic acid.

$$\underset{\text{2-Methyl-2-butene}}{(H_3C)_2C{=}CH(CH_3)} \xrightarrow[\text{ii. } H_2O_2,\ NaOH]{\text{i. } O_3,\ CH_2Cl_2} \underset{\text{Acetone}}{(H_3C)_2C{=}O} + \underset{\text{Acetic acid}}{O{=}C(CH_3)OH}$$

5.7.7 Oxidation of alkynes: preparation of diketones and carboxylic acids

Alkynes are oxidized to diketones by cold, dilute and basic potassium permanganate.

$$R{-}C{\equiv}C{-}R' \xrightarrow[H_2O,\ NaOH]{\text{cold } KMnO_4} \underset{\text{Diketone}}{R{-}\overset{O}{\overset{\|}{C}}{-}\overset{O}{\overset{\|}{C}}{-}R'}$$

When the reaction condition is too warm or basic, the oxidation proceeds further to generate two carboxylate anions, which on acidification yield two carboxylic acids.

$$R{-}C{\equiv}C{-}R' \xrightarrow[\text{KOH, heat}]{KMnO_4} R{-}CO_2^- + {}^-O_2C{-}R' \xrightarrow{H_3O^+} R{-}CO_2H + R'{-}CO_2H$$

Unsubstituted carbon atoms are oxidized to CO_2, and mono-substituted carbon atoms to carboxylic acids. Therefore, oxidation of 1-butyne with hot basic potassium permanganate followed by acidification produces propionic acid and carbon dioxide.

$$\underset{\text{1-Butyne}}{C_2H_5C{\equiv}CH} \xrightarrow[\text{ii. } H_3O^+]{\text{i. } KMnO_4,\ \text{KOH, heat}} \underset{\text{Propionic acid}}{C_2H_5{-}\overset{O}{\overset{\|}{C}}{-}OH} + CO_2$$

5.7.8 Ozonolysis of alkynes: preparation of carboxylic acids

Ozonolysis of alkynes followed by hydrolysis gives similar products to those obtained from permanganate oxidation. This reaction does not require oxidative or reductive work-up. Unsubstituted carbon atoms are oxidized to CO_2, and mono-substituted carbon atoms to carboxylic acids. For example, ozonolysis of 1-butyene followed by hydrolysis gives propionic acid and carbon dioxide.

$$\underset{\text{1-Butyne}}{C_2H_5C{\equiv}CH} \xrightarrow[\text{ii. } H_2O]{\text{i. } O_3,\ -78\ ^\circ C} \underset{\text{Propionic acid}}{C_2H_5{-}\overset{O}{\overset{\|}{C}}{-}OH} + CO_2$$

5.7.9 Oxidation of primary alcohols

Primary alcohols are oxidized either to aldehydes or to carboxylic acids, depending on the oxidizing reagents and conditions used.

Preparation of carboxylic acids

Primary alcohols are oxidized to carboxylic acids using a variety of aqueous oxidizing agents, including $KMnO_4$ in basic solution, chromic acid in aqueous acid (H_2CrO_4) and Jones' reagent (CrO_3 in acetone). Potassium permanganate is most commonly used for oxidation of a 1° alcohol to a carboxylic acid. The reaction is generally carried out in aqueous basic solution. A brown precipitate of MnO_2 indicates that the oxidation has taken place.

$$\underset{\text{Primary alcohol}}{RCH_2CH_2{-}OH} \xrightarrow[NaOH]{KMnO_4} \underset{\text{Carboxylic acid}}{RCH_2{-}\overset{O}{\overset{\|}{C}}{-}OH} + MnO_2$$

$$\underset{\text{Propanol}}{CH_3CH_2CH_2OH} \xrightarrow[NaOH]{KMnO_4} \underset{\text{Propanoic acid}}{CH_3CH_2{-}\overset{O}{\overset{\|}{C}}{-}OH} + MnO_2$$

Chromic acid is produced *in situ* by the reaction of sodium dichromate ($Na_2Cr_2O_7$) or chromic trioxide (CrO_3), sulphuric acid and water.

$$Na_2Cr_2O_7 \text{ or } CrO_3 \xrightarrow[H_2O]{H_2SO_4} \underset{\text{Chromic acid}}{H_2CrO_4}$$

$$\underset{\text{Primary alcohol}}{RCH_2CH_2{-}OH} \xrightarrow{H_2CrO_4} \underset{\text{Carboxylic acid}}{RCH_2{-}\overset{O}{\overset{\|}{C}}{-}OH}$$

$$\underset{\text{Cyclohexyl methanol}}{C_6H_{11}{-}CH_2OH} \xrightarrow{H_2CrO_4} \underset{\text{Cyclohexyl carboxylic acid}}{C_6H_{11}{-}\overset{O}{\overset{\|}{C}}{-}OH}$$

Preparation of aldehydes

A convenient reagent that selectively oxidizes primary alcohols to aldehyde is anhydrous pyridinium chlorochromate, abbreviated to PCC

($C_5H_5NH^+CrO_3Cl^-$). It is made from chromium trioxide and pyridine under acidic conditions in dry dichloromethane (CH_2Cl_2).

$$CrO_3 + HCl + \text{Pyridine} \xrightarrow{CH_2Cl_2} \text{PCC}$$

Chromium trioxide, Pyridine, PCC

$$RCH_2CH_2\text{–}OH \xrightarrow[CH_2Cl_2]{PCC} RCH_2\text{–}C(=O)\text{–}H$$

Primary alcohol, Aldehyde

Cyclohexanol $\xrightarrow[CH_2Cl_2]{PCC}$ Hexanal

5.7.10 Oxidation of secondary alcohols: preparation of ketones

Any oxidizing reagents, including H_2CrO_4, Jones' reagent or PCC, can be used to oxidize 2° alcohols to ketones. However, the most common reagent used for oxidation of 2° alcohols is chromic acid (H_2CrO_4).

$$RCH_2\text{–}CH(CH_3)\text{–}OH \xrightarrow{H_2CrO_4} RCH_2\text{–}C(CH_3)=O \qquad H_3C\text{–}CH(CH_3)\text{–}OH \xrightarrow{H_2CrO_4} H_3C\text{–}C(CH_3)=O$$

Secondary alcohol, Ketone, Isopropanol, Propanone

Mechanism. Chromic acid reacts with isopropanol to produce a chromate ester intermediate. An elimination reaction occurs by removal of a hydrogen atom from the alcohol carbon, and departure of the chromium group with a pair of electrons. The Cr is reduced from Cr (VI) to Cr (IV), and the alcohol is oxidized.

$$H_3C\text{–}CH(CH_3)\text{–}OH + HO\text{–}CrO_2\text{–}OH \longrightarrow H_3C\text{–}CH(CH_3)\text{–}O\text{–}CrO_2\text{–}OH + H_2O$$

Chromic acid (Cr VI)

$$\xrightarrow{H_2O} O{=}Cr(OH)\text{–}O^- + H_3O^+ + H_3C\text{–}C(CH_3)=O$$

5.7.11 Oxidation of aldehydes: preparation of carboxylic acids

Any aqueous oxidizing reagent, e.g. chromic acid (CrO_3 in aqueous acid), Jones' reagent (CrO_3 in acetone) and $KMnO_4$ in basic solution, can oxidize aldehydes to carboxylic acids.

$$R{-}\overset{O}{\overset{\|}{C}}{-}H \xrightarrow{[O]} R{-}\overset{O}{\overset{\|}{C}}{-}OH$$

Aldehydes can also be oxidized selectively in the presence of other functional groups using silver (I) oxide (Ag_2O) in aqueous ammonium hydroxide (*Tollen's reagent*). Since ketones have no H on the carbonyl carbon, they do not undergo this oxidation reaction.

$$\text{(cyclohexenyl)}{-}CHO \xrightarrow[NH_4OH,\ H_2O]{Ag_2O} \text{(cyclohexenyl)}{-}COOH + \underset{\text{Metallic silver}}{Ag}$$

5.7.12 Baeyer–Villiger oxidation of aldehydes and ketones

Aldehyde reacts with peroxyacid (RCO_3H) to yield carboxylic acid. Most oxidizing reagents do not react with ketones. However, a ketone reacts with peroxyacid (RCO_3H) to yield an ester. Cyclic ketones give lactones (cyclic esters). This reaction is known as *Baeyer–Villiger oxidation.* A peroxyacid contains one more oxygen atom than a carboxylic acid. This extra oxygen is inserted between the carbonyl carbon and R group (R=H in an aldehyde, and R = alkyl group in a ketone).

$$\underset{\text{Aldehyde}}{R{-}\overset{O}{\overset{\|}{C}}{-}H} + \underset{\text{Peroxyacid}}{R{-}\overset{O}{\overset{\|}{C}}{-}O{-}OH} \longrightarrow \underset{\text{Carboxylic acid}}{R{-}\overset{O}{\overset{\|}{C}}{-}OH}$$

$$\underset{\text{Ketone}}{R{-}\overset{O}{\overset{\|}{C}}{-}R'} + \underset{\text{Peroxyacid}}{R{-}\overset{O}{\overset{\|}{C}}{-}O{-}OH} \longrightarrow \underset{\text{Ester}}{R{-}\overset{O}{\overset{\|}{C}}{-}OR'}$$

5.7.13 Reduction of alcohols via tosylates: preparation of alkanes

Generally, an alcohol cannot be reduced directly to an alkane in one step, because the —OH group is a poor leaving group.

$$\underset{\text{Alcohol}}{R{-}OH} \xrightarrow[\times]{LiAlH_4} \underset{\text{Alkane}}{R{-}H}$$

However, the hydroxyl group can easily be converted to water, a better leaving group, and this allows the reaction to proceed. One such conversion involves tosyl chloride, and the formation of a tosylate. For example, cyclopentanol reacts with TsCl to form cyclopentyl tosylate, and the corresponding tosylate is reduced conveniently to cyclopentane.

$$\text{Cyclopentanol} \xrightarrow[\text{Py}]{\text{Ts-Cl}} \text{Cyclopentyl tosylate} \xrightarrow[\text{THF}]{LiAlH_4} \text{Cyclopentane}$$

5.7.14 Reduction of alkyl halides: preparation of alkanes

Lithium aluminium hydride ($LiAlH_4$), a strong reducing agent, reduces alkyl halides to alkanes. Essentially, a hydride ion (H^-) acts as a nucleophile displacing the halide. A combination of metal and acid, usually Zn with acetic acid (AcOH), can also be used to reduce alkyl halides to alkanes.

$$\underset{\text{Propyl bromide}}{CH_3CH_2CH_2Br} \xrightarrow[\text{Zn, AcOH}]{LiAlH_4\text{, THF or}} \underset{\text{Propane}}{CH_3CH_2CH_3}$$

5.7.15 Reduction of organometallic reagents: preparation of alkanes

Organometallics are generally strong nucleophiles and bases. They react with weak acids, e.g. water, alcohol, carboxylic acid and amine, to become protonated and yield hydrocarbons. Thus, small amounts of water or moisture can destroy organometallic compounds. For example, ethylmagnesium bromide or ethyllithium reacts with water to form ethane. This is a convenient way to reduce an alkyl halide to an alkane via Grignard and organolithium synthesis.

$$\underset{\text{Ethyl bromide}}{CH_3CH_2{-}Br} \xrightarrow[\text{Dry ether}]{Mg} \underset{\text{Ethyl magnesium bromide}}{CH_3CH_2{-}MgBr} \xrightarrow{H_2O} \underset{\text{Ethane}}{CH_3CH_3} + Mg(OH)Br$$

$$\underset{\text{Ethyl bromide}}{CH_3CH_2{-}Br} \xrightarrow[\text{Ether}]{2\ Li} \underset{\text{Ethyllithium}}{CH_3CH_2{-}Li} + LiBr \xrightarrow{H_2O} \underset{\text{Ethane}}{CH_3CH_3} + LiOH$$

5.7.16 Reduction of aldehydes and ketones

Aldehydes and ketones are reduced to 1° and 2° alcohols, respectively, by hydrogenation with metal catalysts (Raney nickel, Pd—C and PtO_2). They are also reduced to alcohols relatively easily with mild reducing agent, e.g. $NaBH_4$, or powerful reducing agent, e.g. $LiAlH_4$. The key step in the reduction is the reaction of hydride with the carbonyl carbon.

Preparation of alcohols: catalytic hydrogenation

Catalytic hydrogenation using H_2 and a catalyst reduces aldehydes and ketones to 1° and 2° alcohols, respectively. The most common catalyst for these hydrogenations is Raney nickel, although PtO_2 and Pd—C can also be used. The C=C double bonds are reduced more quickly than C=O double bonds. Therefore, it is not possible to reduce C=O selectively in the presence of a C=C without reducing both by this method.

$$H_2C{=}CHCH_2CH_2{-}C({=}O){-}H \xrightarrow[\text{Raney Ni}]{H_2} CH_3CH_2CH_2CH_2CH_2OH$$

Pentanol

$$H_2C{=}CHCH_2{-}C({=}O){-}CH_3 \xrightarrow[\text{Raney Ni}]{H_2} CH_3CH_2CH_2{-}CH(CH_3)OH$$

2-Pentanol

Preparation of alcohols: hydride reduction

The most useful reagents for reducing aldehydes and ketones are the metal hydride reagents. Complex hydrides are the source of hydride ions, and the two most commonly used reagents are $NaBH_4$ and $LiAlH_4$. Lithium aluminium hydride is extremely reactive with water and must be used in an anhydrous solvent, e.g. dry ether.

$$Na^+\ [BH_4]^- \qquad Li^+\ [AlH_4]^-$$

$$R{-}C({=}O){-}Y \xrightarrow[\text{ii. } H_3O^+]{\text{i. } NaBH_4 \text{ or } LiAlH_4} R{-}CH(OH){-}Y$$

Y = H or R

1° or 2° Alcohol

Mechanism. Hydride ions attack carbonyl groups, generating alkoxide ions, and protonation furnishes alcohols. The net result of adding H^- from $NaBH_4$ or $LiAlH_4$, and H^+ from aqueous acids, is the addition of the elements of H_2 to the carbonyl π bond.

$$R{-}C({=}\ddot{O}{:}){-}Y + H^- \longrightarrow R{-}CH(O^-){-}Y \xrightarrow{H_3O^+} R{-}CH(OH){-}Y$$

Y = H or R

Sodium borohydride is the more selective and milder reagent of the two. It cannot reduce esters or carboxylic acids, whereas $LiAlH_4$ reduces esters and carboxylic acids to 1° alcohols (see Sections 5.7.20 and 5.7.22). These hydride sources do not reduce alkene double bonds. Therefore, when a compound contains both a C=O group and a C=C bond, selective reduction of one functional group can be achieved by proper choice of the reagent.

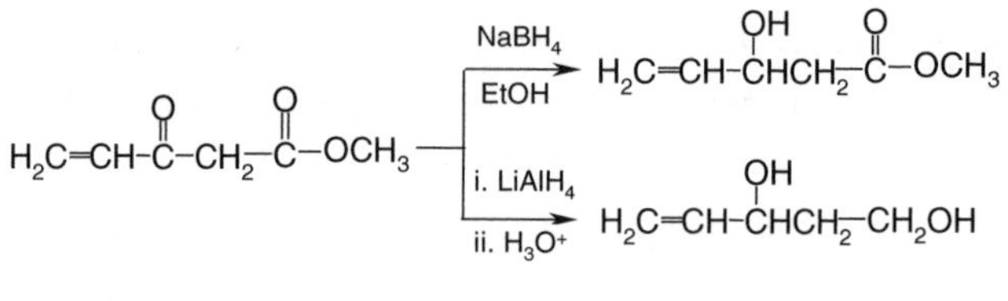

Stereochemistry of hydride reduction Hydride converts a planar sp^2-hybridized carbonyl carbon to a tetrahedral sp^3-hybridized carbon. Thus, hydride reduction of an achiral ketone with $LiAlH_4$ or $NaBH_4$ gives a racemic mixture of alcohol when a new stereocentre is formed.

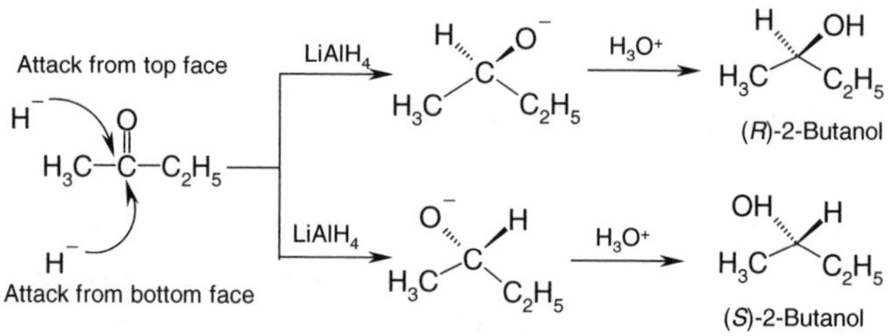

5.7.17 Clemmensen reduction: preparation of alkanes

This method is used for the reduction of acyl benzenes to alkyl benzenes, but it also reduces aldehydes and ketones to alkanes.

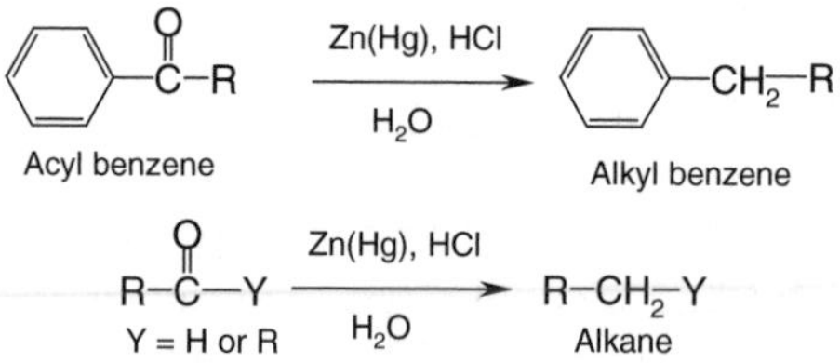

Sometimes the acidic conditions used in the Clemmensen reduction are unsuitable for certain molecules. In these cases, *Wolff–Kishner reduction* is employed, which occurs in basic conditions.

5.7.18 Wolff–Kishner reduction: preparation of alkanes

This method reduces acyl benzenes as well as aldehydes and ketones, but does not reduce alkenes, alkynes or carboxylic acids. Hydrazine reacts with aldehyde or ketone to give hydrazone (see Section 5.3.2). The hydrazone is treated with a strong base (KOH) to generate alkane.

$$\underset{Y = H \text{ or } R}{R{-}C({=}O){-}Y} \xrightarrow{NH_2NH_2} \underset{\text{Hydrazone}}{R{-}C({=}N{-}NH_2){-}Y} \xrightarrow[\text{Heat}]{KOH} \underset{\text{Alkane}}{R{-}CH_2{-}Y} + N_2$$

Mechanism. The aqueous base deprotonates the hydrazone, and the anion produced is resonance stabilized. The carbanion picks up a proton from water, and another deprotonation by the aqueous base generates an intermediate, which is set up to eliminate a molecule of nitrogen (N_2), and produce a new carbanion. This carbanion is quickly protonated by water, giving the final reduced product as alkane.

$$\underset{\text{Hydrazone}}{R{-}C({=}NNH_2){-}Y} \xrightarrow{KOH} \underset{\text{Resonance stabilized}}{\left[R{-}\bar{C}({-}N{=}NH){-}Y \longleftrightarrow R{-}C({=}N{-}\bar{N}H){-}Y\right]} \xrightarrow{H_2O} R{-}CH({-}N{=}NH){-}Y$$

$$R{-}CH({-}N{=}NH){-}Y \xrightarrow{KOH} R{-}CH({-}N{=}\bar{N}){-}Y \longrightarrow R{-}\bar{C}H{-}Y + N_2 \xrightarrow{H_2O} \underset{Y = H \text{ or } R}{R{-}CH_2{-}Y}$$

5.7.19 Reduction of oximes and imine derivatives

The most general method for synthesizing amines involves the reduction of oximes and imine derivatives obtained from aldehydes or ketones (see Sections 5.5.2 and 4.3.11). By catalytic hydrogenation or by $LiAlH_4$ reduction, while 1° amines are prepared from oxime or unsubstituted imine, 2° amines are obtained from substituted imine. Unsubstituted imines are relatively unstable, and are reduced *in situ*.

$$\underset{\substack{\text{Oxime}\\ Y = H \text{ or } R}}{R{-}C({=}N{-}OH){-}Y} \text{ or } \underset{\substack{\text{Imine}\\ Y = H \text{ or } R}}{R{-}C({=}NH){-}Y} \xrightarrow[LiAlH_4]{H_2/Pd\text{-}C \text{ or}} \underset{1^\circ \text{ amine}}{R{-}CH(NH_2){-}Y}$$

$$\underset{\substack{\text{Imine}\\ Y = H \text{ or } R}}{R{-}C({=}N{-}R'){-}Y} \xrightarrow[LiAlH_4]{H_2/Pd\text{-}C \text{ or}} \underset{2^\circ \text{ amine}}{R{-}CH(NHR'){-}Y}$$

Tertiary amines are made from iminium salts by catalytic hydrogenation or by $LiAlH_4$ reduction. The iminium salts are usually unstable, and so are reduced as they are formed by a reducing agent already in the reaction mixture. A mild reducing agent, e.g. sodium cyanoborohydride ($NaBH_3CN$), can also be used.

$$\underset{\substack{\text{Iminium salt}\\ Y = H \text{ or } R}}{R{-}C({=}\overset{+}{N}R'_2){-}Y} \xrightarrow[NaBH_3CN]{H_2/Pd\text{-}C \text{ or}} \underset{3^\circ \text{ amine}}{R{-}CH(NR'_2){-}Y}$$

5.7.20 Reduction of carboxylic acids: preparation of primary alcohols

Carboxylic acids are considerably less reactive than acid chlorides, aldehydes and ketones towards reduction. They cannot be reduced by catalytic hydrogenation or sodium borohydride ($NaBH_4$) reduction. They require the use of a powerful reducing agent, e.g. $LiAlH_4$. The reaction needs two hydrides (H^-) from $LiAlH_4$, since the reaction proceeds through an aldehyde, but it cannot be stopped at that stage. Aldehydes are more easily reduced than the carboxylic acids, and $LiAlH_4$ reduces all the way back to 1° alcohols.

$$R{-}\overset{\displaystyle O}{\overset{\|}{C}}{-}OH \xrightarrow[\text{ii. } H_3O^+]{\text{i. } LiAlH_4} RCH_2OH$$

5.7.21 Reduction of acid chlorides

Preparation of primary alcohols

Acid chlorides are easy to reduce than carboxylic acids and other carboxylic acid derivatives. They are reduced conveniently all the way to 1° alcohols by metal hydride reagents ($NaBH_4$ or $LiAlH_4$), as well as by catalytic hydrogenation (H_2/Pd−C).

$$\underset{1^\circ\text{ Alcohol}}{RCH_2OH} \xleftarrow{H_2/Pd\text{-}C} \underset{\text{Acid chloride}}{R{-}\overset{\displaystyle O}{\overset{\|}{C}}{-}Cl} \xrightarrow[\text{ii. } H_3O^+]{\text{i. } NaBH_4 \text{ or } LiAlH_4} \underset{1^\circ\text{ Alcohol}}{RCH_2OH}$$

Preparation of aldehydes

Sterically bulky reducing agents, e.g. lithium tri-*t*-butoxyaluminium hydride, can selectively reduce acid chlorides to aldehydes at low temperatures (−78 °C). Lithium tri-*t*-butoxyaluminium hydride, $LiAlH(O\text{-}t\text{-}Bu)_3$, has three electronegative oxygen atoms bonded to aluminium, which makes this reagent less nucleophilic than $LiAlH_4$.

$$Li^+\ (CH_3)_3CO{-}\overset{\displaystyle H}{\overset{|}{\underset{\underset{\displaystyle OC(CH_3)_3}{|}}{Al}}}{-}OC(CH_3)_3$$

Lithium tri-*t*-butoxyaluminium hydride

$$R{-}\overset{\displaystyle O}{\overset{\|}{C}}{-}Cl \xrightarrow[\text{ii. } H_3O^+]{\text{i. } LiAlH(O\text{-}t\text{-}Bu)_3,\ -78\ ^\circ C} \underset{\text{Aldehyde}}{R{-}\overset{\displaystyle O}{\overset{\|}{C}}{-}H}$$

5.7.22 Reduction of esters

Preparation of primary alcohols

Esters are harder to reduce than acid chlorides, aldehydes and ketones. They cannot be reduced with milder reducing agents, e.g. $NaBH_4$, or by catalytic hydrogenation. Only $LiAlH_4$ can reduce esters. Esters react with $LiAlH_4$ generating aldehydes, which react further to produce 1° alcohols.

$$\underset{\text{Ester}}{R-\overset{O}{\overset{\|}{C}}-OR} \xrightarrow[\text{ii. } H_3O^+]{\text{i. } LiAlH_4} \underset{1^\circ\ \text{Alcohol}}{RCH_2OH}$$

Preparation of aldehydes

Sterically bulky reducing agents, e.g. diisobutylaluminium hydride (DIBAH), can selectively reduce esters to aldehydes. The reaction is carried out at low temperatures (−78 °C) in toluene. Diisobutylaluminium hydride has two bulky isobutyl groups, which make this reagent less reactive than $LiAlH_4$.

$$\underset{\text{Diisobutylaluminium hydride (DIBAH)}}{(CH_3)_2CHCH_2-\overset{H}{\overset{|}{Al}}-CH_2CH(CH_3)_2} \qquad \underset{\text{Ester}}{R-\overset{O}{\overset{\|}{C}}-OR} \xrightarrow[\text{ii. } H_2O]{\text{i. DIBAH, -78 °C}} \underset{\text{Aldehyde}}{R-\overset{O}{\overset{\|}{C}}-H}$$

5.7.23 Reduction of amides, azides and nitriles

Preparation of amines

Amides, azides and nitriles are reduced to amines by catalytic hydrogenation (H_2/Pd−C or H_2/Pt−C) as well as metal hydride reduction ($LiAlH_4$). They are less reactive towards the metal hydride reduction, and cannot be reduced by $NaBH_4$. Unlike the $LiAlH_4$ reduction of all other carboxylic acid derivatives, which affords 1° alcohols, the $LiAlH_4$ reduction of amides, azides and nitriles yields amines. Acid is not used in the work-up step, since amines are basic. Thus, hydrolytic work-up is employed to afford amines. When the nitrile group is reduced, an NH_2 and an extra CH_2 are introduced into the molecule.

$$\underset{1^\circ\ \text{Amide}}{R-\overset{O}{\overset{\|}{C}}-NH_2} \xrightarrow[\text{i. } LiAlH_4 \text{ ii. } H_2O]{H_2/\text{Pd-C or}} \underset{1^\circ\ \text{Amine}}{RCH_2-NH_2}$$

$$\underset{\text{Alkyl azide}}{RCH_2-N=\overset{+}{N}=\overset{-}{N}} \xrightarrow[\text{i. } LiAlH_4 \text{ ii. } H_2O]{H_2/\text{Pd-C or}} \underset{1^\circ\ \text{Amine}}{RCH_2-NH_2}$$

$$\underset{\text{Nitrile}}{R-C\equiv N} \xrightarrow[\text{i. } LiAlH_4 \text{ ii. } H_2O]{2\ H_2/\text{Pd-C or}} \underset{1^\circ\ \text{Amine}}{RCH_2-NH_2}$$

Preparation of aldehydes

Reduction of nitrile with a less powerful reducing reagent, e.g. DIBAH, produces aldehyde. The reaction is carried out at low temperatures (−78 °C) in toluene.

$$\underset{\text{Nitrile}}{R{-}C{\equiv}N} \xrightarrow[\text{ii. } H_2O]{\text{i. DIBAH}} \underset{\text{Aldehyde}}{R{-}\overset{O}{\overset{\|}{C}}{-}H}$$

5.8 Pericyclic reactions

Pericyclic reactions are concerted reactions that take place in a single step without any intermediates, and involve a cyclic redistribution of bonding electrons. The concerted nature of these reactions gives fine stereochemical control over the generation of the product. The best-known examples of this reaction are the Diels–Alder reaction (cyclo-addition) and sigmatropic rearrangement.

5.8.1 Diels–Alder reaction

In the *Diels–Alder reaction*, a conjugated diene reacts with an α,β-unsaturated carbonyl compound, generally called a *dienophile*. A dienophile is a reactant that loves a diene. The most reactive dienophiles usually have a carbonyl group, but it may also have another electron-withdrawing group, e.g. a cyano, nitro, haloalkene or sulphone group conjugated with a carbon–carbon double bond.

CN NO_2 Cl O O O O S

Dienophiles other than carbonyl group directly linked to the conjugated system

The Diels–Alder reaction is in fact a [4 + 2] cycloaddition reaction, where C-1 and C-4 of the conjugated diene system become attached to the double-bonded carbons of the dienophile to form a six-membered ring. For example, 1,3-butadiene reacts with maleic anhydride to produce tetrahydrophthalic anhydride on heating.

1 4 + O O O Benzene Heat H O O O H

1, 3-Butadiene
A conjugated diene

Maleic anhydride
A dienophile

Tetrahydrophthalic anhydride
95%

Different types of cyclic compound can be produced just varying the structures of the conjugated diene and the dienophile. Compounds containing carbon–carbon triple bonds can be utilized as dienophiles to produce compounds with two bonds as shown below.

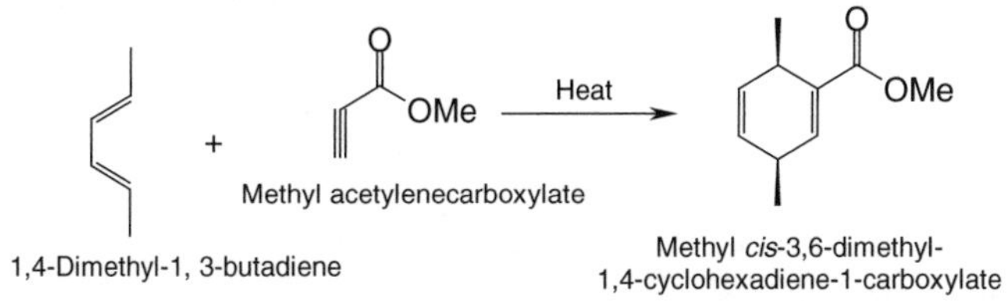

1,4-Dimethyl-1, 3-butadiene + Methyl acetylenecarboxylate —Heat→ Methyl *cis*-3,6-dimethyl-1,4-cyclohexadiene-1-carboxylate

In the case of a cyclic conjugated diene, the Diels–Alder reaction yields a bridged bicyclic compound. A bridged bicyclic compound contains two rings that share two nonadjacent carbons. For example, cyclopentadiene reacts with ethylene to produce norbornene.

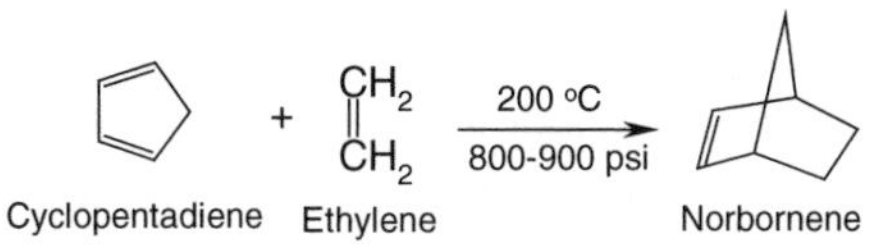

Cyclopentadiene + Ethylene ($CH_2{=}CH_2$) —200 °C, 800-900 psi→ Norbornene

Cyclo-addition is used extensively in the synthesis of chiral natural products and pharmaceutical agents, because the reaction can determine the relative configuration of up to four chiral centres in a single reaction.

Essential structural features for dienes and dienophiles

In the Diels–Alder reaction, the conjugated diene can be cyclic or acyclic, and it may contain different types of substituent. A conjugated diene can exist in two different conformations, an s-*cis* and an s-*trans*. The ‘s’ stands for single bond or σ bond; e.g., s-*cis* means the double bonds are *cis* about the single bond. In the Diels–Alder reaction, the conjugated diene has to be in an s-*cis* conformation. A conjugated diene that is permanently in an s-*trans* conformation cannot undergo this reaction. This s-*cis* feature must also be present in conjugated cyclic dienes for Diels–Alder reaction.

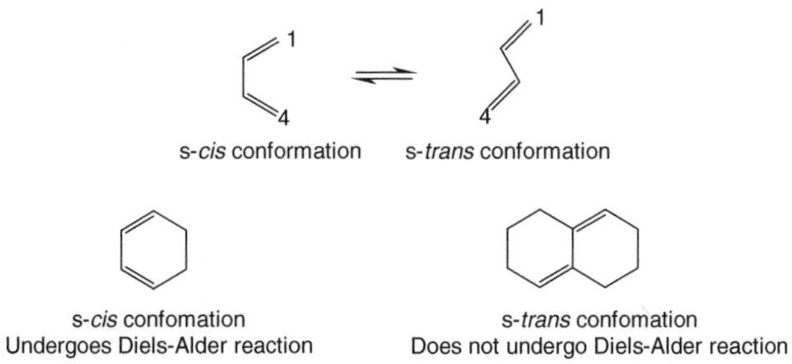

s-*cis* conformation ⇌ s-*trans* conformation

s-*cis* confomation
Undergoes Diels-Alder reaction

s-*trans* confomation
Does not undergo Diels-Alder reaction

Cyclic conjugated dienes that are s-*cis* conformation, e.g. cyclopentadiene and 1,3-cyclohexadiene, are highly reactive in Diels–Alder reactions. In fact, cyclopentadiene is reactive both as a diene and as a dienophile, and forms dicyclopentadiene at room temperature. When dicyclopentadiene is heated to 170 °C, a reverse Diels–Alder reaction takes place and reforms the cyclopentadiene.

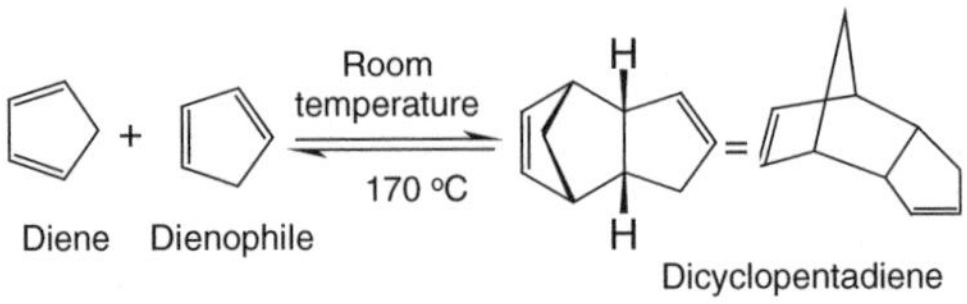

Room temperature
170 °C
Diene Dienophile
Dicyclopentadiene

Stereochemistry of Diels–Alder reaction

The Diels–Alder reaction is stereospecific. The stereochemistry of the dienophile is retained in the product; i.e., *cis* and *trans* dienophiles produce different diastereoisomers in the product. For example, freshly distilled cyclopentadiene, having s-*cis* configuration, reacts with maleic anhydride to give *cis*-norbornene-5,6-endo-dicarboxylic anhydride.

0 °C
Few minutes
Cyclopentadiene Maleic anhydride
cis-norbornene-5,6-endo-dicarboxylic anhydride

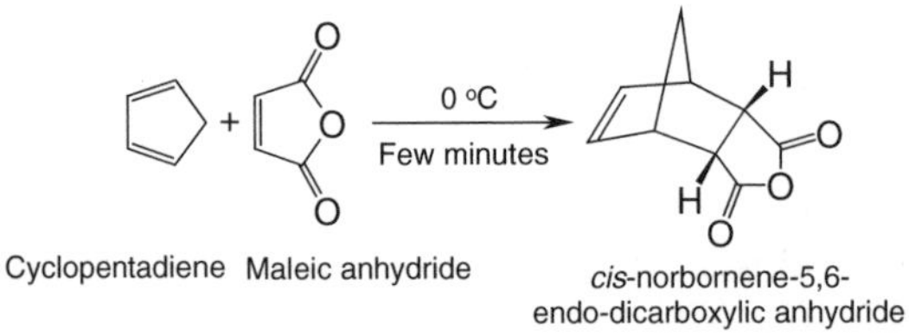

There are two possible configurations, *endo* and *exo*, for bridged bicyclic compounds resulting from the reaction of a cyclic diene and cyclic dienophile. A substituent on a bridge is *endo* if it is closer to the longer of the two other bridges, and it is *exo* if it is closer to the shorter bridge. Most of these reactions result in an *endo* product. However, if this reaction is reversible, and thermodynamically controlled, the *exo* product is formed.

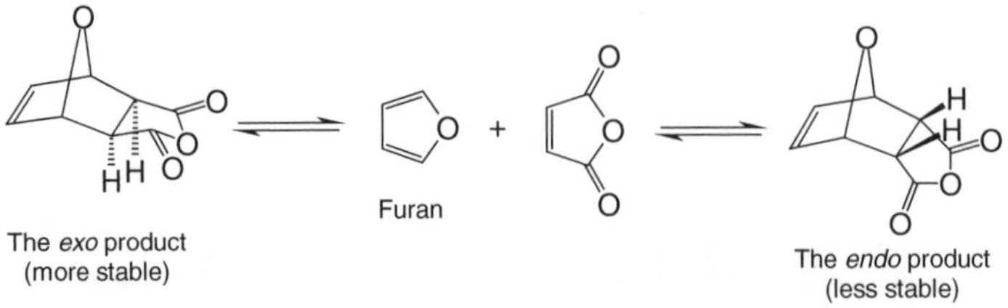

The *exo* product (more stable)
Furan
The *endo* product (less stable)

5.8.2 Sigmatropic rearrangements

Sigmatropic rearrangements are unimolecular processes, and involve the movement of a σ bond with the simultaneous rearrangement of the π system. In this rearrangement reaction, a σ bond is broken in the reactant and a new σ bond is formed in the product, and the π bonds rearrange. However, the number of π bonds does not change, i.e. the reactant and the product possess the same number of π bonds. Sigmatropic reactions are usually uncatalysed, although Lewis acid catalysts are sometimes used. Sigmatropic rearrangement plays an important role in the biosynthesis of vitamin D in our bodies.

Bond broken — Heat → New bond formed

This reaction can occur through hydrogen shift, alkyl shift (Cope rearrangement) or Claisen rearrangement.

Hydrogen shift

A sigmatropic rearrangement involves the migration of a σ bond from one position to another with a simultaneous shift of the π bonds. For example, a hydrogen atom and its σ bond can be migrated in (*Z*)-1,3-pentadiene. This is known as *hydrogen shift*. Hydrogen shifts occur at $4n+1$ positions in a *suprafacial* fashion. It can also take place at $4n+3$ positions in an *antarafacial* fashion. *Antarafacial* means that opposite faces are involved, whereas it is suprafacial when both changes occur at the same face. Many sigmatropic rearrangements and Diels–Alder reactions can be either suprafacial or antarafacial and this dictates the stereochemistry. Antarafacial hydrogen shifts are observed in the conversion of lumisterol to vitamin D.

(*Z*)-1,3-pentadiene ⇌ (200 °C) (*Z*)-1,3-pentadiene

Alkyl shift: Cope rearrangement

In addition to the migration of hydrogen atoms in sigmatropic rearrangements, *alkyl shifts* also take place. A large number of such reactions occur with a migration of a carbon atom and a σ bond, but do not have ionic intermediates. More specifically, these reactions involve methyl shifts at $4n+3$ positions in a suprafacial fashion with inversion of stereochemistry.

Alkyl shift is evident in the *Cope rearrangement*. A Cope rearrangement is a [3,3] sigmatropic rearrangement of a 1,5-diene. This reaction leads to the formation of a six-membered ring transition state. As [3,3] sigmatropic rearrangements involve three pairs of electrons, they take place by a suprafacial pathway under thermal conditions.

Ph Heat Ph

Claisen rearrangement

Sigmatropic rearrangements involving the cleavage of a σ bond at an oxygen atom are called *Claisen rearrangement*. A Claisen rearrangement is a [3,3] sigmatropic rearrangement of an allyl vinyl ether to produce a γ, δ-unsaturated carbonyl compound. Like Cope rearrangement, this reaction also forms a six-membered ring transition state. This reaction is exothermic and occurs by a suprafacial pathway under thermal conditions.

O Ph Heat O Ph

Claisen rearrangement plays an important part in the biosynthesis of several natural products. For example, the chorismate ion is rearranged to the prephenate ion by the Claisen rearrangement, which is catalysed by the enzyme chorismate mutase. This prephenate ion is a key intermediate in the shikimic acid pathway for the biosynthesis of phenylalanine, tyrosine and many other biologically important natural products.

CH_2 ^-OOC O COO^- OH Chorismate mutase COO^- O COO^- HO

Chorismate ion Prephenate

Recommended further reading

Clayden, J., Greeves, N., Warren, S. and Wothers, P. *Organic Chemistry*, Oxford University Press, Oxford, 2001.

6

Natural product chemistry

Learning objectives

After completing this chapter the student should be able to

- provide an overview of the natural product drug discovery process;
- discuss the importance of natural products in medicine;
- describe the origin, chemistry, biosynthesis and pharmaceutical importance of various classes of natural products including alkaloids, carbohydrates, glycosides, iridoids and secoiridoids, phenolics, steroids and terpenoids.

6.1 Introduction to natural product drug discovery process

6.1.1 Natural products

Natural products are products from various natural sources, plants, microbes and animals. Natural products can be an entire organism (e.g. a plant, an animal or a micro-organism), a part of an organism (e.g. leaves or flowers of a plant, an isolated animal organ), an extract of an organism or part of an organism and an exudate, or pure compound (e.g. alkaloids, coumarins, flavonoids, lignans, steroids and terpenoids) isolated from plants, animals or micro-organisms. However, in practice, the term *natural product* refers to secondary metabolites, small molecules (molecular weight < 1500 amu), produced by an organism, but not strictly necessary for the survival of the organism.

Chemistry for Pharmacy Students Satyajit D Sarker and Lutfun Nahar

6.1.2 Natural products in medicine

The use of natural products, especially plants, for healing is as ancient and universal as medicine itself. The therapeutic use of plants certainly goes back to the Sumerian and the Akkadian civilizations in about the third millenium BC. Hippocrates (ca. 460–377 BC), one of the ancient authors who described medicinal natural products of plant and animal origins, listed approximately 400 different plant species for medicinal purposes. Natural products have been an integral part of the ancient traditional medicine systems, e.g. Chinese, Ayurvedic and Egyptian. Even now, continuous traditions of natural product therapy exist throughout the third world, especially in the orient, where numerous minerals, animal substances and plants are still in common use. According to the World Health Organisation (WHO), some 3.4 billion people in the developing world depend on plant-based traditional medicines. This represents about 88 per cent of the world's inhabitants, who rely mainly on traditional medicine for their primary health care. In China alone, 7295 plant species are utilized as medicinal agents.

Nature has been a potential source of therapeutic agents for thousands of years. An impressive number of modern drugs have been derived from natural sources. Over the last century, a number of top selling drugs have been developed from natural products. Anticancer drug vincristine from *Vinca rosea*, narcotic analgesic morphine from *Papaver somniferum*, antimalarial drug artemisinin from *Artemisia annua*, anticancer drug Taxol® from *Taxus brevifolia* and antibiotic penicillins from *Penicillium* ssp. are just a few examples.

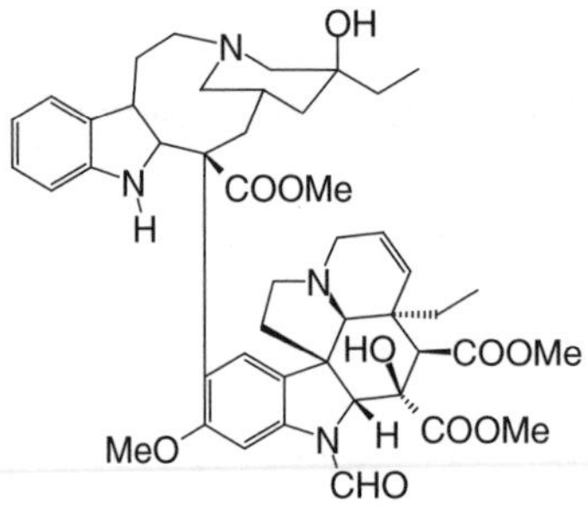

Vincristine

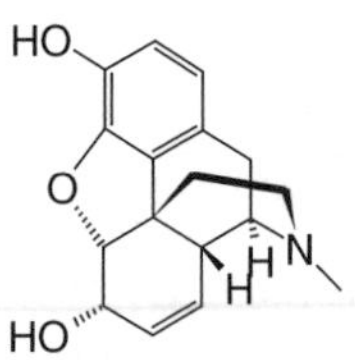

Morphine

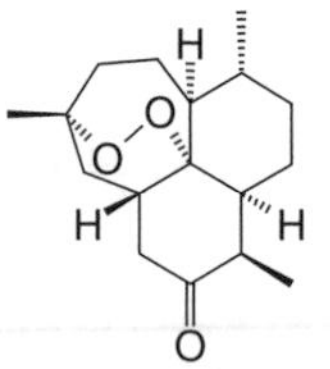

Artemisinin

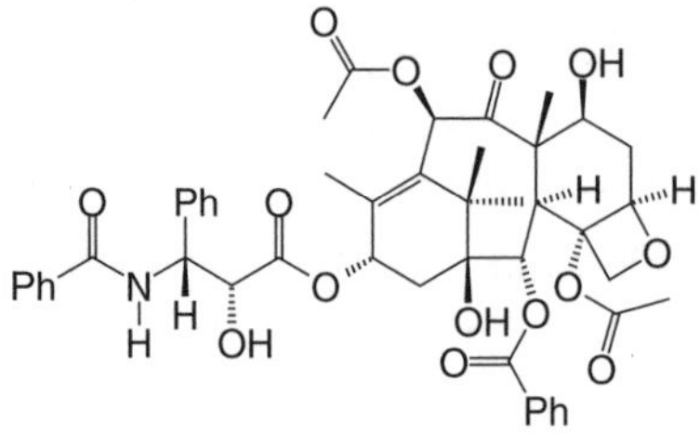

Taxol

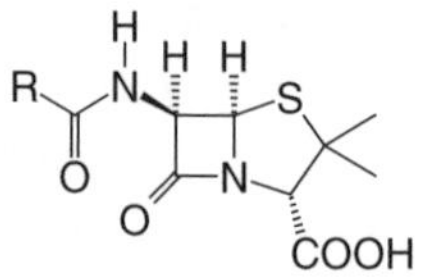

Penicillin G R =

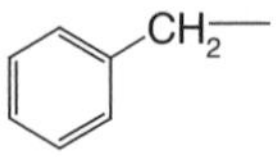

Penicillin V R =

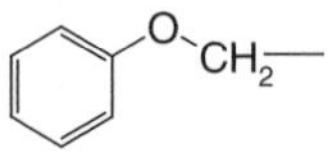

Apart from natural-product-derived modern medicine, natural products are also used directly in the 'natural' pharmaceutical industry that is growing rapidly in Europe and North America, as well as in the traditional medicine programmes being incorporated into the primary health care systems of Mexico, The People's Republic of China, Nigeria and other developing countries.

6.1.3 Drug discovery and natural products

Although *drug discovery* may be considered to be a recent concept that evolved from modern science during the 20th century, in reality the concept of drug discovery dates back many centuries, and has its origins in nature. Time and time again, humans have turned to Mother Nature for cures, and discovered unique drug molecules. Thus, the term *natural product* has become almost synonymous with the concept of drug discovery. In modern drug discovery and development processes, natural products play an important role at the early stage of 'lead' discovery, i.e. discovery of the active (determined by various bioassays) natural molecule, which itself or its structural analogues could be an ideal drug candidate.

Natural products have been a wellspring of drugs and drug leads. It is estimated that 61 per cent of the 877 small molecule new chemical entities introduced as drugs worldwide during 1981–2002 can be traced back to or were developed from natural products. These include natural products (6 per cent), natural product derivatives (27 per cent), synthetic compounds with natural-product-derived pharmacophores (5 per cent) and synthetic compounds designed on the basis of knowledge gained from a natural product, i.e. a natural product mimic (23 per cent). In some therapeutic areas, the contribution of natural products is even greater, e.g. about 78 per cent of antibacterials and 74 per cent of anticancer drug candidates are natural products or structural analogues of natural products. In 2000, approximately 60 per cent of all drugs in clinical trials for the multiplicity of cancers were of natural origins. In 2001, eight (simvastatin, pravastatin, amoxycillin, clavulanic acid, clarithromycin, azithromycin, ceftriaxone, cyclosporin and paclitaxel) of 30 top selling medicines were natural products or derived from natural products, and these eight drugs together totalled US$16 billion in sales.

Despite the outstanding record and statistics regarding the success of natural products in drug discovery, 'natural product drug discovery' has been neglected by many big pharmaceutical companies in the recent past. The declining popularity of natural products as a source of new drugs began in the 1990s, because of some practical factors, e.g. the apparent lack of compatibility of natural products with the modern high throughput screening (HTS) programmes, where significant degrees of automation, robotics and computers are used, the complexity in the isolation and identification of natural products and the cost and time involved in the natural product 'lead' discovery process. Complexity in the chemistry of natural products, especially in the case of novel structural types, also became the rate-limiting step in drug discovery programmes. Despite being neglected by the pharmaceutical companies, attempts to discover new drug 'leads' from natural sources has never stopped, but continued in academia and some semi-academic research organizations, where more traditional approaches to natural product drug discovery have been applied.

Neglected for years, natural product drug discovery appears to be drawing attention and immense interest again, and is on the verge of a comeback in the mainstream of drug discovery ventures. In recent years, a significant revival of interests in natural products as a potential source for new medicines has been observed among academics as well as several pharmaceutical companies. This extraordinary comeback of natural products in drug discovery research is mainly due to the following factors: combinatorial chemistry's promise to fill drug development pipelines with *de novo* synthetic small molecule drug candidates is somewhat unsuccessful; the practical difficulties of natural product drug discovery are being overcome by advances in separation and identification technologies and in the speed and sensitivity of structure elucidation and, finally, the unique and incomparable chemical diversity that natural products have to offer. Moreover, only a small fraction of the world's biodiversity has ever been explored for bioactivity to date. For example, there are at least 250 000 species of higher plants that exist on this planet, but merely five to 10 per cent of these terrestrial plants have been investigated so far. In addition, re-investigation of previously investigated plants has continued to produce new bioactive compounds that have the potential for being developed as drugs. While several biologically active compounds have been found in marine organisms, e.g. antimicrobial compound cephalosporin C from marine organisms (*Cephalosporium acremonium* and *Streptomyces* spp.) and antiviral compounds such as avarol and avarone from marine sponges, e.g. *Dysidea avara*, research in this area is still in its infancy.

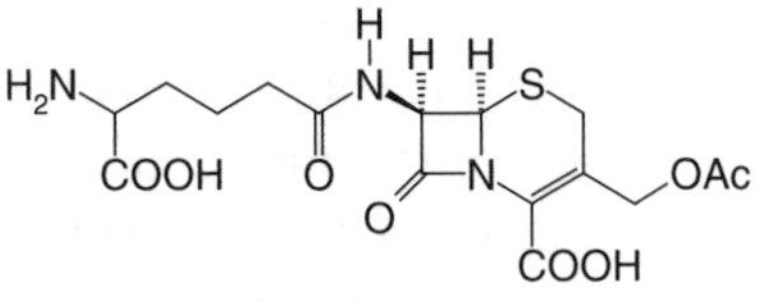

Cephalosporin C

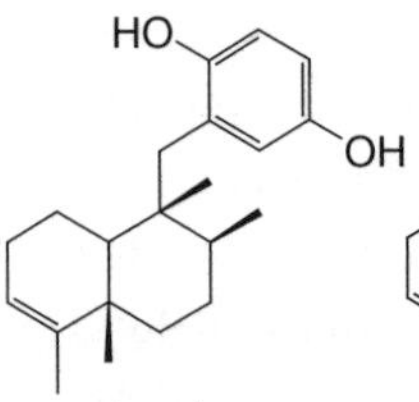

Avarol

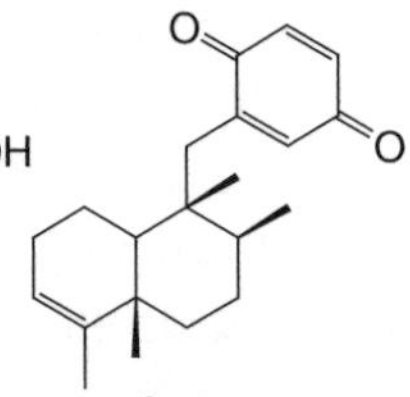

Avarone

Now, let us have a look at the summary of the traditional as well as the modern drug discovery processes involving natural products.

Natural product drug discovery: the traditional way

In the traditional, rather more academic, method of drug discovery from natural products, drug targets are exposed to crude extracts, and in the case of a hit, i.e. any evidence of activity, the extract is fractionated and the active compound is isolated and identified. Every step of fractionation and isolation is usually guided by bioassays, and the process is called *bioassay-guided isolation*. The following scheme presents an overview of a bioassay-guided traditional natural product drug discovery process.

Sometimes, a straightforward natural product isolation route, irrespective of bioactivity, is also applied, which results in the isolation of a number of natural compounds (*small compound library*) suitable for undergoing any bioactivity screening. However, the process can be slow, inefficient and labour intensive, and it does not guarantee that a 'lead' from screening would be chemically workable or even patentable.

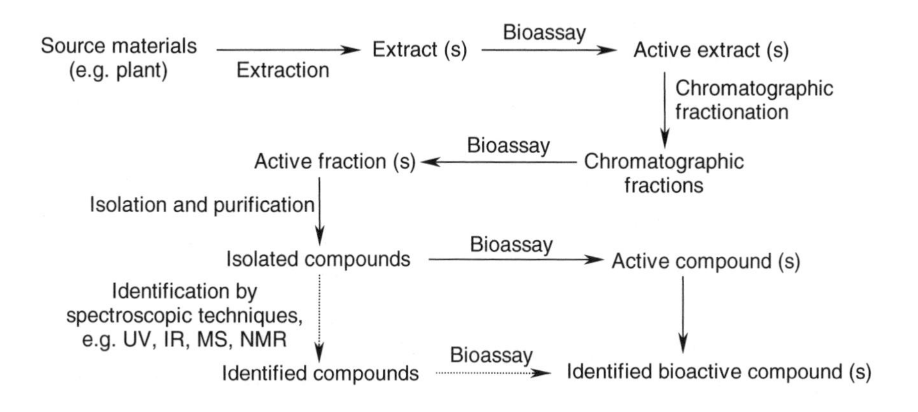

Natural product drug discovery: the modern processes

Modern drug discovery approaches involve HTS, where, applying full automation and robotics, hundreds of molecules can be screened using several assays within a short time, and with very little amounts of compounds. In order to incorporate natural products in the modern HTS programmes, a natural product library (a collection of *dereplicated* natural products) needs to be built. *Dereplication* is the process by which one can eliminate recurrence or re-isolation of same or similar compounds from various extracts. A number of hyphenated techniques are used for dereplication, e.g. LC-PDA (liquid chromatography–photo-diode-array detector),

LC-MS (liquid chromatography–mass detector) and LC-NMR (liquid chromatography–nuclear magnetic resonance spectroscopy).

While in the recent past it was extremely difficult, time consuming and labour intensive to build such a library from purified natural products, with the advent of newer and improved technologies related to separation, isolation and identification of natural products the situation has improved remarkably. Now, it is possible to build a 'high quality' and 'chemically diverse' natural product library that can be suitable for any modern HTS programmes. Natural product libraries can also be of crude extracts, chromatographic fractions or semi-purified compounds. However, the best result can be obtained from a fully identified pure natural product library as it provides scientists with the opportunity to handle the 'lead' rapidly for further developmental work, e.g. total or partial synthesis, dealing with formulation factors, *in vivo* assays and clinical trials.

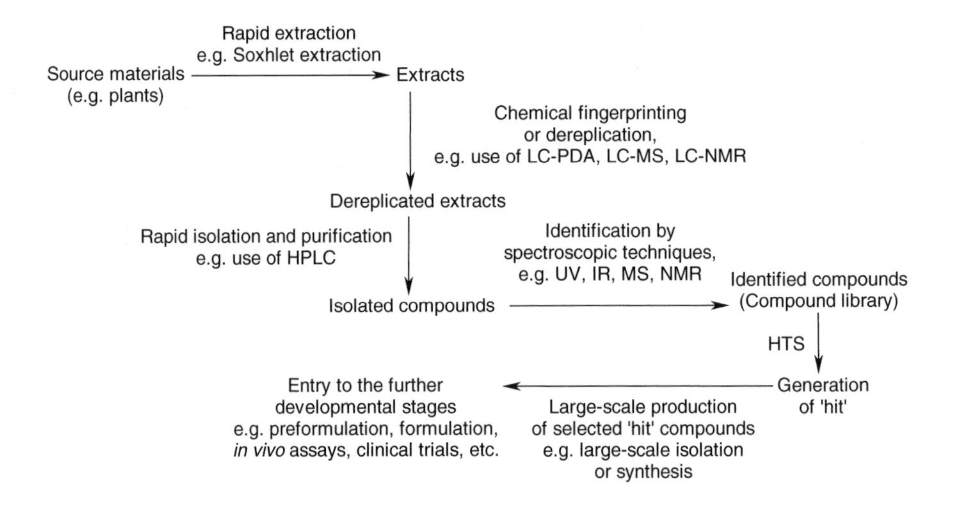

To continue to exploit natural sources for drug candidates, the focus must be on exploiting newer approaches for natural product drug discovery. These approaches include the application of genomic tools, seeking novel sources of organisms from the environment, new screening technologies and improved processes of sample preparation for screening samples. In addition, the recent focus on the synthesis of diversity-oriented combinatorial libraries based on natural-product-like compounds is an attempt to enhance the productivity of synthetic chemical libraries.

6.2 Alkaloids

Alkaloids are a naturally occurring large group of pharmacologically active nitrogen-containing secondary metabolites of plants, microbial or animal origin. In most alkaloids, the nitrogen atom is a part of the ring. Alkaloids are biosynthetically derived from amino acids. The name 'alkaloid' derives from the word 'alkaline', which means a water soluble base. A number of natural alkaloids and their derivatives have been developed as drugs to treat various diseases, e.g. morphine, reserpine and taxol.

6.2.1 Properties

Alkaloids are basic in nature, and form water soluble salts with mineral acids. In fact, one or more nitrogen atoms that are present in an alkaloid, typically as 1°, 2° or 3° amines, contribute to the basicity of the alkaloid. The degree of basicity varies considerably, depending on the structure of the molecule, and presence and location of the functional groups. Most alkaloids are crystalline solids and are bitter in taste.

6.2.2 Classification of alkaloids

Alkaloids are generally classified according to the amino acid that provides both the nitrogen atom and the fundamental alkaloidal skeleton. However, alkaloids can also be grouped together on the basis of their generic structural similarities. The following table shows different major types of alkaloid, their generic skeletons and specific examples.

Class/structural types	Generic structure	Examples
Aporphine (Tyrosine derived)	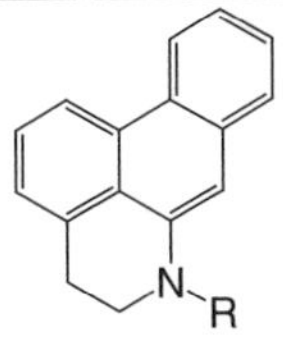 N, R Aporphine R = Me Noraporphine R = H	Boldine
Betaines	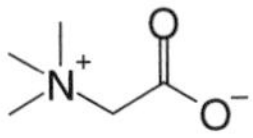 N+, O, O− Betaine	Choline, muscarine and neurine
Imidazole	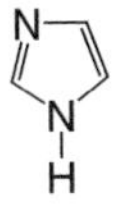 N, N, H Imidazole	Pilocarpine
Indole (Tryptophan derived)	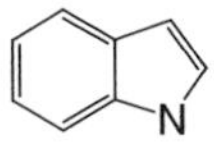 N Indole	

Class/structural types	Generic structure	Examples
Tryptamines	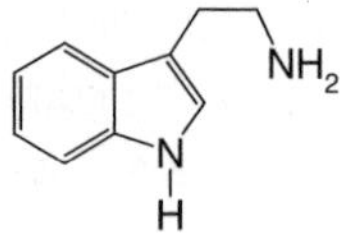 Tryptamine	Moschamine, moschamindole, psilocybin and serotonin
Ergolines	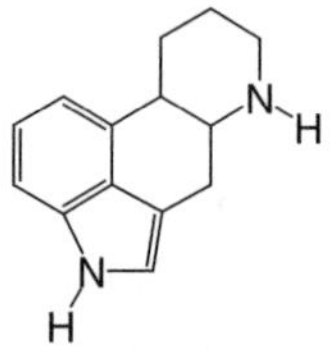 Ergoline	Ergine, ergotamine and lysergic acid
β-carbolines	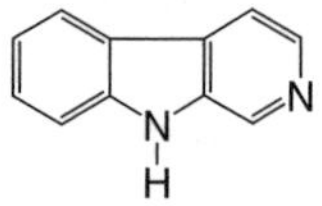 β-Carboline	Emetine, harmine and reserpine
Indolizidine	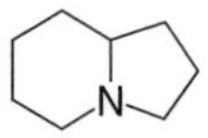 Indolizidine	Swainsonine and castanospermine
Isoquinoline (Tyrosine derived)	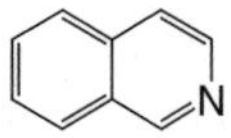 Isoquinoline	Codeine, berberine, morphine, papaverine, sanguinarine and thebaine
Macrocyclic spermines and spermidines	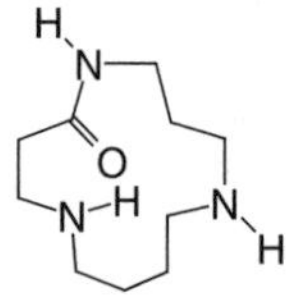 Spermine	Celabenzine
Norlupinane (Lysine derived)	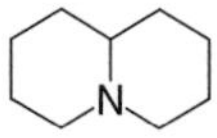 Norlupinane	Cytisine and lupanine
Phenethylamine (Phenylalanine derived)	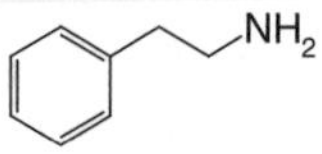 Phenylethylamine	Ephedrine and mescaline
Purine	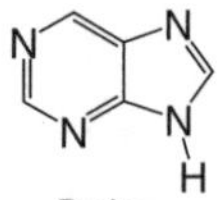 Purine	Caffeine, theobromine and theophylline
Pyridine and (Nicotinic acid derived)	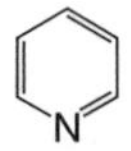 Pyridine 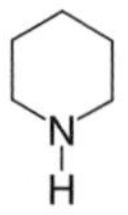 Piperidine	Arecoline, coniine, nicotine, piperine, sparteine and trigonelline

Class/structural types	Generic structure	Examples
Pyrrole and pyrrolidine (Ornithine derived)	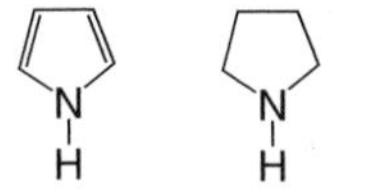 Pyrrole Pyrrolidine	Hygrine, cuscohygrine and nicotine
Pyrrolizidine	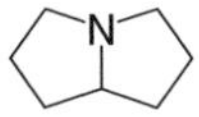 Pyrrolizidine	Echimidine and symphitine
Quinoline (Tryptophan/anthranilic acid derived)	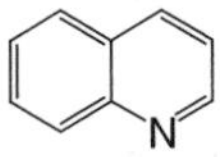 Quinoline	Cinchonine, brucine, quinine and quinidine
Terpenoidal/steroidal Terpenoidal	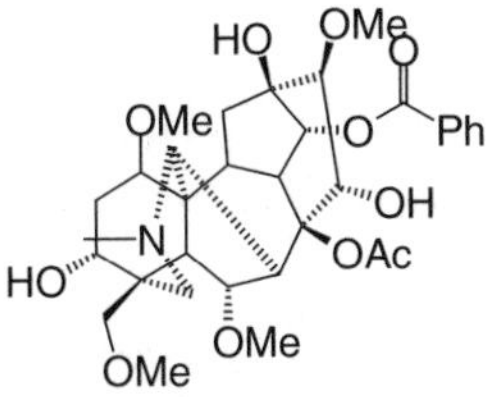 Aconitine	Aconitine
Steroidal	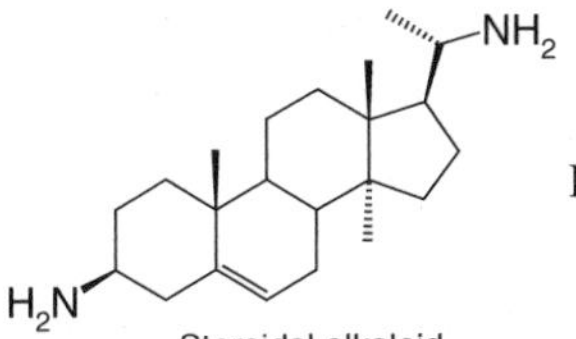 Steroidal alkaloid	Batrachotoxin, conanine, irehdiamine A, solanine, samandarine and tomatillidine
Tropane (Ornithine derived)	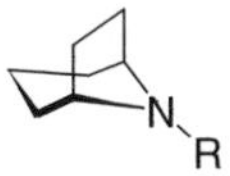 Tropane R = Me Nortropane R = H	Atropine, cocaine, ecgonine, hyoscine and scopalamine

Pyridine and piperidine alkaloids

Alkaloids, e.g. piperine, coniine, trigonelline, pilocarpine, nicotine and sparteine, possess a pyridine or modified pyridine heterocyclic ring system (e.g. piperidine ring).

Piperine Piperine, molecular formula $C_{17}H_{19}NO_3$, a component of black pepper (*Piper nigrum*), has been used in various traditional medicine preparations, and also as an insecticide. Piperine has various effects on human drug metabolizing enzymes, and is marketed under the brand name, Bioperine®, as an adjunct for increasing bioavailability of various dietary supplements, especially curcumin, one of the active ingredients of turmeric (*Curcuma longa*).

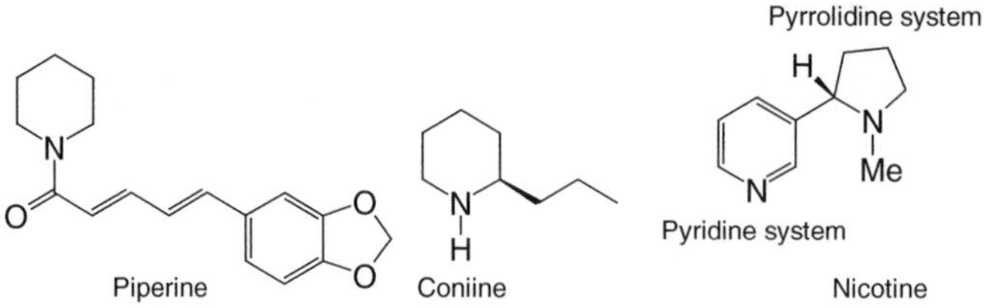

Pyrrolidine system

Pyridine system

Piperine Coniine Nicotine

Coniine Coniine or (*S*)-2-propylpiperidine, molecular formula $C_8H_{17}N$, is a poisonous alkaloid found in the hemlock poison and the 'yellow pitcher plant' (*Sarracenia flava*). Coniine contributes to the foul smell of hemlock. It is a neurotoxin, causes respiratory paralysis and is toxic to all classes of livestock and humans. In 399 BC, Socrates was put to death by this poison.

Nicotine Nicotine, molecular formula $C_{10}H_{14}N_2$, is the major pharmacologically active component of tobacco, *Nicotiana tabacum*, and is also found extensively in other species of the family Solanaceae, e.g. tomato, potato, aubergine and green pepper. Nicotine is a hygroscopic oily liquid, and miscible with water in its base as well as its salt form. Nicotine possesses two nitrogenous ring systems: one is pyridine, but the other is a pyrrolidine ring system. Thus, this alkaloid can be classified either under pyridine or pyrrolidine.

Nicotine is a potent nerve poison, and is included in many insecticide preparations. In lower concentrations, nicotine is a stimulant, i.e. it increases activity, alertness and memory, and this is one of the main factors that contribute to the dependence-forming properties of tobacco smoking. Nicotine increases heart-rate and blood pressure, and reduces appetite. In higher doses, nicotine acts as a depressant. In large doses, it can cause nausea and vomiting. The main symptoms of the withdrawal of nicotine intake include irritability, headaches, anxiety, cognitive disturbances and sleep disruption.

Pyrrole and pyrrolidine alkaloids

These alkaloids contain pyrrole or modified pyrrole, e.g. pyrrolidine, ring system. The simplest example of this class is nicotine. A pyrrolidine ring is the central structure of the amino acids proline and hydroxyproline. These alkaloids are also found in many drug preparations, e.g. procyclidine hydrochloride, which is an anticholinergic drug mainly used for the treatment of drug-induced Parkinsonism, akathisia and acute dystonia.

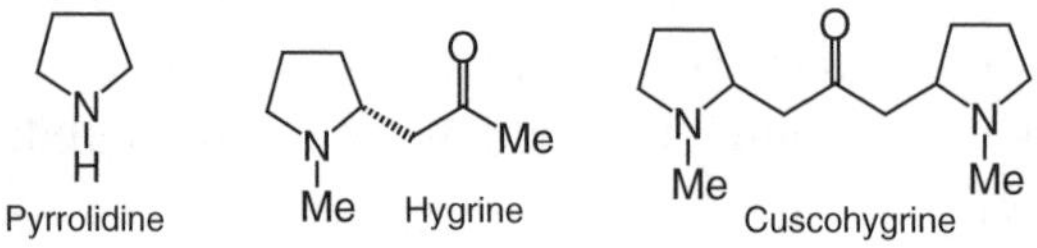

Pyrrolidine Hygrine Cuscohygrine

Hygrine Hygrine, molecular formula $C_8H_{15}NO$, is found mainly in coca leaves (*Erythroxylum coca*). It is a thick yellow oil, having a pungent taste and odour.

Cuscohygrine Cuscohygrine, molecular formula $C_{13}H_{24}N_2O$, is a dimeric pyrrolidine alkaloid found in coca, and also in many species of the Solanaceae. It is an oil, but soluble in water.

Tropane alkaloids

These are the group of alkaloids that possess a 8-methyl-8-aza-bicyclo [1,2,3]octane or tropane skeleton, e.g. atropine, cocaine and scopolamine. Tropane alkaloids occur mainly in plants from the families Solanaceae and Erythroxylaceae. 8-Aza-bicyclo[1,2,3]octane, i.e. tropane without the 8-methyl group, is known as *nortropane*.

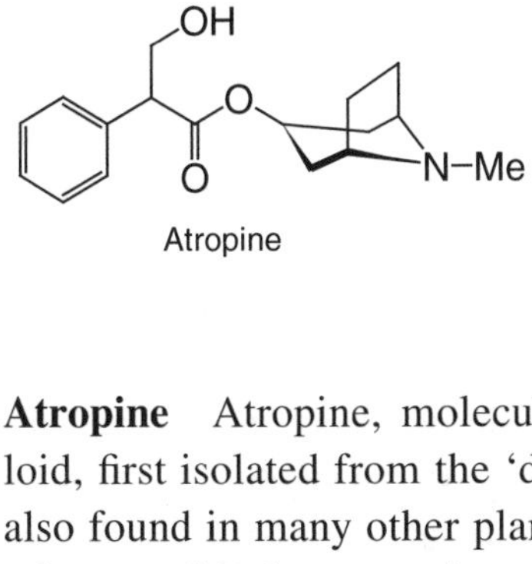

Atropine

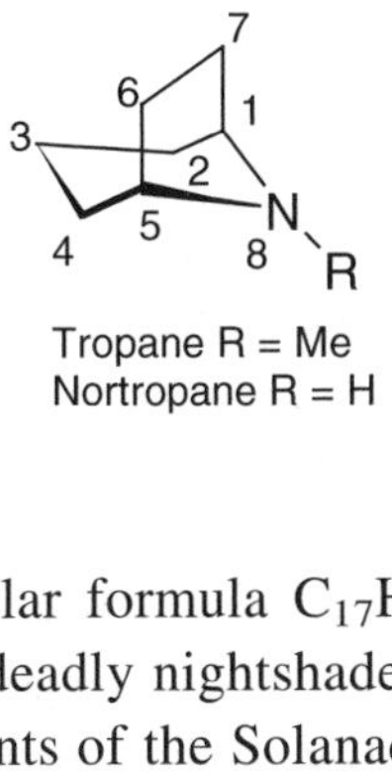

Tropane R = Me
Nortropane R = H

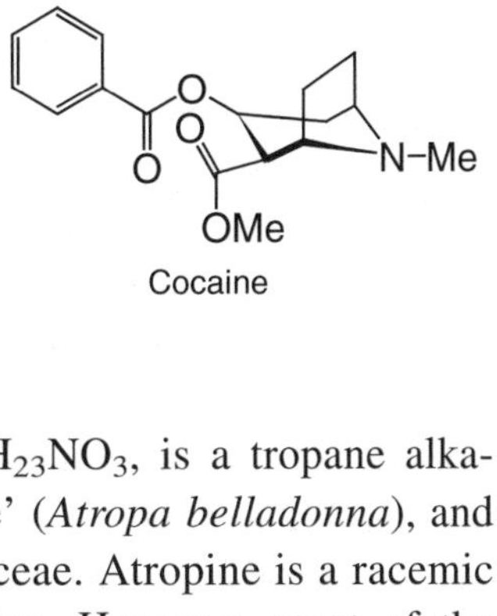

Cocaine

Atropine Atropine, molecular formula $C_{17}H_{23}NO_3$, is a tropane alkaloid, first isolated from the 'deadly nightshade' (*Atropa belladonna*), and also found in many other plants of the Solanaceae. Atropine is a racemic mixture of D-hyoscyamine and L-hyoscyamine. However, most of the pharmacological properties of atropine are due to its L-isomer, and due to its binding to muscarinic acetylcholine receptors. Atropine is a *competitive antagonist* of the muscarinic acetylcholine receptors. The main medicinal use of atropine is as an opthalmic drug. Usually a salt of atropine, e.g. atropine sulphate, is used in pharmaceutical preparations. Atropine is used as an acycloplegic to paralyse accommodation temporarily, and as a mydriatic to dilate the pupils. It is contraindicated in patients predisposed to narrow angle glaucoma. Injections of atropine are used in the treatment of bradycardia (an extremely low heart-rate), asystole and pulseless electrical activity (PEA) in cardiac arrest. It is also used as an antidote for poisoning by organophosphate insecticides and nerve gases.

The major adverse effects of atropine include ventricular fibrillation, tachycardia, nausea, blurred vision, loss of balance and photophobia. It also produces confusion and hallucination in elderly patients. Overdoses of atropine can be fatal. The antidote of atropine poisoning is physostigmine or pilocarpine.

Cocaine Cocaine, molecular weight $C_{17}H_{21}NO_4$, is a white crystalline tropane alkaloid found mainly in coca plant. It is a potent central nervous system (CNS) stimulant and appetite suppressant. For its euphoretic effect, cocaine is often used recreationally, and it is one of the most common drugs of abuse and addiction. Cocaine is also used as a topical anaesthetic in eye, throat and nose surgery. Possession, cultivation and distribution of cocaine is illegal for non-medicinal and non-government sanctioned purposes virtually all over the world. The side-effects of cocaine include twitching, paranoia and impotence, which usually increase with frequent usage. With excessive dosage it produces hallucinations, paranoid delusions, tachycardia, itching and formication. Cocaine overdose leads to tachyarrhythmias and elevated blood pressure, and can be fatal.

Quinoline alkaloids

The chemistry of the quinoline heterocycle has already been discussed in Chapter 4. Any alkaloid that possesses a quinoline, i.e. 1-azanaphthalene, 1-benzazine, or benzo[b]pyridine, skeleton is known as a quinoline alkaloid, e.g. quinine. Quinoline itself is a colourless hygroscopic liquid with strong odour, and slightly soluble in water, but readily miscible with organic solvents. Quinoline is toxic. Short term exposure to the vapour of quinoline causes irritation of the nose, eyes, and throat, dizziness and nausea. It may also cause liver damage.

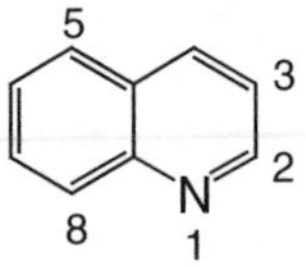

5 3 2 N 8 1

Quinoline

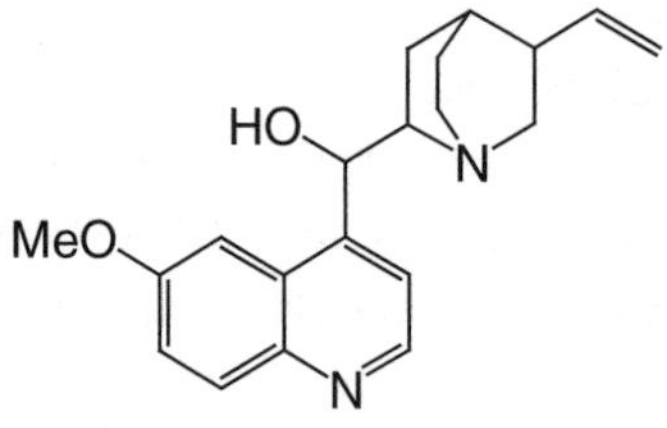

HO N MeO N

Quinine

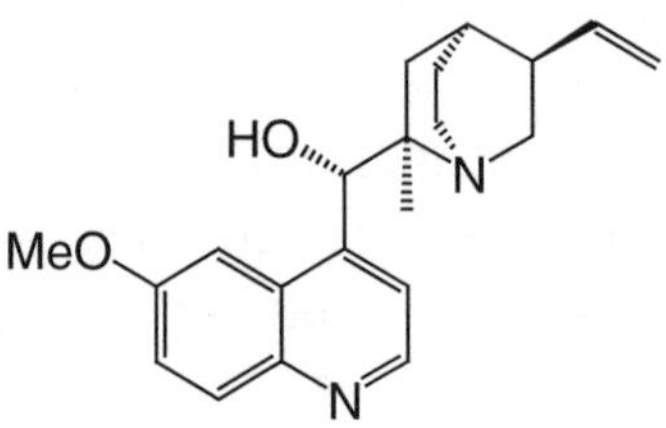

HO N MeO N

Quinidine

Quinine Quinine, molecular formula $C_{20}H_{24}N_2O_2$, is a white crystalline quinoline alkaloid, isolated from Cinchona bark (*Cinchona succirubra*), and is well known as an antimalarial drug. Quinine is extremely bitter, and also possesses antipyretic, analgesic and anti-inflammatory properties. While quinine is still the drug of choice for the treatment of *Falciparum* malaria, it can be also used to treat nocturnal leg cramps and arthritis. Quinine is an extremely basic compound, and is available in its salt forms, e.g. sulphate, hydrochloride and gluconate.

Despite being a wonder drug against malaria, quinine in therapeutic doses can cause various side-effects, e.g. nausea, vomiting and cinchonism, and in some patients pulmonary oedema. It may also cause paralysis if accidentally injected into a nerve. An overdose of quinine may have fatal consequences. Non-medicinal uses of quinine include its uses as a flavouring agent in tonic water and bitter lemon.

Quinidine Quinidine, molecular formula $C_{20}H_{24}N_2O_2$, is a stereoisomer of quinine found in Cinchona bark. Chemically, it is known as (2-ethenyl-4-azabicyclo[2.2.2]oct-5-yl)-(6-methoxyquinolin-4-yl)-methanol, or 6′-methoxycinchonan-9-ol. It is used as a Class 1 anti-arrhythmic agent. Intravenous injection of quinidine is also used in the treatment of *P. falciparum* malaria. Among the adverse effects, quinidine induces thrombocytopenia (low platelet counts) and may lead to thrombocytic purpurea.

Isoquinoline alkaloids

Isoquinoline is in fact an isomer of quinoline, and chemically known as benzo[c]pyridine or 2-benzanine. Any alkaloids that possess an isoquinoline skeleton are known as isoquinoline alkaloids, e.g. papaverine and morphine. The isoquinoline backbone is biosynthesized from the aromatic amino acid tyrosine.

Isoquinoline itself is a colourless hygroscopic liquid at room temperature. It has an unpleasant odour. It is slightly soluble in water but well soluble in ethanol, acetone, ether and other common organic solvents. Isoquinoline is a weak base with a pK_a of 8.6.

Isoquinoline alkaloids play an important part in medicine. A number of these alkaloids are available as drugs. Some examples of isoquinoline derivatives with medicinal values are summarized in the following table. In addition to their medicinal uses, isoquinolines are used in the manufacture of dyes, paints and insecticides, and as a solvent for the extraction of resins.

Isoquinoline alkaloids	Medicinal uses
Dimethisoquin	Anaesthetic
Quinapril	Antihypertensive agent
2,2′-hexadecamethylenediisoquinolinium dichloride	Topical antifungal agent
Papaverine	Vasodilator
Morphine	Narcotic analgesic

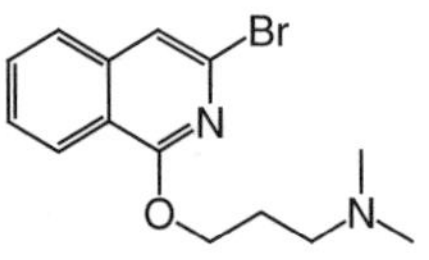

Dimethisoquin

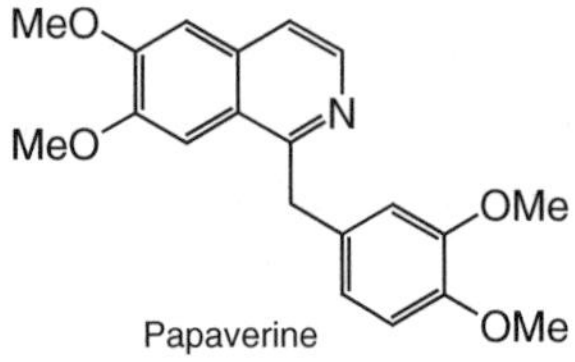

Papaverine

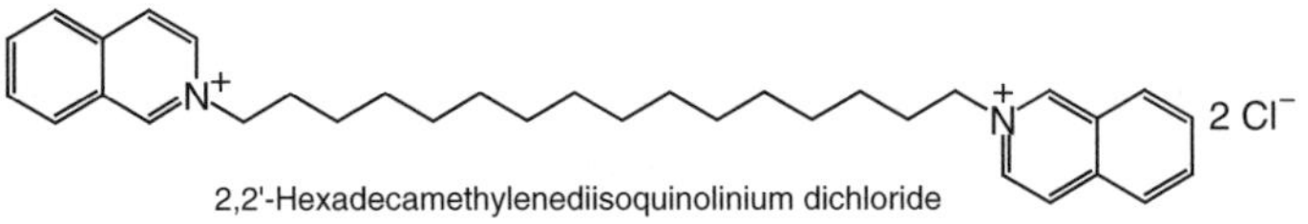

2,2'-Hexadecamethylenediisoquinolinium dichloride

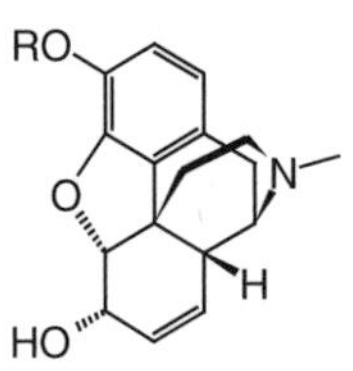

Morphine R = H
Codeine R = OMe

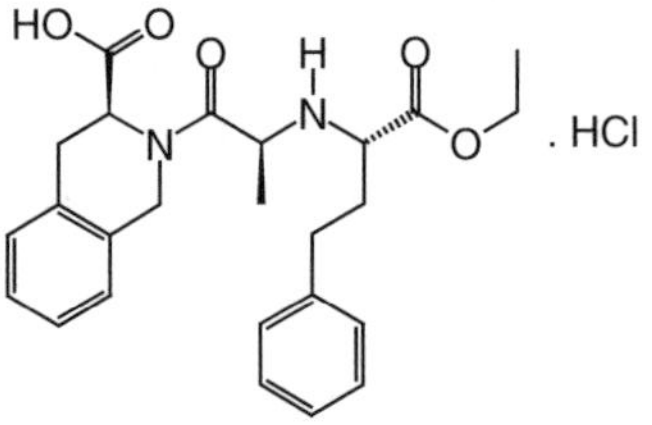

Quinapril hydrochloride

Papaverine Papaverine, molecular formula $C_{20}H_{21}NO_4$, is an isoquinoline alkaloid isolated from poppy seeds (*Papaver somniferum*, family Papaveraceae). This alkaloid is used mainly in the treatment of spasms and of erectile dysfunction. It is also used as a cerebral and coronary vasodilator. Papaverine may be used as a smooth muscle relaxant in microsurgery. In pharmaceutical preparations, papaverine is used in its salt form, e.g. hydrochloride, codecarboxylate, adenylate and teprosylate. The usual side-effects of papaverine treatment include polymorphic ventricular tachycardia, constipation, increased transaminase levels, hyperbilirunemia and vertigo.

Morphine Morphine ($C_{17}H_{19}NO_3$), a habit forming Class A analgesic drug, is the major bioactive constituent of opium poppy seeds. Like other opium constituents (opiates), e.g. heroin, morphine acts directly on the CNS to relieve pain. Morphine is used for the treatment of post-surgical pain and chronic pain (e.g. cancer pain), and as an adjunct to general anaesthesia, and an antitussive for severe cough. Side-effects of morphine treatment generally include impairment of mental performance, euphoria, drowsiness, loss of appetite, constipation, lethargy and blurred vision.

Phenylethylamines

Phenylethylamine, a neurotransmitter or neuromodulator, is a monoamine. Although the nitrogen is not a part of the ring, phenylethylamine and its derivatives are classified as alkaloids. Phenylethylamine itself is a colourless liquid, and it is biosynthesized from phenylalanine through enzymatic decarboxylation. The phenylethylamine moiety can be found in various complex ring systems, e.g. the ergoline system in lysergic acid diethylamide (LSD) or the morphinan system in morphine. A number of alkaloids of this class are used as neurotransmitters, stimulants (e.g. ephedrine, cathinone and amphetamine), hallucinogens (e.g. mescaline), bronchodilators (e.g. ephedrine and salbutamol) and antidepressants (e.g. bupropion).

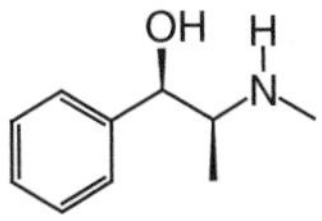

Ephedrine
A constituent of *Ephedra sinica*

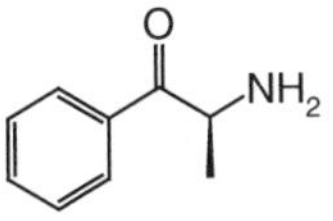

Cathinone
A constituent of *Catha edulis*

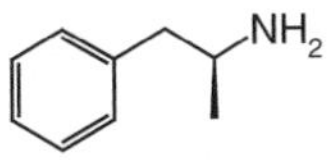

Amphetamine
Colloquially known as Speed

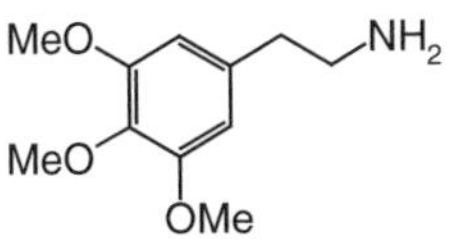

Mescaline
A hallucinogen from the cactus
Lophophora williamsii

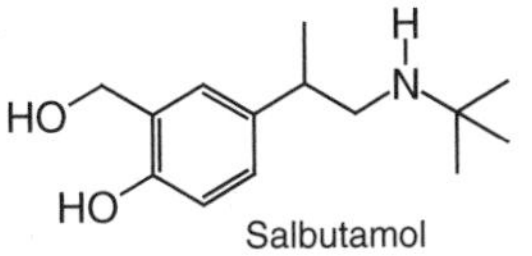

Salbutamol
A bronchodilator

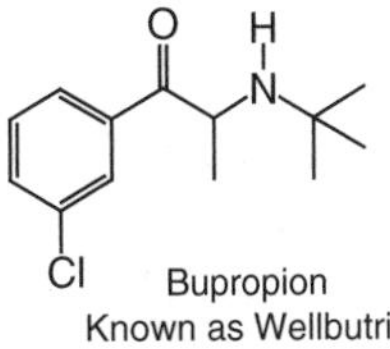

Bupropion
Known as Wellbutrin
An antidepressant

Indole alkaloids

Indole chemistry has already been discussed in Chapter 4. This is one of the major groups of naturally occurring bioactive alkaloids, and can be classified into three main categories: tryptamine and its derivatives, ergoline and its derivatives, and β-carboline and its derivatives.

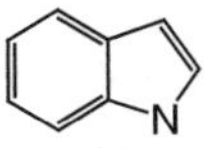

Indole

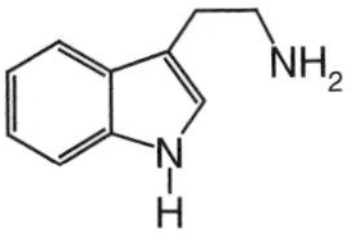

Tryptamine

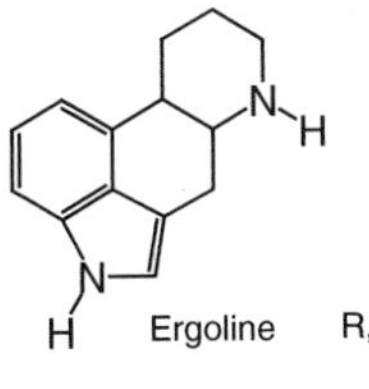

Ergoline

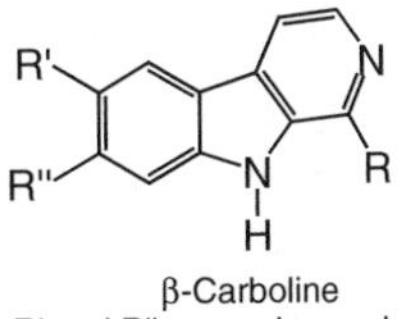

β-Carboline
R, R' and R'' are various substituents

Tryptamine derivatives Tryptamine, chemically known as 3-(2-aminoethyl)indole, is widespread in plants, fungi and animals. Biosynthetically, tryptamine derives from the amino acid tryptophan. Tryptamine acts as the precursor of many other indole alkaloids. Substitutions to the tryptamine skeleton give rise to a group of compounds collectively known as *tryptamines*: e.g. serotonin, an important neurotransmitter, is the 5-hydroxy derivative of tryptamine; melatonin, a hormone found in all living creatures, is 5-methoxy-*N*-acetyltryptamine. Some of the pharmacologically active natural tryptamines are psilocybin (4-phosphoryloxy-*N*,*N*-dimethyltryptamine) from 'magic mushrooms' (*Psilocybe cubensis* and *P. semilanceata*), DMT (*N*,*N*-dimethyltryptamine) from a number of plants and DET (*N*,*N*-diethyltryptamine), an orally active hallucinogenic drug and psychedelic compound of moderate duration. Many synthetic tryptamines, e.g. sumatripan (5-methylaminosulphonyl-*N*,*N*-dimethyltryptamine), a drug used for the treatment of migraine, are also available.

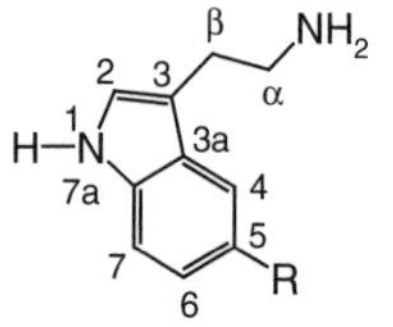

Tryptamine R = H
Serotonin R = OH

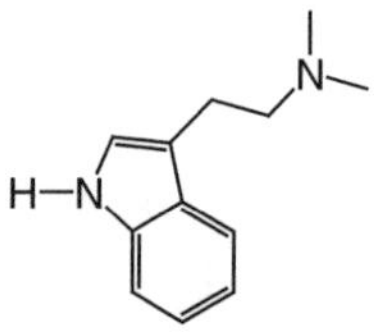

N,*N*-Dimethyltryptamine (DMT)

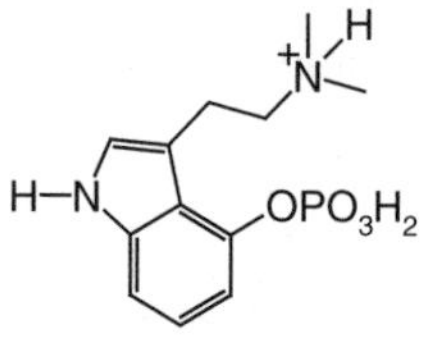

Psilocybin
A psychedelic tryptamine alkaloid

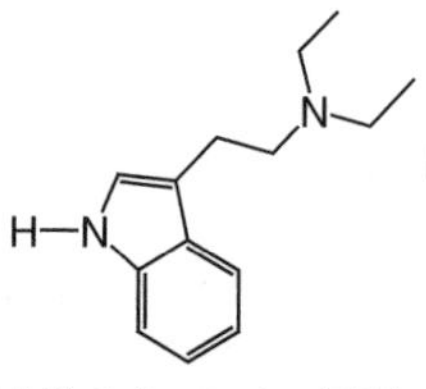

N,*N*-Diethyltryptamine (DET)
A hallucinogenic drug

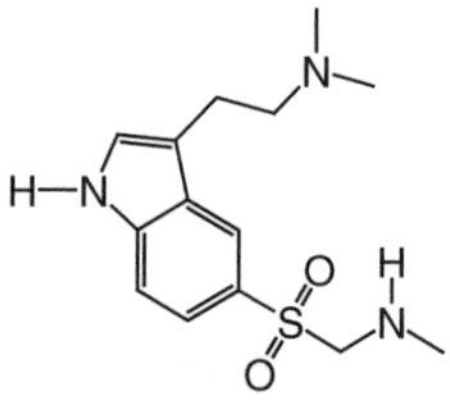

Sumatriptan
A drug for migraine treatment

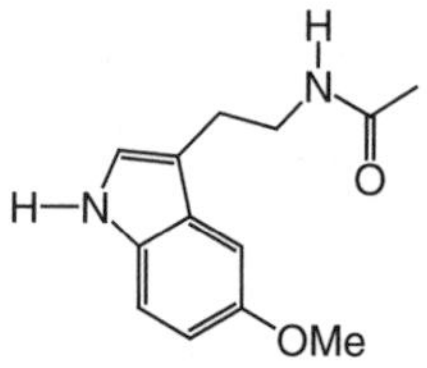

Melatonin
A hormone found in all living beings

Ergolines Alkaloids that contain an ergoline skeleton are called ergoline alkaloids, and some of them are psychedelic drugs, e.g. LSD. A number of ergoline derivatives are used clinically as vasoconstrictors (e.g. 5-HT 1 agonists, ergotamine), and in the treatment of migraine and Parkinson's disease, and some are implicated in the disease ergotism.

Ergine, molecular formula $C_{16}H_{17}N_3O$, is the amide of D-lysergic acid, and commonly known as LSA or LA-111. It is an ergoline alkaloid that occurs in various species of the Convolvulaceae, and in some species of fungus. *Rivea corymbosa* (ololiuqui), *Argyreia nervosa* (Hawaiian baby woodrose) and *Ipomoea violacea* (tlitliltzin) are three major sources of this alkaloid.

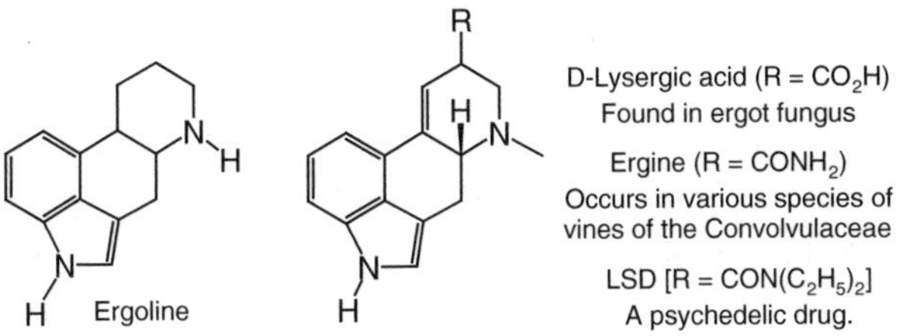

Ergoline

D-Lysergic acid (R = CO_2H)
Found in ergot fungus
Ergine (R = $CONH_2$)
Occurs in various species of vines of the Convolvulaceae
LSD [R = $CON(C_2H_5)_2$]
A psychedelic drug.

Lysergic acid diethylamide, molecular formula $C_{20}H_{25}N_3O$, also known as LSD or LSD-25, is a semi-synthetic psychedelic drug, synthesized from the natural precursor lysergic acid found in ergot, a grain fungus. It is a colourless, odourless and mildly bitter compound. LSD produces altered experience of senses, emotions, memories, time and awareness for 8 to 14 h. Moreover, LSD may cause visual effects, e.g. moving geometric patterns, 'trails' behind moving objects and brilliant colours.

β-carbolines Alkaloids that possess a 9*H*-pyrid-[3,4-b]-indole skeleton are called *β-carboline alkaloids*, and are found in several plants and animals. The structure of β-carboline is similar to that of tryptamine, with the ethylamine chain re-connected to the indole ring via an extra carbon atom, to produce a three-membered ring structure. The biosynthesis of β-carboline alkaloids follows a similar pathway to tryptamine. The β-carbolines, e.g. harmine, harmaline and tetrahydroharmine, play an important role in the pharmacology of the psychedelic brew ayahuasca. Some β-carbolines, notably tryptoline and pinoline, are formed naturally in the human body.

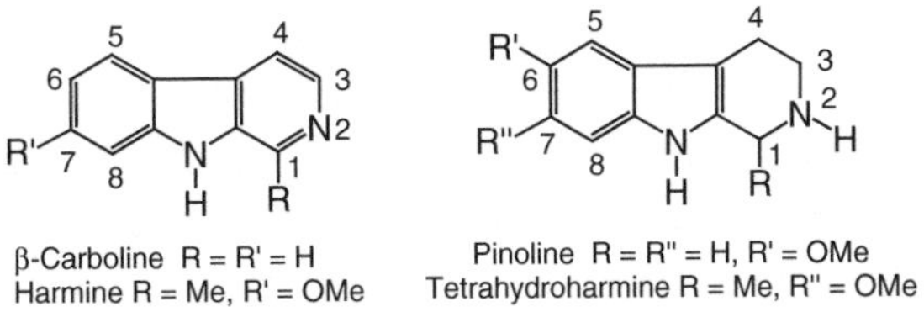

β-Carboline R = R' = H
Harmine R = Me, R' = OMe

Pinoline R = R" = H, R' = OMe
Tetrahydroharmine R = Me, R" = OMe

The major sources of β-carboline alkaloids with their medicinal or pharmacological properties are summarized below.

β-carboline alkaloids	Natural sources	Medicinal or pharmacological properties
Harmine and harmaline	Seeds of 'harmal' (*Peganum harmala*) and *Banisteriopsis caapi*	CNS stimulant, acts by inhibiting the metabolism of serotonin and other monoamines

Purine alkaloids

Alkaloids that contain a purine skeleton (see Section 4.7) are commonly known as *purine alklaoids*, e.g. caffeine and theobromine. We have already learnt that two of the bases in nucleic acids, adenine and guanine, are purines.

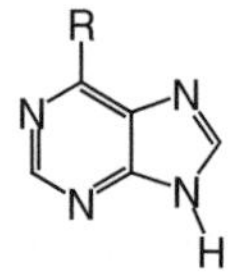

Purine R = H
Adenine R = NH_2

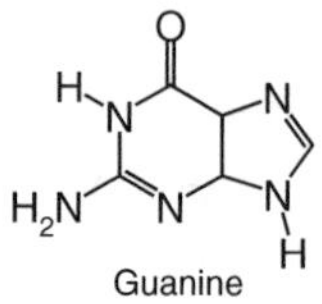

Guanine

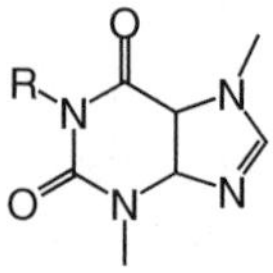

Caffeine R = Me
Theobromine R = H

Caffeine Caffeine [1,3,7-trimethyl-1*H*-purine-2,6(3*H*,7*H*)-dione], molecular formula $C_8H_{10}N_4O_2$, is a xanthine (purine) alkaloid, found mainly in tea leaves (*Camellia sinensis)* and coffee beans (*Coffea arabica*). Caffeine is sometimes called guaranine when found in guarana (*Paullinia cupana)*, mateine when found in mate (*Ilex paraguariensis*) and theine when found in tea. Caffeine is found in a number of other plants, where it acts as a natural pesticide. It is odourless white needles or powder. Apart from its presence in the tea and coffee that we drink regularly, caffeine is also an ingredient of a number of soft drinks.

Caffeine is a potent CNS and metabolic stimulant, and is used both recreationally and medically to reduce physical fatigue, and to restore mental alertness. It stimulates the CNS first at the higher levels, resulting in increased alertness and wakefulness, faster and clearer flow of thought, increased focus, and better general body coordination, and later at the spinal cord level at higher doses. Caffeine is used in combination with a number of painkillers. Caffeine is also used with ergotamine in the treatment of migraine and cluster headaches as well as to overcome the drowsiness caused by antihistamines.

Terpenoidal alkaloids

Aconite alkaloids Aconitine, molecular formula $C_{34}H_{47}NO_{11}$, is an example of an aconite alkaloid. It is soluble in organic solvents, e.g. $CHCl_3$ and C_6H_6, and slightly soluble in alcohol or ether, but insoluble in water. Aconitine is an extremely toxic substance obtained from the plants of the genus *Aconitum* (family Ranunculaceae), commonly known as 'aconite' or 'monkshood'. It is a neurotoxin, and used for creating models of cardiac arrhythmia.

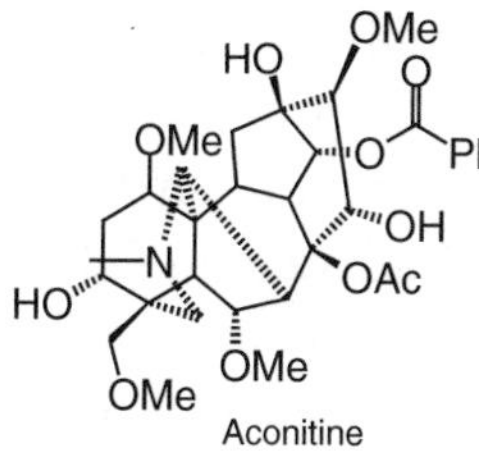

Aconitine

Steroidal alkaloids These alkaloids have a core steroidal skeleton as part of the molecule, e.g. solanine. There are a number of structural varieties that exist in steroidal alkaloids. Following discussion is just on a few selected steroidal alkaloids.

Solanine is a poisonous steroidal alkaloid, also known as glycoalkaloid, found in the nightshades family (Solanaceae). It is extremely toxic even in small quantities. Solanine has both fungicidal and pesticidal properties, and it is one of the plant's natural defences.

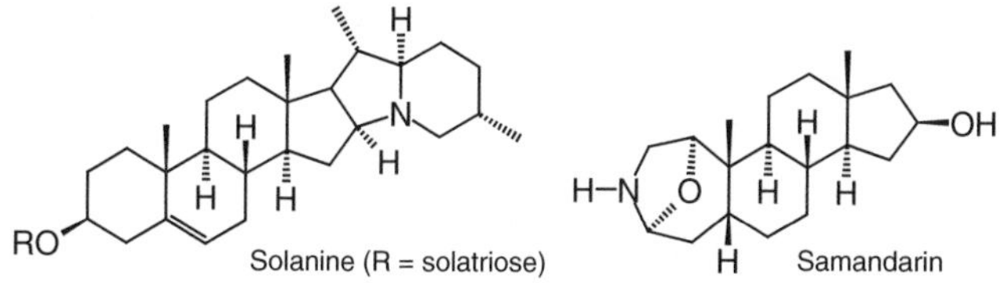

Solanine (R = solatriose)

Samandarin

Solanine hydrochloride has been used as a commercial pesticide. It has sedative and anticonvulsant properties, and has sometimes been used for the treatment of asthma, as well as for cough and common cold. However, gastrointestinal and neurological disorders result from solanine poisoning. Symptoms include nausea, diarrhoea, vomiting, stomach cramps, burning of the throat, headaches and dizziness. Other adverse reactions, in more severe cases, include hallucinations, loss of sensation, paralysis, fever, jaundice, dilated pupils and hypothermia. Solanine overdose can be fatal.

Samandarin, molecular formula $C_{19}H_{31}NO_2$, is the major steroidal alkaloid of the skin glands of the fire salamander (*Salamandra salamandra*), and is extremely toxic. The toxicities of samandarin include muscle convulsions, raised blood pressure and hyperventilation.

Betaines

Alkaloids that contain the betaine (*N,N,N*-trimethylglycine or TMG) skeleton are included in this class, e.g. muscarine. Betaine itself is used to treat high homocysteine levels, and sometimes as a mood enhancer.

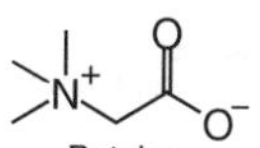

Betaine

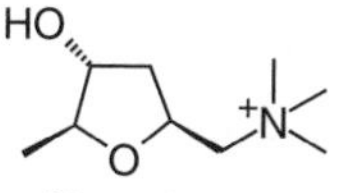

Muscarine

Muscarine Muscarine, molecular formula $C_9H_{20}NO_2^+$, first isolated from fly agaric *Amanita muscaria*, occurs in certain mushrooms, especially in the species of the genera *Inocybe* and *Clitocybe*. It is a parasympathomimetic substance. It causes profound activation of the peripheral parasympathetic nervous system, which may result in convulsions and death. Muscarine mimics the action of the neurotransmitter acetylcholine at the muscarinic acetylcholine receptors.

Macrocyclic alkaloids

This group of alkaloids possess a macrocycle, and in most cases nitrogen is a part of the ring system. Macrocyclic spermine group of alkaloids is one of such examples. These polyamine alkaloids are found in a number of plant families, e.g. Acanthaceae, Scrophulariaceae, Leguminosae, Ephedraceae and possess various biological properties, for example budmunchiamines L4 and L5, two antimalarial spermine alkaloids isolated from *Albizia adinocephala* (Leguminosae).

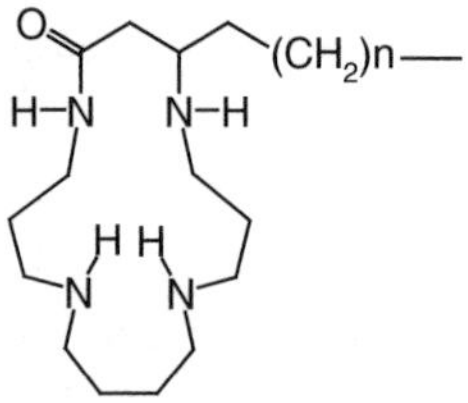

Budmunchiamine L4 n = 11
Budmunchiamine L5 n = 13

6.2.3 Tests for alkaloids

Reagent/test	Composition of the reagent	Result
Meyer's reagent	Potassiomercuric iodide solution	Cream precipitate
Wagner's reagent	Iodine in potassium iodide	Reddish-brown precipitate
Tannic acid	Tannic acid	Precipitation
Hager's reagent	A saturated solution of picric acid	Yellow precipitate
Dragendorff's reagent	Solution of potassium bismuth iodide	Orange or reddish-brown precipitate (except with caffeine and a few other alkaloids)

Caffeine and other purine derivatives can be detected by the *Murexide test*. In this test the alkaloids are mixed with a tiny amount of potassium chlorate and a drop of hydrochloric acid and evaporated to dryness, and the resulting residue is exposed to ammonia vapour. Purine alkaloids produce pink colour in this test.

6.3 Carbohydrates

Carbohydrates are the primary fuel for our muscles and the brain. Eating a high carbohydrate diet will ensure maintenance of muscle and liver glycogen (storage forms of carbohydrate), improve performance and delay fatigue. The word *carbohydrate* means 'hydrate of carbon'. Thus, carbohydrates are a group of polyhydroxy aldehydes, ketones or acids or their derivatives, together with linear and cyclic polyols. Most of these compounds are in the form $C_nH_{2n}O_n$ or $C_n(H_2O)_n$, for example glucose, $C_6H_{12}O_6$ or $C_6(H_2O)_6$. Sometimes, carbohydrates are referred to simply as sugars and their derivatives.

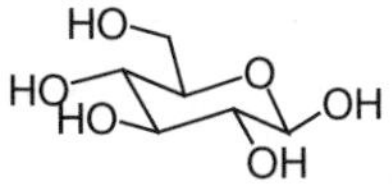

β-D-Glucose

Carbohydrates are found abundantly in nature, both in plants and animals, and are essential constituents of all living matter. Photosynthesis is the means by which plants produce sugars from CO_2 and water.

6.3.1 Classification

General classification

Generally, carbohydrates are classified into four different categories, *monosaccharides*, *di-*, *tri-* and *tetrasaccharides*, *oligosaccharides* and *polysaccharides*.

Monosaccharides These carbohydrates, commonly referred to as 'sugars', contain from three to nine carbon atoms. Most common monosaccharides in nature possess five (*pentose*, $C_5H_{10}O_5$) or six (*hexose*, $C_6H_{12}O_6$) carbon atoms. For example, glucose, a six-carbon-containing sugar, is the most common monosaccharide that is metabolized in our body to provide energy, and fructose is also a hexose found in many fruits.

Di-, tri- and tetrasaccharides These carbohydrates are dimers, trimers and tetramers of monosaccharides, and are formed from two, three or four monosaccharide molecules, with the elimination of one, two or three molecules of water. For example, sucrose is a disaccharide composed of two monosaccharides, glucose and fructose.

Oligosaccharides The name 'oligosaccharide' refers to saccharides containing two to 10 monosaccharides. For example, raffinose, found in beans and pulses, is an oligosaccharide composed of three monosaccharide units, i.e. galactose, glucose and fructose.

Polysaccharides Polysaccharides are composed of a huge number of monosaccharide units, and the number forming the molecule is often approximately known. For example, cellulose and starch are polysaccharides composed of hundreds of glucose units.

Classification of monosaccharides according to functional groups and carbon numbers

The two most common functional groups found in monosaccharides (in open chain form) are aldehyde and ketone. When a monosaccharide contains an aldehyde, it is known as an *aldose*, e.g. glucose, and in the case of ketone, it is called a *ketose* or *keto sugar*, e.g. fructose.

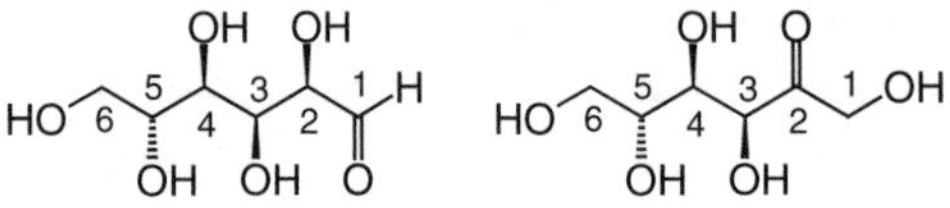

D-Glucose, an aldose
Contains an aldehyde group

D-Fructose, a ketose
Contains a ketone group

Depending on the number of carbon atoms present, monosaccharides are classified as *triose*, *tetrose*, *pentose* or *hexose*, containing three, four, five or six carbon atoms, respectively. Glucose is a hexose as it contains six carbon atoms. Sometimes, monosaccharides are classified more precisely to denote the functional group as well as the number of carbon atoms. For example, glucose can be classified as an *aldohexose*, as it contains six carbon atoms as well as an aldehyde group.

If any monosaccharide lacks the usual numbers of hydroxyl groups, it is often called a *deoxy sugar*. For example, 2-deoxyribose, which is a component of DNA nucleosides, has one less hydroxyl group than its parent sugar, ribose.

D-Ribose, an aldopentose
A component of RNA nucleosides

D-2-Deoxyribose, a deoxy aldopentose
A component of DNA nucleosides

Hydroxyls, aldehyde and keto groups are not the only functional groups that are present in monosaccharides. Monosaccharides containing carboxylic acid (–COOH) and amino (–NH_2) groups are common structural units in biologically important carbohydrates. For example, 2-amino-2-deoxy-D-glucose, also known as glucosamine, is an *amino sugar*, and glucuronic acid is a *sugar acid*.

Glucosamine, an amino sugar   Glucuronic acid, a sugar acid

6.3.2 Stereochemistry of sugars

With monosaccharides, the configuration of the highest numbered chiral carbon is compared with that of D- or L-glyceraldehyde (the simplest aldose); e.g., D-sugar has the same configuration as D-glyceraldehyde and L-sugar has the same configuration as L-glyceraldehyde. It can be noted that D- and L-notations have no relation to the direction in which a given sugar rotates the plane-polarized light i.e. (+) or (−).

(+)-D-Glyceraldehyde
The hydroxyl group
on the chiral carbon
is on the right hand side

(−)-L-Glyceraldehyde
The hydroxyl group
on the chiral carbon
is on the left hand side

Glucose, fructose, and many other natural monosaccharides have the same configuration as D-glyceraldehyde at the chiral centre farthest from the carbonyl group. In Fischer projections, most natural sugars have the hydroxyl group at the highest numbered chiral carbon pointing to the right. All these sugars are referred to as *D-sugars*, e.g. D-glucose.

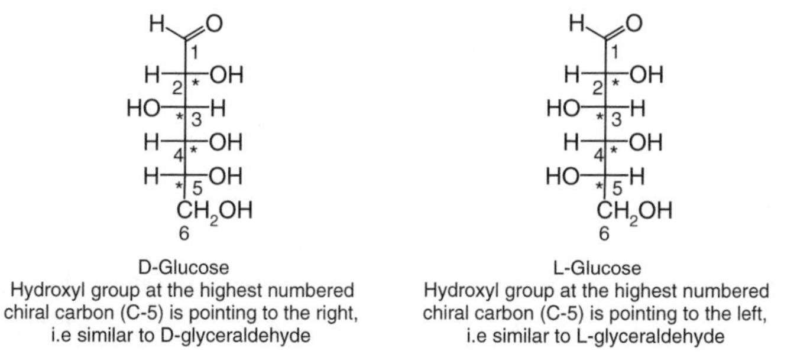

D-Glucose
Hydroxyl group at the highest numbered chiral carbon (C-5) is pointing to the right, i.e similar to D-glyceraldehyde

L-Glucose
Hydroxyl group at the highest numbered chiral carbon (C-5) is pointing to the left, i.e similar to L-glyceraldehyde

All *L-sugars* have the configuration as L-glyceraldehyde at the chiral centre farthest from the carbonyl group, e.g. L-glucose. In Fischer projections, L-sugars have the hydroxyl group at the highest numbered chiral carbon pointing to the left. Thus, an L-sugar is the mirror image (enantiomer) of the corresponding D-sugar.

6.3.3 Cyclic structures of monosaccharides

Monosaccharides not only exist as open chain molecules (acyclic), but also as cyclic compounds. Cyclization leading to the formation of a cyclic *hemiacetal* or *hemiketals* occurs due to intramolecular nucleophilic addition reaction between a –OH and a C=O group. Many monosaccharides exist in an equilibrium between open chain and cyclic forms.

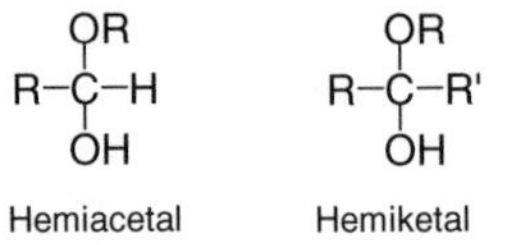

Hemiacetal Hemiketal

Hemiacetals and hemiketals have the important structural feature of –OH and –OR attached to the same carbon as shown above. Through cyclization, sugars attain a pyranose and/or a *furanose* form.

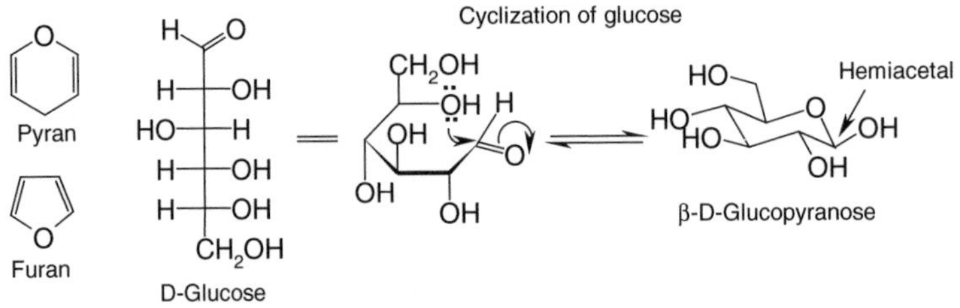

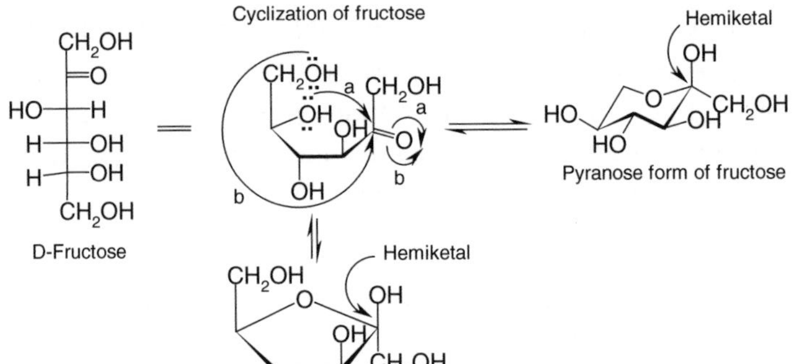

Pyran Furan D-Glucose

Cyclization of glucose

β-D-Glucopyranose

D-Fructose

Cyclization of fructose

Pyranose form of fructose

Furanose form of fructose

Cyclization produces a new chiral centre at C-1 in the cyclic form. This carbon is called the *anomeric carbon*. At the anomeric carbon, the –OH group can project upwards (β configuration) or downwards (α configuration).

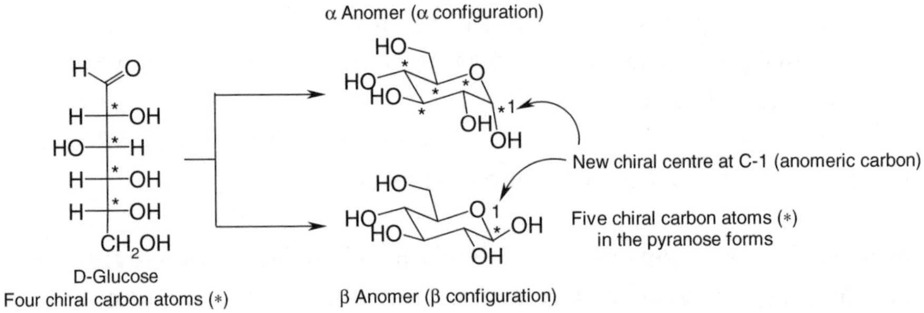

α Anomer (α configuration)

New chiral centre at C-1 (anomeric carbon)

Five chiral carbon atoms (*) in the pyranose forms

D-Glucose
Four chiral carbon atoms (*)

β Anomer (β configuration)

6.3.4 Mutarotation

The term *mutarotation* means the variation of optical rotation with time, observed in a solution of sugar on standing. Let us have a look at this phenomenon in a glucose solution. The pure α anomer of glucose has an m.p. of 146 °C and a specific rotation $[\alpha]_D + 112.2°$, and the specific rotation on standing is $+52.6°$, while pure β anomer has an m.p. of 148–155 °C and a specific rotation $[\alpha]_D + 18.7°$, and the specific rotation on standing is $+52.6°$. When a sample of either pure anomer is dissolved in water, its optical rotation slowly changes and ultimately reaches a constant value of $+52.6°$. Both anomers, in solution, reach an equilibrium with fixed amounts of α (35 per cent), β (64 per cent) and open chain (~1 per cent) forms.

6.3.5 Acetal and ketal formation in sugars

We have already learnt that *hemiacetal* or *hemiketal* exists in the cyclic structure of a sugar. For example, the anomeric carbon (C-1) in glucose is a hemiacetal, and that in fructose is a hemiketal. When the hydroxyl group in a hemiacetal or hemiketal is replaced by a –OR group, *acetal* or *ketal*, respectively, is formed. R can be an alkyl group or another sugar. The following example shows the acetal formation in glucopyranose.

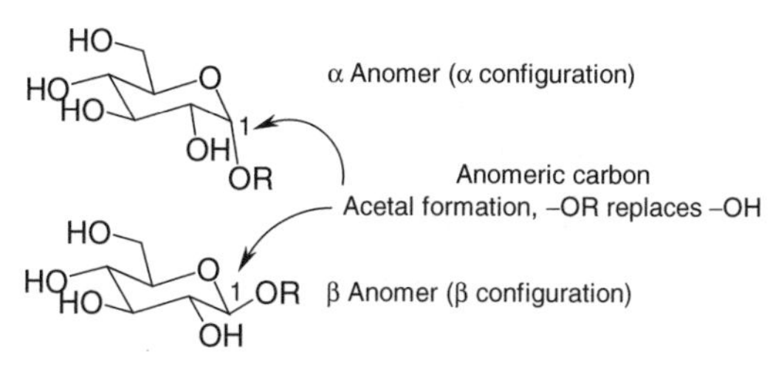

Acetal formation in glucopyranose

Acetals and ketals are also called *glycosides*. Acetals and ketals (glycosides) are not in equilibrium with any open chain form. Only hemi-acetals and hemiketals can exist in equilibrium with an open chain form. Acetals and ketals do not undergo mutarotation or show any of the reactions specific to the aldehyde or ketone groups. For example, they cannot be oxidized easily to form sugar acids. As an acetal, the carbonyl group is effectively protected.

When glucose is treated with methanol containing hydrogen chloride, and prolonged heat is applied, acetals are formed. In this reaction the hemi-acetal function is converted to the monomethyl acetal.

Hemi-acetal

D-Glucopyranose

MeOH, 4% HCl

Stereochemistry is not specified

Acetal

Methyl-β-D-glucopyranoside
Minor product

Methyl-α-D-glucopyranoside
Major product

6.3.6 Oxidation and reduction of monosaccharides

For most of the reactions of monosaccharides that involve the aldehyde or ketone functional group, the presence of open chain form is crucial, as only in this form do these functional groups exist. A sugar solution contains two cyclic anomers and the open chain form in an equilibrium. Once the aldehyde or ketone group of the open chain form is used up in a reaction, the cyclic forms open up to produce more open chain form to maintain the equilibrium.

Reduction

Reduction with sodium borohydride Treatment of monosaccharide with sodium borohydride ($NaBH_4$) reduces it to a polyalcohol called an *alditol*.

β-D-Glucopyranose

$NaBH_4$, H_2O

D-Glucose in open chain form

D-Glucitol or D-sorbitol, an alditol

The reduction occurs by interception of the open chain form present in the aldehyde/ketone–hemi-acetal/hemiketal equilibrium. Although only a small amount of the open chain form is present at any given time, that small amount is reduced. Then more is produced by opening of the pyranose form, and that additional amount is reduced, and so on until the entire sample has undergone reaction.

Reaction (reduction) with phenylhydrazine (osazone test) The open chain form of the sugar reacts with phenylhydrazine to produce a *phenylosazone*. Three moles of phenylhydrazine are used, but only two moles taken up at C-1 and C-2.

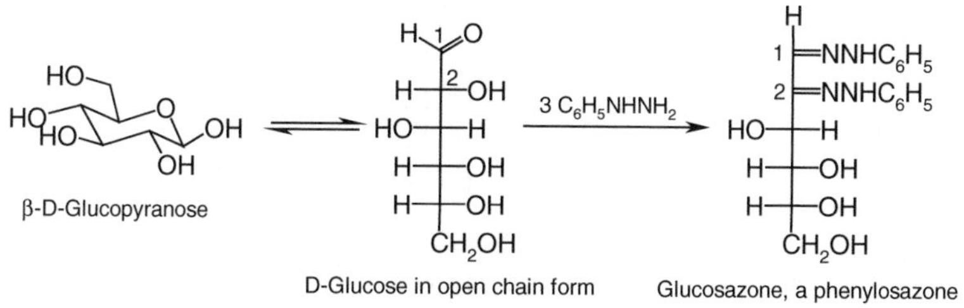

β-D-Glucopyranose ⇌ D-Glucose in open chain form $\xrightarrow{3\ C_6H_5NHNH_2}$ Glucosazone, a phenylosazone

In monosaccharides where structures differ at C-1 and C-2, but are the same in the rest of the molecule, we get the same phenylosazone. If we examine the structures of glucose and mannose, the only structural difference we can identify is the orientation of the hydroxyl group at C-2. The rest of the molecules are exactly the same. Therefore, glucose and mannose produce the same phenylosazone. Phenylosazones are highly crystalline solids with characteristic shaped crystals. Shapes are diagnostic of phenylosazone type.

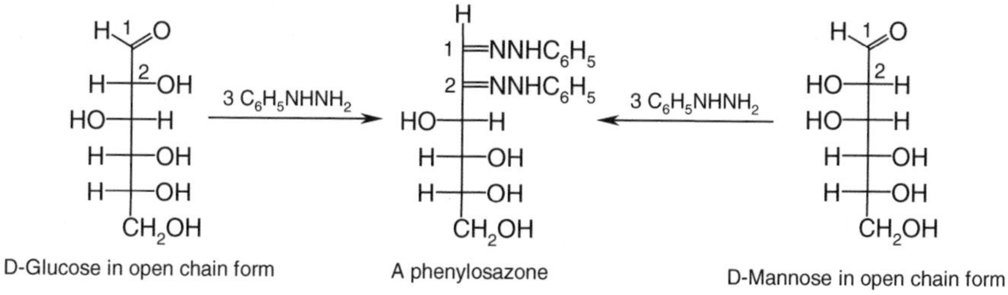

D-Glucose in open chain form $\xrightarrow{3\ C_6H_5NHNH_2}$ A phenylosazone $\xleftarrow{3\ C_6H_5NHNH_2}$ D-Mannose in open chain form

Oxidation

Aldoses are easily oxidized to produce aldonic acid. Aldoses react with *Tollens'* (Ag^+ in aqueous NH_3), *Fehling's* (Cu^{2+} in aqueous sodium tartrate) and *Benedict's reagents* (Cu^{2+} in aqueous sodium citrate), and produce characteristic colour changes. All these reactions produce oxidized sugar and a reduced metallic species. These reactions are simple chemical tests for *reducing sugars* (sugars that can reduce an oxidizing agent).

Reaction with Fehling's (and Benedict's) reagent, aldehydes and ketones (i.e. with sugars – aldoses and ketoses) can reduce Fehling's (and Benedict's) reagents, and they themselves are oxidized.

$$Cu^{2+}\ (\text{blue}) + \text{aldose or ketose} \rightarrow Cu_2O\ (\text{red/brown}) + \text{oxidized sugar}$$

Although majority of sugar molecules are in cyclic form, the small amounts of open chain molecules are responsible for this reaction.

Therefore, glucose (open chain is an aldose) and fructose (open chain is a ketose) give positive test and are *reducing sugars.*

When an oxidizing agent, e.g. nitric acid, is used, a sugar is oxidized at both ends of the chain to the dicarboxylic acid, called *aldaric acid.* For example, galactose is oxidized to galactaric acid by nitric acid.

D-Galactose → (HNO_3) → Galactaric acid, a *meso* compound (Plane of symmetry)

6.3.7 Reactions of monosaccharides as alcohols

Ester formation

Monosaccharides contain a number of alcoholic hydroxyl groups, and thus can react with acid anhydrides to yield corresponding esters. For example, when glucose is treated with acetic anhydride and pyridine, it forms a pentaacetate. The ester functions in glucopyranose pentaacetate undergo the typical ester reactions.

β-D-Glucopyranose → (Ac_2O, Pyridine) → β-D-glucopyranose pentaacetate

Monosaccharides also form phosphate esters with phosphoric acid. Monosaccharide phosphate esters are important molecules in biological system. For example, in the DNA and RNA nucleotides, phosphate esters of 2-deoxyribose and ribose are present, respectively. Adenosine triphosphate (ATP), the triphosphate ester at C-5 of ribose in adenosine, is found extensively in living systems.

Ether formation

When methyl α-D-glucopyranoside (an acetal) is treated with dimethyl sulphate in presence of aqueous sodium hydroxide, the methyl ethers of the alcohol functions are formed. The methyl ethers formed from monosaccharides are stable in bases and dilute acids.

Methyl-α-D-Glucopyranoside → ($(CH_3)_2SO_4$, NaOH, H_2O) → Methyl-2,3,4,6-tetra-*O*-methyl-α-D-Glucopyranoside

6.3.8 Pharmaceutical uses of monosaccharides

Pharmaceutically, glucose is probably the most important of all regular monosaccharides. A solution of pure glucose has been recommended for use by subcutaneous injection as a restorative after severe operations, or as a nutritive in wasting diseases. It has also been used to augment the movements of the uterus. Glucose is added to nutritive enemata for rectal alimentation. Its use has also been recommended for rectal injection and by mouth in delayed chloroform poisoning.

Glucose is used as a pharmaceutical additive. Liquid glucose is used mainly as a pill or tablet additive. For coloured pills many dispensers prefer a mixture of equal weights of extract of gentian and liquid glucose. Liquid glucose is particularly suitable for the preparation of pills containing ferrous carbonate. It preserves the ferrous salt from oxidation, and will even reduce any ferric salt present. Conversely, it should not be used where such reduction is to be avoided, as in the preparation of pills containing cupric salts. Apart from the pharmaceutical or medicinal uses, glucose is also used in large quantities in the food and confectionery industries, often in the form of thick syrup.

Fructose, another common monosaccharide found in fruits and honey, is more soluble in water than glucose and is also sweeter than glucose. It is used as a sweetener for diabetic patients, and in infusion for parenteral nutrition.

6.3.9 Disaccharides

Disaccharides contain a *glycosidic acetal bond* between the anomeric carbon of one sugar and an –OH group at any position on the other sugar. A glycosidic bond between C-1 of the first sugar and the –OH at C-4 of the second sugar is particularly common. Such a bond is called a 1,4′-link, for example maltose, where two glucose units are linked between C-1 and C-4 via oxygen. A glycosidic bond to the anomeric carbon can be either α or β.

The most common naturally occurring disaccharides are sucrose (table sugar) and lactose (milk sugar). While sucrose is derived from plants and is prepared commercially from sugar cane and sugar beet, lactose is found in the milk of animals. Other common disaccharides that are produced by breaking down polysaccharides include maltose (obtained from starch) and cellobiose (obtained from cellulose).

Maltose and cellobiose

Maltose is a disaccharide, composed of two units of glucose linked (α linkage) between C-1 of one and C-4 of the other via oxygen.

Chemically, it can be called 4-*O*-α-D-glucopyranosyl-D-glucopyranose. Cellobiose is also composed of two units of glucose, but the 1,4′- link is β, instead of α. Thus, it can be called 4-*O*-β-D-glucopyranosyl-D-glucopyranose. 'Linkage' always refers to 'left-hand' sugar. For example, in maltose, since the 'linkage' is α, and is in between C-1 of one glucose and C-4 of the other, the 'linkage' is called α 1,4′.

Both maltose and cellobiose exist as α and β anomers, and undergo mutarotation. These are reducing sugars. They react with Benedict's and Fehling's reagents, and also react with phenylhydrazine to yield the characteristic phenylosazone. If you have a closer look at the following structures of maltose and cellobiose, you will see that the left-hand glucose possesses an acetal link (glycosidic link) but the right-hand glucose still has the hemiacetal at C-1′. The right-hand glucose can exist in an equilibrium of α and β anomers, and the open chain form. This is why maltose and cellobiose behave like glucose in chemical reactions.

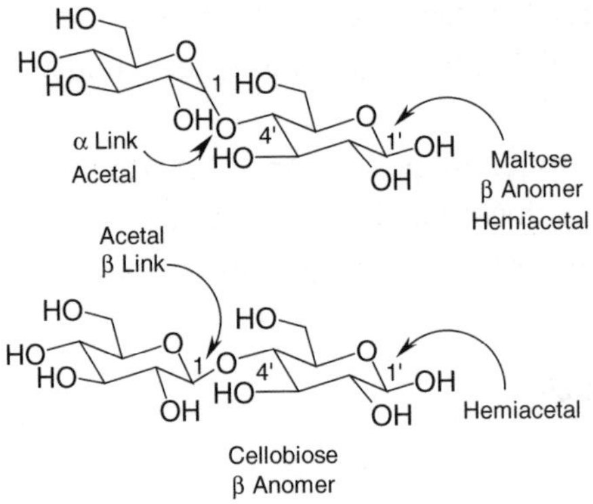

Maltose is hydrolyzed by the enzyme maltase (specific for α-glycosidic linkage) to two units of glucose, but for the hydrolysis of cellobiose the enzyme emulsin (specific for β-glycosidic linkage) is necessary. While maltose is the building block of the polysaccharide starch, cellobiose is the building block of another polysaccharide, cellulose.

Malt consists of the grain of barley, *Hordeum distichon* (family Gramineae), partially germinated and dried. Maltose is the major carbohydrate of malt and malt extracts. Pharmaceutically, extract of malt is used as a vehicle for the administration of cod-liver oil, and the liquid extract is given with haemoglobin, extract of cascara and various salts.

Lactose

Lactose, found in milk and a major component of whey, is a disaccharide that is composed of a unit of glucose and a unit of galactose through a β 1,4′-linkage. Chemically, it can be called 4-*O*-β-D-galactopyranosyl-D-glucopyranose. Like maltose and cellobiose, lactose is a reducing sugar because of the presence of hemiacetal on the right-hand sugar (glucose). Therefore, it also undergoes similar reactions to those of cellobiose and maltose, and shows mutarotation.

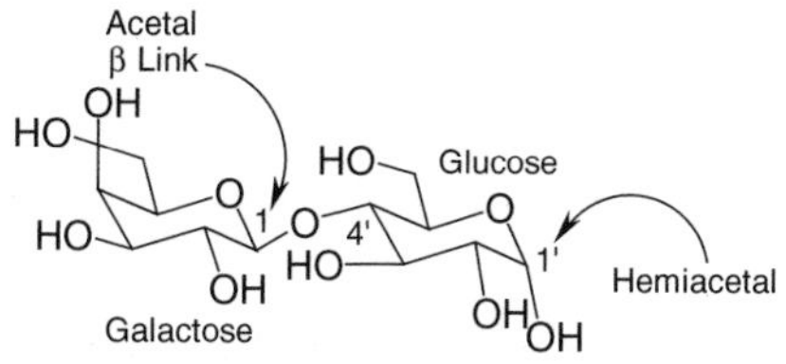

Lactose
α Anomer

Lactose has a sweetish taste, and is used extensively in the pharmaceutical industry. It is the second most widely used compound and is employed as a diluent, filler or binder in tablets, capsules and other oral product forms. α-lactose is used for the production of lactitol, which is present in diabetic products, low calorie sweeteners and slimming products. As lactose is only 30 per cent as sweet as sugar it is used as a sugar supplement, and also in food and confectionery. It is used in infant milk formulas.

Sucrose

Sucrose is a disaccharide that is composed of a unit of glucose (acetal form) and a unit of fructose (ketal form) linked through C-1 of glucose and C-2 of fructose, i.e. a 1,2′ link. In sucrose, neither glucose nor fructose can exist in open chain form because of the formation of acetal and ketal as shown below. As a result, sucrose is not a reducing sugar, and does now exhibit mutarotation. The specific rotation $[\alpha]_D$ of sucrose is $+66°$.

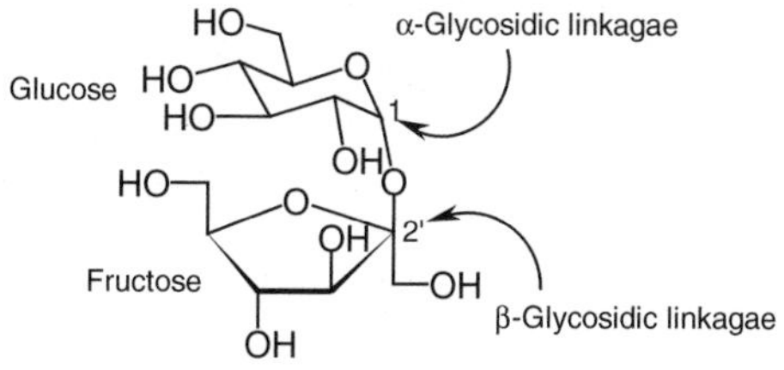

Sucrose molecule

Hydrolysis of sucrose yields glucose and fructose with specific rotations $[\alpha]_D + 52.5°$ and $-92°$, respectively, and makes the resulting mixture laevorotatory (−). This phenomenon of sucrose is called the *inversion of sucrose*, and the resulting mixture is known as *invert sugar*, which is the main component of honey, and is sweeter than sucrose itself.

6.3.10 Polysaccharides

A number of monosaccharide units combine together to form a polysaccharide, e.g. starch, cellulose and inulin. Starch and cellulose are the two most important polysaccharides from biological as well as economical viewpoints.

Starch

Starch, an essential component of our diet, is a high molecular weight polymer of glucose where the monosaccharide (glucose) units are linked mainly by 1,4′-α-glycoside bonds, similar to maltose. Plants are the main source of starch. Starch is obtained from wheat (*Triticum sativum*), rice (*Oryza sativa*) and maize (*Zea mays*), all from the plant family Gramineae. Potato (*Solanum tuberosum*; family Solanaceae) and maranta (*Maranta arundinacea*; family Marantaceae) are also good sources of starch.

Starch consists of two main components: *amylose* (insoluble in cold water) and *amylopectin* (soluble in cold water). Amylose, which accounts for about 20 per cent by weight of starch, has an average molecular weight of over 10^6. It is a polymer of glucopyranose units linked together through α 1,4′-linkages in a linear chain. Hydrolysis of amylose produces maltose. Amylose and iodine form a colour complex, which is blue/black. This is the colour reaction of iodine in starch, a confirmatory test for the presence of starch.

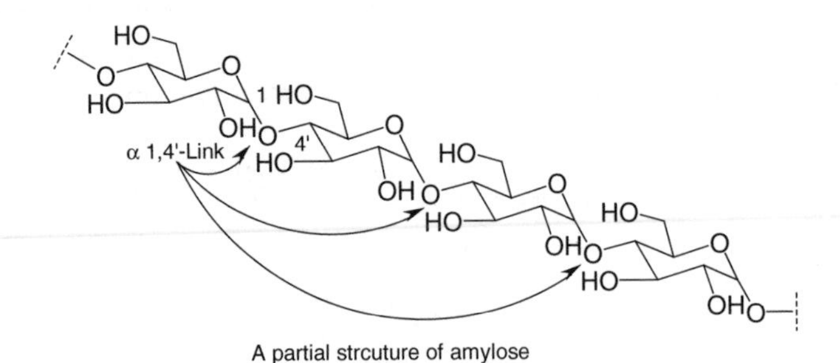

A partial strcuture of amylose

On the other hand, amylopectin accounts for about 80 per cent by weight of starch and consists of hundreds of glucose molecules linked together by 1,4′-α- and also 1,6′-α-glycoside bonds. Amylopectin contains branches (nonlinear), approximately one in every 20 to 25 glucose units. Hydrolysis of amylopectin yields maltose.

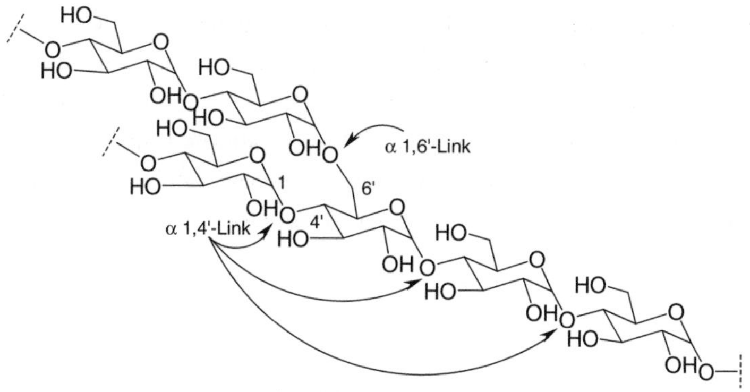

A partial strcuture of amylopectine

The pharmaceutical and cosmetic uses of starch include its use as dusting powder, binder, dispersing agent, thickening agent, coating agent and diluent. Starch soaks up secretions and helps to render injured parts less liable to bacterial infections. As a dusting powder for application to chafings and excoriations, it is used either alone or mixed with zinc oxide, boric acid and other similar substances. It also forms the basis of violet powder. Boiled with water it may be employed as an emollient for the skin. Starch is the best antidote for poisoning by iodine. Some examples of commercial preparations of starch are presented here.

Products	Composition	Applications
Amylum Iodisatum BPC	Iodized starch	It is administered internally in syphilis and other cachexias, and may be given in milk, water, gruel or arrowroot. Externally, it is used as a dry dressing, being a good substitute for iodoform.
Cataplasma Amyli BPC	Starch poultice	Used as a substitute for the domestic bread poultice for application to small superficial ulcerations.
Cataplasma Amyli et Acidi Borici BPC	Starch and boric acid poultice	Starch, 10; boric acid, 6; water, 100. An antiseptic poultice for application to ulcerated wounds.
Glycerinum Amyli BP and Glycerinum Amyli USP	Glycerin of starch	It is a soothing and emollient application for the skin, and is used for chapped hands and chilblains.
Mucilago Amyli BPC	Mucilage of starch	This mucilage is used as a basis for enemata

Glycogen

Glycogen, a homopolymer of glucose, is the major form of stored carbohydrate in animals and serves the energy storage function. Dietary carbohydrates that are not needed for immediate energy are converted by the body to glycogen for long term storage. It can release glucose units if cellular glucose levels are low. Glycogen 'mops up' excess glucose in cells. Like amylopectin, glycogen contains a complex branching structure with both 1,4′ and 1,6′ links, but it is larger than amylopectin (up to 100 000 glucose units) and much more branched. It has one end glucose unit (where glucose can be added or released) for every 12 units and a branch in every six glucose units.

Cellulose

Cellulose, the most abundant natural organic polymer, consists of several thousands of D-glucose units linked by 1,4′-β-glycoside bonds as in cellobiose. Cellulose has a linear chain structure. Different cellulose molecules can then interact to form a large aggregate structure held together by *hydrogen bonds*. On hydrolysis, cellulose produces cellobiose, and finally glucose.

Nature uses cellulose mainly as a structural material to provide plants with strength and rigidity. Human digestive enzymes contain α-glucosidase, but not β-glucosidase. Therefore, human digestive enzymes cannot hydrolyze β-glycosidic links between glucose units. In human beings, starch (but not cellulose), is hydrolyzed enzymatically to produce glucose. Therefore, cellulose does not have any dietary importance. While there is no food value in cellulose for humans, cellulose and its derivatives are commercially important. Cellulose is used as raw material for the manufacture of cellulose acetate, known commercially as *acetate rayon*, and cellulose nitrate, known as *guncotton*. Commercially important fibres, e.g. cotton and flax, consist almost completely of cellulose.

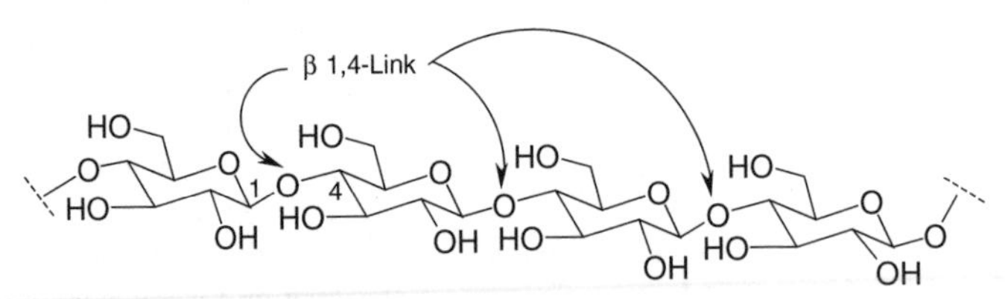

Partial structure of cellulose

Water soluble, high viscosity grade cellulose ether compositions are useful for the reduction of serum lipid levels, particularly total serum cholesterol, serum triglycerides and low density lipoprotein (LDL) levels, and/or attenuate the rise of blood glucose levels. The compositions may be in the form of a prehydrated ingestible composition, e.g. a gelatin, or a comestible, e.g. a biscuit.

Cellulose derivatives, e.g. hydroxyethylcellulose, are used in the formulation of sustained release tablets and suspensions. Natrosol (hydroxyethylcellulose) is a nonionic water-soluble polymer that is extensively used as a thickener, protective colloid, binder, stabilizer and suspending agent, particularly in applications where a nonionic material is desired. Natrosol is also used in cosmetic preparations as a thickening agent for shampoos, conditioners, liquid soaps and shaving creams.

6.3.11 Miscellaneous carbohydrates

Sugar phosphates

These sugars are formed by *phosphorylation* with ATP, e.g. glucose 6-phosphate. They are extremely important in carbohydrate metabolism. We have already seen that nucleotides contain sugar phosphates.

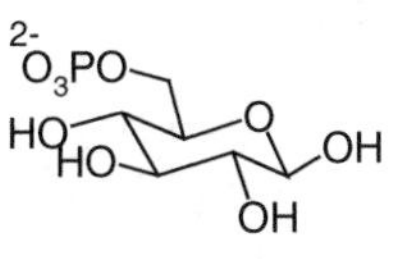

Glucose 6-phosphate

Nitrogen-containing sugars

Glycosylamines In these sugars, the anomeric –OH group (of common sugars) is replaced by an amino ($-NH_2$) group: for example, adenosine.

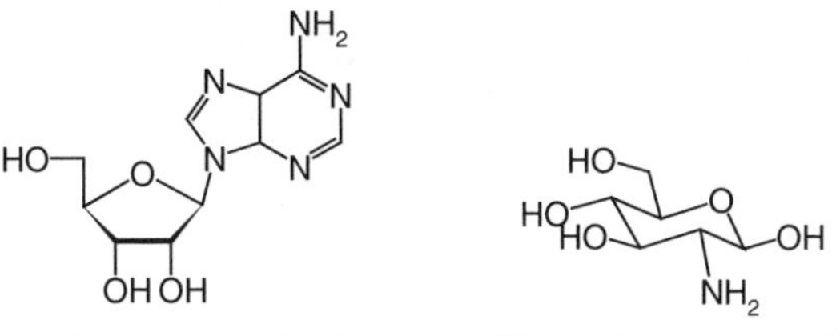

Adenosine, a nucleoside

Glucosamine, an amino sugar

Amino sugars In amino sugars, a non-anomeric –OH group (of common sugars) is replaced by an amino ($-NH_2$) group: for example, glucosamine, which is found in exoskeletons of insects and crustacea, and also isolated from heparin (anticoagulant in mast cells in arterial cell walls). Other amino sugars are found in antibiotics such as streptomycin and gentamicin.

Carbohydrate antibiotics Antibiotics that contain one or more *amino sugars* within the molecule are called *carbohydrate antibiotics*. For example, gentamicin is composed of three different units: purpurosamine, 2-deoxystreptamine and garosamine. Other examples include streptomycin and neomycin.

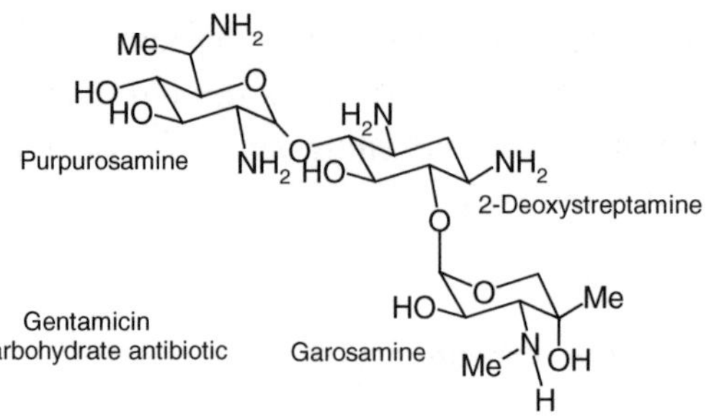

Gentamicin
A carbohydrate antibiotic

Sulphur-containing carbohydrate

In these sugars, the anomeric –OH group (of common sugars) is replaced by a sulphur-containing group, e.g. lincomycin.

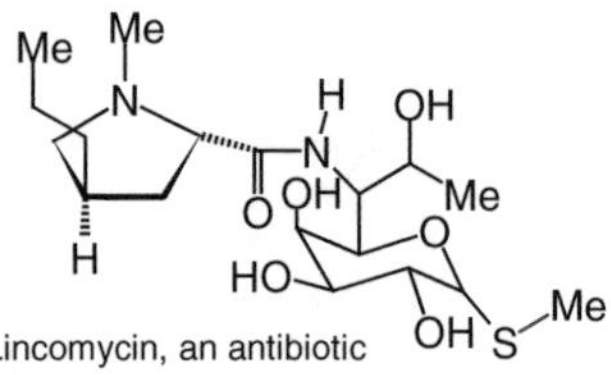

Lincomycin, an antibiotic

Ascorbic acid (vitamin C)

Ascorbic acid, commonly known as vitamin C, is a sugar acid, biosynthesized in plants, and also found in the livers of most vertebrates, except human beings. Therefore, human beings need an external supply of this vitamin, mainly from fresh vegetables and fruits. In many pharmaceutical preparations ascorbic acid is used as an antioxidant preservative. Ascorbic acid is highly susceptible to oxidation, and oxidized easily to dehydroascorbic acid.

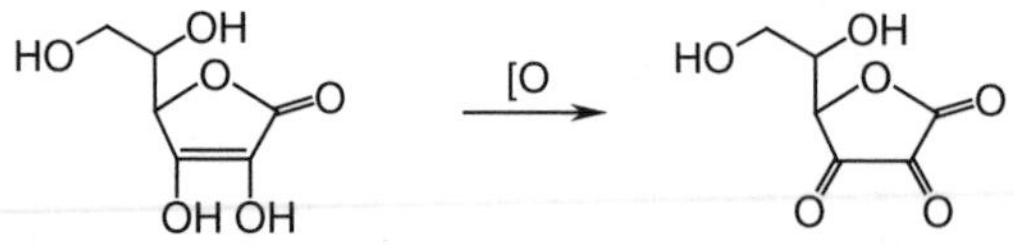

Ascorbic acid (Vitamin C) → [O] → Dehydroascorbic acid

Glycoprotein and glycolipids

Glycoproteins and glycolipids are formed when sugars combine, respectively, with proteins and lipids. Biologically these are important compounds as they are an integral part of cell membranes. Biological membranes are composed of proteins, lipids and carbohydrates. The carbohydrates in the membrane are covalently bonded to proteins (*glycoproteins*) or with lipids (*glycolipids*).

6.3.12 Cell surface carbohydrates and blood groupings

Small polysaccharide chains, covalently bonded by glycosidic links to hydroxyl groups on proteins (glycoproteins), act as biochemical markers (i.e. *antigenic determinants*) on cell surfaces. The membrane of the red blood cells (RBCs) contains glycoproteins/glycolipids, and the type of sugar that combines with these proteins/lipids varies from person to person. This gives rise to different *blood groups* (A, B, AB and O). Human blood group compatibilities are presented in the following table.

Donor blood type	Acceptor blood type			
	A	B	AB	O
A	Compatible	Incompatible	Compatible	Incompatible
B	Incompatible	Compatible	Compatible	Incompatible
AB	Incompatible	Incompatible	Compatible	Incompatible
O	Compatible	Compatible	Compatible	Compatible

6.4 Glycosides

Compounds that yield one or more sugars upon hydrolysis are known as glycosides. A glycoside is composed of two moieties: sugar portion (*glycone*) and non-sugar portion (*aglycone* or *genin*). For example, the hydrolysis of salicin produces a glucose unit and salicyl alcohol.

Glycosidic link

Hydrolysis

Salicin, a glycoside Glucose, a glycone (sugar) Salicyl alcohol, an aglycone

Glycosides of many different aglycones are extensively found in the plant kingdom. Many of these glycosides are formed from phenols, polyphenols, steroidal and terpenoidal alcohols through glycosidic attachment to sugars. Among the sugars found in natural glycosides, D-glucose is the most prevalent one, but L-rhamnose, D- and L-fructose and L-arabinose also occur quite frequently. Of the pentoses, L-arabinose is more common than D-xylose and the sugars often occur as oligosaccharides.

The sugar moiety of a glycoside can be joined to the aglycone in various ways, the most common being via an oxygen atom (*O-glycoside*). However, this bridging atom can also be a carbon (*C-glycoside*), a nitrogen (*N-glycoside*) or a sulphur atom (*S-glycoside*). By virtue of the aglycone and/or sugar, glycosides are extremely important pharmaceutically and medicinally. For example, digitoxin is a cardiac glycoside found in the foxglove plant (*Digitalis purpurea*).

6.4.1 Biosynthesis

The biosynthetic pathways are widely variable depending on the type of aglycone as well as the glycone units present in the glycosides. The aglycone and the sugar parts are biosynthesized separately, and then coupled to form a glycoside. The coupling of the sugar and aglycone takes place in the same way, irrespective of the structural type of the aglycone. Phosphorylation of a sugar yields a sugar 1-phosphate, which reacts with a uridine triphosphate (UDP) to form a uridine diphosphate sugar (UDP-sugar) and inorganic phosphate. This UDP-sugar reacts with the aglycone to form the glycoside and a free UDP.

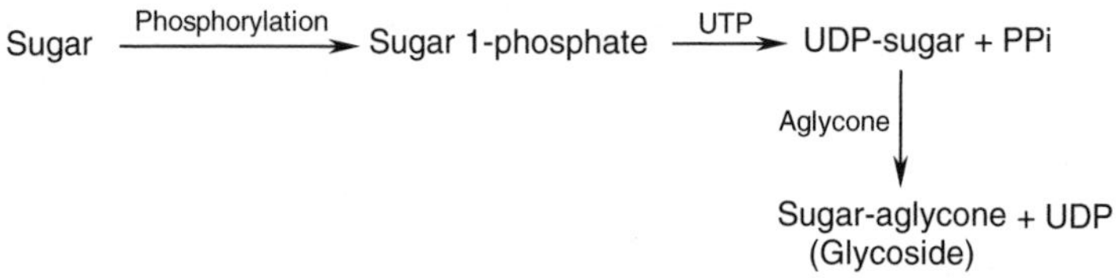

Sugar —(Phosphorylation)→ Sugar 1-phosphate —(UTP)→ UDP-sugar + PPi —(Aglycone)→ Sugar-aglycone + UDP
(Glycoside)

6.4.2 Classification

Based on sugar component

Glycosides that contain glucose are called *glucoside*. Similarly, when the sugars are fructose or galactose, the glycosides are called *fructoside* or *galactoside*, respectively.

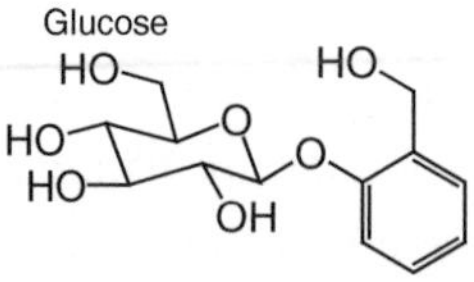

Glucose

Salicin, a glucoside

Based on aglycone

Glycosides can be classified on the basis of the structural types of aglycone present in the glycoside. For example, in anthraquinone, flavonoid, iridoid, lignan or steroid glycosides, the aglycones are anthraquinone, flavonoid, iridoid, lignan or steroid, respectively.

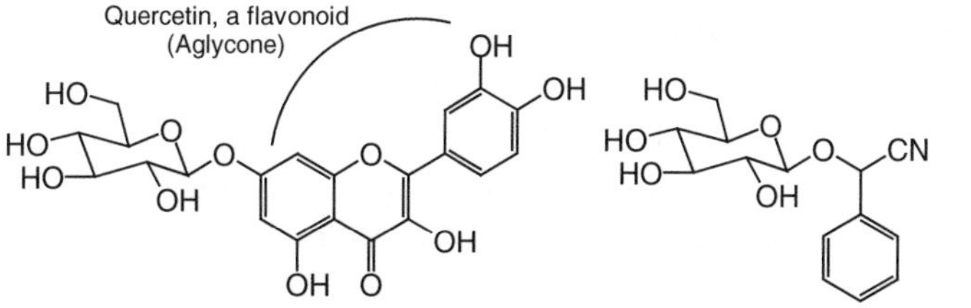

Quercetin, a flavonoid (Aglycone)

Quercetin 7-*O*-β-D-glucopyranoside
A flavonoid glycoside

Prunasin, a cyanogenic glycoside

Based on properties or functions

Glycosides that have 'soaplike' properties are called *saponins*. Similarly, glycosides that liberate hydrocyanic acid (HCN) on hydrolysis are known as *cyanogenic glycosides*, and glycosides that have an effect on heart muscle are called *cardiac glycosides*.

6.4.3 Cyanogenic glycosides

Amygdalin, prunasin and a number of other related glycosides belong to this class of glycosides, which liberate hydrocyanic acid upon hydrolysis. Biosynthetically, the aglycones of cyanogenic glycosides are derived from L-amino acids, e.g. amygdalin from L-phenylalanine, linamarin from L-valine and dhurrin from L-tyrosine.

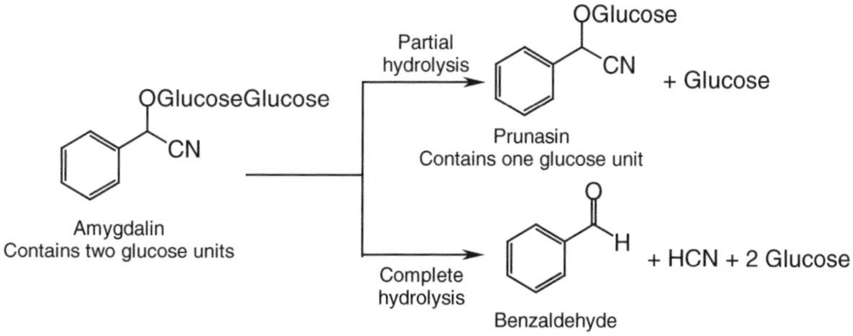

Amygdalin
Contains two glucose units

Partial hydrolysis → Prunasin (Contains one glucose unit) + Glucose

Complete hydrolysis → Benzaldehyde + HCN + 2 Glucose

Cyanogenic glycosides, particularly amygdalin and prunasin, are found in the kernels of apricots, bitter almonds, cherries, plums and peaches. The following are a few other sources of cyanogenic glycosides.

Common name	Botanical name	Family	Major cyanogenic glycoside present
Almond	*Prunus amygdalus*	Rosaceae	Amygdalin
Cassava	*Manihot utilissima*	Euphorbiaceae	Manihotoxin
Linseed	*Linum usitatissimum*	Linaceae	Linamarin

Test for hydrocyanic acid (HCN)

The liberation of hydrocyanic acid due to complete hydrolysis of cyanogenic glycosides can be determined by a simple colour test using sodium picrate paper (yellow), which turns red (sodium isopurpurate) in contact with HCN.

Pharmaceutical uses and toxicity

The extracts of plants that contain cyanogenic glycosides are used as flavouring agents in many pharmaceutical preparations. Amygdalin has been used in the treatment of cancer (HCN liberated in stomach kills malignant cells), and also as a cough suppressant in various preparations.

Excessive ingestion of cyanogenic glycosides can be fatal. Some foodstuffs containing cyanogenic glycosides can cause poisoning (severe gastric irritations and damage) if not properly handled.

6.4.4 Anthracene/anthraquinone glycosides

The aglycones of anthracene glycosides belong to structural category of anthracene derivatives. Most of them possess an anthraquinone skeleton, and are called *anthraquinone glycosides*, e.g. rhein 8-*O*-glucoside and aloin (a C-glucoside). The most common sugars present in these glycosides are glucose and rhamnose.

Anthracene

9,10-Anthraquinone

Rhein 8-*O*-glucoside

Aloin, an anthraquinone C-glucoside

Anthraquinone glycosides are coloured substances, and are the active components in a number of crude drugs, especially with laxative and purgative properties. Anthraquinone aglycone increases peristaltic action of large intestine. A number of 'over the counter' laxative preparations contain anthraquinone glycosides. The use of anthraquinone drugs, however, should be restricted to short term treatment of constipation only, as frequent or long term use may cause intestinal tumours.

Anthraquinones are found extensively in various plant species, especially from the families Liliaceae, Polygonaceae, Rhamnaceae, Rubiaceae and Fabaceae. They are also biosynthesized in micro-organisms, e.g. *Penicillium* and *Aspergillus* species. The following structural variations within anthraquinone aglycones are most common in nature.

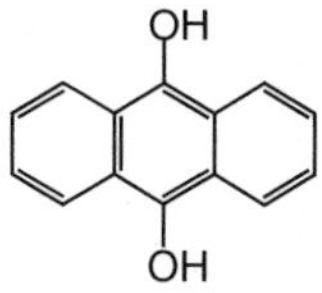

Anthrahydroquinone 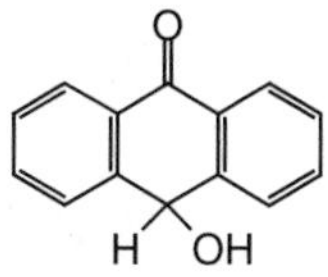 Oxanthrone 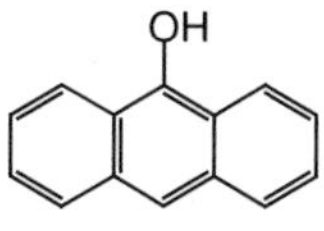 Anthranol 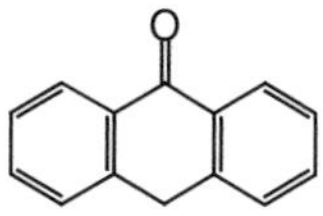 Anthrone

Dimeric anthraquinone and their derivatives are also present as aglycones in anthraquinone glycoside found in the plant kingdom.

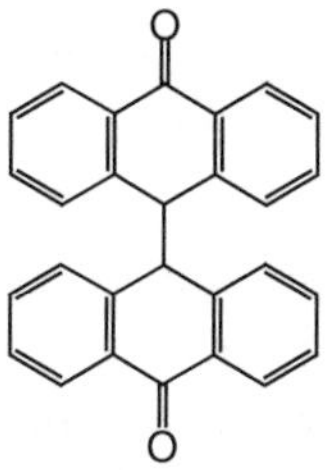

Dianthrone 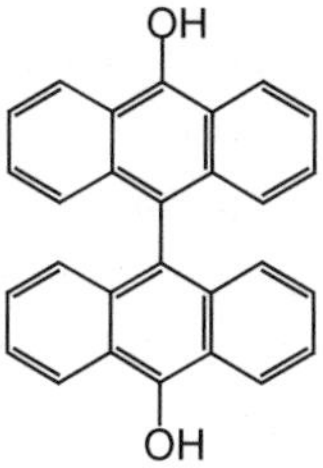 Dianthranol

Sennosides

The most important anthraquinone glycosides are sennosides, found in the senna leaves and fruits (*Cassia senna* or *Cassia angustifolia*). These are, in fact, dimeric anthraquinone glycosides. However, monomeric anthraquinone glycosides are also present in this plant.

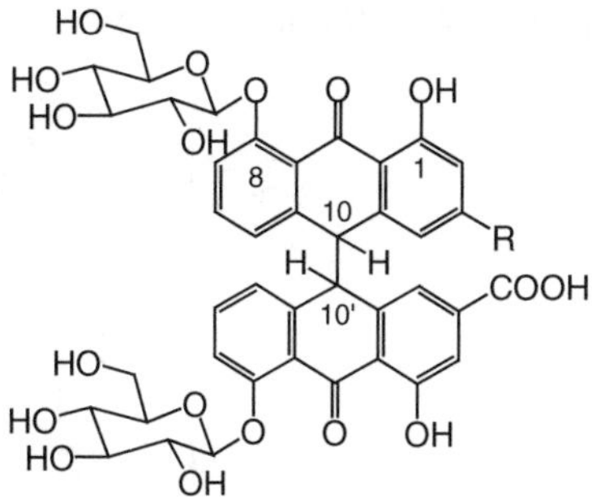

Sennoside A	R = COOH	10,10'-*trans*
Sennoside B	R = COOH	10,10'-*cis*
Sennoside C	R = CH_2OH	10,10'-*trans*
Sennoside D	R = CH_2OH	10,10'-*cis*

Cascarosides

Cascara bark (*Rhamnus purshianus*) contains various anthraquinone *O*-glycosides, but the main components are the C-glycosides, which are known as cascarosides. Rhubarb (*Rheum palmatum*) also contains several different *O*-glycosides and cascarosides. *Aloe vera* mainly produces anthraquinone C-glycosides, e.g. aloin.

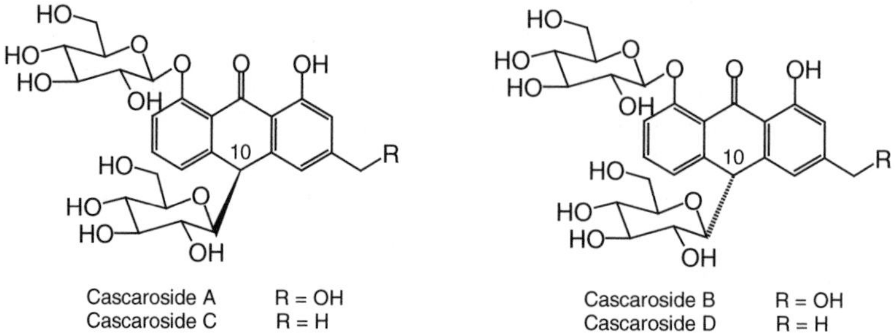

Cascaroside A R = OH
Cascaroside C R = H

Cascaroside B R = OH
Cascaroside D R = H

Test for anthraquinone glycosides

For *free anthraquinones*, powdered plant material is mixed with organic solvent and filtered, and an aqueous base, e.g. NaOH or NH_4OH solution, is added to it. A pink or violet colour in the base layer indicates the presence of anthraquinones in the plant sample.

For *O-glycosides*, the plant samples are boiled with HCl/H_2O to hydrolyse the anthraquinone glycosides to respective aglycones, and the then the above method for free anthraquinones is carried out.

For *C-glycosides*, the plant samples are hydrolysed using $FeCl_3/HCl$, and then the above method for free anthraquinones is carried out.

Biosynthesis of anthraquinone glycosides

In higher plants, anthraquinones are biosynthesized either via acylpolymalonate (as in the plants of the families Polygonaceae and Rhamnaceae) or via shikimic acid pathways (as in the plants of the families Rubiaceae and Gesneriaceae) as presented in the following biosynthetic schemes.

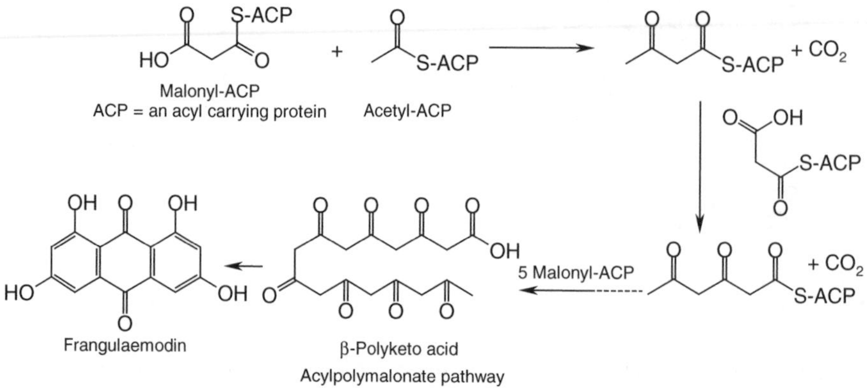

Malonyl-ACP
ACP = an acyl carrying protein
Acetyl-ACP
5 Malonyl-ACP
Frangulaemodin
β-Polyketo acid
Acylpolymalonate pathway

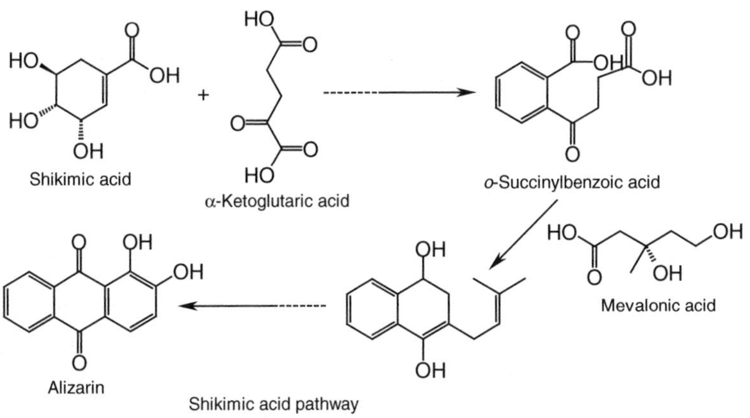

Shikimic acid pathway

6.4.5 Isoprenoid glycosides

The aglycone of this type of glycoside is biosynthetically derived from isoprene units. There are two major classes of isoprenoid glycosides: saponins and cardiac glycosides.

Saponins

Saponin glycosides possess 'soaplike' behaviour in water, i.e. they produce foam. On hydrolysis, an aglycone is produced, which is called *sapogenin*. There are two types of sapogenin: steroidal and triterpenoidal. Usually, the sugar is attached at C-3 in saponins, because in most sapogenins there is a hydroxyl group at C-3.

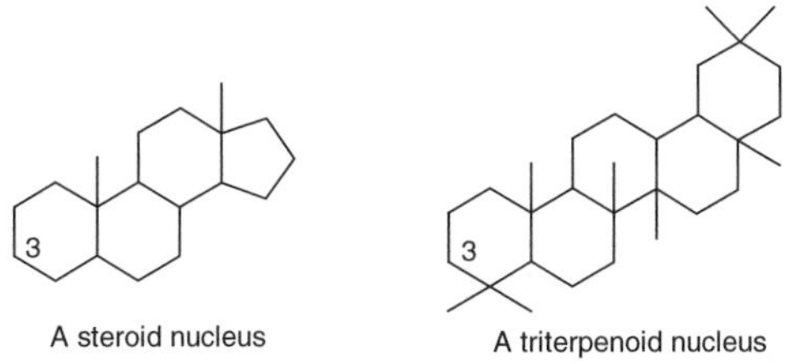

A steroid nucleus

A triterpenoid nucleus

The two major types of steroidal sapogenin are *diosgenin* and *hecogenin*. Steroidal saponins are used in the commercial production of sex hormones for clinical use. For example, progesterone is derived from diosgenin.

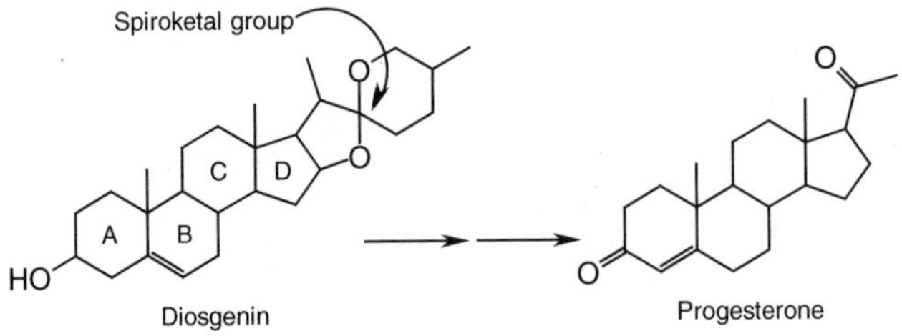

Spiroketal group

Diosgenin

Progesterone

The most abundant starting material for the synthesis of progesterone is diosgenin isolated from *Dioscorea* species, formerly supplied from Mexico, and now from China. The spiroketal group attached to the D ring of diosgenin can easily be removed. Other steroidal hormones, e.g. cortisone and hydrocortisone, can be prepared from the starting material hecogenin, which can be isolated from Sisal leaves found extensively in East Africa.

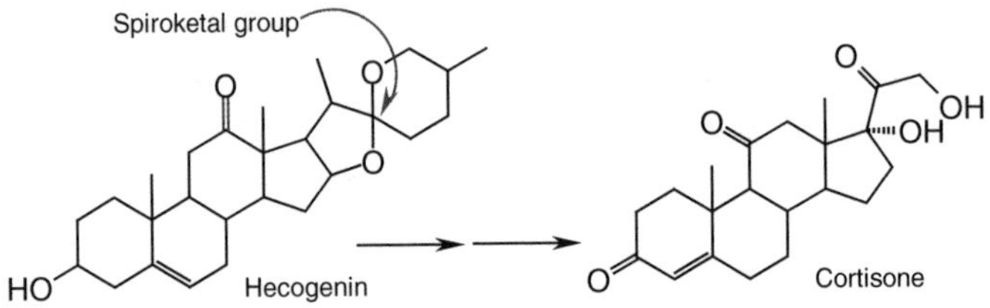

Spiroketal group

Hecogenin

Cortisone

In *triterpenoidal saponins*, the aglycone is a triterpene. Most aglycones of triterpenoidal saponins are pentacyclic compounds derived from one of the three basic structural classes represented by α-amyrin, β-amyrin and lupeol. However, tetracyclic triterpenoidal aglycones are also found, e.g. ginsenosides. These glycosides occur abundantly in many plants, e.g. liquorice and ginseng roots contain glycyrrhizinic acid derivatives and ginsenosides, respectively. Most crude drugs containing triterpenoid saponins are usually used as expectorants. Three major sources of triterpenoidal glycosides along with their uses are summarized below.

Plants Botanical names (Family)	Main constituents	Uses
Liquorice root *Glycyrrhiza glabra* (Fabaceae)	Glycyrrhizinic acid derivatives	In addition to expectorant action, it is also used as a flavouring agent.
Quillaia bark *Quillaja saponaria* (Rosaceae)	*Several complex* triterpenoidal saponins, e.g. senegin II	*Tincture of this plant is* used as an emulsifying agent.
Ginseng *Panax ginseng* (Araliaceae)	Ginsenosides	As a tonic, and to promote the feeling of well being.

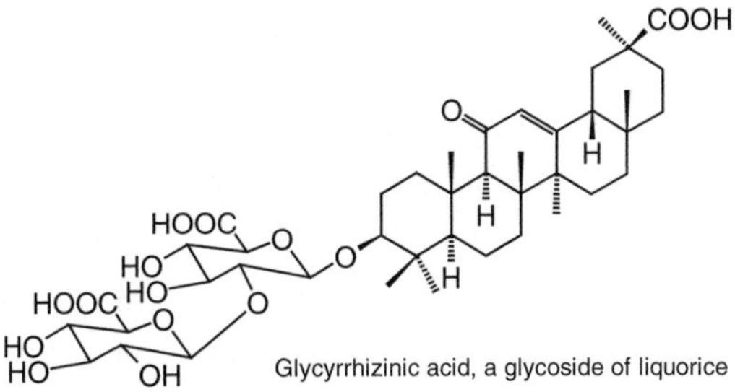

Glycyrrhizinic acid, a glycoside of liquorice

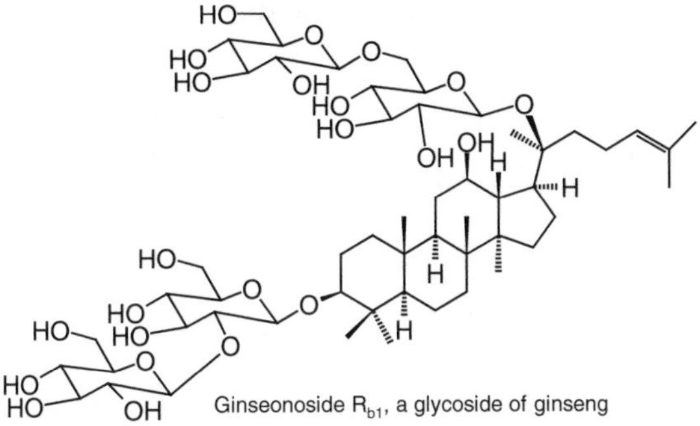

Ginseonoside R_{b1}, a glycoside of ginseng

Cardiac glycosides

Glycosides that exert a prominent effect on heart muscle are called cardiac glycosides, e.g. digitoxin from *Digitalis purpurea*. Their effect is specifically on myocardial contraction and atrioventricular conduction. The aglycones of cardiac glycosides are steroids with a side-chain containing an unsaturated lactone ring, either five membered γ-lactone (called *cardenolides*) or six membered δ-lactone (called *bufadienolides*). The sugars present in these glycosides are mainly digitoxose, cymarose, digitalose, rhamnose and sarmentose. Digitoxose, cymarose and sarmentose are 2-deoxysugars.

Cardiac glycosides are found only in a few plant families, e.g. Liliaceae, Ranunculaceae, Apocynaceae and Scrophulariaceae are the major sources of these glycosides. Among the cardiac glycosides isolated to date, digitoxin and digoxin, isolated from *Digitalis purpurea* and *Digitalis lanata*, respectively, are the two most important cardiotonics. Digitoxin and digoxin are also found in in *Strophanthus* seeds and squill. Both these cardiac glycosides are cardenolides, and the sugar present is the 2-deoxysugar digitoxose.

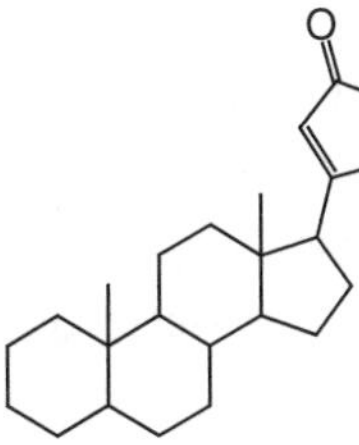

Cardenolide

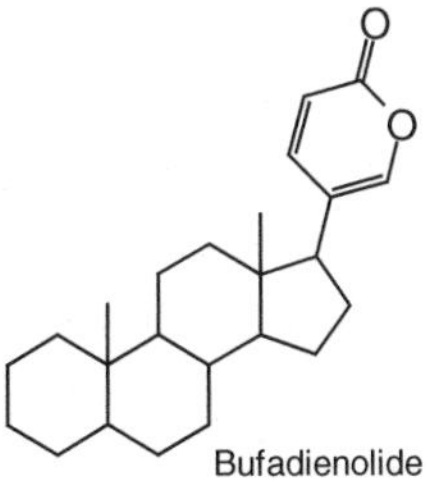

Bufadienolide

Both sugar and aglycone parts are critical for biological activity. The sugar part possibly is responsible for binding the glycoside to heart muscle, and the aglycone moiety has the desired effect on heart muscle once bound. It has been found that the lactone ring is essential for the pharmacological action. The orientation of the 3-OH groups is also important, and for more prominent activity this has to be β. In large doses these glycosides lead to cardiac arrest and can be fatal, but at lower doses these glycosides are used in the treatment of congestive heart failure.

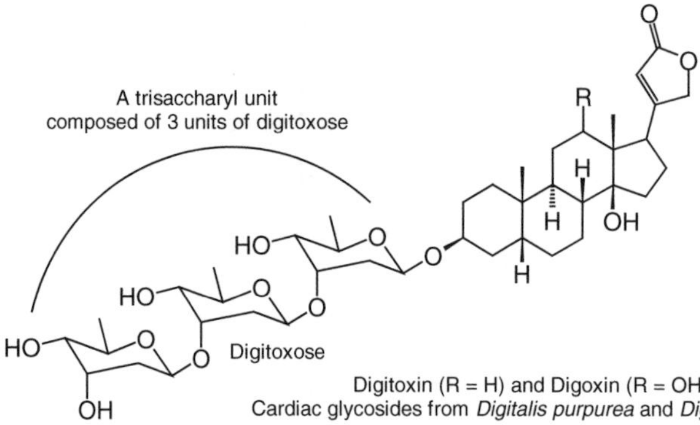

A trisaccharyl unit composed of 3 units of digitoxose

Digitoxose

Digitoxin (R = H) and Digoxin (R = OH)
Cardiac glycosides from *Digitalis purpurea* and *Digitalis lanata*

Cardiac glycosides with bufadienolide skeleton, e.g. proscillaridin A, have been found in plants (e.g. squill, *Drimia maritima*).

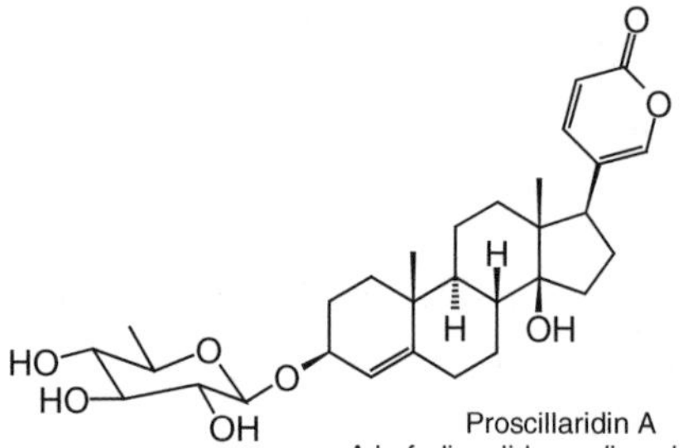

Proscillaridin A
A bufadienolide cardiac glycoside

Iridoid and secoiridoid glycosides

The iridoids and secoiridoids form a large group of plant constituents that are found usually, but not invariably, as glycosides. For example, harpagoside, an active constituent of *Harpagophytum procumbens*, is an iridoid glycoside. Plant families, e.g. Lamiaceae (especially genera *Phlomis*, *Stachys* and *Eremostachys*), Gentianaceae, Valerianaceae and Oleaceae, are good sources of these glycosides.

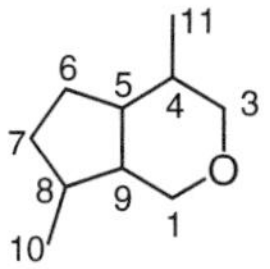

Iridoid 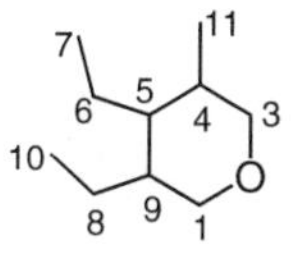 Secoiridoid 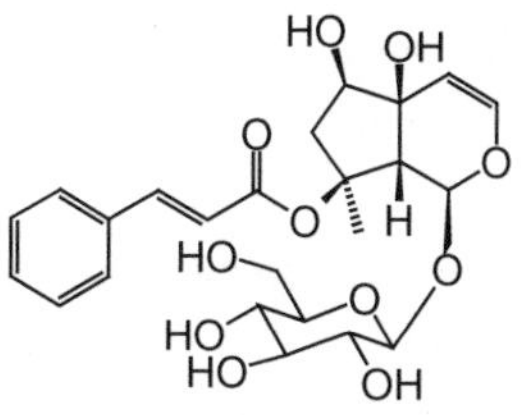 Harpagoside

In most natural iridoids and secoiridoids, there is an additional oxygenation (hydroxy) at C-1, which is generally involved in the glycoside formation.

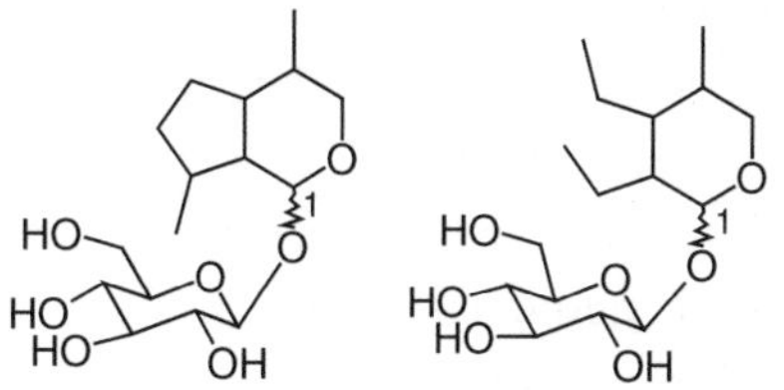

Iridoid and secoiridoid glycosides
Glycosylation is at C-1

It is also extremely common amongst natural iridoids and secoiridoids, to have a double bond between C-3 and C-4, and a carboxylation at C-11. Changes in functionalities at various other carbons in iridoid and secoiridoid skeletons are also found in nature, as shown below.

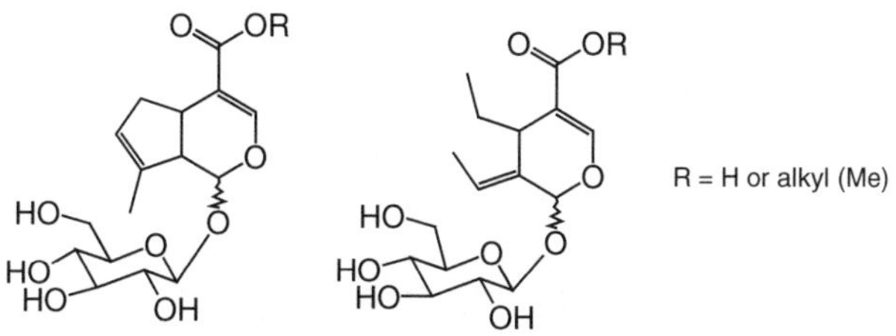

R = H or alkyl (Me)

Iridoid and secoiridoid glycosides
with modified functionalities

Some examples of plants that produce irirdoid or secoiridoid glycosides, and their medicinal uses, are summarized below.

Devil's claw (*Harpagophytum procumbens*) *Harpagophytum procumbens* is native to South Africa, Namibia and Madagascar, and traditionally used in the treatment of osteoarthritis, rheumatoid arthritis, indigestion and low back pain. This plant contains 0.5–3 per cent iridoid glycosides, harpagoside, harpagide and procumbine being the major active iridoid glycosides present.

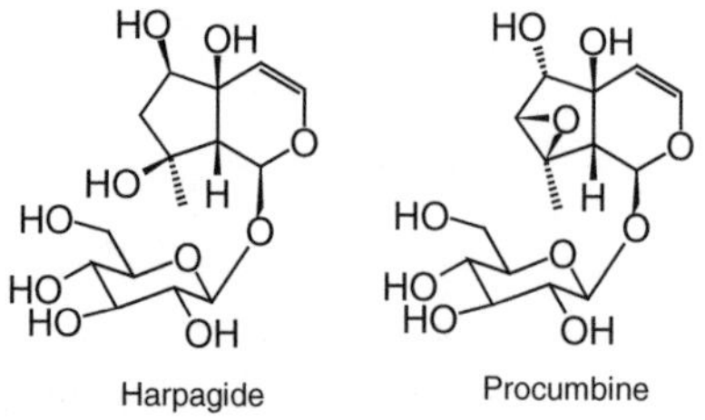

Harpagide

Procumbine

The toxicity of *H. procumbens* is considered extremely low. To date, there have been no reported side-effects following its use. However, this plant is said to have oxytocic properties and should be avoided in pregnancy. In addition, due to its reflex effect on the digestive system, it should be avoided in patients with gastric or duodenal ulcers.

Picrorhiza (*Picrorhiza kurroa*) *Picrorhiza kurroa* is a small perennial herb that grows in hilly parts of India, particularly in the Himalayas between 3000 and 5000 m. The bitter rhizomes of this plant have been used for thousands of years in *Ayurvedic traditional medicine* to treat indigestion, dyspepsia, constipation, liver dysfunction, bronchial problems and fever. It is, in combination with various metals, useful in the treatment of acute viral hepatitis. The active constituents of picrorhiza are a group of iridoid glycosides known as picrosides I–IV and kutkoside.

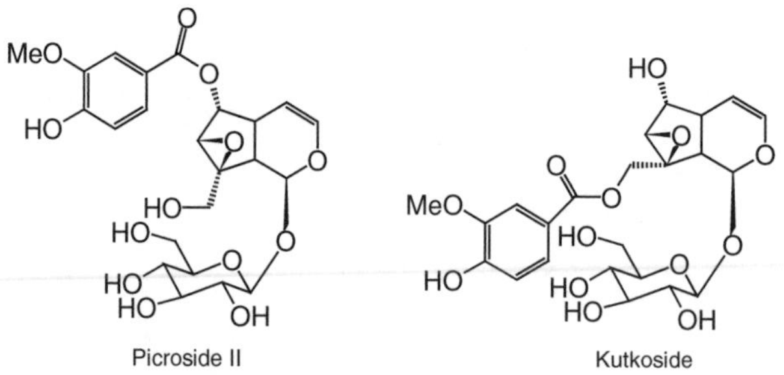

Picroside II

Kutkoside

Picrorhiza has been used widely in India, and no significant adverse reactions have been reported to date. The oral LD_{50} of *Picrorhiza* iridoid glycosides (known as 'kutkin') is greater than 2600 mg/kg in rats.

Oleuropein, a secoiridoid glycoside *Fraxinus excelsior* (ash tree), *Olea europaea* (olive tree) and *Ligustrum obtusifolium* from the family Oleaceae are the major sources of oleuropein. This compound has hypotensive, antioxidant, antiviral and antimicrobial properties. There is no known toxicity or contraindications for this compound.

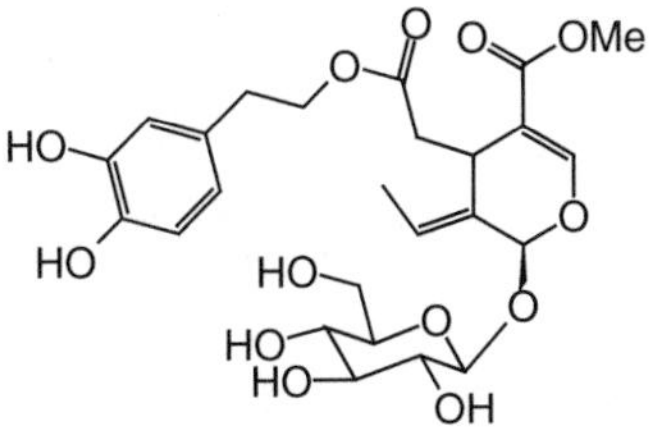

Oleuropein, a bioactive secoiridoid glycoside

6.5 Terpenoids

Terpenoids are compounds derived from a combination of two or more *isoprene* units. Isoprene is a five carbon unit, chemically known as 2-methyl-1,3-butadiene. According to the *isoprene rule* proposed by Leopold Ruzicka, terpenoids arise from *head-to-tail* joining of isoprene units. Carbon 1 is called the 'head' and carbon 4 is the 'tail'. For example, myrcene is a simple 10-carbon-containing terpenoid formed from the head-to-tail union of two isoprene units as follows.

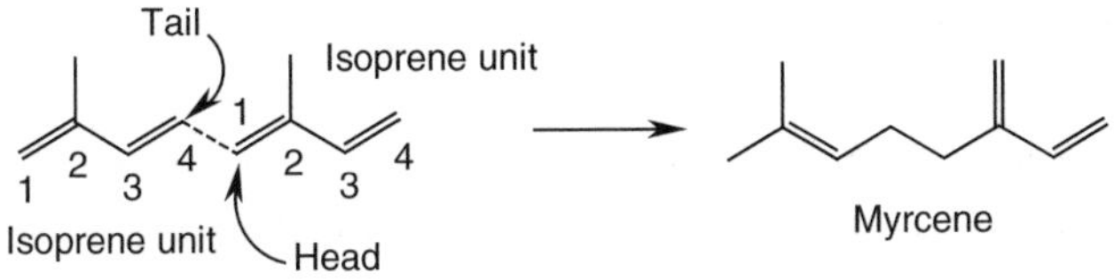

Myrcene

Terpenoids are found in all parts of higher plants and occur in mosses, liverworts, algae and lichens. Terpenoids of insect and microbial origins have also been found.

6.5.1 Classification

Terpenoids are classified according to the number of isoprene units involved in the formation of these compounds.

Type of terpenoids	Number of carbon atoms	Number of isoprene units	Example
Monoterpene	10	2	Limonene
Sesquiterpene	15	3	Artemisinin
Diterpene	20	4	Forskolin
Triterpene	30	6	α-amyrin
Tetraterpene	40	8	β-carotene
Polymeric terpenoid	Several	Several	Rubber

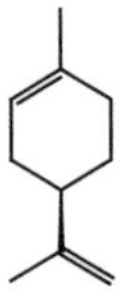

(+)-Limonene, a monoterpene

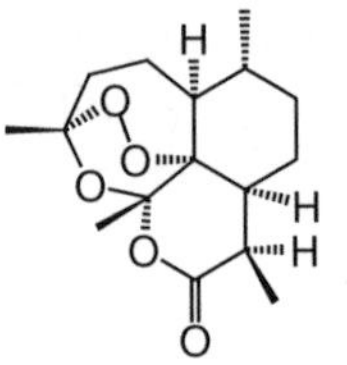

Artemisinin,
An antimalarial sesquiterpene

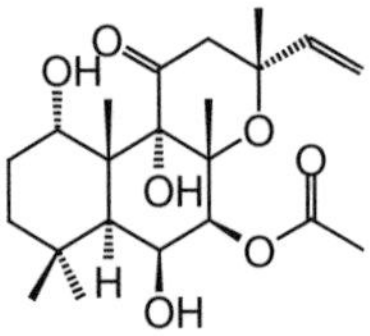

Forskolin
An antihypertensive diterpene

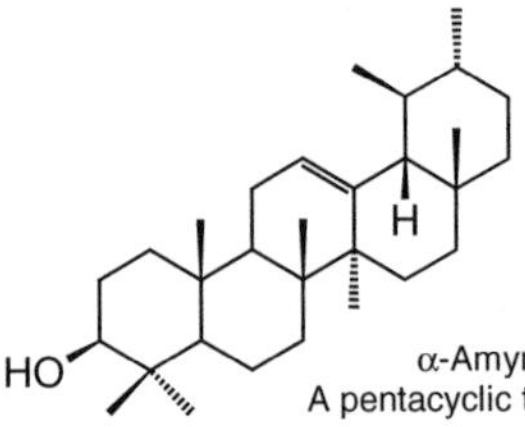

α-Amyrin
A pentacyclic triterpene

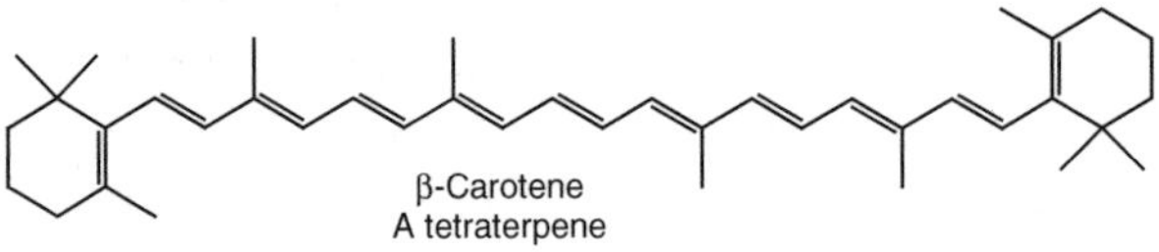

β-Carotene
A tetraterpene

6.5.2 Biosynthesis

3*R*-(+)-mevalonic acid is the precursor of all terpenoids. The enzymes mevalonate kinase and phosphomevalonate kinase catalyse phosphorylation of mevalonic acid to yield 3*R*-(+)-mevalonic acid 5-diphosphate, which is finally transformed to isopentenyl diphosphate, also known as isopentenyl pyrophosphate (IPP), by the elimination of a carboxyl and a hydroxyl group mediated by mevalonate 5-diphosphate decarboxylase. Isopentenyl pyrophosphate is isomerized by isopentenyl isomerase to dimethylallylpyrophosphate (DMAPP). A unit of IPP and a unit of DMAPP are combined together head to tail by dimethylallyl transferase to form geranyl pyrophosphate, which is finally hydrolysed to geraniol, a simple monoterpene. Geranyl pyrophosphate is the precursor of all monoterpenes.

In similar fashions, the core pathway up to C_{25} compounds (five isoprene units) is formed by sequential addition of C_5 moieties derived from IPP to a starter unit derived from DMAPP. Thus, sesquiterpenes are formed form the precursor 2*E*, 6*E*-farnesyl pyrophosphate (FPP), and diterpenes from 2*E*, 6*E*, 10*E*-geranylgeranyl pyrophosphate (GGPP). The parents of triterpenes and tetraterpenes are formed by reductive coupling of two FPPs or GGPPs, respectively. Rubbers and other polyisoprenoids are produced from repeated additions of C_5 units to the starter unit GGPP.

3R (+)-Mevalonic acid
ATP ADP
3R (+)-Mevalonic acid 5-phosphate
ATP ADP
P = phosphate group
3R (+)-Mevalonic acid 5-diphosphate
ADP ATP
$-CO_2$, -HO-P
Isopentenyl diphosphate
Or isopentenyl pyrophosphate
Isopentenyl isomerase
Dimethylallylpyrophosphate
Isopentenyl pyrophosphate + Dimethylallyl pyrophosphate
Dimethylallyl transferase
Geranyl pyrophosphate
Hydrolysis
Geraniol

6.5.3 Monoterpenes

Monoterpenes, 10-carbon-containing terpenoids, are composed of two isoprene units, and found abundantly in plants, e.g. (+)-limonene from lemon oil, and (−)-linalool from rose oil. Many monoterpenes are the constituents of plant volatile oils or essential oils. These compounds are particularly important as flavouring agents in pharmaceutical, confectionery and perfume products. However, a number of monoterpenes show various types of bioactivity and are used in medicinal preparations. For example, camphor is used in liniments against rheumatic pain, menthol is used in ointments and liniments as a remedy against itching, bitter-orange peel is used as an aromatic bitter tonic and as a remedy for poor appetite and thymol and carvacrol are used in bactericidal preparations.

Types of monoterpene

Monoterpenes occur in plants in various structural forms; some are cyclic while the others are acyclic. They also contain various types of functional group, and depending on their functional groups they can be classified as simple hydrocarbons, alcohols, ketones, aldehydes, acids or phenols. Some examples are cited below.

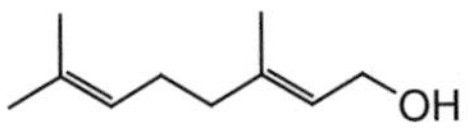

Geraniol
An acyclic monoterpene

(+)-α-Pinene
A cyclic monoterpene

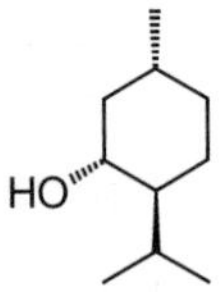

(−)-Menthol
A monoterpene alcohol

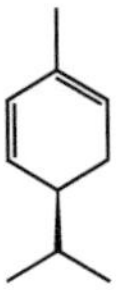

(+)-α-Phellandrene
A monoterpene hydrocarbon

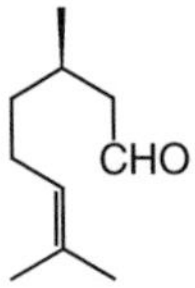

(+)-Citronellal
A monoterpene aldehyde

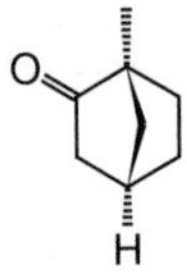

(+)-Camphor
A monoterpene ketone

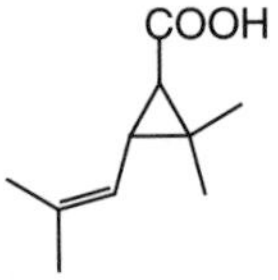

(+)-*trans*-Chrysanthemic acid
A monoterpene acid

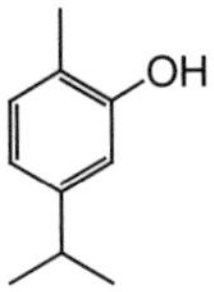

Carvacrol
A phenolic sesquiterpene

Botanical sources

A number of plants produce a variety of monoterpenes. The following table lists just a few of these sources, and their major monoterpene components.

Source		Major monoterpenes
Common name	Botanical name (Family)	
Black pepper	*Piper nigrum* (Piperaceae)	α- and β-pinene, phellandrene
Peppermint leaf	*Mentha piperita* (Lamiaceae)	Menthol, menthone
Oil of rose	*Rosa centifolia* (Rosaceae)	Geraniol, citronellol, linalool
Cardamom	*Elettaria cardamomum* (Zingiberaceae)	α-terpineol, α-terpinene
Rosemary	*Rosmarinus officinalis* (Lamiaceae)	Borneol, cineole, camphene
Bitter orange	*Citrus aurantium* (Rutaceae)	(+)-limonene, geranial
Camphor	*Cinnamomum camphora* (Lauraceae)	(+)-camphor
Caraway	*Carum carvi* (Apiaceae)	(+)-carvone, (+)-limonene
Thyme	*Thymus vulgaris* (Lamiaceae)	Thymol, carvacrol

6.5.4 Sesquiterpenes

Sesquiterpenes, 15-carbon-containing terpenoids, are composed of three isoprene units, and found abundantly in plants, e.g. artemisinin from

Artemisia annua and (−)-α-bisabolol from *Matricaria recutita* (German chamomile). Addition of IPP to GPP produces 2*E*,6*E*-farnesylpyrophosphate (FPP), the precursor for all sesquiterpenes. Farnesylpyrophosphate can cyclize by various cyclase enzymes in various ways, leading to the production of a variety of sesquiterpenes. Some of these sesquiterpenes are medicinally important bioactive compounds. For example, (−)-α-bisabolol and its derivatives have potent anti-inflammatory and spasmolytic properties, and artemisinin is an antimalarial drug.

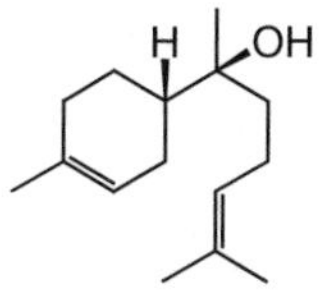

(–)-α-Bisabolol

Structural types

Sesquiterpenes can be of various structural types, some of which are presented with specific examples in the following table.

Major structural types	Specific examples
Simple farnesane-type acyclic sesquiterpenes	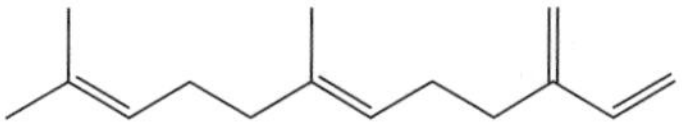 *trans*-β-Farnesene A potent aphid-repellant found in hops and sweet potatoes
Furanoid farnesane sesquiterpenes	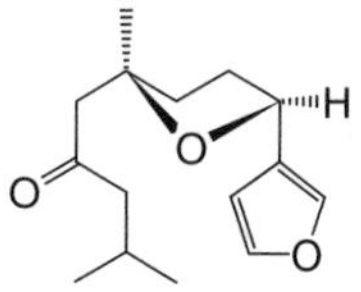 Ipomeamarone A phytoalexin
Cyclobutane and cyclopentane sesquiterpenes	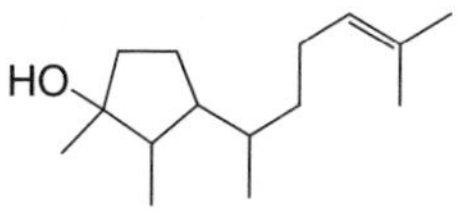 Cyclonerodiol A fungal metabolite
Cyclofarnesane-type sesquiterpenes	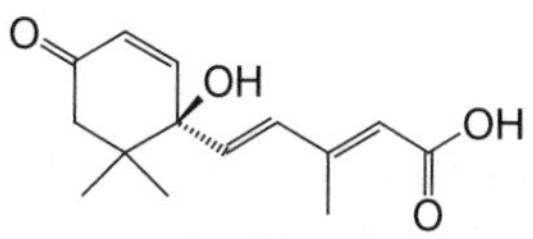 Abscisic acid A plant-growth regulatory agent

(Continued)

Major structural types	Specific examples
Bisabolane sesquiterpenes	(–)-α-Bisabolol; (–)-β-Bisabolene
Cyclobisabolane sesquiterpenes	Sesquicaren
Elemane sesquiterpenes	Curzerenone A component of turmaric; β-Elemene A component of turmaric
Germacranes	Germacrene A; Parthenolide; Germacrene D
Lepidozanes and bicyclogermacrane sesquiterpenes	Bicyclogermacrene
Humulane sesquiterpenes	Humulene A sesquiterpene from hops
Caryophyllane sesquiterpenes	β-Caryophyllene A bioactive component of *Cinnamomum zeylanicum*
Cuparane and cyclolaurane sesquiterpenes	Cuparene; 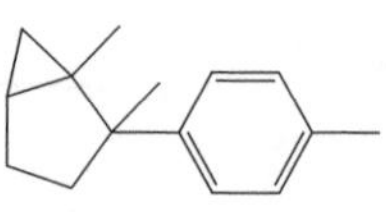 Cyclolaurene

(Continued)

Major structural types	Specific examples
Laurane sesquiterpenes	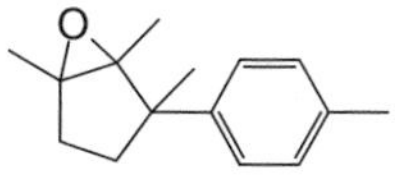 Curcumene ether; 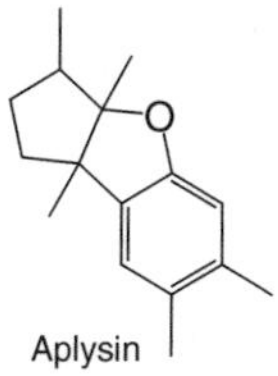 Aplysin
Trichothecane sesquiterpenes	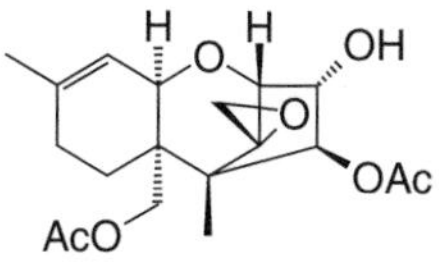 Diacetoxyscirpenol A fungal toxin
Eudesmane sesquiterpenes	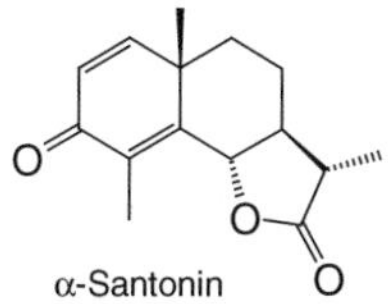 α-Santonin An anthelmintic component of *Artemisia* species; 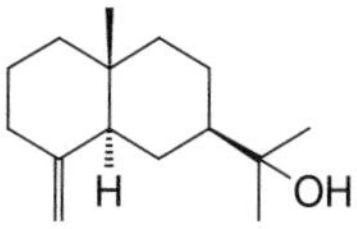 β-Eudesmol
Emmotin sesquiterpenes	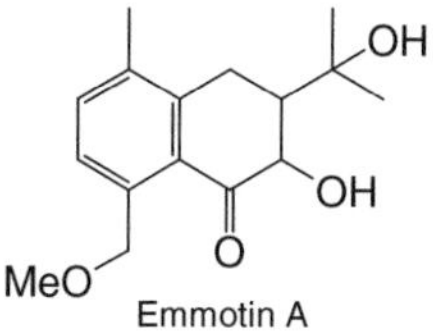 Emmotin A; 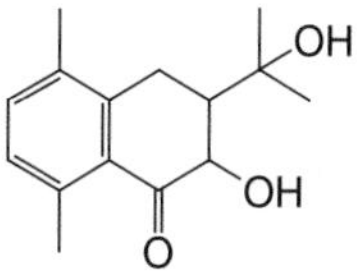 Emmotin F
Oppositane sesquiterpenes	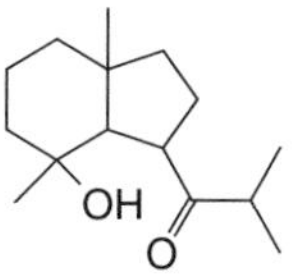 4-Hydroxy-7-oppositanone; 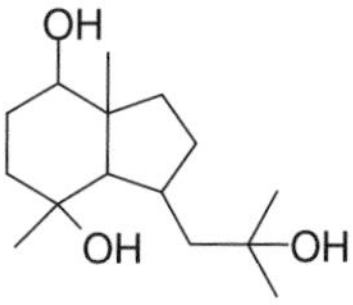 1β,4β,11-Oppositanetriol
Cycloeudesmane sesquiterpenes	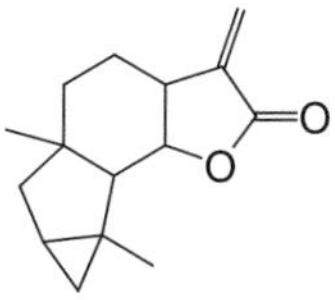 Brothenolide; 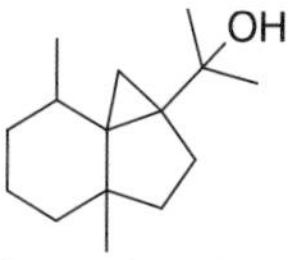 Cycloeudesmol
Eremophilane sesquiterpenes	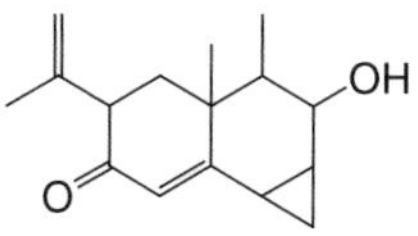 Eremofortin A; 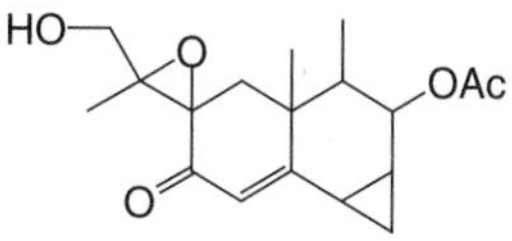 Eremofortin C

(Continued)

Major structural types	Specific examples	
Aristolane sesquiterpenes	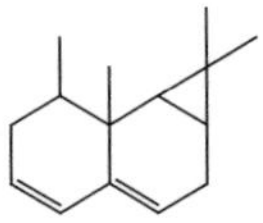 1,9-Aristoladiene	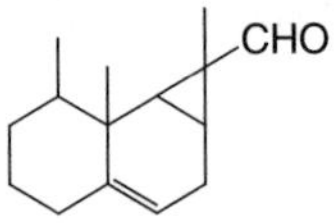 9-Aristolen-12-al
Nardosinane sesquiterpenes	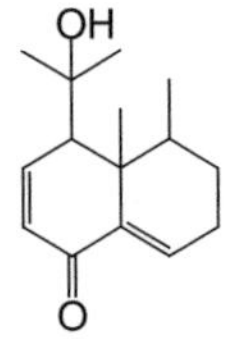 Kanshone A	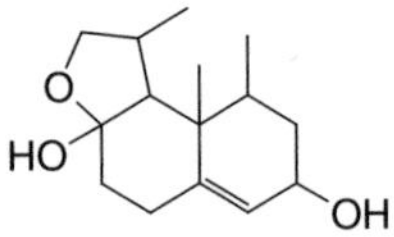 Lemnacarnol
Cacalol sesquiterpenes	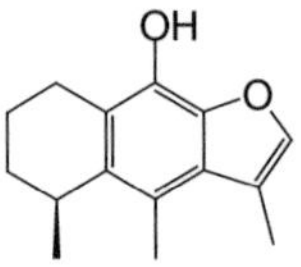 Cacalol	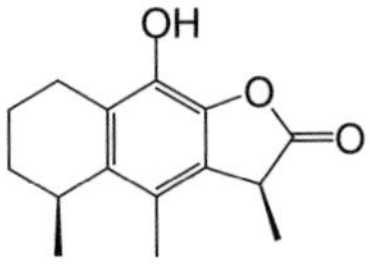 Cacalolide
Cadinane sesquiterpenes	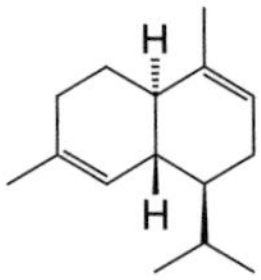 α-Cadinene	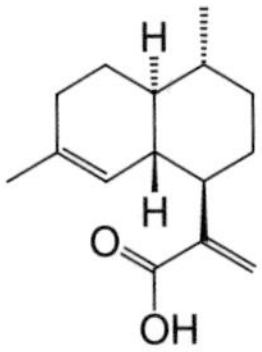 Artemisinic acid
Alliacane sesquiterpenes	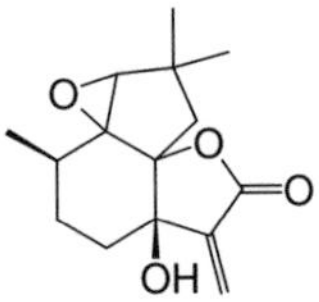 Alliacol A	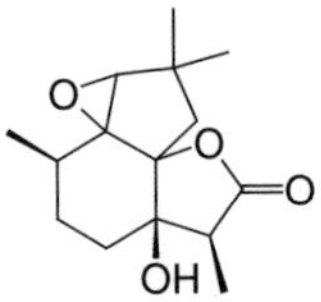 Alliacolide
Oplopane sesquiterpenes	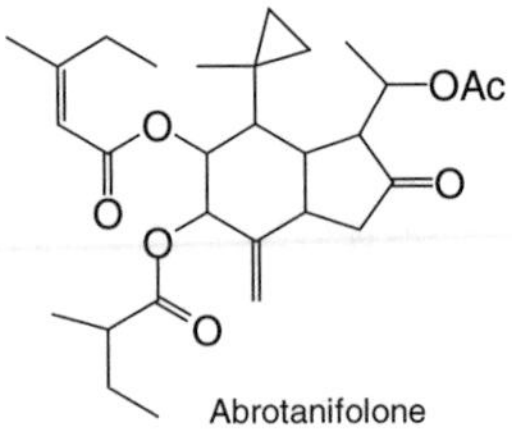 Abrotanifolone	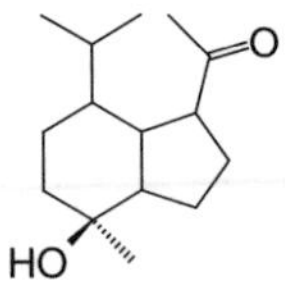 10-Hydroxy-4-oplopanone
Drimane sesquiterpenes	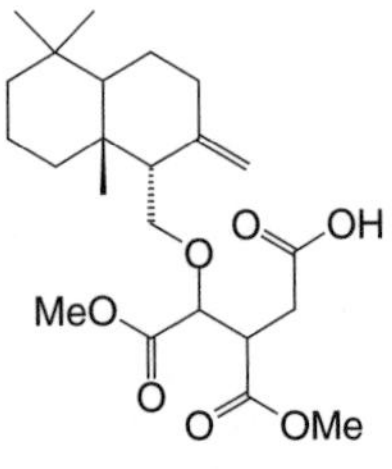 Cryptoporic acid A	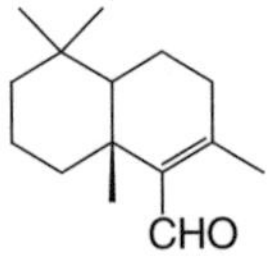 8-Drimen-11-al 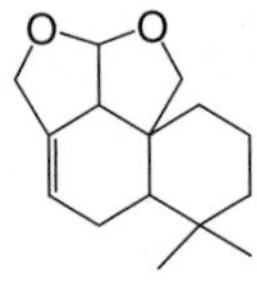 Marasmene

(Continued)

Major structural types	Specific examples
Xanthane sesquiterpenes	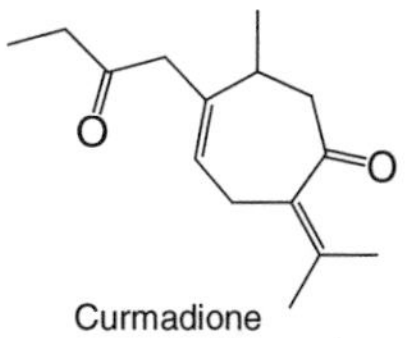 Curmadione A component of turmaric
Carabrane sesquiterpenes	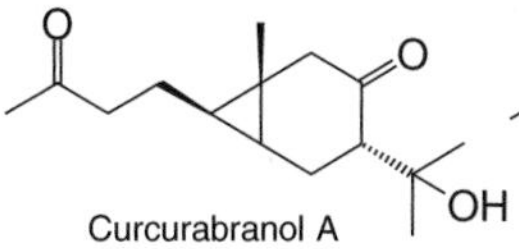 Curcurabranol A; 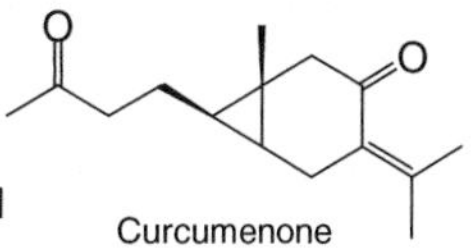 Curcumenone
Guaiane sesquiterpenes	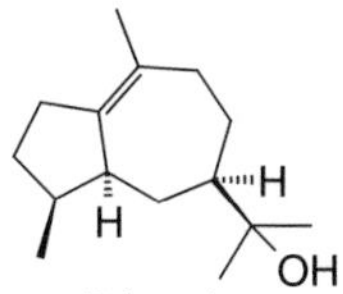 Bulnesol; 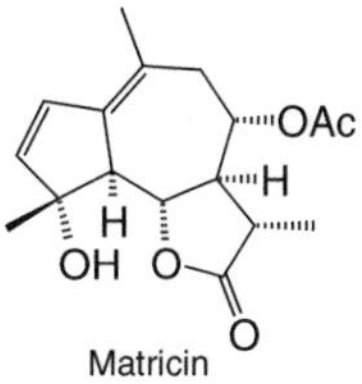 Matricin
Aromadendrane sesquiterpenes	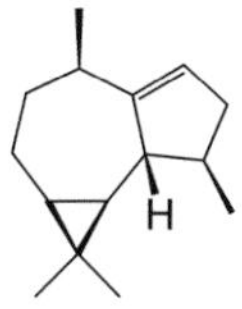 1-Aromadendrene; 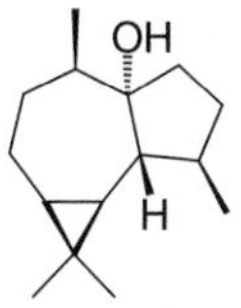 1-Aromadendranol
Patchoulane and rearranged patchoulane sesquiterpenes	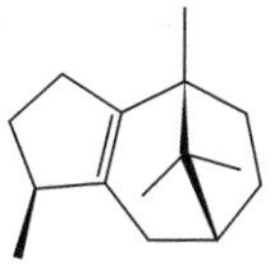 β-Patchoulene
Valerenane sesquiterpenoids	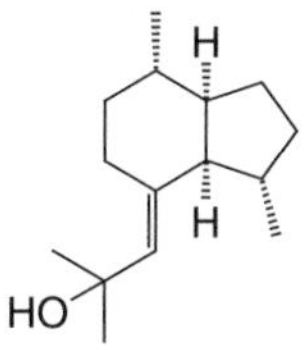 6-Valerenen-11-ol; 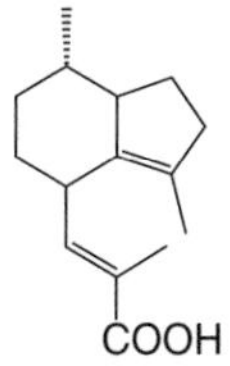 Valerenic acid
Africanane sesquiterpenes	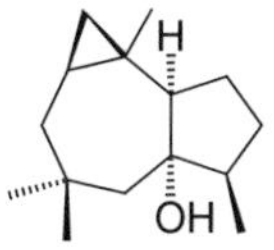 2-Africananol; 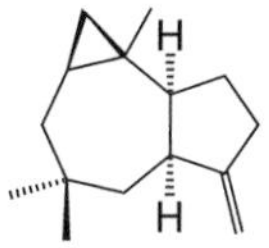 3(15)-Africanene
Lippifoliane and himachalane sesquiterpenes	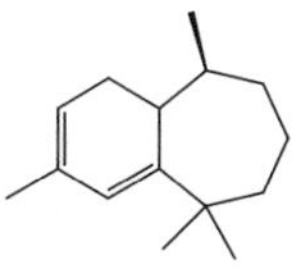 1,3-Himachaladiene; 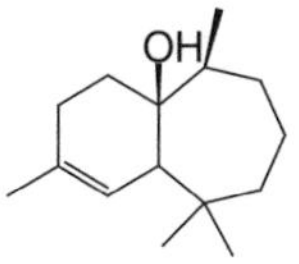 2-Himachalen-6-ol

(Continued)

Major structural types	Specific examples
Longipinane sesquiterpenes	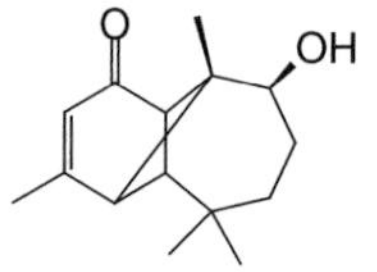 8β-Hydroxy-3-longipinen-5-one; 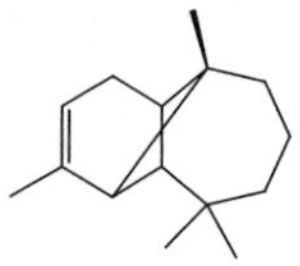 3-Longipinene
Longifolane sesquiterpenes	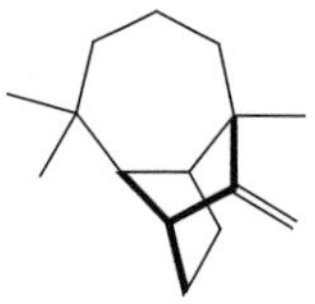 (+)-Longifolene
Pinguisane sesquiterpenes	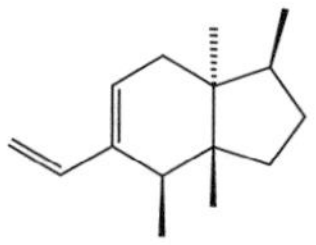 5,10-Pinguisadiene; 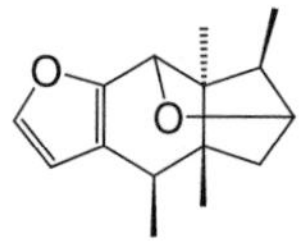 Pinguisanin
Picrotoxane sesquiterpenes	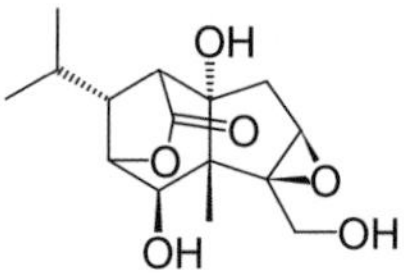 Amoenin; 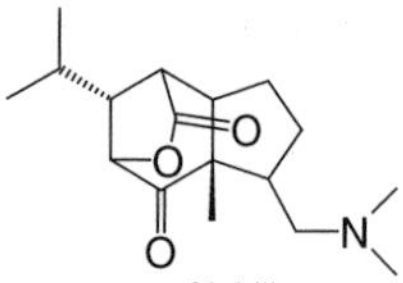 Nobiline
Daucane and isodaucane sesquiterpenes	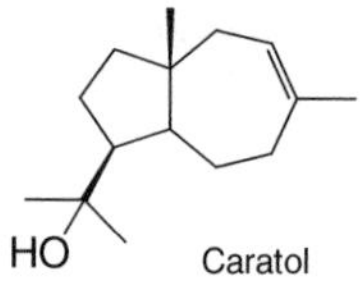 Caratol
Illudane and protoilludane sesquiterpenes	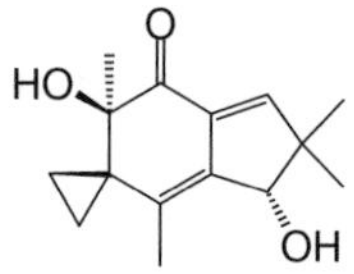 Illudin S A component of luminescent mushrooms
Sterpurane sesquiterpenes	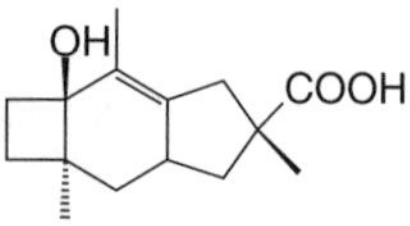 6-Hydroxy-6-sterpuren-12-oic acid
Illudalane sesquiterpenes	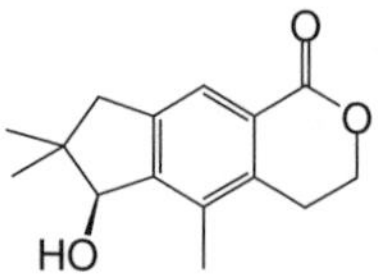 Calomelanolactone; 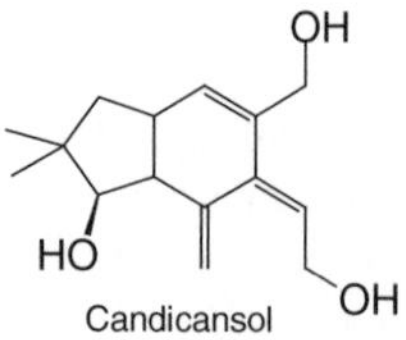 Candicansol

(Continued)

Major structural types	Specific examples
Isolactarane, merulane, lactarane and marasmane sesquiterpenes	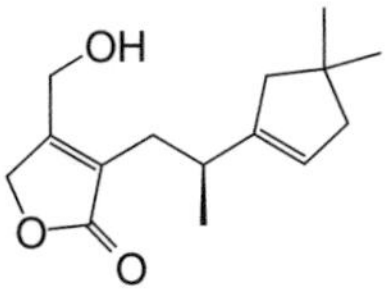 Blennin C; 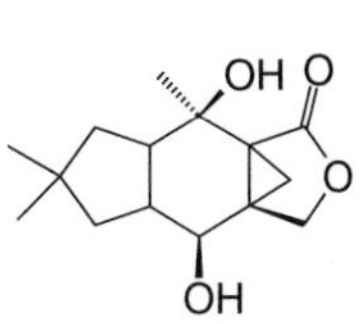 Lacterorufin C
Botrydial sesquiterpenes	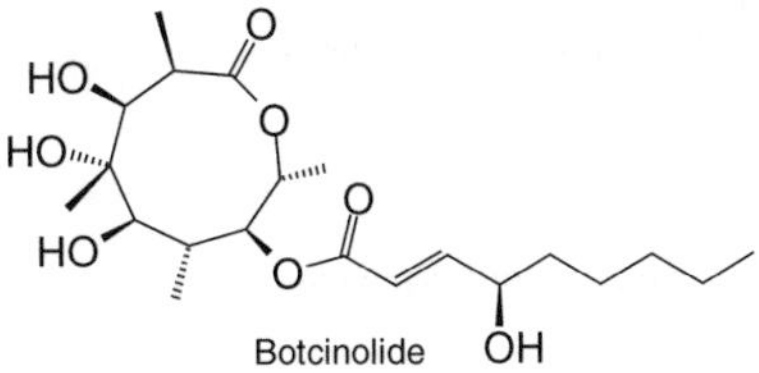 Botcinolide
Spirovetivane sesquiterpenes	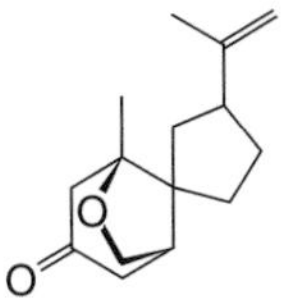 Cyclodehydroisolubimin; 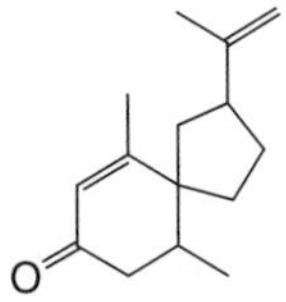 1(10), 7(11)-Spirovetivadien-2-one
Acorane sesquiterpenes	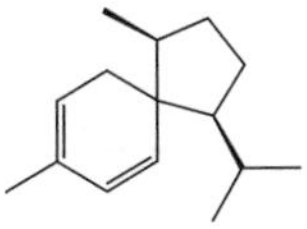 3,5-Acoradiene; 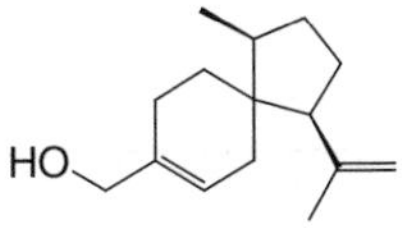 3,11-Acoradien-15-ol
Chamigrane sesquiterpenes	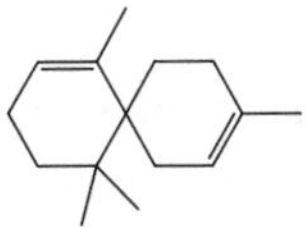 2,7-Chamigradiene;  Majusculone
Cedrane and isocedrane sesquiterpenes	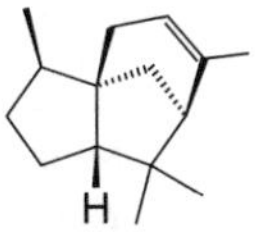 α-Cedrene; 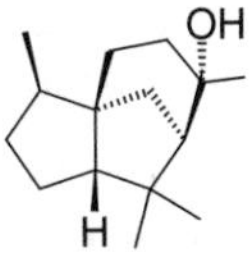 Cedrol
Precapnellane and capnellane sesquiterpenes	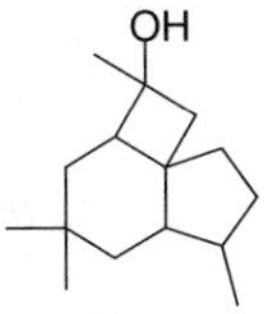 Viridianol; 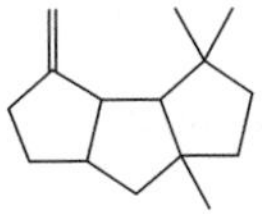 9,12-Capnellene
Hirsutane and rearranged hirsutane sesquiterpenes	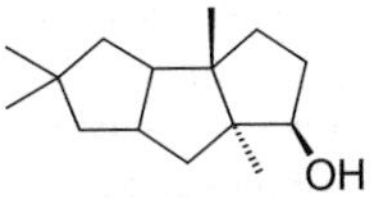 Ceratopicanol; 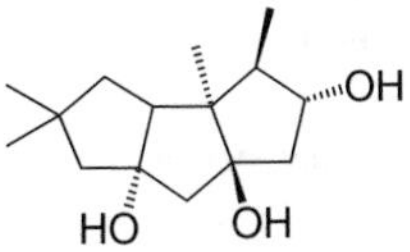 5α,7β,9α-Hirsutanetriol

(Continued)

Major structural types	Specific examples
Pentalenane sesquiterpenes	 Pentalenene; 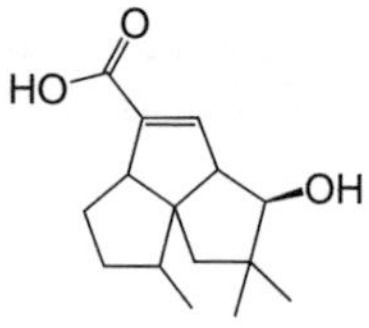 Pentalenic acid
Campherenane, α-santalane and β-santalane sesquiterpenes	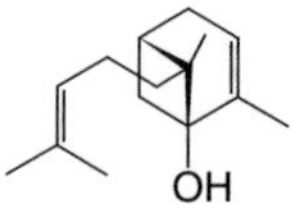 α-Bergamotene; 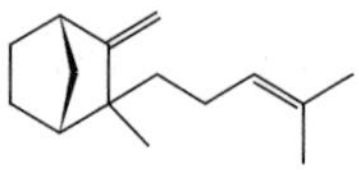 α-Santalene
Copaane sesquiterpenes	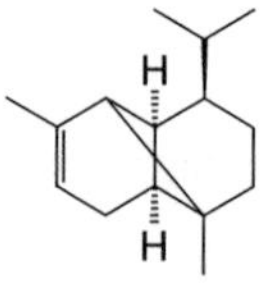 3-Copaene; 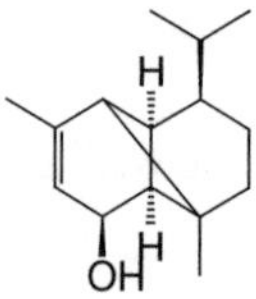 3-Copaen-2-ol
Gymnomitrane sesquiterpenes	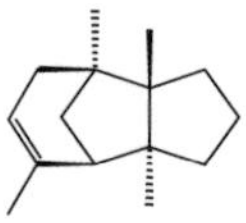 3-Gymnomitrene; 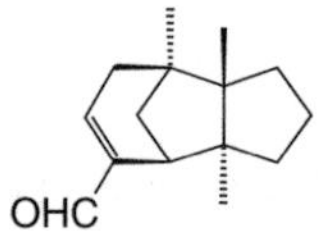 3-Gymnomitren-15-al

Botanical sources

Plants produce a variety of sesquiterpenes. The following table lists just a few of these sources, and their major sesquiterpene components.

Source		Major sesquiterpenes
Common name	Botanical name (Family)	
German chamomile	*Matricaria recutita* (Asteraceae)	−α-bisabolol and its derivatives
Feverfew	*Tanacetum parthenium* (Asteraceae)	Farnesene, germacrene D, parthenolide
Qinghao	*Artemisia annua* (Asteraceae)	Artemisinin and its derivatives
Holy thistle	*Cnicus benedictus* (Asteraceae)	Cnicin
Cinnamon	*Cinnamomum zeylanicum* (Lauraceae)	β-caryophyllene
Cloves	*Syzygium aromaticum* (Myrtaceae)	β-caryophyllene
Hop	*Humulus lupulus* (Cannabaceae)	Humulene
Wormseed	*Artemisia cinia* (Asteraceae)	α-santonin
Valerian	*Valeriana officinalis* (Valerianaceae)	Valeranone
Juniper berries	*Juniperus communis* (Cupressaceae)	α-cadinene
Curcuma or turmeric	*Cucuma longa* (Zingiberaceae)	Curcumenone, curcumabranol A, curcumabranol B, β-elemene, curzerenone

6.5.5 Diterpenes

The diterpenoids constitute a large group of 20-carbon-containing compounds derived from 2*E*,6*E*,10*E*-geranylgeranyl pyrophosphate (GGPP) or its allylic geranyl linaloyl isomer through condensation of IPP with 2*E*,6*E*-FPP. They are found in higher plants, fungi, insects and marine organisms.

One of the simplest and most significant of the diterpenes is phytol, a reduced form of geranylgeraniol, which constitutes the lipophilic side-chain of the chlorophylls. Phytol also forms a part of vitamin E (tocopherols) and K molecules. Vitamin A is also a 20-carbon-containing compound, and can be regarded as a diterpene. However, vitamin A is formed from a cleavage of a tetraterpene. Among the medicinally important diterpenes, paclitaxel, isolated from *Taxus brevifolia* (family Taxaceae), is one of the most successful anticancer drugs of modern time.

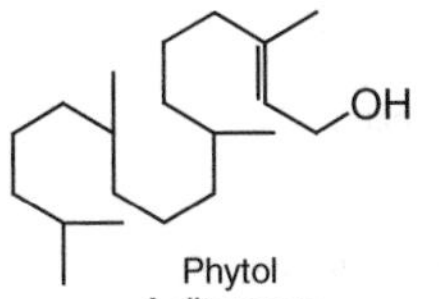

Phytol
A diterpene

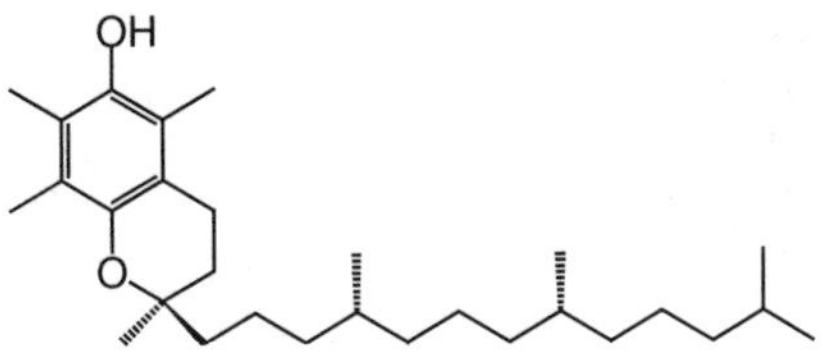

α-Tocopherol
A member of the vitamin E group

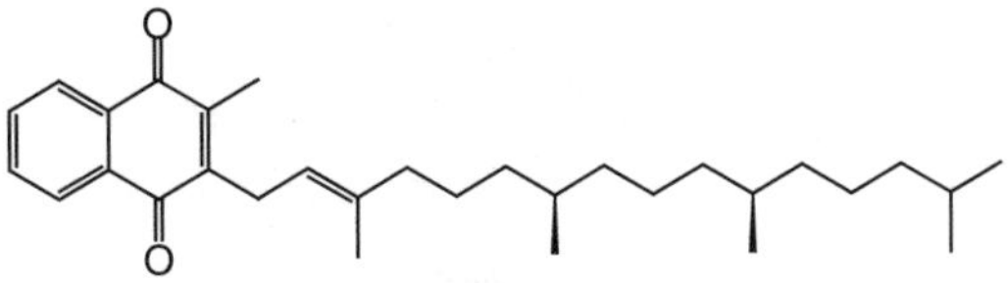

Vitamin K_1
Contains a diterpenoidal part

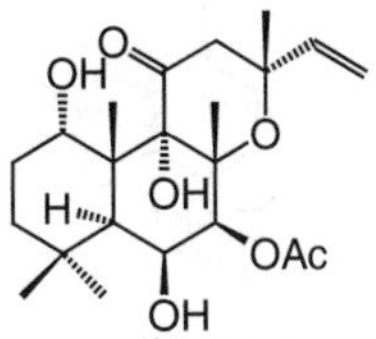

Forskolin
An antihypertensive agent

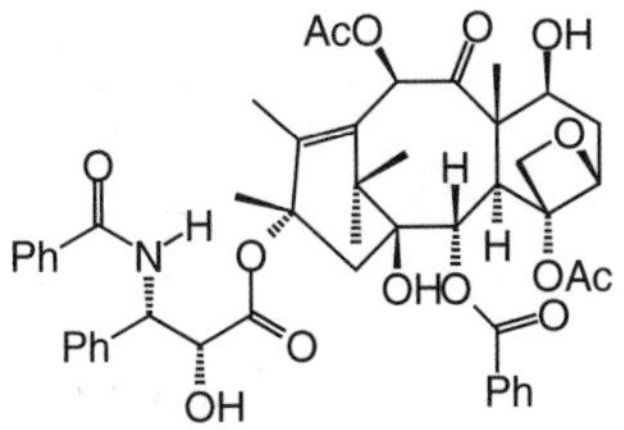

Paclitaxel or Taxol
An anticancer drug

Major structural types

While there are a number of acyclic diterpenes such as phytol, cyclization of these acyclic diterpenes, driven by various enzymes, leads to the formation of cyclic diterpenes. A number of other biogenetic reactions, e.g. oxidation, also bring in variation among these cyclic diterpenes.

Some of the major structural types encountered in diterpenes are shown below.

Abietane diterpenoids

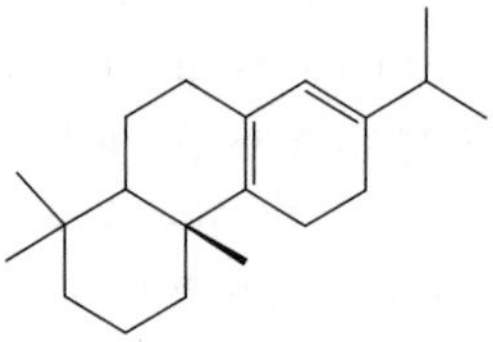

8,13-Abietadiene

Amphilectan diterpenoids

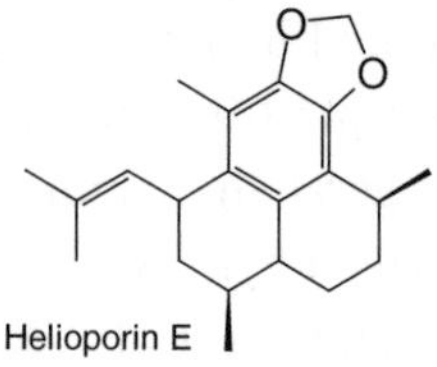

Helioporin E

Beyerane diterpenoids

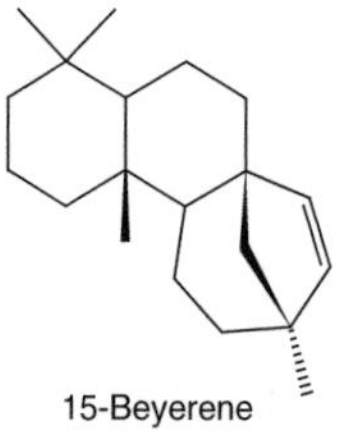

15-Beyerene

Briarane diterpenoids

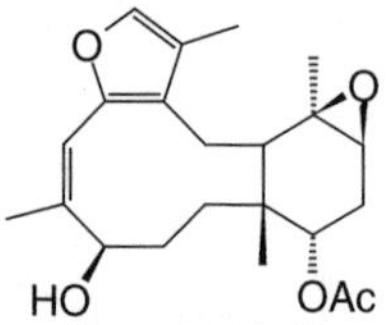

HO OAc
Verecynarmin G

Cassane diterpenoids

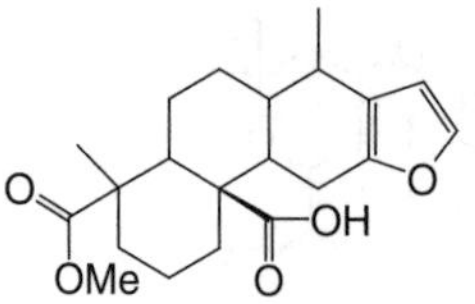

O OMe O OH O O
Caesaljapin

Cembrane diterpenoids

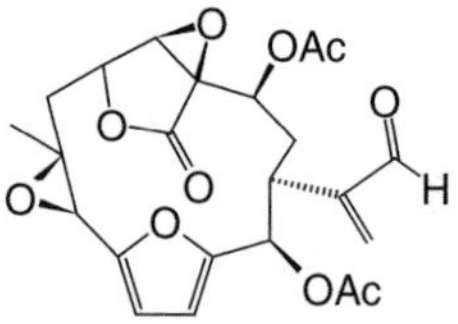

O OAc O O O O H O O OAc
Bippinatin B

Cleistanthane diterpenoids

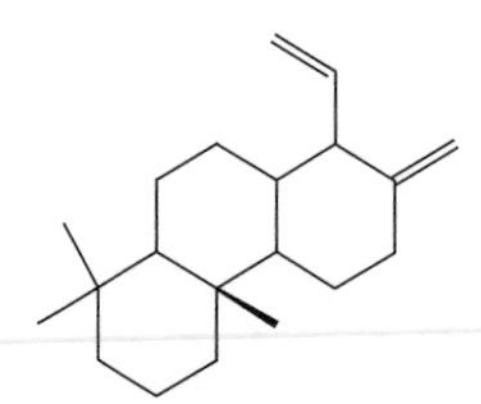

13(17), 15-Cleistanthadiene

Cyathane diterpenoids

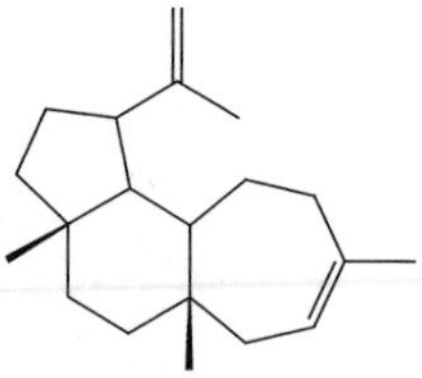

12, 18-Cyathadiene

Daphnane diterpenoids

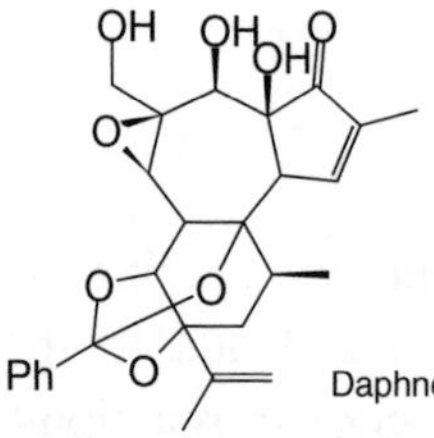

OH OH OH O O O O Ph O
Daphnetoxin

Dolabellane diterpenoids

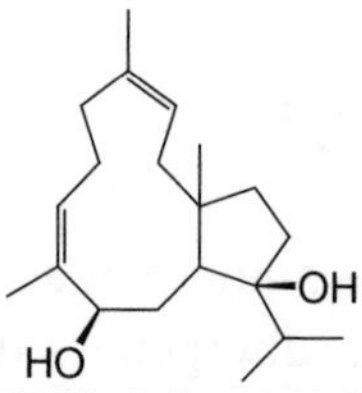

OH HO
3,7-Dolabelladiene-9,12-diol

Dolastane diterpenoids

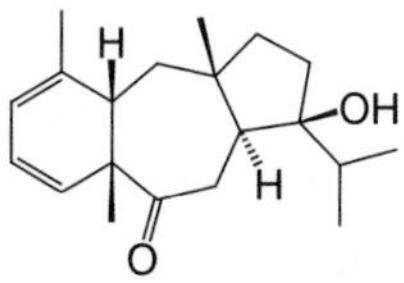

9-Hydroxy-1,3-dolastadien-6-one

Eunicellane and asbestinane diterpenoids

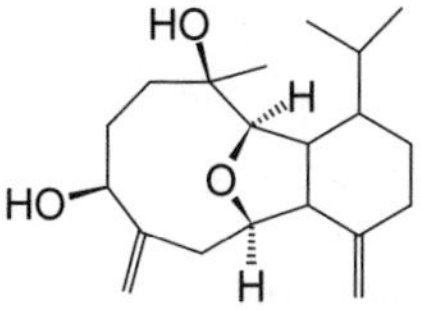

6,13-Epoxy-4(18), 8(19)-eunicelladiene-9,12-diol

Fusicoccane diterpenoids

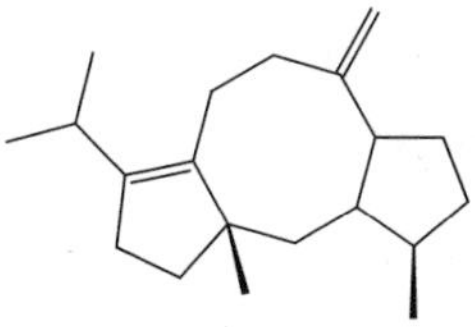

7(17), 10(14)-Fusicoccadiene

Gibberellins

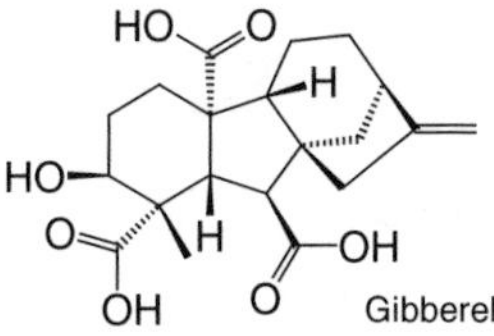

Gibberellin A_{13}

Isocopalane diterpenoids

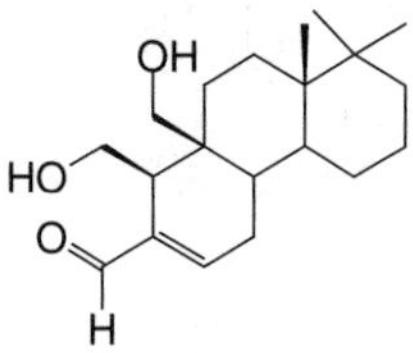

15,17-Dihydroxy-12-isocopalen-16-al

Jatrophane diterpenoids

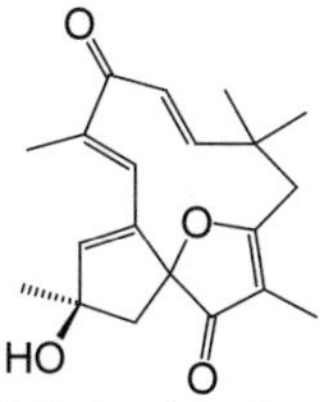

2β-Hydroxyjatrophone

Kaurane and phyllocladane diterpenoids

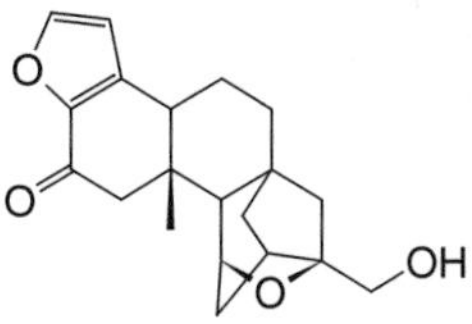

Bengalensol

Labdane diterpenoids

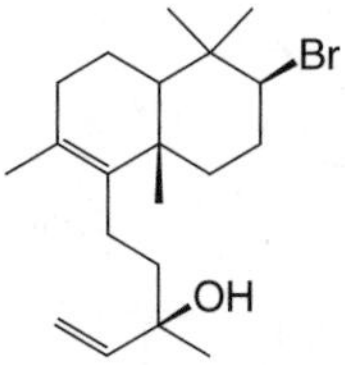

3-Bromo-7,14-labdadien-13-ol

Lathyrane diterpenoids

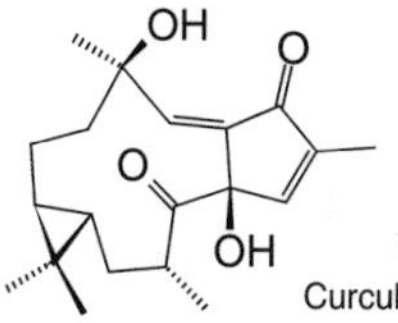

Curculathyrane A

Lobane diterpenoids

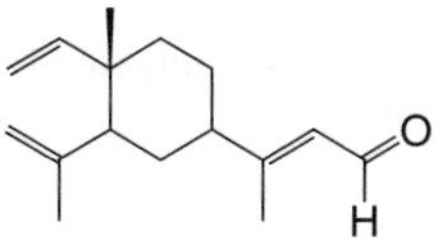

Lobophytal

Pachydictyane diterpenoids

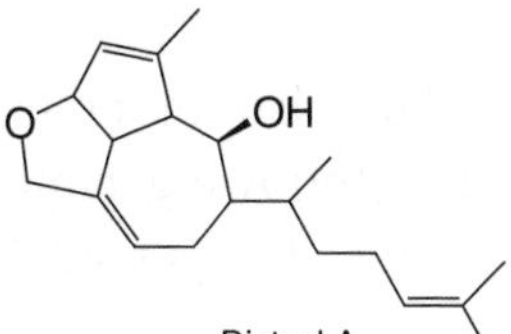

Dictyol A

Phytane diterpenoids

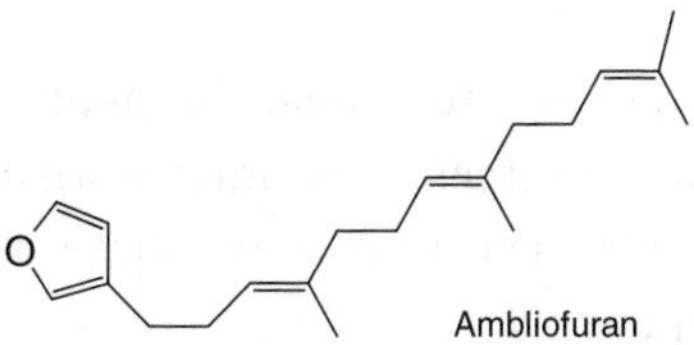

Ambliofuran

Pimarane diterpenoids

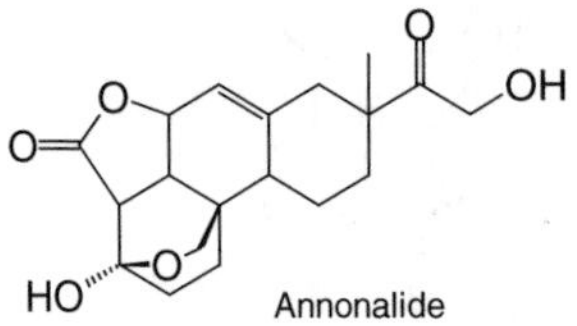

Annonalide

Podocarpane diterpenoids

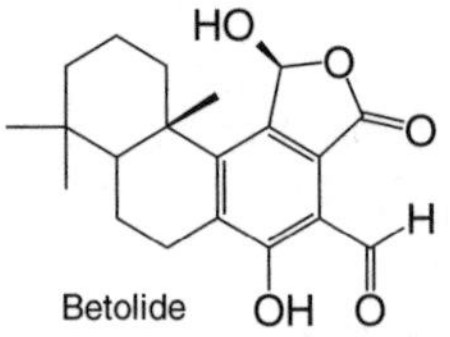

Betolide

Prenylgermacrane diterpenoids

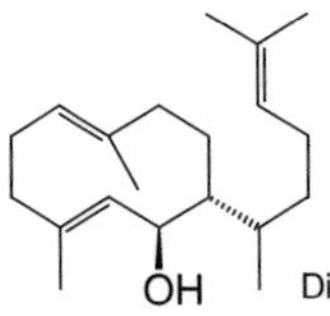

Dilophol

Serrulatane and biflorane diterpenoids

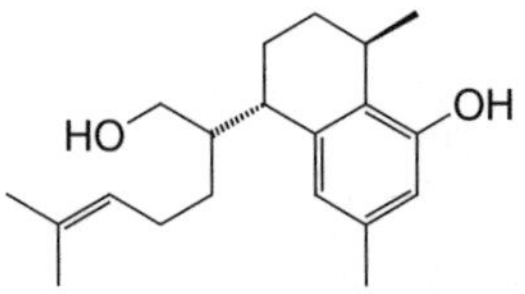

14-Serrulatene-8,18-diol

Sphenolobane diterpenoids

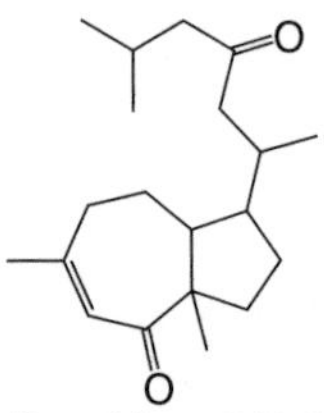

3-Spenolobene-5,16-dione

Taxane diterpenoids

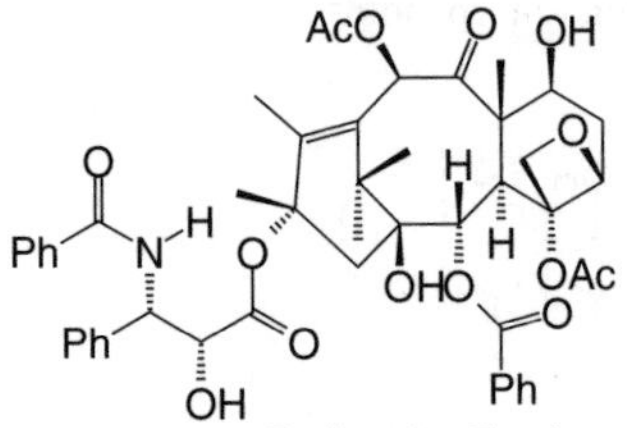

Paclitaxel or Taxol

Tigliane and ingenane diterpenoids

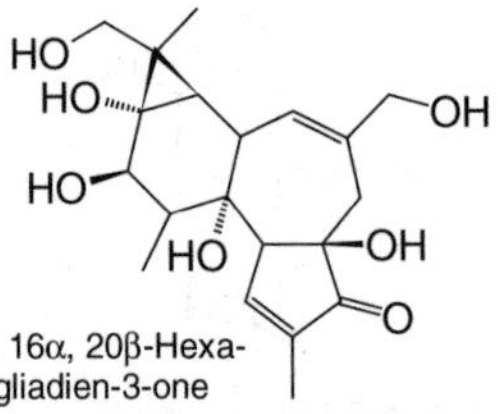

4β, 9α, 12β, 13α, 16α, 20β-Hexa-
-hydroxy-1,6-tigliadien-3-one

Verrucosane diterpenoids

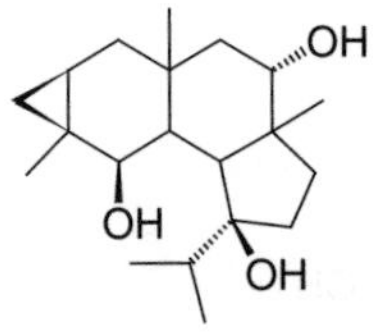

2β, 9α, 13β-Verrucosanetriol

Xenicane and xeniaphyllane diterpenoids

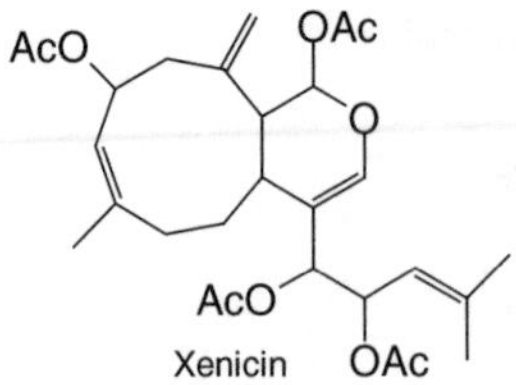

Xenicin

Botanical sources

Diterpenes are found in nature, mainly in plants, but also in other natural sources, e.g. micro-organisms and insects. The following table presents just a few of these sources, and their major diterpenoidal components.

	Source	Major diterpenes
Common name	Botanical name (Family)	
Yew tree	*Taxus brevifolia* (Taxaceae)	Paclitaxol
Sawada fungus	*Gibberella fujikuroi*	Gibberellins
Coleus	*Coleus forskohlii* (Lamiaceae)	Forskolin
Stevia	*Stevia rebaudiana* (Asteraceae)	Stevioside
Ginkgo	*Ginkgo biloba* (Ginkgoaceae)	Ginkgolides

6.5.6 Triterpenes

The triterpenoids encompass a large and diverse group of naturally occurring 30-carbon-atom-containing compounds derived from squalene or, in the case of 3β-hydroxytriterpenoids, the 3*S*-isomer of squalene 2,3-epoxide. Two molecules of farnesyl pyrophosphate are joined tail-to-tail to yield squalene. The conformation that all-*trans*-squalene 2,3-epoxide adopts, when the initial cyclization takes place, determines the stereochemistry of the ring junctions in the resulting triterpenoids. The initially formed cation intermediate may undergo a series of 1,2-hydride and methyl migrations, commonly called backbone rearrangements, to provide a variety of skeletal types.

A number of triterpenoids are bioactive compounds and are used in medicine. For example, fusidic acid is an antimicrobial fungal metabolite, isolated from *Fusidium coccineum*, and cytotoxic dimeric triterpenoids, crellastatins, are isolated from marine sponges *Crella* species.

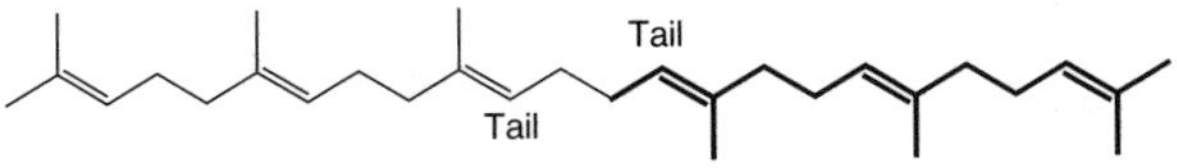

Squalene
Formed from tail-to-tail combination of 2 FPP

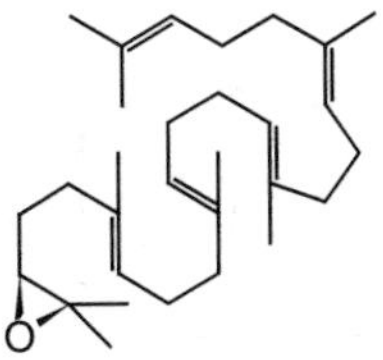

Squalene 2,3-epoxide

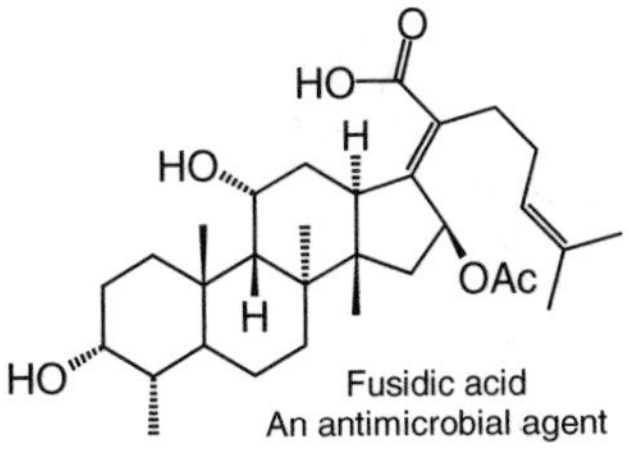

Fusidic acid
An antimicrobial agent

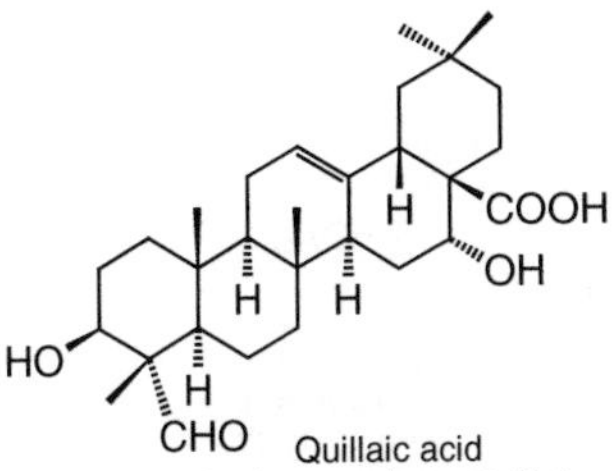

Quillaic acid
A triterpene from *Quillaja*

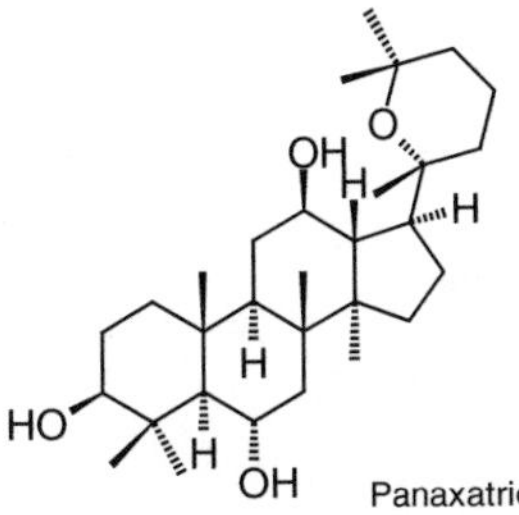

Panaxatriol
A triterpene from *Panax ginseng*

Major structural types

While squalene, the parent of all triterpenoids, is a linear acyclic compound, the majority of triterpeneoids exist in cyclic forms, penta- and tetracyclic triterpenes being the major types. Within these cyclic triterpenoids distinct structural variations lead to several structural classes of triterpenoids. Some of the major structures types of triterpenoids are shown below.

Tetracyclic triterpenes

Apotirucallane

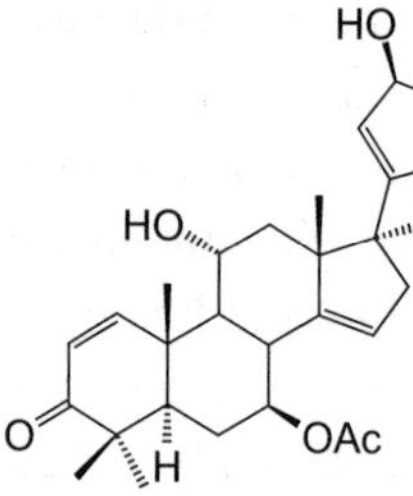

Azadirachtol

Cucurbitane

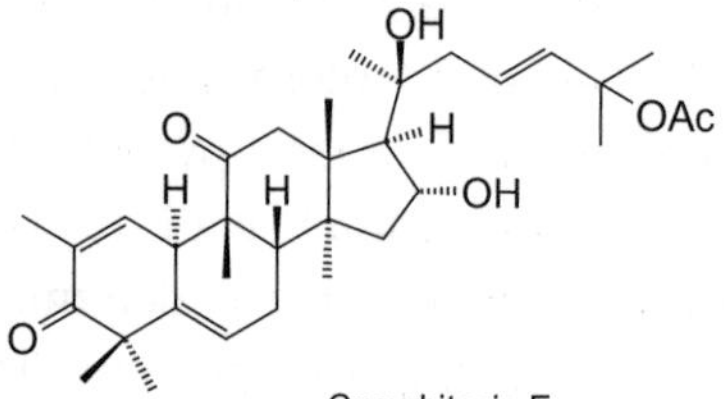

Cucurbitacin E
Occurs in the family Cucurbitaceae

Cycloartane

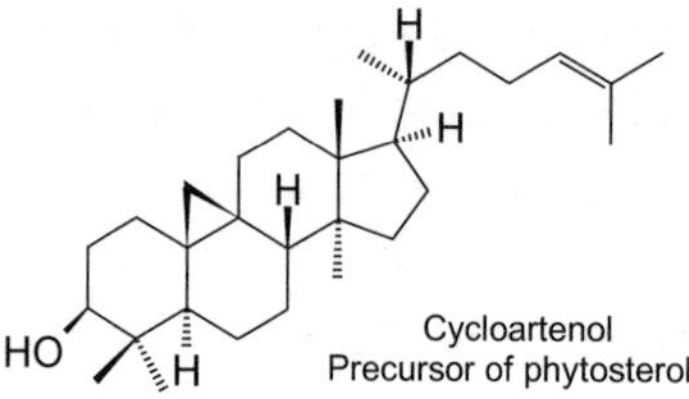

Cycloartenol
Precursor of phytosterols

Dammarane

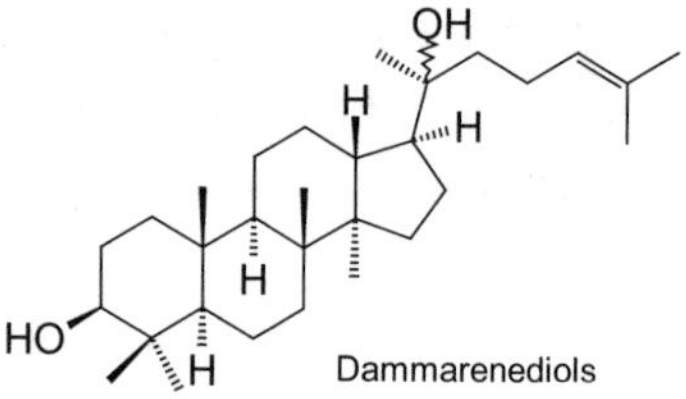

Dammarenediols

Euphane

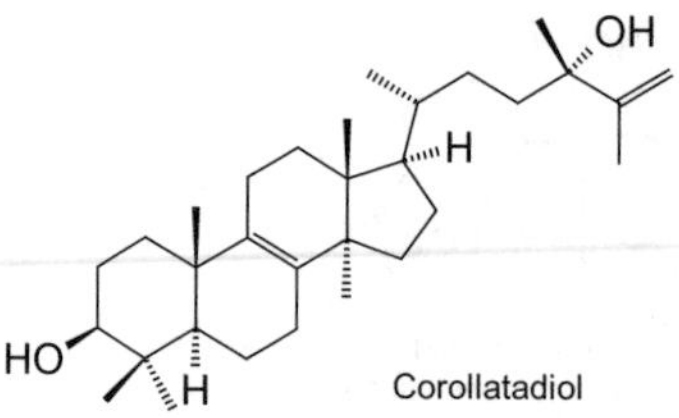

Corollatadiol

Lanostane

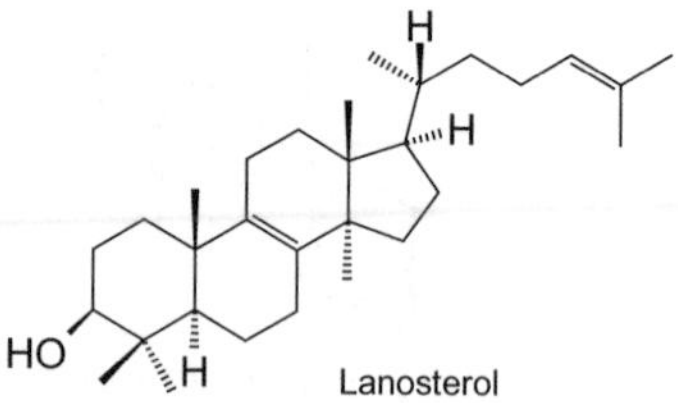

Lanosterol

Prostostane and fusidane

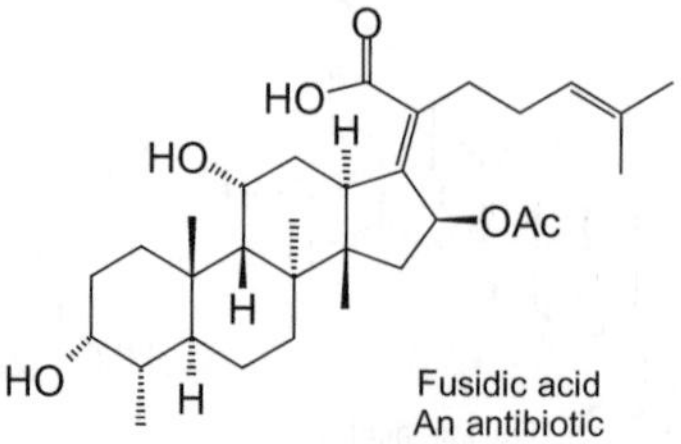

Fusidic acid
An antibiotic

Pentacyclic triterpenes

Friedelane

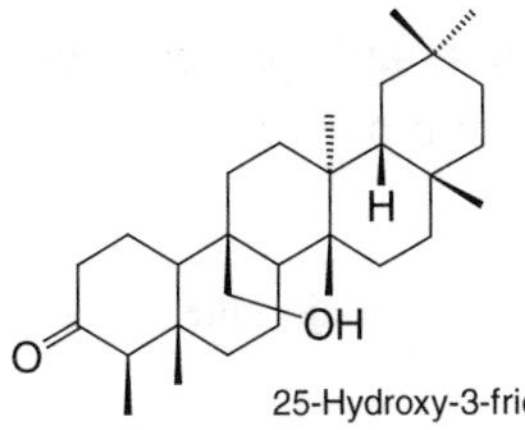

25-Hydroxy-3-friedelanone

Hopane

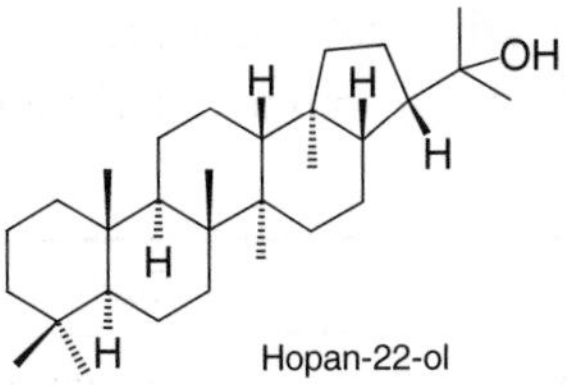

Hopan-22-ol

Lupane

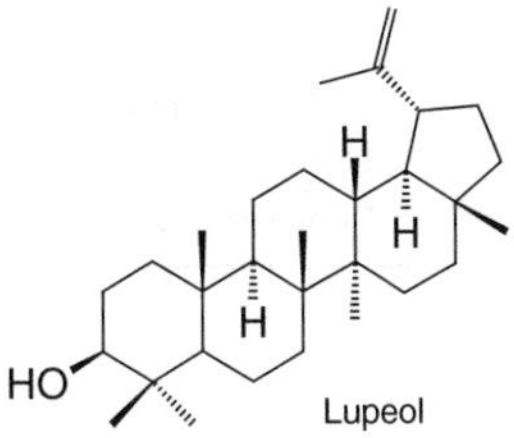

Lupeol

Oleanane

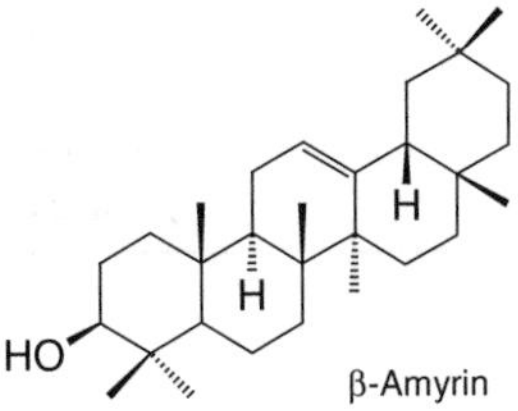

β-Amyrin

Serratane

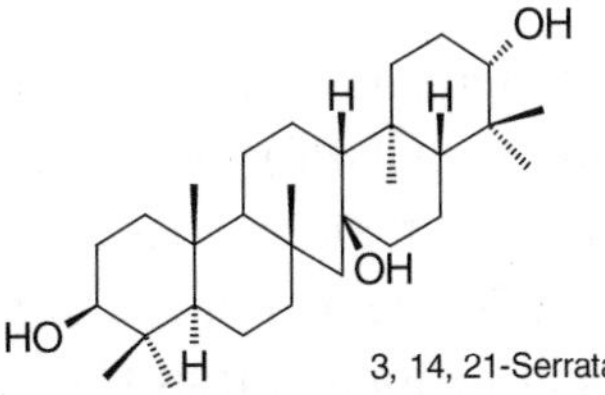

3, 14, 21-Serratanetriol

Modified triterpenes

Limonoids

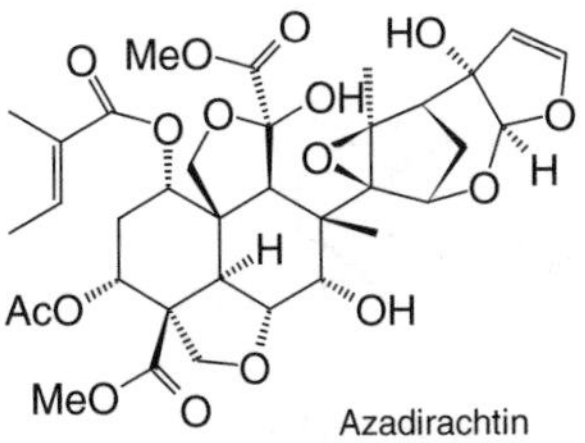

Azadirachtin
A limonoid from *Azadirachta indica* (Neem)

Quassinoids

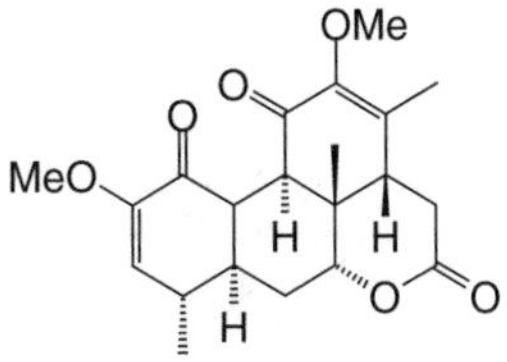

Quassin
A quassinoid from *Quassia amara*

Steroids

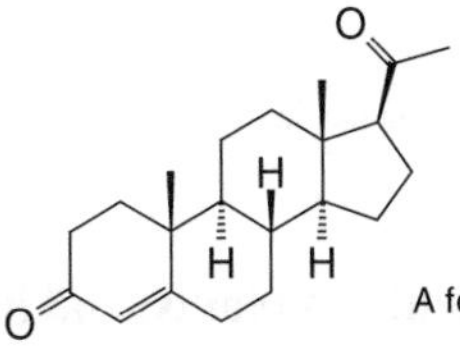

Progesterone
A female sex hormone

Botanical sources

Plants are the main sources of natural triterpenes. However, they are also found in other natural sources, e.g. fungus. The following table presents just a few of these sources, and their major triterpenoidal components.

Source		Major diterpenes
Common name	Botanical name (Family)	
Fusidium	*Fusidium coccineum*	Fusidic acid
Ganoderma	*Ganoderma lucidum*	Lanosterol
Dammar resin	*Balanocarpus heimii* (Dipterocarpaceae)	Dammarenediols
Ginseng	*Panax ginseng* (Araliaceae)	Dammarenediols
Lupin	*Lupinus luteus* (Fabaceae)	Lupeol
Quillaia	*Quillaja saponaria* (Rosaceae)	Quillaic acid

6.5.7 Tetraterpenes

The tetraterpenes arise by tail-to-tail coupling of two geranylgeranylpyrophosphate (GGPP) molecules. Tetraterpenes are represented by the carotenoids and their analogues, e.g. β-carotene, an orange colour pigment of carrots (*Daucus carota*, family Apiaceae), lycopene, a characteristic pigment in ripe tomato fruit (*Lycopersicon esculente*, family Solanaceae), and capsanthin, the brilliant red pigment of peppers (*Capsicum annuum*, family Solanaceae).

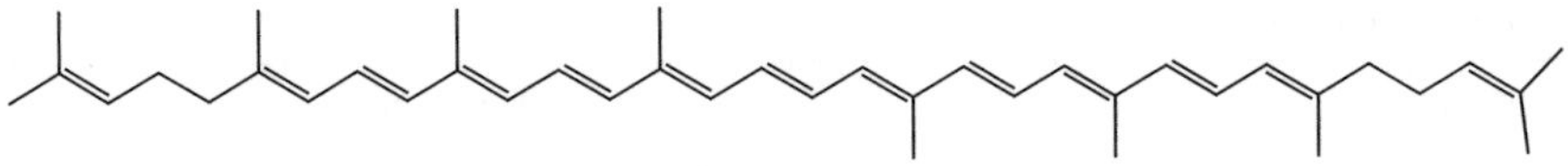

Lycopene

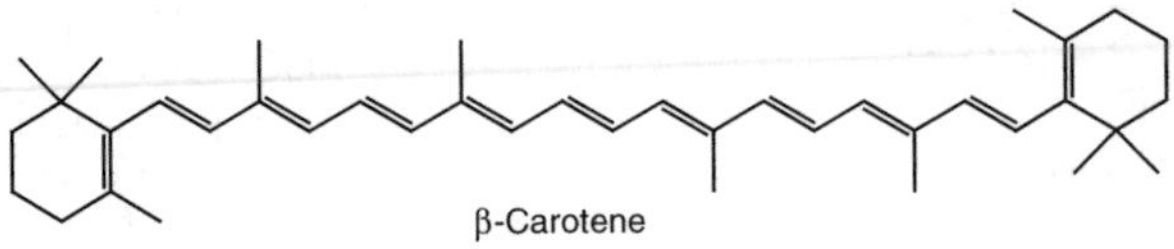

β-Carotene

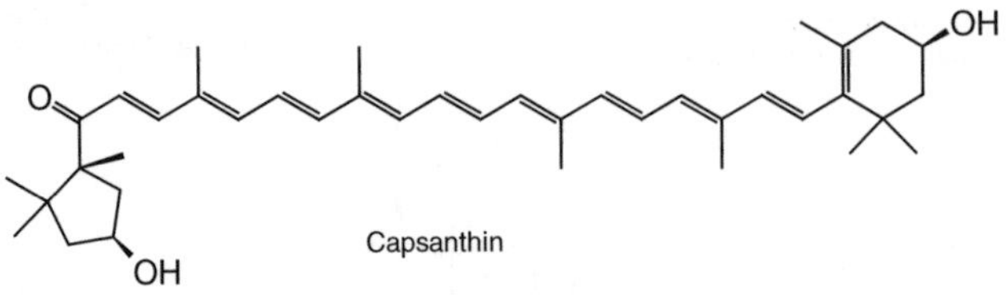

OH

O

OH

Capsanthin

Carotenoids are found abundantly in plants, and have been used as colouring agents for foods, drinks, confectionery and drugs. The vitamin A group of compounds are important metabolites of carotenoids, e.g. vitamin A_1 (retinol).

Chemistry of vision: role of vitamin A

β-carotene is converted to vitamin A_1 (retinol) in our liver. Vitamin A_1 is a fat-soluble vitamin found in animal products, e.g. eggs, dairy products, livers and kidneys. It is oxidized to an aldehyde called all-*trans*-retinal, and then isomerized to produce 11-*cis*-retinal, which is the light-sensitive pigment present in the visual systems of all living beings.

Rod and *cone* cells are the light sensitive receptor cells in the retina of the human eye. About three million rod cells are responsible for our vision in dim light, whereas the hundred million cone cells are responsible for our vision in the bright light and for the perception of bright colours. In the rod cells, 11-*cis*-retinal is converted to *rhodopsin.*

When light strikes the rod cells, isomerization of the C-11/C-12 double bond takes place, and *trans*-rhodopsin (metarhodopsin II) is formed. This *cis–trans* isomerization is accompanied by an alteration in molecular geometry, which generates a nerve impulse to be sent to the brain, resulting in the perception of *vision.* Metarhodopsin II is recycled back to rhodopsin by a multi-step sequence that involves the cleavage to all-*trans*-retinal and *cis–trans* isomerization back to 11-*cis*-retinal.

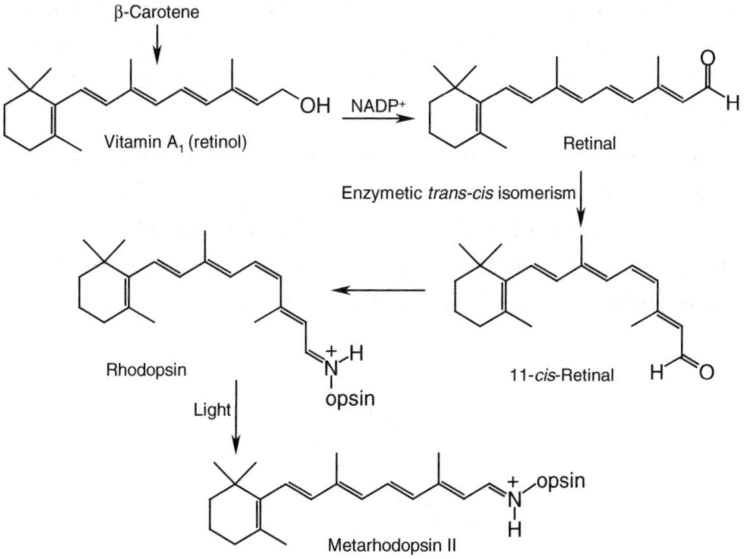

A deficiency of vitamin A leads to vision defects, e.g. night blindness. Vitamin A is quite unstable and sensitive to oxidation and light. Excessive intake of vitamin, however, can lead to adverse effects, e.g. pathological changes in the skin, hair loss, blurred vision and headaches.

6.6 Steroids

You have surely come across the items of news that appear quite frequently in the media, related to world-class athletes and sports personalities abusing *anabolic steroids*, e.g. nandrolone, to enhance performance and also to improve physical appearance. What are these substances? Well, all these drugs and many other important drugs belong to the class of compounds called steroids.

Steroids are chemical messengers, also known as hormones. They are synthesized in glands and delivered by the bloodstream to target tissues to stimulate or inhibit some process. Steroids are nonpolar and therefore lipids. Their nonpolar character allows them to cross cell membranes, so they can leave the cells in which they are synthesized and enter their target cells.

6.6.1 Structure

Structurally, a steroid is a lipid characterized by a carbon skeleton with four fused rings. All steroids are derived from the acetyl CoA biosynthetic pathway. Hundreds of distinct steroids have been identified in plants, animals and fungi, and most of them have interesting biological activity. They have a common basic ring structures, three-fused cyclohexane rings, together the phenanthrene part, fused to a cyclopentane ring system, known as *cyclopentaphenanthrene*.

The four rings are lettered A, B, C and D, and the carbon atoms are numbered beginning in the A ring as shown in gonane. These fused rings can be *trans* fused or *cis* fused. In steroids, the B, C and D rings are always *trans* fused. In most naturally occurring steroids, rings A and B are also *trans* fused. Different steroids vary in the functional groups attached to these rings.

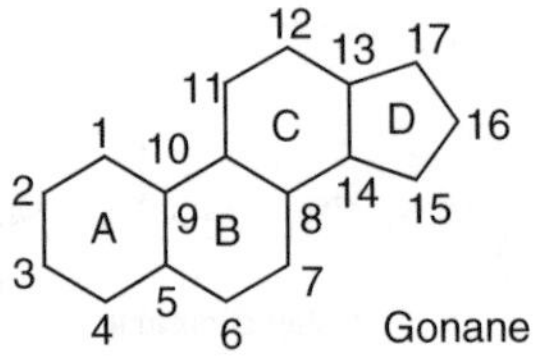

Gonane

All steroids possess at least 17 carbons. Many steroids have methyl groups at C-10 and C-13 positions. These are called angular methyl groups. Steroids may also have a side chain attached to C-17 and a related series of steroids are named after their fundamental ring systems, which are shown in the following structures.

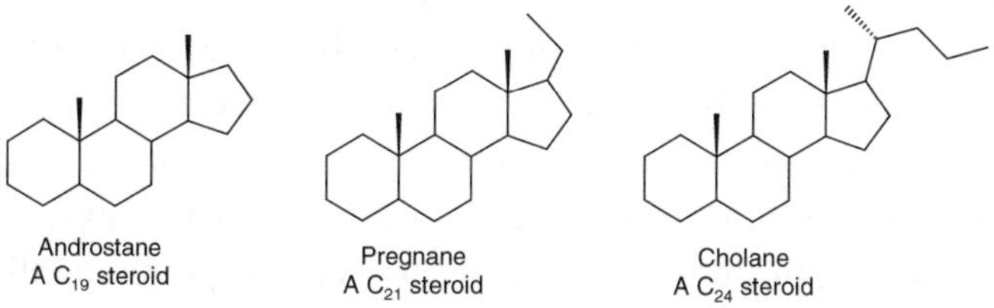

Androstane
A C_{19} steroid

Pregnane
A C_{21} steroid

Cholane
A C_{24} steroid

Many steroids have an alcoholic hydroxyl attached to the ring system, and are known as *sterols*. The most common sterol is cholesterol, which occurs in most animal tissues. There are many different steroid hormones, and cholesterol is the precursor for all of them. Cholesterol is also the precursor of vitamin D.

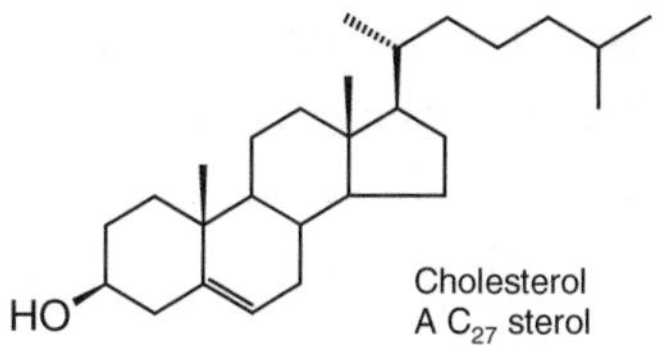

HO

Cholesterol
A C_{27} sterol

6.6.2 Stereochemistry of steroids

Sir Derek H. R. Barton of Great Britain received the Nobel Prize in 1969 for recognizing that functional groups could vary in reactivity depending on whether they occupied an *axial* or an *equatorial* position on a ring (see Chapter 3). The steroid skeleton shows a specific stereochemistry. All three of the six-membered rings can adopt strain-free chair conformations as shown below. Unlike simple cyclohexane rings, which can undergo chair–chair interconversions, steroids, being the large rigid molecules cannot undergo ring-flips. Steroids can have either *cis* or *trans* fusion of the A and B rings; both kind of steroid are relatively long, flat molecules but the A, B *trans*-fused steroids are by far the more common, though *cis*-fused steroids are found in bile. Furthermore, the presence of two angular methyl groups at C-10 and C-13 positions is characteristic in cholesterol. Substituents on the steroid ring system may be either *axial* or *equatorial*, and as usual *equatorial* substitution is more favourable than *axial* substitution for *steric* reasons. Thus, the hydroxyl group at C-3 of cholesterol has the more stable *equatorial* orientation.

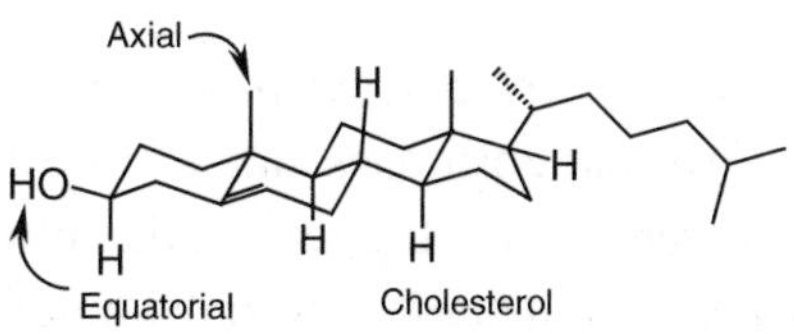

Axial
H
HO
H
H
H
H
Equatorial
Cholesterol

In most steroids the B–C and C–D rings are fused, usually in a *trans* manner. The lower side of the steroid is denoted α, the upper side of the steroid is denoted β, usually drawn as projected below the plane of the paper, which is shown as broken lines, and above the plane of the paper, which is drawn as solid lines. Thus, substituents attached to the steroid are also characterized as α and β. Cholesterol has eight *chiral* centres, therefore 256 stereoisomers are theoretically possible, but only one exists in nature! Stereogenic centres in steroid side chains are denoted preferentially with the *R* and *S* nomenclature.

6.6.3 Physical properties of steroids

The main feature, as in all lipids, is the presence of a large number of carbon–hydrogens that makes steroids nonpolar. The solubility of steroids in nonpolar organic solvents, e.g. ether, chloroform, acetone and benzene, and general insolubility in water, results from their significant hydrocarbon components. However, with the increase in number of hydroxyl or other polar functional groups on the steroid skeleton, the solubility in polar solvents increases.

6.6.4 Types of steroid

On the basis of the physiological functions, steroids can be categorized as follows.

(a) *Anabolic steroids* or *anabolic androgenic steroids* are a class of natural and synthetic steroids that interact with androgen receptors to promote cell growth and division, resulting in growth of several types of tissue, especially muscle and bone. There are natural and synthetic anabolic steroids. Examples: testosterone, nandrolone and methandrostenolone.

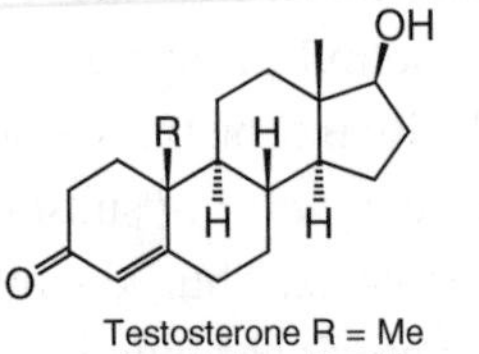

Testosterone R = Me
Nandrolone R = H
Natural anabolic steroids

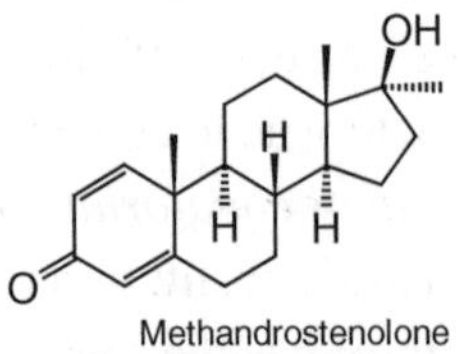

Methandrostenolone
A synthetic anabolic steroid

(b) *Corticosteroids (glucocorticoids and mineralocorticoids).* Glucocorticoids are a class of steroid hormones characterized by an ability to bind with the *cortisol receptor* and trigger similar effects. Glucocorticoids regulate many aspects of metabolism and immune functions, and are often prescribed as a remedy for inflammatory conditions such as asthma and arthritis. Example: cortisol.

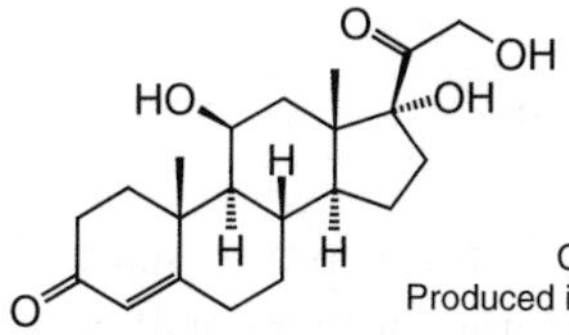

Cortisol
Produced in adrenal cortex

Mineralocorticoids are corticosteroids that help maintain blood volume and control renal excretion of electrolytes. Example: aldosterone.

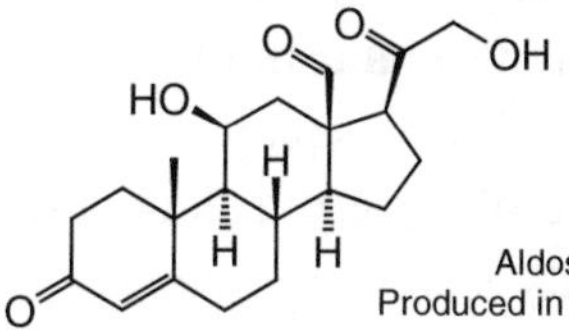

Aldosterone
Produced in adrenal cortex

(c) *Sex steroids* or *gonadal steroids* are a subset of sex hormones that interacts with vertebrate androgen or oestrogen receptors to produce sex differences (primary and secondary sex characters) and support reproduction. They include androgens, oestrogens and progestagens. Examples: testosterone, oestradiol and progesterone.

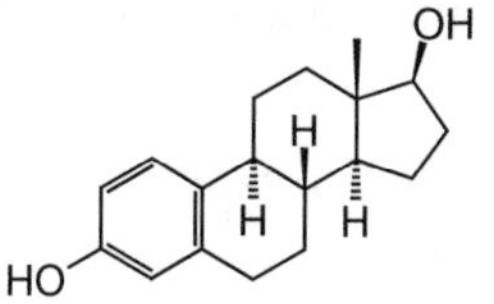

Oestradiol (estradiol)
The major oestrogen in humans

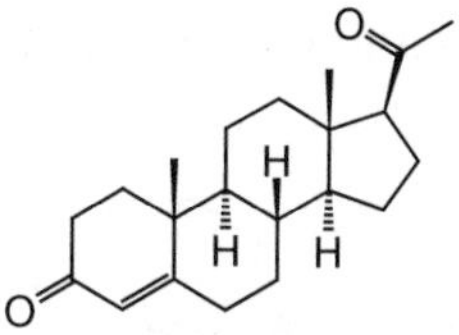

Progesterone
Supports gestation

(d) *Phytosterols* or *plant sterols* are steroid alcohols that occur naturally in plants. Example: β-sitosterol.

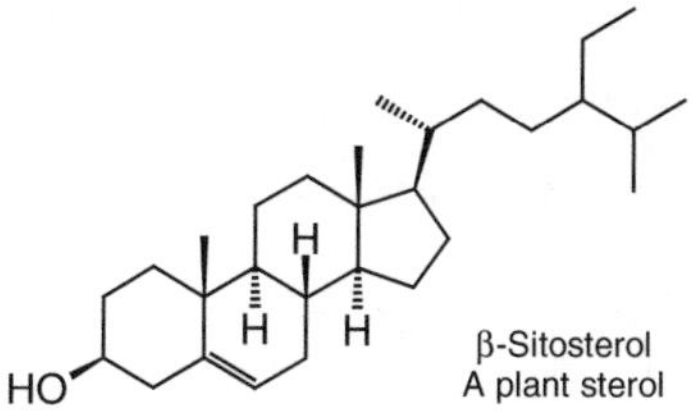

β-Sitosterol
A plant sterol

(e) *Ergosterols* are steroids that occur in fungi, and include some vitamin D supplements.

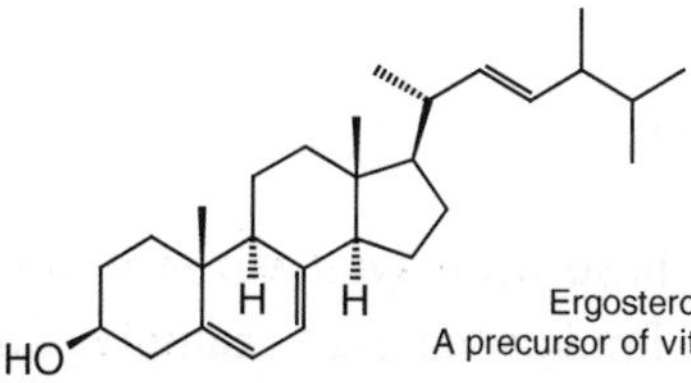

Ergosterol
A precursor of vitamin D_2

However, broadly steroids can be classified only in to two main classes: *sex or reproductive hormones* and *adrenocorticoid or adrenocortical hormones.*

6.6.5 Biosynthesis of steroids

The all-*trans*-squalene ($C_{30}H_{50}$), discovered in shark liver oil in the 1920s, is a triterpene, but one in which the isoprene rule at violated in one point. Rather than a head-to-tail arrangement of six units of isoprene, there appear to be farnesyl units that have been connected tail to tail. Almost all steroids are biosynthesized from cholesterol. Cholesterol is biosynthesized from squalene, which is first converted to lanosterol. The conversion of squalene to the steroid skeleton is an oxirane, squalene-2,3-oxide, which is transformed by enzymes into lanosterol, a steroid alcohol naturally found in wool fat. The whole process is highly stereoselective.

Squalene is an important biological precursor of many triterpenoids, one of which is cholesterol. The first step in the conversion of squalene to lanosterol is epoxidation of the 2,3-double bond of squalene. Acid-catalysed ring opening of the epoxide initiates a series of cyclizations, resulting in the formation of protesterol cation. Elimination of a C-9 proton leads to the 1,2-hydride and 1,2-methyl shifts, resulting in the formation of lanosterol, which in turn converted to cholesterol by enzymes in a series of 19 steps.

Tail
Tail
Squalene
Formed from tail-to-tail combination of 2 FPP
O_2 Squalene epoxidase
Squalene-2,3-oxide
- H^+ Lanosterol cyclase
HO H Lanosterol
Enzyme
19 steps
HO Cholesterol
A C_{27} sterol

6.6.6 Synthetic steroids

Several synthetic steroids have been synthesized in an effort to investigate their physiological effects. Prednisone is an example of a synthetic drug. The oral contraceptives and anabolic steroids are the best known of all steroids.

It was discovered a relatively long time ago (the 1930s) that injections of progesterone were effective as a contraceptive in preventing pregnancies. The pill is an oral contraceptive containing synthetic derivatives of the female sex hormones, progesterone and oestrogen. These synthetic hormones prevent ovulation and thus prevent pregnancy. The two most important birth-control pills are norethindrone and ethynyloestradiol. Many synthetic steroids have been found to be much more potent than natural steroids. For example, the contraceptive drug, norethindrone is better than progesterone in arresting (terminating) ovulation.

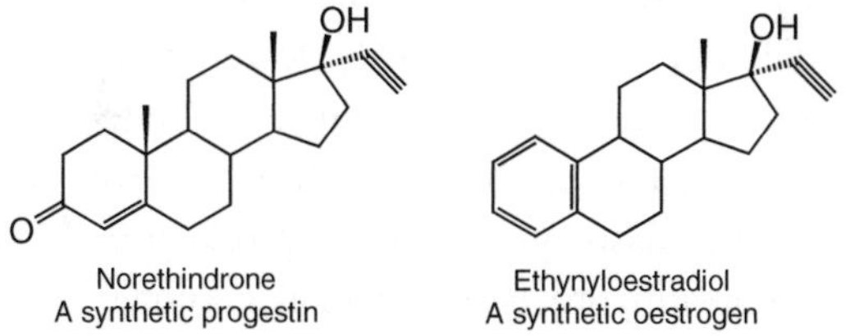

Norethindrone
A synthetic progestin

Ethynyloestradiol
A synthetic oestrogen

Steroids that aid in muscle development are called *anabolic steroids*. They are synthetic derivatives of testosterone, thus have the same muscle-building effect as testosterone. There are more than 100 different anabolic steroids which, vary in structure, duration of action, relative effects and toxicities. Androstenedione, stanozolol and dianabol are anabolic steroids. They are used to treat people suffering from traumas accompanied by muscle deterioration. The use of anabolic steroid can lead to a number of dangerous side-effects, including lowered levels of high density lipoprotein cholesterol, which benefits the heart, and elevated levels of harmful low density lipoprotein, stimulation of prostate tumours, clotting disorders and liver problems.

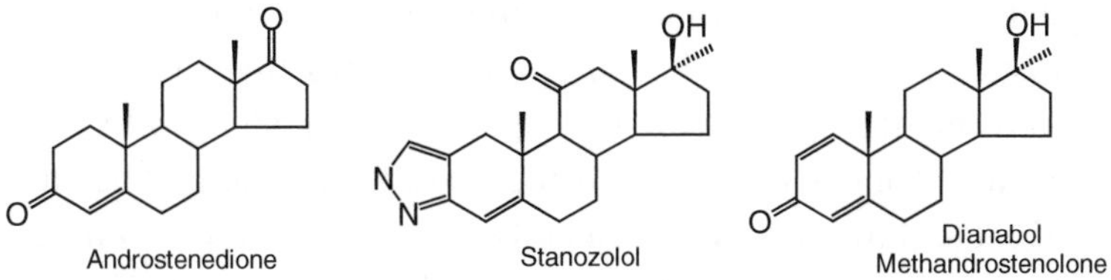

Androstenedione

Stanozolol

Dianabol
Methandrostenolone

6.6.7 Functions

The most important function of steroids in most living systems is as hormones. Steroid hormones produce their physiological effects by binding to steroid hormone receptor proteins. The binding of steroids to their receptors causes changes in gene transcription and cell function. From biological and physiological viewpoints, probably the most important steroids are cholesterol, the steroid hormones, and their precursors and metabolites. Cholesterol, a common component of animal cell membranes, is an important steroid alcohol. Cholesterol is formed in brain tissue, nerve tissue and blood stream. It is the major compound found in gallstones and bile salts.

Cholesterol also contributes to the formation of deposits on the inner walls of blood vessels. However, these deposits harden and obstruct the flow of blood. This condition results in various heart diseases, strokes and high blood pressure, and a high level of cholesterol can be life-threatening. A number of vertebrate hormones, which govern a number of physiological functions, from growth to reproduction, are biosynthesized from cholesterol.

Much research is currently underway to determine whether a correlation exists between cholesterol levels in the blood and diet. Cholesterol not only comes from the diet, but is also synthesized in the body from carbohydrates and proteins as well as fat. Therefore, the elimination of cholesterol rich foods from the diet does not necessarily lower blood cholesterol levels. Some studies have found that if certain unsaturated fats and oils are substituted for saturated fats the blood cholesterol level decreases.

Sex hormones control tissue growth and reproduction. Male sex hormones are testosterone and 5α-dihydrotestosterone, also known as androgens, which are secreted by the testes. The primary male hormone, testosterone, is responsible for the development of secondary sex characteristics during puberty. They also promote muscle growth. The two most important female sex hormones are oestradiol and oestrone, also known as *oestrogens* (*estrogens*). They are responsible for the development of female secondary sex characteristics.

Testosterone

5α-Dihydrotestosterone

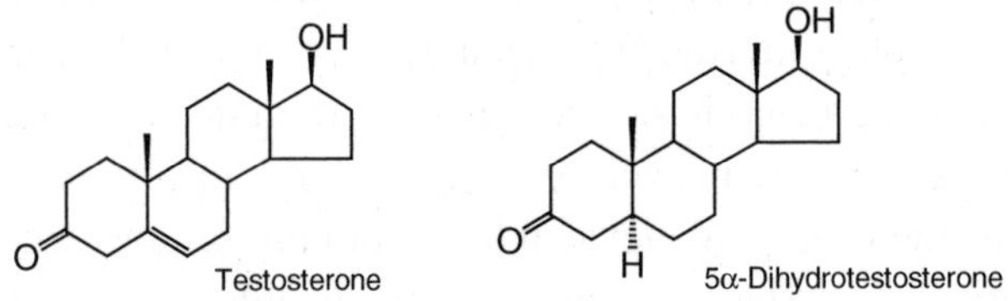

Oestrogen (estrogen) is biosynthesized from testosterone by making the first ring aromatic, which results in more double bonds, the loss of a methyl group and formation of an alcohol group. Oestrogen, along with progesterone, regulates changes occurring in the uterus and ovaries known as the *menstrual cycle*. Progesterone is a member of the class called *progestins*. It is also the precursor of sex hormones and adrenal cortex steroids. Progesterone is an essential component for the maintenance of pregnancy. It also prevents ovulation during pregnancy. Many of the steroid hormones are ketones, including testosterone and progesterone. The male and female hormones have only slight differences in structure, but yet have very different physiological effects. For example, the only difference between testosterone and progesterone is the substituent at C-17.

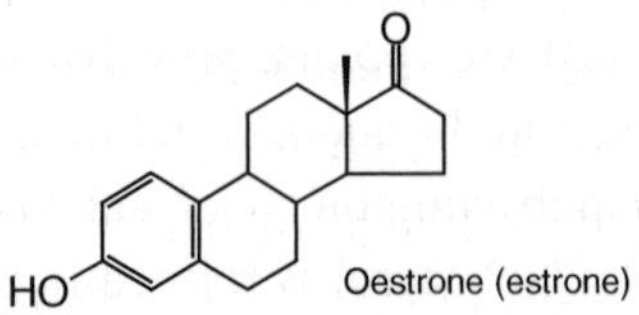

Oestrone (estrone)

Adrenocorticoid hormones are produced in the adrenal glands. They regulate a variety of metabolic processes. The most important mineralocorticoid is aldosterone, an aldehyde as well as a ketone, which regulates the reabsorption of sodium and chloride ions in the kidney, and increases the loss of potassium ions. Aldosterone is secreted when blood sodium ion levels are too low to cause the kidney to retain sodium ions. If sodium levels are elevated, aldosterone is not secreted, so some sodium will be lost in the urine and water. Aldosterone also controls swelling in the tissues.

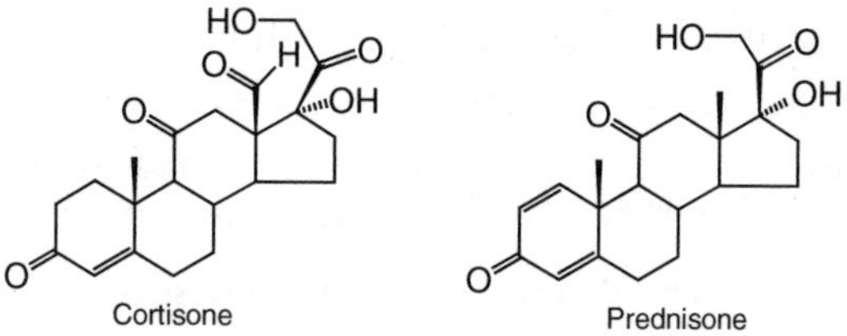

Cortisone

Prednisone

Cortisol or hydrocortisone, the most important glucocorticoid, has the function of increasing glucose and glycogen concentrations in the body. These reactions are completed in the liver by taking fatty acids from lipid storage cells and amino acids from body proteins to make glucose and glycogen. Cortisol and its ketone derivative, cortisone, are potent anti-inflammatory agents. Cortisone or similar synthetic derivatives such as prednisolone, the active metabolite of prednisone, are used to treat inflammatory diseases, rheumatoid arthritis and bronchial asthma. There are many side-effects with the use of cortisone drugs, so their use must be monitored carefully. Prednisolone is designed to be a substitute for cortisone, which has much greater side-effects than prednisolone.

Phytosterols found in plants have many applications as food additives and in medicine and cosmetics. Ergosterol is a component of fungal cell membranes, serving the same function that cholesterol serves in animal cells. The presence of ergosterol in fungal cell membranes coupled with its absence from animal cell membranes makes it a useful target for antifungal drugs. Ergosterol is also used as a fluidizer in the cell membranes of some protists, such as trypanosomes. This explains the use of some antifungal agents against West African sleeping sickness.

6.7 Phenolics

This is a large group of structurally diverse naturally occurring compounds that possess at least a phenolic moiety in their structures. For example, umbelliferone, a coumarin, has a phenolic hydroxyl functionality at C-7; quercetin is a flavonoid that has four phenolic hydroxyls at C-5, C-7, C-3′ and C-4′. Although the phenolic group of compounds encompasses various structural types, in this section we will mainly focus our discussion on phenyl propanoids, coumarins, flavonoid and isoflavonoids, lignans and tannins.

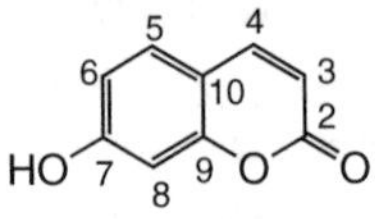

Umbelliferone
The most common natural coumarin

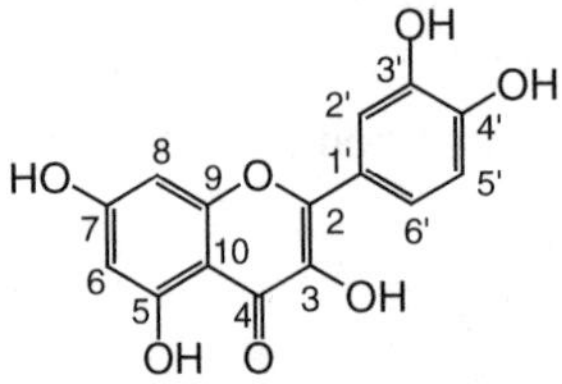

Quercetin
A natural antioxidant

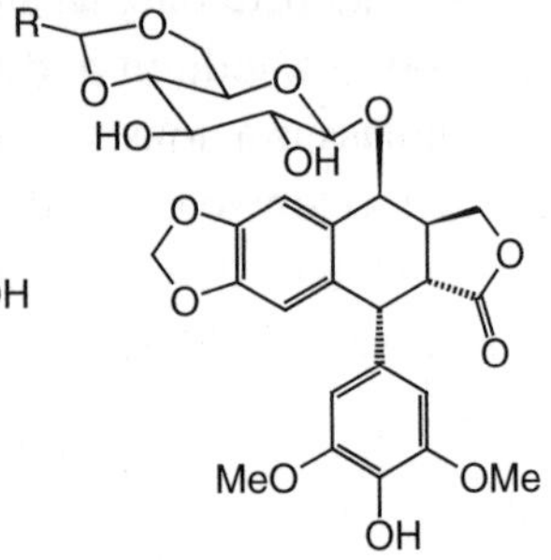

Etoposide R = Me
Teniposide R = thiophenyl
Anticancer lignans

Most of these compounds, e.g. quercetin, possess various degrees of antioxidant or free radical scavenging properties. A number of phenolic compounds have medicinal properties and have long been used as drugs. For example, etoposide and teniposide, two lignans, are anticancer drugs.

6.7.1 Phenylpropanoids

Phenylpropanes are aromatic compounds with a propyl side chain attached to the benzene ring, which can be derived directly from phenylalanine. Naturally occurring phenylpropanoids often contain oxygenated substituents, e.g. OH, OMe or methylenedioxy, on the benzene ring. Phenylpropanoids with hydroxyl substituent(s) on the benzene ring belongs to the group of phenolics, e.g. caffeic acid and coumaric acid.

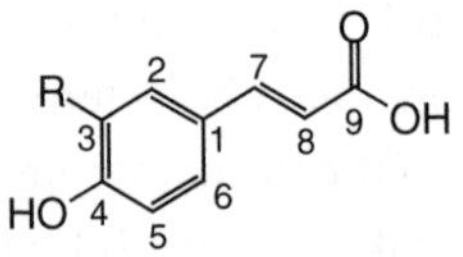

Coumaric acid (or 4-hydroxycinnamic acid) R = H
Caffeic acid (or 3,4-dihydroxycinnamic acid) R = OH

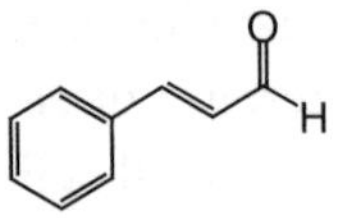

Cinnamaldehyde

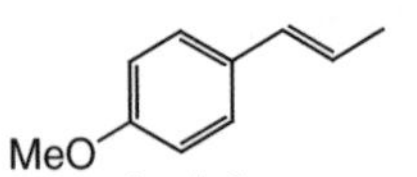

Anethole

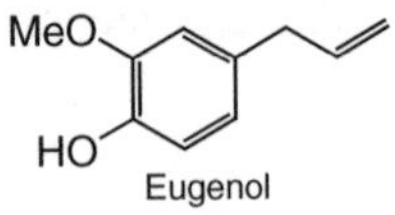

Eugenol

Phenylpropanoids are widespread in higher plants, especially in the plants that produce essential oils, e.g. plants of the families, Apiaceae, Lamiaceae, Lauraceae, Myrtaceae and Rutaceae. For example, Tolu balsam (*Myroxylon balsamum*, family Fabaceae) yields a high concentration of cinnamic acid esters, cinnamon (*Cinnamomum verum*, family Lauraceae) produces cinnamaldehyde, fennel (*Foeniculum vulgare*, family Apiaceae) is a good source of eugenol and star anise (*Illicium verum*, family Illiaceae) produces high amounts of anethole. The biosynthesis of phenylpropanoids follows the shikimic acid pathway, and the immediate precursor of cinnamic acid is phenylalanine. Other phenylpropanoids, and a number of other phenolics, e.g. coumarins, flavonoids and lignans, originate from cinnamic acid.

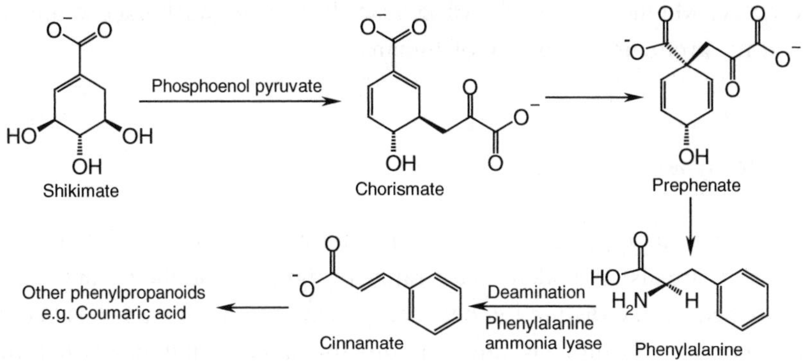

Shikimate
Phosphoenol pyruvate
Chorismate
Prephenate
Phenylalanine
Deamination
Phenylalanine ammonia lyase
Cinnamate
Other phenylpropanoids e.g. Coumaric acid

6.7.2 Lignans

The lignans are a large group of plant phenolics, biosynthesized from the union of two phenylpropane molecules; e.g., both matairesinol (*Centaurea* species, family Asteraceae) and podophyllotoxin (*Podophyllum peltatum*, family Berberidaceae) are formed from the phenylpropane coniferyl alcohol. Lignans are essentially cinnamoyl alcohol dimers, though further cyclization and other structural modifications result in various structural types, e.g. dibenzylbutyrolactone and epoxy lignan.

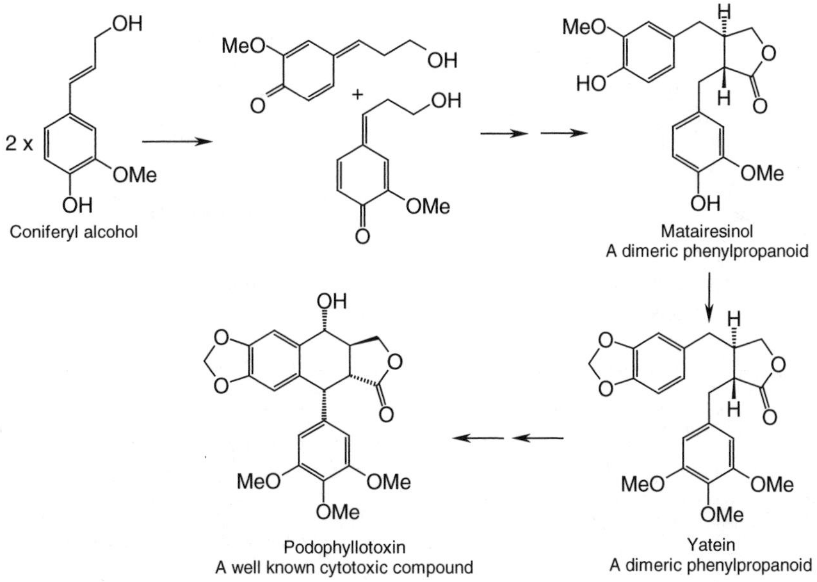

2 x
Coniferyl alcohol
Matairesinol
A dimeric phenylpropanoid
Yatein
A dimeric phenylpropanoid
Podophyllotoxin
A well known cytotoxic compound

Natural lignans are optically active, although a few *meso*-compounds exist in nature. Like any other optically active compounds, important physiological or pharmacological properties of lignans are generally associated with a particular absolute configuration, e.g. the antitumour agent podophyllotoxin. Lignans, including neolignans, are quite widespread in the plant kingdom, and plants from, e.g., the families Asteraceae, Berberidaceae, Piperaceae, Magnoliaceae, Phytolaccacae, Rutaceae and Pinaceae are well known for producing a variety of lignans.

Structural types

Major structural types encountered in natural lignans are shown below. *Neolignans* are also included, as the range of lignoids and their plant sources has widened, so the distinction between lignans and neolignans has become less important. Neolignans are also dimers of cinnamyl units, but their structures are obtained by coupling of mesomeric radicals other than the β–β link typical of the lignans.

Simple dibenzylbutane lignans

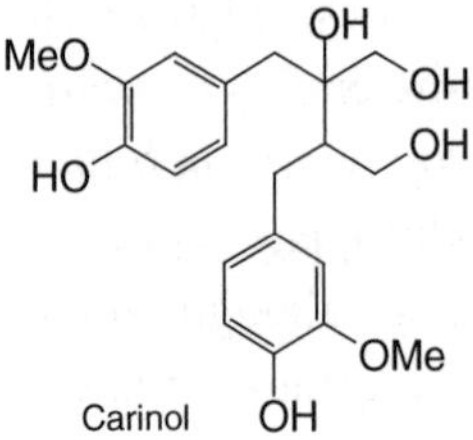

Carinol

Dibenzylbutyrolactone lignans

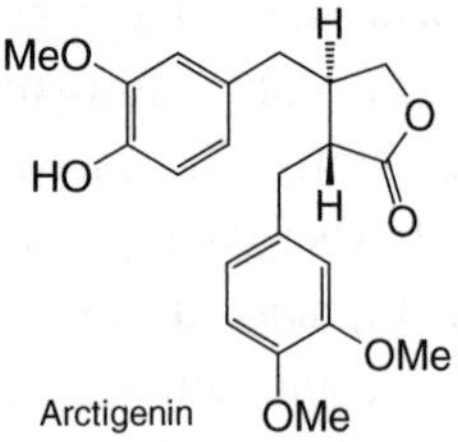

Arctigenin

Epoxy and diepoxy lignans

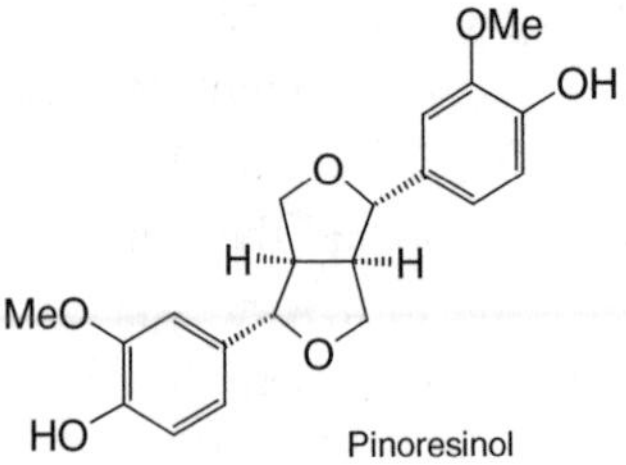

Pinoresinol

Simple aryltetralin lignans (2,7′-cyclolignans)

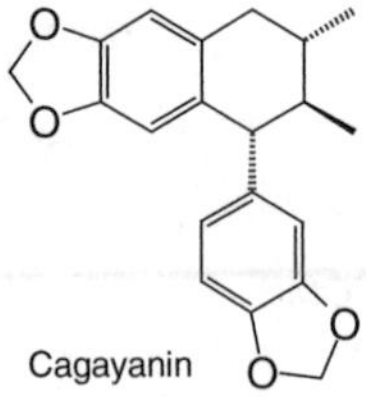

Cagayanin

Dibenzocycloctadiene lignans (2,2′-cyclolignans)

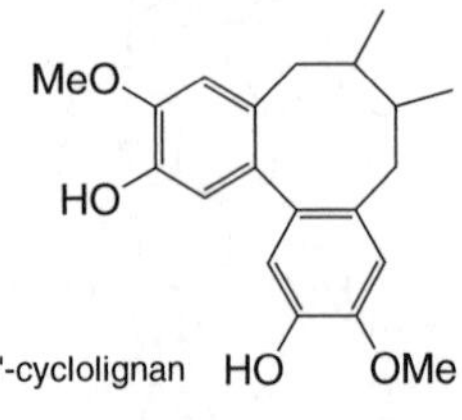

A 2,2'-cyclolignan

Neolignans

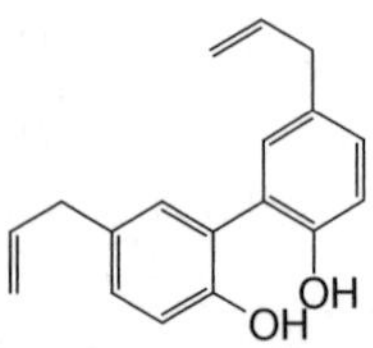

Magnolol

A bioactive neolignan of *Magnolia* species

6.7.3 Coumarins

The coumarins (2*H*-1-benzopyran-2-one) are the largest class of *1-benzopyran* derivatives, found mainly in higher plants. Most natural coumarins are oxygenated at C-7, e.g. umbelliferone (7-hydroxycoumarin). Umbelliferone is considered as the structural and biogenetic parent of the more highly oxygenated coumarins, e.g. scopoletin. C- and O-prenylations are common in a large number of coumarins, e.g. imperatorin. The prenyl groups found in coumarins exhibit the greatest number of biogenetic modifications, including cyclization to dihydropyrans, pyrans, dihydrofurans and furans.

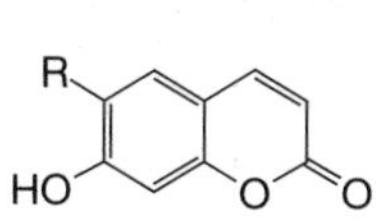

Umbelliferone R = H
Scopoletin R = OMe

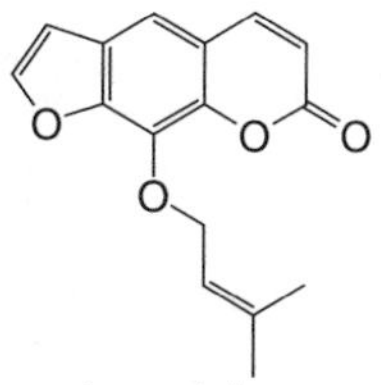

Imperatorin
An *O*-prenylated furanocoumarin

Coumarins occur abundantly in various plant families, e.g. Apiaceae, Asteraceae, Fabaceae, Lamiaceae, Moraceae, Poaceae, Rutaceae and Solanaceae. However, the Apiaceae (*alt.* Umbelliferae) and the Rutaceae are the two most important coumarin-producing plant families.

Many coumarins are used in sunscreen preparations for the protection against the sunlight, because these compounds absorb short-wave UV radiation (280–315 nm), which is harmful for human skin, but transmits the long-wave UV radiation (315–400 nm) that provides the brown sun-tan. Dicoumarol, a dimeric coumarin, occurs in mouldy sweet clover, *Melilotus officinalis* (family Fabaceae), has a prominent anticoagulant property and has been used in medicine as an anti-blood-clotting agent for the prevention of thrombosis. Psoralen, a linear furanocoumarin, isolated from *Psoralea corylifolia* (family Fabaceae) and also found in the families Rutaceae, Apiaceae and Moraceae, has long been used in the treatment of vertigo. A number of coumarins also possess antifungal and antibacterial properties.

Biosynthesis

The biosynthesis of coumarins begins with *trans*-4-cinnamic acid, which is oxidized to *ortho*-coumaric acid (2-hydroxy cinnamic acid) followed by formation of the glucoside. This glucoside isomerizes to the corresponding *cis*-compound, which finally through ring closure forms coumarin. However, as most natural coumarins contains an oxygenation at C-7, the biosynthesis proceeds through 4-hydroxylation of cinnamic acid.

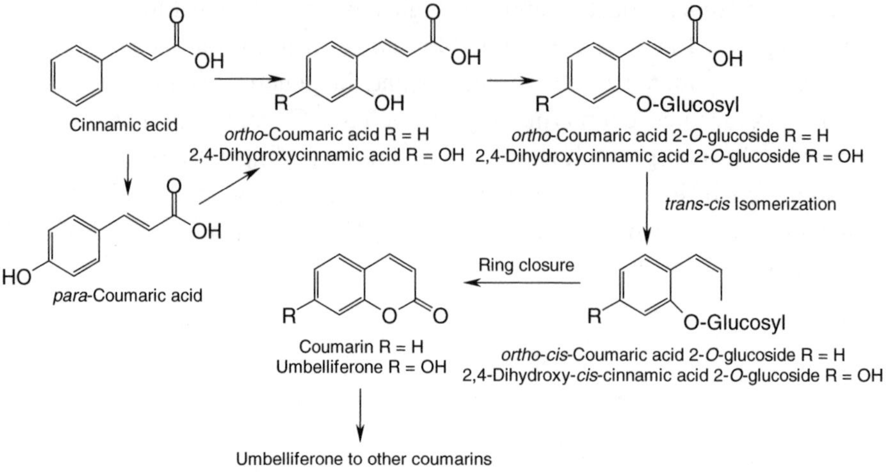

Cinnamic acid

para-Coumaric acid

ortho-Coumaric acid R = H
2,4-Dihydroxycinnamic acid R = OH

ortho-Coumaric acid 2-*O*-glucoside R = H
2,4-Dihydroxycinnamic acid 2-*O*-glucoside R = OH

trans-cis Isomerization

ortho-cis-Coumaric acid 2-*O*-glucoside R = H
2,4-Dihydroxy-*cis*-cinnamic acid 2-*O*-glucoside R = OH

Ring closure

Coumarin R = H
Umbelliferone R = OH

Umbelliferone to other coumarins

Structural types

Simple coumarins

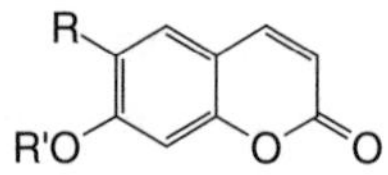

Umbelliferone R = R' = H
Aesculetin R = OH, R' = H
Scopoletin R = OMe, R' = H
Scopolin R = OMe, R' = glucosyl

Simple prenylated coumarins

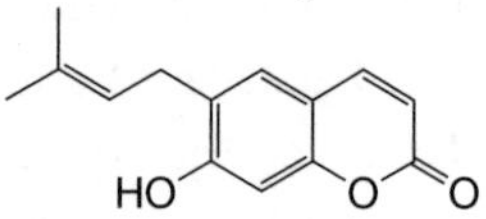

Demethylsuberosin

Linear furanocoumarins

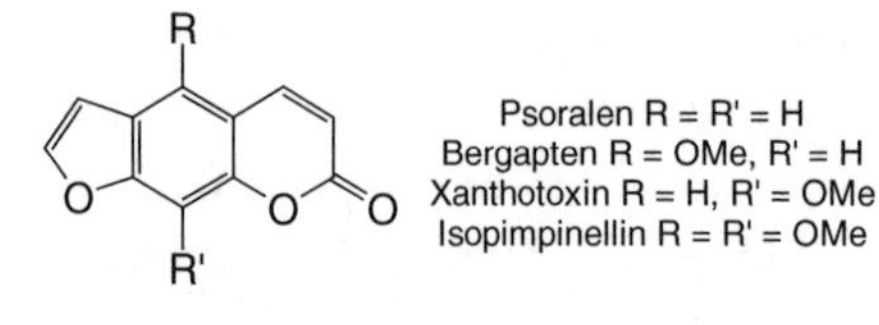

Psoralen R = R' = H
Bergapten R = OMe, R' = H
Xanthotoxin R = H, R' = OMe
Isopimpinellin R = R' = OMe

Angular furanocoumarins

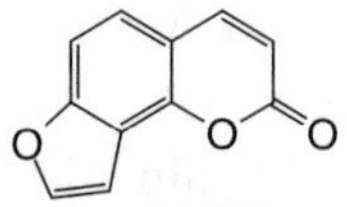

Angelicin
A typical coumarin of *Angelica* species

Linear dihydrofuranocoumarins

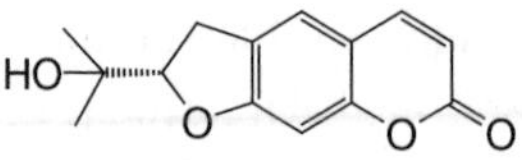

Marmesin

Angular dihydrofuranocoumarins

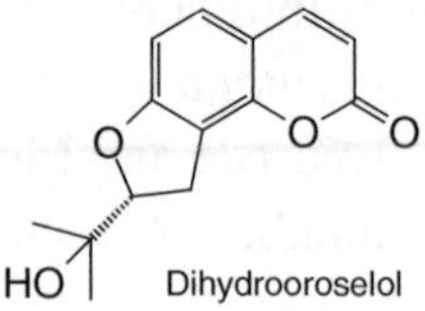

Dihydrooroselol

Linear pyranocoumarins

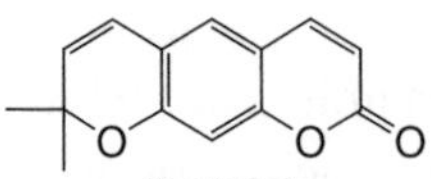

Xanthyletin

Angular pyranocoumarins

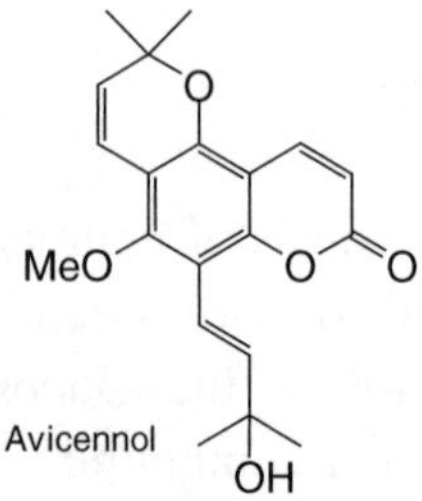

Avicennol

Linear dihydropyranocoumarins

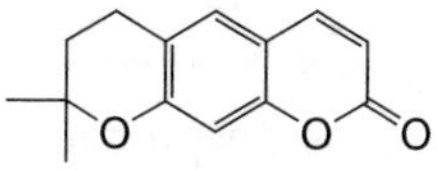

1',2'-Dihydroxanthyletin

Angular dihydropyranocoumarins

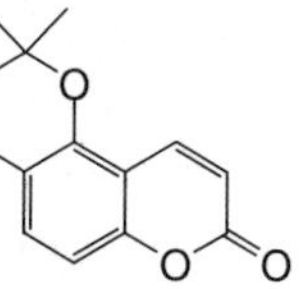

Libanotin A

Sesquiterpenyl coumarins

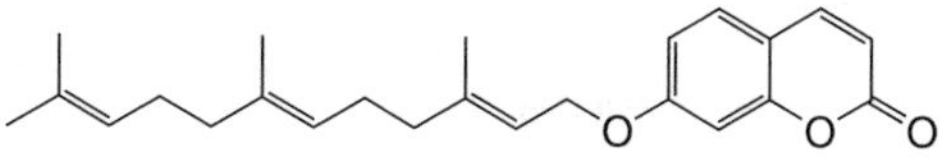

Umbelliprenin

Dimeric coumarins

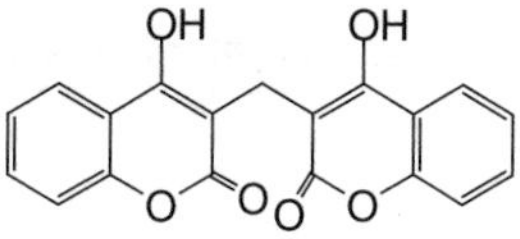

Dicoumarol

6.7.4 Flavonoids and isoflavonoids

The flavonoids, the derivatives of 1,3-diphenylpropane, are a large group of natural products, which are widespread in higher plants but also found in some lower plants, including algae. Most flavonoids are yellow compounds, and contribute to the yellow colour of the flowers and fruits, where they are usually present as glycosides.

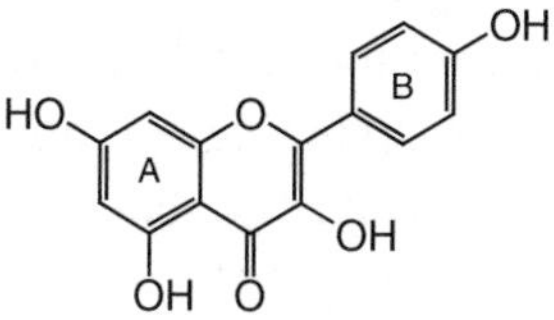

Kaempferol

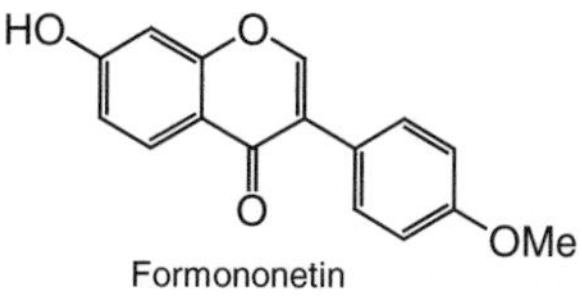

Formononetin

Most flavonoids occur as glycosides, and within any one class may be characterized as monoglycosidic, diglycosidic and so on. There are over 2000 glycosides of the flavones and flavonols isolated to date. Both *O*- and *C*-glycosides are common in plant flavonoids; e.g., *rutin* is an *O*-glycoside, whereas *isovitexin* is a *C*-glycoside. Sulphated conjugates are also common in the flavone and flavonol series, where the sulphate conjugation may be on a phenolic hydroxyl and/or on an aliphatic hydroxyl of a glycoside moiety.

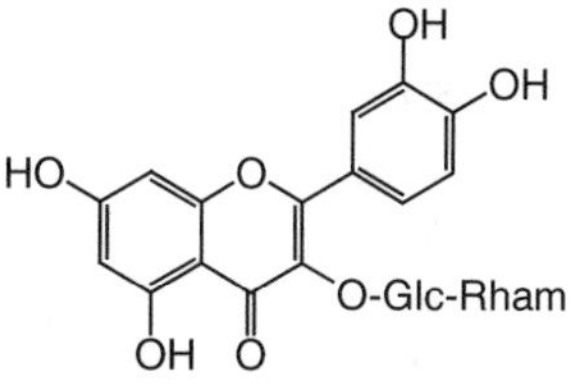

Rutin

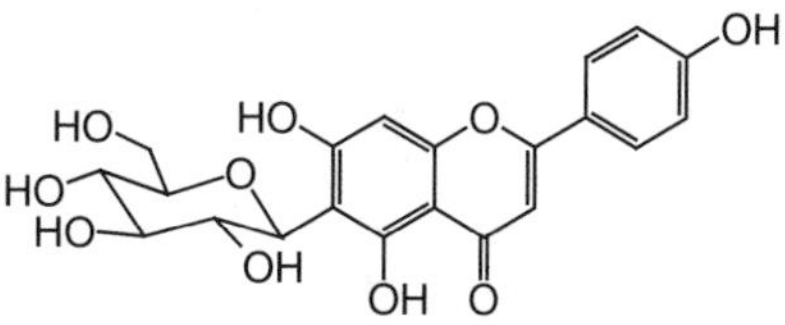

Isovitexin

Most flavonoids are potent antioxidant compounds. Several flavonoids possess anti-inflammatory, antihepatotoxic, antitumour, antimicrobial and antiviral properties. Many traditional medicines and medicinal plants contain flavonoids as the bioactive compounds. The antioxidant properties of flavonoids present in fresh fruits and vegetables are thought to contribute to their preventative effect against cancer and heart diseases. Rutin, a flavonoid glycoside found in many plants, e.g. *Sophora japonica* (Fabaceae), buckwheat (*Fagopyrum esculentum*, family Polygonaceae) and rue (*Ruta graveolens*, family Rutaceae), is probably the most studied of all flavonoids, and is included in various multivitamin preparations. Another flavonoid glycoside, hesperidin from *Citrus* peels, is also included in a number of dietary supplements, and claimed to have a beneficial effect for the treatment of capillary bleeding.

Hesperidin

Biosynthesis

Structurally, flavonoids are derivatives of 1,3-diphenylpropane, e.g. kaempferol. One of the phenyl groups, ring B, originates from the shikimic acid pathway, while the other ring, ring A, is from the acetate pathway through ring closure of a polyketide. One hydroxyl group in ring A is always situated in the *ortho* position to the side chain, and involved in the formation of the third six-membered ring or a five-membered ring (only found in *aurones*). The 2-phenyl side-chain of the flavonoid skeleton isomerizes to the 3-position, giving rise to isoflavones, e.g. formononetin. The biosynthesis of flavonoids can be summarized as follows.

Phenylalanine

Cinnamic acid

4-Hydroxycinnamic acid (*para*-Coumaric acid)

para-Coumaroyl-CoA

Naringenin chalcone

Naringenin

A = Phenylalanine-ammonia lyase
B = Cinamate 4-hydroxylase
C = 3 x Malonyl-CoA
D = Chalcone synthase
E = Chalcone isomerase

Classification

Flavonoids can be classified according to their biosynthetic origins. Some flavonoids are both intermediates in biosynthesis and end-products, e.g. chalcones, flavanones, flavanon-3-ols and flavan-3,4-diols. Other classes are only known as the end-products of biosynthesis, e.g. anthocyanins, flavones and flavonols. Two further classes of flavonoids are those in which the 2-phenyl side-chain of flavonoid isomerizes to the 3-position (giving rise to isoflavones and related isoflavonoids) and then to the 4-position (giving rise to the neoflavonoids). The major classes of flavonoids, with specific examples, are summarized below.

Chalcone

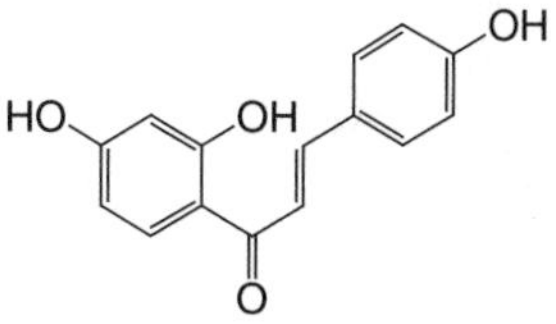

Isoliquiritigenin

Dihydrochalcone

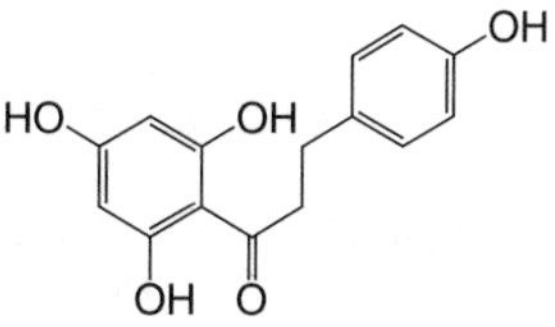

Dihydronaringenin chalcone

Flavanone

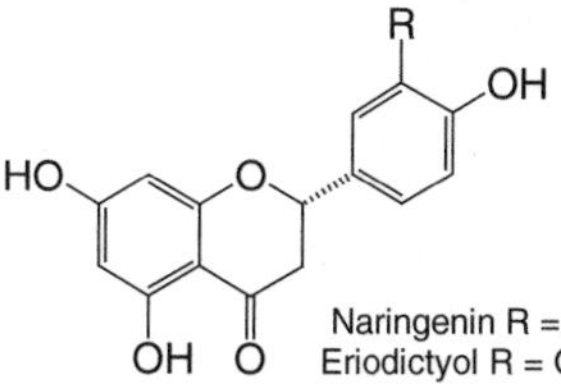

Naringenin R = H
Eriodictyol R = OH

Flavone

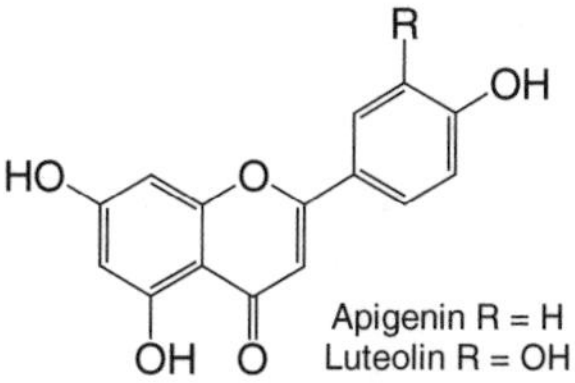

Apigenin R = H
Luteolin R = OH

Flavanon-3-ol

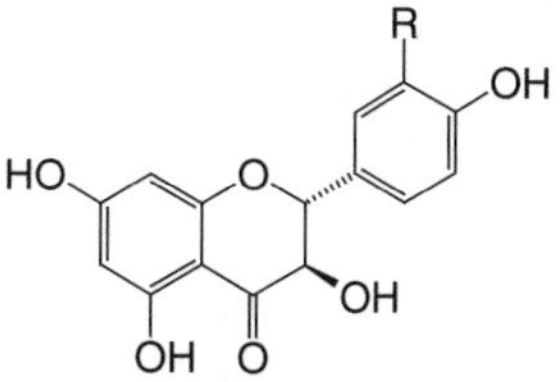

Dihydrokaempferol R = H
Dihydroquercetin R = OH

Flavonol

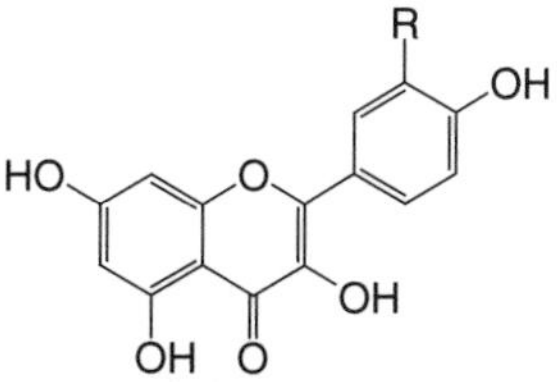

Kaempferol R = H
Quercetin R = OH

Flavan-3,4-diol

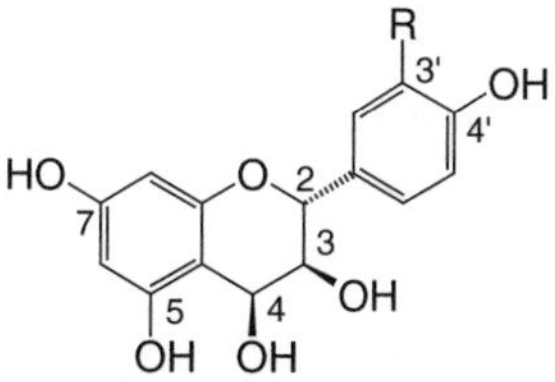

Leucopelargonidin R = H
Leucocyanidin R = OH

Flavan-3-ol

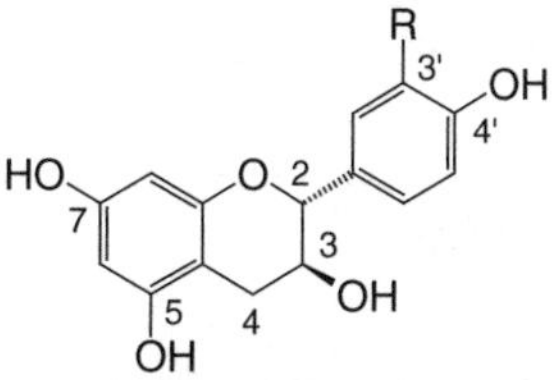

Afzalechin R = H
(+)-Catechin R = OH

Flavan

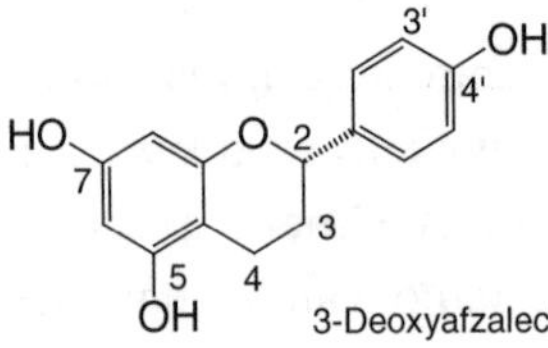

3-Deoxyafzalechin

Anthocyanidin

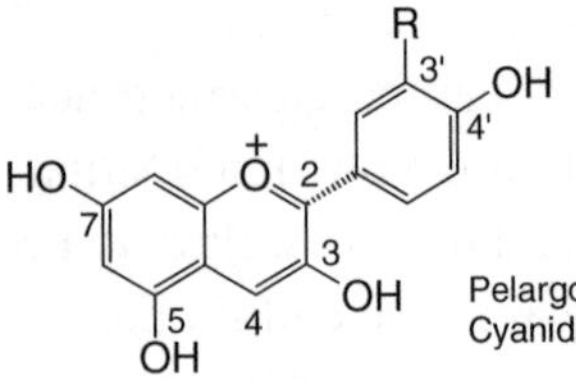

Pelargonidin R = H
Cyanidin R = OH

Flavonoid O-glycoside

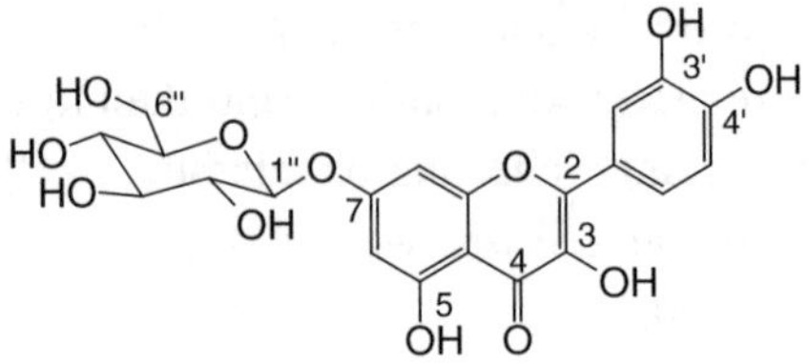

Quercetin 7-*O*-β-D-glucopyranoside

Flavonoid C-glycoside

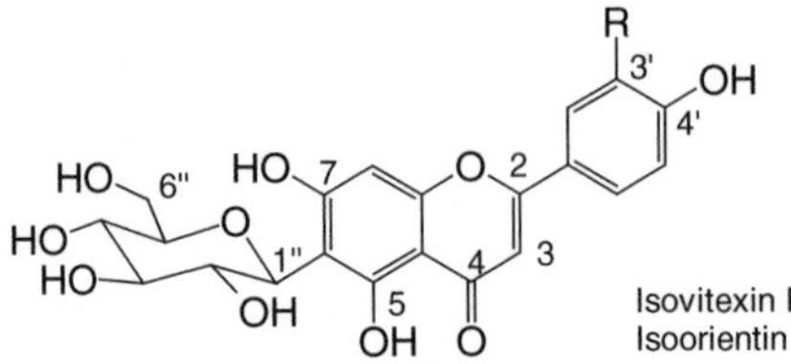

Isovitexin R = H
Isoorientin R = OH

Aurone

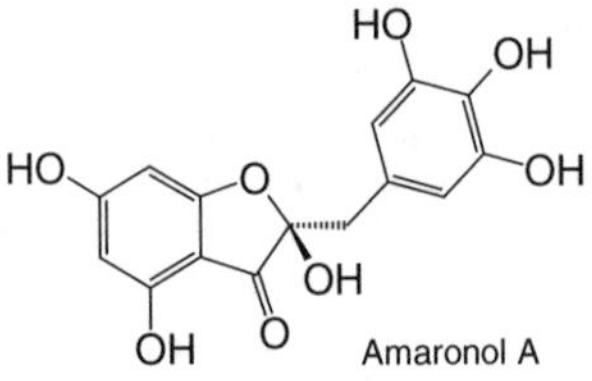

Amaronol A

Proanthocyanidin

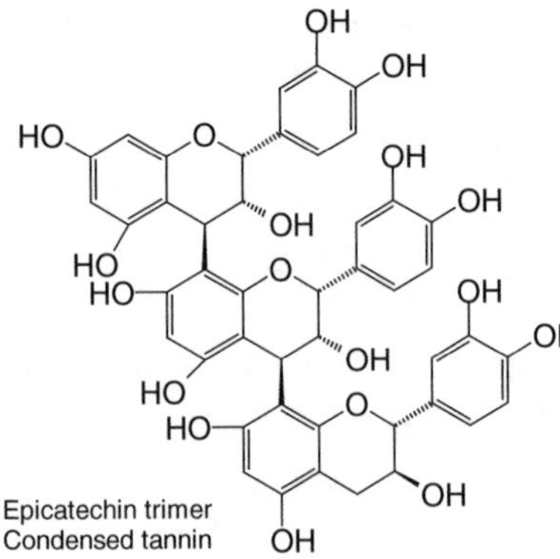

Epicatechin trimer
Condensed tannin

Isoflavonoid

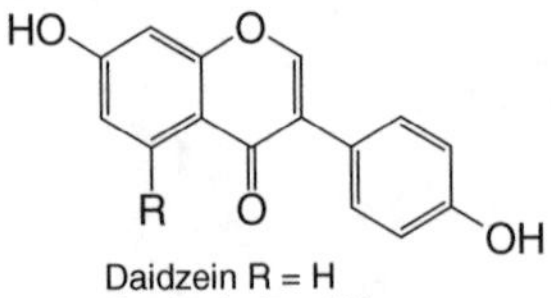

Daidzein R = H
Genistein R = OH

Biflavonoid

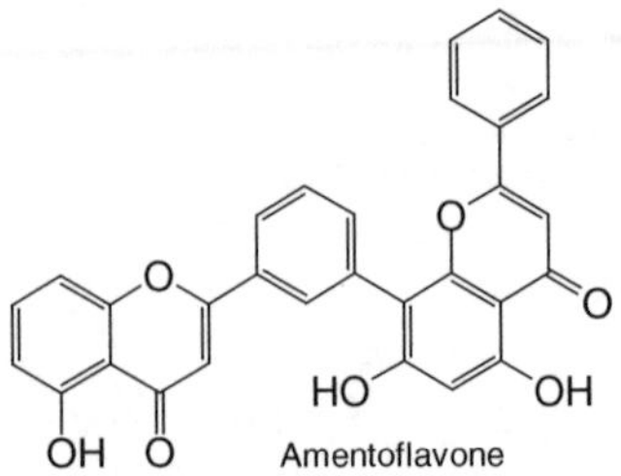

Amentoflavone

6.7.5 Tannins

Plant polyphenols, also known as vegetable *tannins*, are a heterogenous group of natural products widely distributed in the plant kingdom. Tannins are often present in unripe fruits, but disappear during ripening. It is believed that tannins may provide plants with protection against microbial attacks. Tannins are of two broad structural types: condensed proanthocyanidins in which the fundamental structural unit is the phenolic flavan-3-ol (catechin) nucleus, and galloyl and hexahydroxydiphenoyl esters and their derivatives.

Tannins are amorphous substances, which produce colloidal acidic aqueous solutions with astringent taste. With iron salts ($FeCl_3$) they form dark blue or greenish black water-soluble compounds. Tannins form insoluble and indigestible compounds with proteins, and this is the basis of their extensive use in the leather industry (tanning process), and for the treatment of diarrhoea, bleeding gums and skin injuries.

Classification

Tannins can be classified into two major classes: *hydrolysable tannins* and *condensed tannins*. On treatment with acids or enzymes, while hydrolysable tannins are split into simpler molecules, condensed tannins produce complex water-insoluble products.

Hydrolysable tannins are subdivided into *gallotannins* and *ellagitannins*. Gallotannins, on hydrolysis, yield sugar and gallic acid, whereas hydrolysis of ellagitannins results in sugar, gallic acid and ellagic acid. Pentagalloylglucose, which has long been used in the tanning industry, is an example of a gallotannin.

Condensed tannins are complex polymers, where the building blocks are usually catechins and flavonoids, esterified with gallic acid. Example: epicatechin trimer.

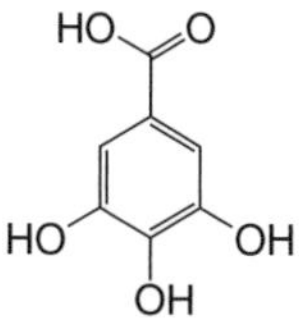

Gallic acid

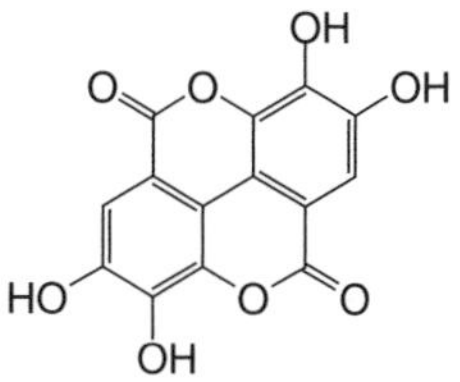

Ellagic acid

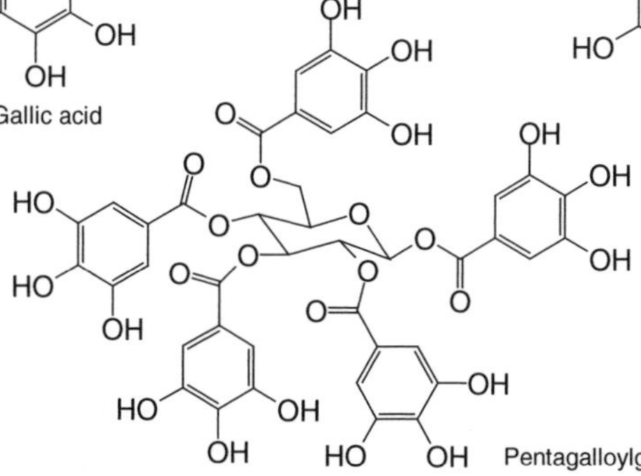

Pentagalloylglucose

Recommended further reading

Hanson, J. R. *Natural Products: the Secondary Metabolites*, The Royal Society of Chemistry, London, 2003.

Dewick, P. M. *Medicinal Natural Products: a Biosynthetic Approach*, 2nd edn, Wiley, London, 2002.

Index